# The Chemistry of Diamondoids

The Chemistry of Diamondoids

Building Blocks for Ligands, Catalysts, Materials, and Pharmaceuticals

Andrey A. Fokin
Marina Šekutor
Peter R. Schreiner

WILEY-VCH

Authors

Prof. Dr. Andrey A. Fokin
Igor Sikorsky Kyiv Polytechnic Institute
Department of Organic Chemistry
Beresteiskyi Ave. 37
03056 Kiev
Ukraine

Dr. Marina Šekutor
Ruđer Bošković Institute
Department of Organic Chemistry and
Biochemistry
Bijenička 54
10 000 Zagreb
Croatia

Prof. Dr. Peter R. Schreiner
Justus Liebig University
Institute of Organic Chemistry
Heinrich-Buff-Ring 17
35392 Giessen
Germany

Cover Image: © Formgeber, Mannheim

All books published by **WILEY-VCH** are carefully produced. Nevertheless, authors, editors, and publisher do not warrant the information contained in these books, including this book, to be free of errors. Readers are advised to keep in mind that statements, data, illustrations, procedural details or other items may inadvertently be inaccurate.

Library of Congress Card No.: applied for

British Library Cataloguing-in-Publication Data
A catalogue record for this book is available from the British Library.

Bibliographic information published by the Deutsche Nationalbibliothek
The Deutsche Nationalbibliothek lists this publication in the Deutsche Nationalbibliografie; detailed bibliographic data are available on the Internet at <http://dnb.d-nb.de>.

© 2024 WILEY-VCH GmbH, Boschstr. 12, 69469 Weinheim, Germany

All rights reserved (including those of translation into other languages). No part of this book may be reproduced in any form – by photoprinting, microfilm, or any other means – nor transmitted or translated into a machine language without written permission from the publishers. Registered names, trademarks, etc. used in this book, even when not specifically marked as such, are not to be considered unprotected by law.

Print ISBN: 978-3-527-34391-1
ePDF ISBN: 978-3-527-81294-3
ePub ISBN: 978-3-527-81293-6
oBook ISBN: 978-3-527-81295-0

Typesetting: Straive, Chennai, India
Printing and Binding: CPI Group (UK) Ltd, Croydon, CR0 4YY

C9783527343911_200125

The manufacturer's authorized representative according to the EU General Product Safety Regulation is Wiley-VCH GmbH, Boschstr. 12, 69469 Weinheim, Germany, e-mail: Product_Safety@wiley.com.

Contents

Preface *ix*
Abbreviations *xi*
Acknowledgments *xv*
Author Biographies *xvii*

1 **Description of Diamondoids** *1*
1.1 Nomenclature *2*
1.2 Strain *5*
1.3 Preparation of Diamondoids *8*
1.4 Physical Properties of Diamondoids *14*
1.5 Spectroscopy of Diamondoids *18*
1.6 Ionization Potentials *23*
1.7 Electron Affinities *24*
1.8 Vibrational Spectroscopy *25*
References *28*

2 **Naturally Occurring Diamondoids** *35*
2.1 Diamondoid Occurrence in the Earth's Crust *38*
2.2 Diamondoids in Geochemical Studies *40*
2.3 Diamondoid Formation in the Earth's Crust *42*
2.4 Large-scale Isolation of Natural Diamondoids *44*
2.5 Alternative Natural Sources of Diamondoids *48*
2.6 Other Diamondoid Derivatives in Nature *50*
References *52*

3 **Diamondoids as Alkane CH Activation Models** *63*
References *69*

4 **Preparative Diamondoid Functionalizations** *75*
4.1 Halogenations *75*
4.2 Diamondoid Alcohols and Ketones *83*
4.3 Carboxylic Acids and Their Derivatives *98*
4.4 Nitrogen-Containing Compounds *103*

4.5	Phosphorous- and Sulfur-Containing Compounds	*108*
4.6	Single Electron Oxidations of Diamondoids	*113*
4.7	Other Diamondoid Derivatives	*118*
	References	*121*

5 Diamondoid Self-Assembly *137*
5.1	Adamantane-Containing SAMs on Surfaces	*137*
5.2	Higher Diamondoids for SAM Formation	*146*
5.3	Pristine Diamondoids on Surfaces	*157*
5.4	Other Applications of Diamondoid SAM Materials	*157*
5.5	Adamantane-stabilized Metal Nanoparticles	*160*
	References	*162*

6 Growing Diamond Structures from Diamondoids Via Seeding *171*
6.1	Diamondoid-Promoted Growth of Diamond Under HT-HP or CVD Conditions	*171*
6.2	Higher Diamondoids for Diamond Nucleation	*177*
6.3	Diamond Growth Inside Nanotubes	*180*
	References	*186*

7 Diamondoid Polymers *193*
7.1	Polymers Based on 1-Adamantyl-1-adamantane (1ADAD)	*194*
7.2	Polymers Based on Monofunctionalized Diamantanes	*197*
7.3	Polymers Based on Difunctionalized Diamantanes	*199*
	References	*207*

8 Diamondoids in Catalysis *213*
References *232*

9 Medicinal Compounds *239*
References *245*

10 Supramolecular Architectures *251*
References *270*

11 Diamondoid Oligomers *279*
11.1	Saturated Diamondoid Oligomers	*279*
11.2	Unsaturated Oligomers	*287*
	References	*293*

12 Doped Diamondoids *305*
12.1	@-Doping	*306*
12.2	Internal Doping	*308*

12.3	External Doping *318*	
	References *329*	
13	**Perspective** *339*	
	References *340*	
	Index *343*	

Preface

Diamondoids are a prominent and large family of nanoscale hydrocarbons that are composed of carbon atoms arranged in a cubic diamond lattice structure. These molecules range in size from ten (adamantane) to several dozen carbon atoms and have unique properties that make them valuable for a large variety of applications. Since the discovery of diamondoids in petroleum deposits in the 1930s, they have been studied extensively, and their potential for use in materials science, nanotechnology, and medicine has been explored.

The origin of diamondoids can be traced back to the geological processes that create petroleum deposits. Petroleum formed from the remains of ancient (marine) organisms, which are buried deep underground and subjected to high temperatures and pressures over millions of years. During this process, some of the carbon in the organic matter transforms into diamondoids because they constitute the thermodynamically most stable hydrocarbons.

In recent decades, researchers have also discovered that diamondoids beyond adamantane can be synthesized in the laboratory using a variety of methods, including chemical vapor deposition, solution-phase synthesis, and high-pressure, high-temperature techniques. This has opened up new avenues for research and development, as they can be tailored to have specific properties and functionalities that are not found in natural diamondoids.

Even though the parent diamondoid, adamantane, was discovered in petroleum as early as 1933, it was not until the 1970s that researchers started to study these molecules in earnest. Advances in synthetic accessibility, analytical techniques, and computational modeling have since enabled us to study the properties and behavior of diamondoids in detail, leading to a better understanding of their potential applications. For example, their exceptional thermodynamic stability and mechanical strength make them ideal for use in nanoscale electronic devices, such as rectifiers and electron emitters. Additionally, their biocompatibility and low toxicity make diamondoid derivatives attractive candidates for use in pharmaceutical applications. Furthermore, diamondoids have been proposed as building blocks for the construction of new materials, such as diamond-like nanowires and coatings.

This comprehensive book on diamondoids aims to provide a thorough overview of the current state of research on these fascinating molecules. It covers a wide range of topics, from the synthesis and characterization of diamondoids to their applications in various fields. The book is divided into several sections, each focusing on a different aspect of diamondoid research.

The first chapter introduces the basic concepts of diamondoid chemistry, including their structures, nomenclature, (spectroscopic) properties, and synthesis. It covers the various methods to synthesize diamondoids, including techniques that also help understand how they abundantly form in nature. The strain energies of increasingly larger diamondoids are also assessed to aid in this endeavor.

The second section of the book focuses on naturally occurring diamondoids and the role they play in the petroleum industry and in geosciences. Ways for their formation also are described including man-made approaches to prepare them on a large scale.

The third and fourth chapters cover C—H-bond functionalization, as this is a precondition for their use in many applications. As diamondoids are alkanes with strong and multiple, similarly reactive C—H-bonds, this is an exercise in practically applicable and scalable methods for alkane functionalization. This also enables the fine-tuning of the diamondoid properties, as they react sensitively to substitution.

Chapters 5–12 are devoted to the applications of (functionalized) diamondoids, including self-assembly, growing diamonds from diamondoids via seeding, polymers, supramolecular architectures, diamondoid oligomers, and doped diamondoids. The applications in catalysis and as medicinal compounds are also outlined to demonstrate the huge and, for the most part, untapped potential these building blocks offer.

We close by providing a brief perspective on where diamondoids could continue to leave their mark and what future directions in diamondoid research might entail. We hope to provide a thorough overview of the current state of research on these fascinating molecules and to highlight their potential for use in a wide range of applications.

February 2024

Andrey A. Fokin
Marina Šekutor
Peter R. Schreiner

Abbreviations

6-31G(d,p)	split-valence basis sets with d and p polarization functions on the heavy atom and on hydrogen, respectively
AFM	atomic force microscopy
AIBN	azobis(isobutyronitrile)
AIM	atoms-in-molecules
B3LYP	Becke 3-parameter Lee–Yang–Parr exchange-correlation DFT functional
B3LYP-D3	Becke 3-parameter Lee–Yang–Parr exchange-correlation DFT dispersion-corrected functional
cc-pVDZ	double-zeta correlation-consistent basis set
cc-pVTZ	triple-zeta correlation-consistent basis set
CB	cucurbituril
CD	cyclodextrin
CNT	carbon nanotube
CVD	chemical vapor deposition
DAST	diethylaminosulfur trifluoride
DBD	dielectric barrier discharge
DED	dispersion energy donor
DFT	density functional theory
DMSO	dimethyl sulfoxide
DMF	dimethyl formamide
EA	electron affinity
EAS	Engler–Andose–Schleyer (force field)
ECEC	electron transfer-chemical reaction-electron transfer-chemical reaction
EOM	equation of motion
EPD	electronic photodissociation
ESI	electrospray ionization
FWHM	full-width half-max
GCD	graphite-cluster diamond
GC–GC-TOFMS	2D gas chromatography coupled with time-of-flight mass spectrometry
GC–MS	gas chromatography coupled with mass spectrometry

GC–MS/MS	gas-chromatography-tandem mass spectrometry
GOR	gas/oil ratio
HCET	H-coupled electron transfer
HOMO	highest occupied molecular orbital
HPLC	high-performance liquid chromatography
HT-HP	high temperature and high pressure
IETS	inelastic electron tunneling spectroscopy
IP	ionization potential
IR	infrared
IRPD	infrared photodissociation
IR-STM	infrared scanning tunneling microscopy
ITO	indium–tin oxide
IUPAC	International Union of Pure and Applied Chemistry
JT	Jahn–Teller
KIE	kinetic isotope effect
LD	London dispersion
LUMO	lowest unoccupied molecular orbital
MALDI	matrix-assisted laser desorption ionization
MANSE	microwave-assisted nonionic surfactant extraction
MD	molecular dynamics
MP	microwave plasma
MP2	Møller–Plesset second-order pertubation
NAPS	non-aqueous phase pollution liquids
NEA	negative electron affinity
NEXAFS	near-edge X-ray absorption fine structure
NHPI	N-hydroxyphthalimide
NMR	nuclear magnetic resonance
PBE	Perdew–Burke–Ernzerhof exchange DFT functional
PINO	phthalimide-N-oxyl (radical)
PTC	phase-transfer catalysis
PMC	polymethacrylate
R_o	vitrinite reflectance
SAM	self-assembled monolayer
SCX	single crystal X-ray
SET	single-electron transfer
SOMO	single-occupied molecular orbital
SWCNT	single-wall carbon nanotube
TBDMS	*tert*-butyldimethylsilyl
TBHP	*tert*-butylhydroperoxide
TCB	tetracyanobenzene
TD	time-dependent
TEM	transmission electron microscopy
TFA	trifluoroacetic acid
TGA	thermogravimetric analysis
TMAD	tetramethyl adamantane

TOF	time-of-flight
TPPE	two-photon photoemission
TSR	thermochemical sulfate reduction
UHV	ultrahigh vacuum
UV	ultraviolet
X-PEEM	X-ray photoemission electron microscopy
XPS	photoelectron microscopy

Acknowledgments

Our work on diamondoids was supported through exchanges with many very knowledgeable and helpful colleagues, including, *inter alia*, Robert M.K. Carlson, Jeremy E.P. Dahl, Matthew A. Gebbie, Jean-Cyrille Hierso, Nicholas A. Melosh, Thomas Möller, Hisanori Shinohara, Z.-X. Shen, and Trevor Willey.

Even more importantly, we are grateful to our many co-workers who have executed all the work on diamondoids both experimentally and computationally. We are deeply indebted to their dedication, hard work, and inspiration.

Finally, this work was generously supported by the Alexander von Humboldt Foundation, the Bundesministerium für Bildung und Forschung, the Deutsche Forschungsgemeinschaft, the Fonds der Chemischen Industrie, the German Academic Exchange Service, the Ministry of Education of Ukraine, the NATO Science Program, the Ukrainian Basic Research Fund, the US Department of Energy, and the Volkswagen Foundation.

Author Biographies

Andrey A. Fokin is currently Head of the Department of Organic Chemistry and Technology at National Technical University "Igor Sikorsky Kiev Polytechnic Institute," Ukraine. He studied chemistry at the Kiev Polytechnic Institute, where he received the degrees of Candidate of Chemical Sciences (1985) in diamondoid chemistry under the supervision of Prof. Pavel A. Krasutsky and Doctor of Chemical Sciences in pesticide chemistry and technology (1995). He was a DAAD (1996) and Alexander von Humboldt (1997–1998) research fellow in the groups of Prof. P. v. R. Schleyer at the University of Erlangen-Nürnberg and Prof. P. R. Schreiner at Göttingen University (Germany). He has been a visiting professor at the universities of Minnesota and Georgia (USA) as well as in Giessen (Germany). His research mostly concentrates on the combination of computational and experimental studies with applications in organic, physical-organic, medical, agricultural chemistry, and nanotechnology and has been published in over 200 peer-reviewed papers and books.

Marina Šekutor is currently a senior research associate at the Ruđer Bošković Institute, Zagreb, Croatia. She was born in 1986 in Zagreb and received her B.Sc.Eng. degree in 2008 and PhD (under the mentorship of Prof. Kata Majerski) in 2013, both from the Faculty of Science, University of Zagreb. She obtained a Humboldt postdoctoral fellowship (research in the Schreiner group, 2015–2017) and upon returning to Zagreb continued to pursue her interests in the chemistry of diamondoids. M. Šekutor received the annual award to young scientists and artists in 2014, the annual award of the Ruđer Bošković Institute in 2017, and the award for organic chemistry "Vladimir Prelog" in 2019 (awarded by the Croatian Chemical Society). She has been a visiting scientist at the University of Maryland, USA, and is a lecturer at the Faculty of Science, University of Zagreb, and the Faculty of Medicine, Josip Juraj Strossmayer

University of Osijek. She is a Croatian Chemical Society member, where she currently holds the position of Secretary from 2021, and a Croatian Humboldt-Club member. Her research interests include supramolecular host–guest chemistry of polycyclic compounds, cluster assemblies in helium nanodroplets driven by London dispersion, and use of diamondoid derivatives in materials science.

Peter R. Schreiner is professor of organic chemistry and Liebig-Chair at the Institute of Organic Chemistry at the Justus Liebig University Giessen, Germany. He studied chemistry in his native city at the University of Erlangen-Nürnberg, Germany, where he received his Dr. rer. nat. (1994) in organic chemistry. Simultaneously, he obtained a PhD (1995) in computational chemistry from the University of Georgia, USA. He completed his habilitation (assistant professorship) at the University of Göttingen (1999) before becoming associate professor at the University of Georgia (Athens, USA) and head of the institute in Giessen in 2002. P. R. Schreiner is an elected member of the Leopoldina – German National Academy of Sciences, the North Rhine-Westphalian Academy of Sciences, Humanities, and the Arts, the Academy of Science and Literature (Mainz), the Berlin-Brandenburg Academy of Sciences, and is a Fellow of the Royal Society of Chemistry. He received the Dirac Medal (2003, WATOC), the Adolf-von-Baeyer Memorial Award of the German Chemical Society in 2017, the RSC Award in Physical Organic Chemistry of the RSC in 2019, the Academy Award of the Berlin-Brandenburg Academy of Science in 2020, the ACS Arthur C. Cope Scholar Award in 2021, and the Gottfried–Wilhelm–Leibniz–Award 2024 of the German Research Council (DFG). He has been a visiting professor at the CNRS in Bordeaux, the Technion in Haifa, the Australian National University in Canberra, and the University of Florida in Gainesville. His research interests include organic reaction dynamics and reactive intermediates, quantum mechanical tunneling, as well as London dispersion interactions as probed in the realm of nanodiamonds and organocatalysis.

1

Description of Diamondoids

Shortly after the 3D structure of diamond was determined, the German chemist Hermann Decker recognized the connection between diamond and saturated hydrocarbons with "the 6-ring system built out into the third dimension" and suggested the term "diamondoid" for such molecules [1]. As the spatial arrangement of the carbon atoms resembles the diamond crystal lattice, diamondoids can therefore be viewed as hydrogen-terminated nanometer-sized diamonds with distinctive properties determined by their sizes and topologies. The smallest diamondoid is adamantane (**AD**, Figure 1.1), which has a cage skeleton consisting of ten carbons. Formal addition of further isobutyl fragments to the **AD** parent structure in a cyclohexane ring-forming manner results in higher diamondoid homologues. Diamondoids are classified as lower and higher homologues: lower diamondoids have only one isomeric form and include **AD** ($C_{10}H_{16}$), diamantane (**DIA**, $C_{14}H_{20}$), and triamantane (**TRIA**, $C_{18}H_{24}$), while higher diamondoids start with tetramantane (**TET**, $C_{22}H_{28}$) and possess isomers. Among three possible **TET** isomers, one is chiral (**123TET**) and is viewed as the parent of a new family of σ-helicenes [2]. As the cage grows, the number of isomers increases and beginning from pentamantane (**PENT**) spreads into different molecular weight subgroups, i.e., **PENT** has nine isomers with the $C_{26}H_{32}$ formula and one isomer with the $C_{25}H_{30}$ formula. Some other hydrocarbons also satisfy the structural criteria of partial or complete superposition on the diamond lattice, e.g., cyclohexane and decalin, so diamondoids are more precisely defined as "hydrocarbons containing at least one adamantane unit wholly or largely superimposable on the diamond lattice" [3]. Due to this definition, higher diamondoids bridge the gap between saturated hydrocarbons and diamond and are sometimes called nanodiamonds (in plural form to differentiate them from heterogeneous mixtures of nanodiamond material obtained by chemical vapor deposition, detonation, or shock-wave techniques [4]).

The molecular symmetry of diamondoids also plays a role in their self-assembly, readily producing crystals or serving as nucleation centers for bigger nanomaterial architectures. The smallest diamondoid **AD** is highly symmetric (T_d point group), whereas symmetry is generally (but not always) reduced as the diamondoids become larger, e.g., **DIA** and **TRIA** belong to the D_{3d} and C_{2v} point groups, respectively.

The Chemistry of Diamondoids: Building Blocks for Ligands, Catalysts, Materials, and Pharmaceuticals, First Edition. Andrey A. Fokin, Marina Šekutor, and Peter R. Schreiner.
© 2024 WILEY-VCH GmbH. Published 2024 by WILEY-VCH GmbH.

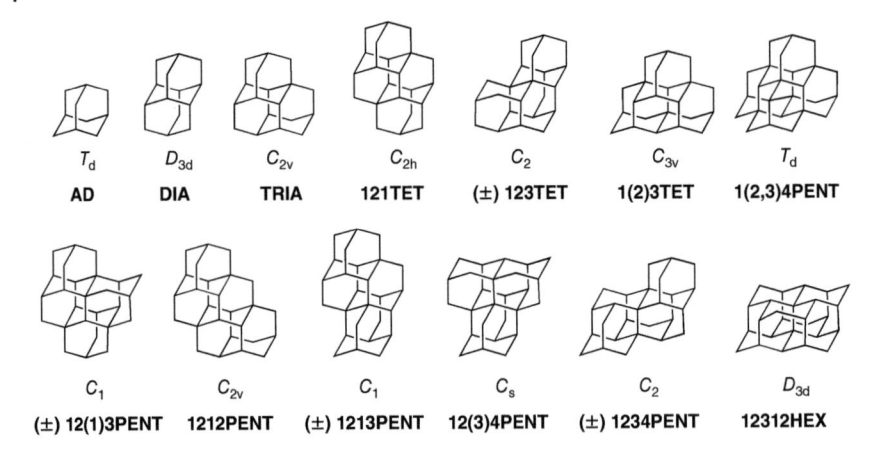

Figure 1.1 Structures and symmetry of diamondoid homologues up to cyclohexamantane (12312HEX).

Isomers of the first higher diamondoid **TET** display C_{2h} (**121TET**), C_2 (**123TET**), and C_{3v} (**1(2)3TET**) symmetry (Figure 1.1).

Note that **TET**s exemplify a common occurrence for higher diamondoids: different isomers have markedly different symmetries, which can be useful in material design by tailoring both the solubility of the material and the shape of the used building blocks. Moreover, diamondoids have one important advantage over bulk diamonds: they are "knowable," that is, their shapes are precisely determined by their molecular structure (rods, disks, helices, prisms, pyramids, cubes, etc.; Figure 1.2) and stoichiometry, and they can be obtained in homogeneous forms because they are single-molecule, nanometer-scale-sized building blocks. For instance, **DIA** and **121TET** are rod-shaped nanodiamond particles, **TRIA** and **1212PENT** have triangular shape, **1(2,3)4PENT** and **1231241(2)3DEC** are tetrahedron and cube, respectively, and **1(2)3TET** and **12312HEX** are prisms.

According to the definition of polymantanes, face-fused **AD** cage structures that are the focus of this book are not the only existing diamondoids. For example, **AD** dimers and higher single-bonded oligomers also belong to the class of diamondoid hydrocarbons (Figure 1.3). The first step in determining whether a saturated hydrocarbon belongs to the class of polymantane compounds is to check whether it has at least one **AD** subunit. If yes, then the next condition is that all cage atoms of the molecule need to be part of an **AD** unit; if that is also true, then the final condition is that two or more **AD** cages need to have at least six common carbon atoms, meaning that they share one face. When these conditions are met, a structure can be classified as a true diamondoid (Figure 1.3) [3].

1.1 Nomenclature

Before going further, we first define the diamondoid classification and nomenclature. As can be anticipated, von Baeyer's IUPAC names for these polycyclic

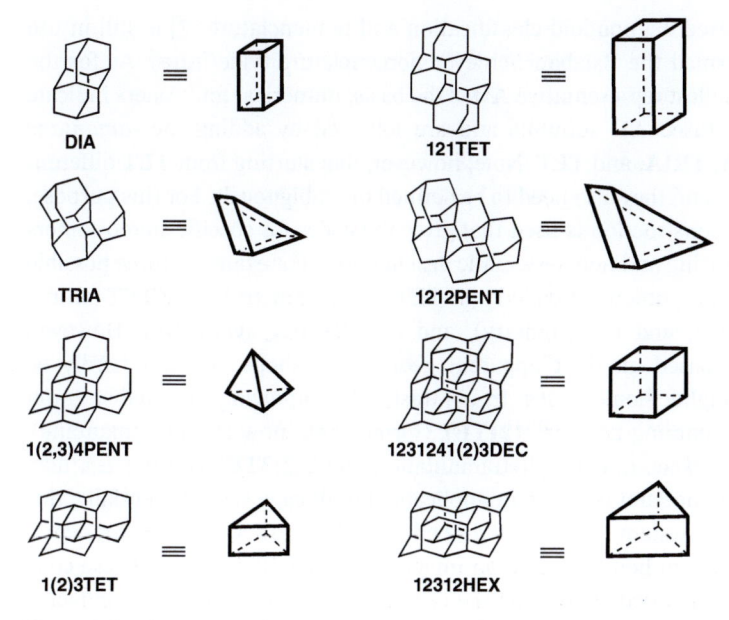

Figure 1.2 Structures of selected diamondoids linked to their geometrical representations. Source: Adapted from Ref. [5].

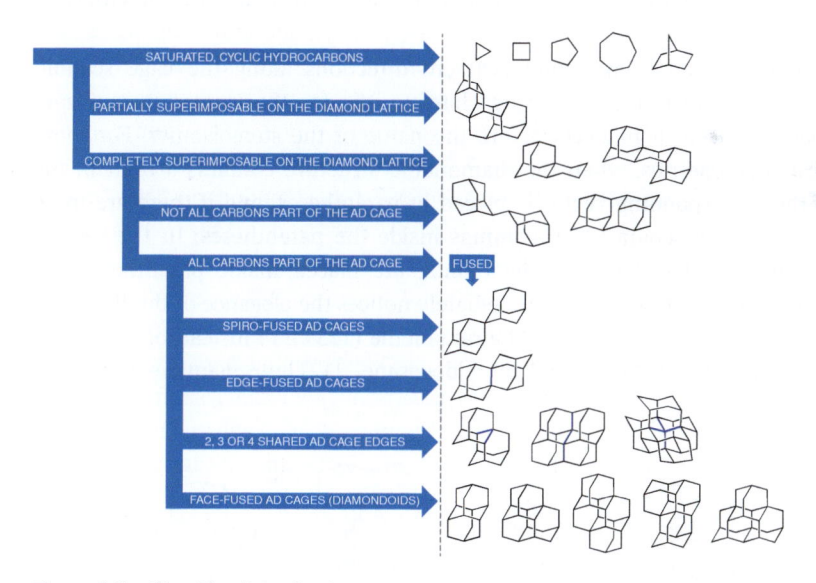

Figure 1.3 Classification of polymantanes.

compounds become quite cumbersome, and precise structural assignments require representing such molecular structures in terms of planar graphs [6]. Note that some programs, such as ChemDoodle, are quite useful for automatic IUPAC naming. As the cages grow larger and become more complex, the need to develop a special nomenclature for diamondoids emerges [3]. The initially proposed graph-theory-based diamondoid classification and nomenclature [7] is still in use today and is termed the Balaban–Schleyer nomenclature (vide infra). As for the naming, the smallest representative **AD** is the basis: numerical multipliers indicate the number of fused **AD** subunits and are followed by adding the -*amantane* suffix, e.g., **DIA**, **TRIA**, and **TET**. Note, however, that starting from **TET** different isomers emerge, and they also need to be defined unambiguously. For this purpose, a dualist graph construction is used that gives the codes for specific stereoisomers and avoids confusing and non-systematic designations. For example, three possible isomers of **TET** are sometimes called *anti*-**TET** (C_{2h}-symmetry), *skew*-**TET** (chiral, enantiomeric pair, and C_2-symmetry), and *iso*-**TET** (C_{3v}-symmetry). However, this naming is based on their apparent geometrical shape and can hardly be transferred to higher homologues. In contrast, when applying the dualist graph convention, the naming becomes **121TET** (former *anti*, now [121]tetramantane), **123TET** (former *skew*, now [123]tetramantane), and **1(2)3TET** (former *iso*, now [1(2)3]tetramantane) and is a system applicable for all cage sizes. Essentially, this Balaban–Schleyer system uses four-digit codes (1, 2, 3, and 4) for the tetrahedral directions of covalent bonds around an imaginary center (Figure 1.4). These code descriptors are generated as follows: the center of the first **AD** cage is connected with the adjacent **AD** moieties in one of the four possible directions (**AD** has four faces), center-to-center. This direction is assigned number 1; the process is repeated until all **AD** subunits are accounted for and the whole molecule is traced with such vectors.

The digits emerge from taking different directions along the cage scaffold and, in the end, give a unique code characteristic for the isomer in question. This code is placed in brackets before the name of the stereoisomer. For more complicated geometries, when the diamondoid structure contains a branch, the digit of the corresponding vector is placed in parentheses, and if there are more branches, they are separated by commas inside the parentheses. In the case of longer branches, the chains of the branch are placed inside parentheses but without comma separation. One immediately notices the elegance of the Balaban–Schleyer approach as we have [123]tetramantane (**123TET**) instead of nonacyclo [11.8.1.0^{1,20}.0^{2,7}.0^{4,21}.0^{6,19}.0^{9,18}.0^{11,16}.0^{15,20}]docosane, [121]tetramantane (**121TET**)

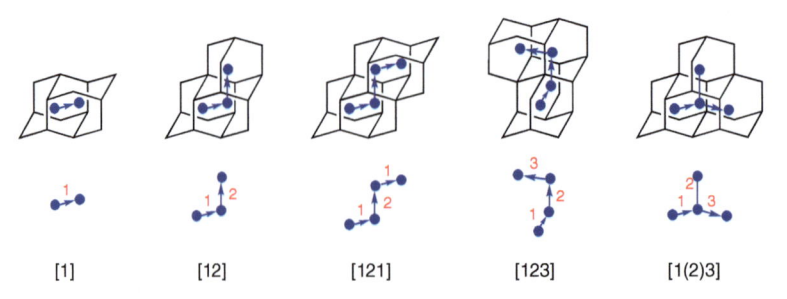

Figure 1.4 Examples of dualist graph construction for diamondoid naming suggested by Balaban and Schleyer.

instead of nonacyclo[11.7.1.1^{6,18}.0^{1,16}.0^{2,11}.0^{3,8}.0^{4,19}.0^{8,17}.0^{10,15}]docosane, [1212]pentamantane (**1212PENT**) instead of undecacyclo[11.11.1.1^{5,21}.0^{1,16}.0^{2,11}.0^{3,8}.0^{4,23}.0^{6,19}.0^{8,17}.0^{10,15}.0^{18,23}]hexacosane, and so on. Note that in the older literature, **DIA** is sometimes called "congressane" since it was proposed [8] as a synthetic challenge for the participants of the XIXth 1963 IUPAC meeting in London.

The dualist graph convention for the nomenclature of diamondoids enables a more straightforward way to designate diamondoid cages but as the cage size increases even such a naming system encounters complications. With increasing cage fusion, it becomes difficult to account for all possible stereoisomers and one way to resolve this type of complexity is to generate partitioned-formula tables based on the distribution of all the present carbon atoms according to them being quaternary (Q), tertiary (T), or secondary (S) [7a, 9]. By following this procedure, one obtains valence isomers of the same molecular formula $C_Q(CH)_T(CH_2)_S$ that are then shortened and denoted as Q – T – S, where the total amount of carbons is C = Q + T + S. For example, by using this convention, the formula for **AD** can be written as $(CH)_4(CH_2)_6$ or as 0–4–6, since the **AD** cage possesses no quaternary, four tertiary, and six secondary carbon atoms. Isomeric diamondoids with the same molecular formula, $C_Q(CH)_T(CH_2)_S$, can be divided into valence isomers by partitioning the number C into Q + T + S. Each [n]diamondoid has a dualist with n vertices and edges connecting vertices of adjacent **AD** units. Such a dualist is characterized by a quadruplet of indices (denoted as p, s, t, and q for primary, secondary, tertiary, and quaternary, respectively) specifying the connectivity of each vertex by assimilating it with a virtual carbon atom. Dualists help in classifying diamondoids as catamantanes with acyclic dualists, perimantanes with dualists having chair-shaped six-membered rings, or coronamantanes with dualists having only higher-membered rings.

An extension of the Schleyer–Balaban nomenclature that is more feasible for computer modeling was suggested recently [10]. This nomenclature is based on the numbering of centers of the diamond lattice starting from the origin (yellow, Figure 1.5) with further expansion along the chains and branches so that the structures can be systematically constructed. For instance, in accordance with this nomenclature, higher diamondoids [12312]hexamantane ($C_{26}H_{30}$), [121321]heptamantane ($C_{30}H_{34}$), and ([1 231 241(2)3]decamantane ($C_{35}H_{36}$) are now [1,2,3,6,7,18], [1,2,3,4,8,9,17], and [1,2,3,6,7,8,9,18,30] diamondoids that simplifies distinguishing and screening of large isomeric structures.

1.2 Strain

Strain reflects the electronic properties, and very recently it was found that the broad σ^*_{C-C} resonances in the near-edge X-ray absorption fine structure spectra of **AD** split into two narrow and intense resonances proportionally to the strain in a series from twistane and octahedrane to cubane [11]. It is often assumed that diamondoids are almost strain-free molecules, and indeed they are the most stable hydrocarbons of given brutto formulae. Strain can be seen as a relationship between

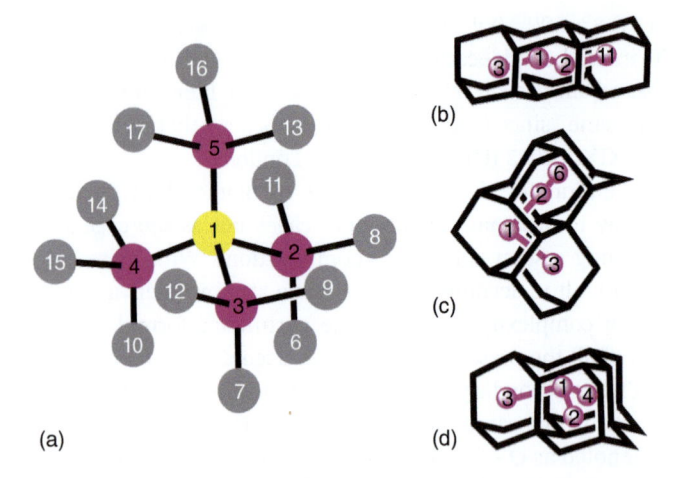

Figure 1.5 Nomenclature of diamondoids. (a) Parts of the diamond lattice with labeled numbers where the yellow atom represents the original atom and magenta and gray atoms are its first and second adjacent atoms, respectively; (b)–(d) are tetramantane isomers with numbering of their corresponding center atoms.

energy and structure that enables the evaluation of the structural feasibility of compounds. Straight-chain hydrocarbons in their linear, non-staggered conformations serve as the structural basis upon which strain is evaluated. Note, however, that even alkanes need not be absolutely strain-free but rather are taken as a point of reference for strain comparisons. Common criteria used when determining strain in structures are the presence of bond angles markedly deviating from the standard values, eclipsed conformations present in the structures, and atoms that approach each other too closely. Since the **AD** cage does not have these indicative features, it has long been considered strain-free (even though one should not forget about cyclohexane *gauche* interactions that lead to small strain, vide infra). While it is certainly true that diamondoids have little strain, there are some nuances to consider. Schleyer recommended using CH_3, CH_2, CH, and C group increments as well as force field calculations [12] for strain energy evaluations, in particular for diamondoids (Table 1.1) [13]. He concluded that the primary source of strain in diamondoids arises from slight deviations from the ideal C–C–C, C–C–H, and H–C–H angles when considering the cage angles where C–CH_2–C and C–CH–C fragments are interconnected. For instance, while the C–C–C angles in **AD** are 109.5°, these values are 112.4° and 111.3° in propane and isobutane, respectively. With size growing, the strain energy of diamondoids increases: it amounts to ca. 6 kcal mol^{-1} for **AD**, 11 and 13 kcal mol^{-1} for **DIA** and **TRIA**, respectively. However, the normalized strain energy (per carbon) remains almost constant (vide infra). Another source of strain is repulsive nonbonding C•••C interactions (akin to *gauche* interactions in *synclinal* n-butane and cyclohexane) throughout the cage itself. However, we also stress that 1,3-nonbonding intramolecular H•••H interactions are in fact acting beneficially due to London dispersion [14] (rediscovered much later as "protobranching" [15]) that is seen as an interplay

Table 1.1 Computed enthalpies of formation (gas) and strain energies (E_{str}) in kcal mol^{-1} for selected saturated hydrocarbons from molecular mechanics computations (EAS force field [12]).

Hydrocarbon	ΔH_f°	E_{str}
AD	−32.5	6.9
DIA	−37.4	10.7
TRIA	−44.4	13.4
Cyclopentane	−18.4	7.3
Cyclohexane	−29.4	1.4
Cycloheptane	−28.3	7.6
Cyclooctane	−29.2	11.9
Cyclononane	−30.7	15.5
Cyclodecane	−34.9	16.4
Norbornane	−13.0	17.0
ADAD	−53.7	21.5
Twistane	−13.3	26.1
Dodecahedrane	−0.2	43.0
Cubane	148.6	165.9

between medium-range correlation and steric repulsion [16]. The enthalpies of formation and evaluated strain energies for selected hydrocarbons presented in Table 1.1 are illustrative of the strain propagation in different cage molecules and are numerically still relevant [17] despite initially being computed using a simple Engler-Andose-Schleyer (EAS) force field method [12].

While the computational group equivalent approach is method-dependent [18], the errors largely cancel if homodesmotic equations are used [19]. The latter allows to compute the strain energies of hydrocarbons directly utilizing conventional strain-free references, such as ethane, propane, isobutane, and neopentane. This again gives the E_{str} of ca. 6 kcal mol^{-1} for **AD** (Eq. 1.1) and 2.2 kcal mol^{-1} for 1,3,5,7-tetramethyladamantane (**TMAD**) (Eq. 1.2) [15]. The fact that formally less crowded **AD** is more strained than **TMAD** may also be associated with additional electron delocalizations due to CC σ → σ* hyperconjugation in the latter [20].

$$\text{AD} + 12\,\text{—} = 6\,\wedge + 4\,\curlywedge \quad -5.8 \text{ kcal mol}^{-1} \tag{1.1}$$

$$\text{TMAD} + 12\,\text{—} = 6\,\wedge + 4\,\curlyvee \quad -2.2 \text{ kcal mol}^{-1} \tag{1.2}$$

Table 1.2 Absolute (E_{str}) and normalized (per one carbon atom, $E_{str}°$) strain energies in kcal mol^{-1} of selected diamondoids $C_{4n+6}H_{4n+12}$ evaluated through homedesmotic equation $C_{4n+6}H_{4n+12} + k\ H_3CCH_3 = l\ H_3CCH_2CH_3 + m\ (CH_3)_3CH + p\ (CH_3)_4C$ at the B3LYP/6-31G(d) level (Source: From [21]).

Diamondoid	E_{str}	$E_{str}°$
AD	6.0	0.60
DIA	9.1	0.63
TRIA	10.8	0.60
121TET	12.3	0.56
1(2)3TET	11.5	0.52
123TET	17.5	0.79
1(2,3)4PENT	9.3	0.36
12312HEX	16.2	0.62

Computations predict [10, 21] the lowest values of the formation enthalpies for the most symmetric diamondoid structures. Accordingly, C_{3v}-tetramantane (**1(2)3TET**) and T_d-pentamantane (**1(2,3)4PENT**) are the most stable isomers. The strain energies of higher diamondoids were calculated utilizing Eq. 1.1, namely, $C_{4n+6}H_{4n+12} + k\ H_3CCH_3 = l\ H_3CCH_2CH_3 + m\ (CH_3)_3CH + p\ (CH_3)_4C$ at the B3LYP/6-31G(d) level [21]. The calculated strain energy of **DIA** (9.1 kcal mol^{-1}) and **TRIA** (10.8 kcal mol^{-1}) is higher than that for **AD** (6.0 kcal mol^{-1}); however, the normalized values ($E_{str}°$) are virtually identical to that of **AD** (ca. 0.6 kcal mol^{-1}, Table 1.2).

Additional strain in higher diamondoids may arise from destabilizing transannular HH contacts, such as in **123TET**, where some distances are ca. 2.1–2.2 Å (Figure 1.6); these are shorter than the sum of the van der Waals radii of two H-atoms (ca. 2.4 Å) [22], which causes destabilization of the structure. Such interactions are not present in **121TET** and **1(2)3TET,** which almost level their strain energies. Note that while the ideal tetrahedral **1(2,3)4PENT** is almost strain-free (0.36 kcal mol^{-1} per carbon), the strain energy of **12312HEX** (0.62 kcal mol^{-1} per carbon) is close to that of **AD**.

1.3 Preparation of Diamondoids

The construction of the **AD** core in tetraester (**TE**) was first achieved in 1937 by Oskar Böttger [23] through methylenation of Meerwein's ester (**ME**, Scheme 1.1) [24], readily available from formaldehyde and malonic ester [25]. The first successful synthesis of **AD** was achieved by Prelog [24, 26] as early as 1941, starting from **ME** through diester (**DE**) [27], however, with a very low yield (0.16% based on

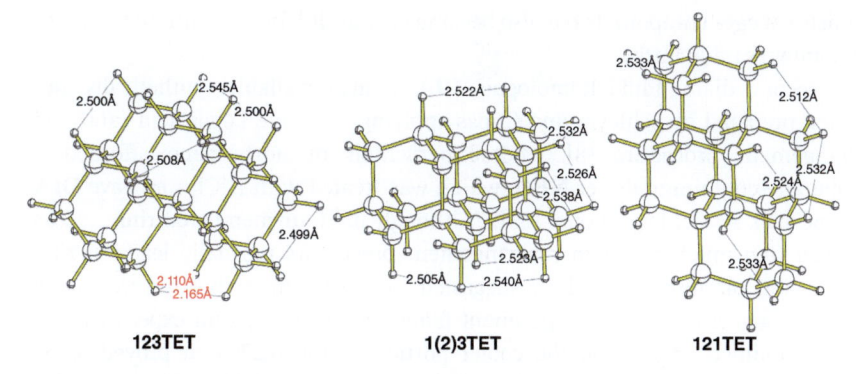

Figure 1.6 Additional strain in the *skew*-diamondoid **123TET** arose from several destabilizing transannular HH contacts (in red). The 1,3-HH contact distances of 2.5 Å are close to the sum of the van der Waals radii of two H-atoms (ca. 2.4 Å) and are stabilizing (protobranching).

ME). Further improvements of the decarboxylation steps, either through reductive decarboxylation of diacid **DA** [28] or tetraacid **TA** [29], utilizing the Hunsdiecker reaction, allowed for an increase in the overall yield of **AD** up to 1.5 and 6.5%, respectively. The bottleneck is the methylenation of **ME**, which determines the generally low yields. Schleyer, one of the most influential organic chemists of the twentieth century, simplified and perfected the synthesis of **AD** [30], making it available on an unprecedented large scale [31]. Schleyer's synthesis (Scheme 1.1, bottom) is based on the thermodynamically favorable Lewis acid-catalyzed rearrangement of tetrahydrodicyclopentadiene (**THCPD**) to **AD** [32]. Such "stabilomeric synthesis" not only made **AD** available in large quantities and thus significantly promoted the

Scheme 1.1 Final steps in the synthesis of **AD** from Meerwein's ester (**ME**, top) and improved Schleyer's synthesis of **AD** from dicyclopentadiene (**DCPD**, bottom) through hydrogenation to tetrahydrodicyclopentadiene (**THCPD**) followed by Lewis acid-catalyzed rearrangement.

chemistry of cage compounds but also became very useful for the synthesis of larger polymantanes (vide infra).

The second diamondoid homologue, **DIA**, is also available synthetically and was first prepared by Schleyer and Cupas applying the same Lewis acid-catalyzed rearrangement procedure [8]. Photodimerization of norbornene afforded a cyclobutane-containing dimer **NBD**, which was treated with AlCl$_3$ and gave **DIA** (Scheme 1.2). As Schleyer noted [33], "Although the four-membered ring ... was not ideal (too much strain in rearrangement precursors normally leads to ring opening and tar formation), I ... suggested to Chris Cupas that he investigate the rearrangement. The first experiment failed but in the second experiment he noticed sublimed crystals on the cooler portion of the flask that proved to be congressane." This method unfortunately gave very low yields and produced a significant amount of tar, although it enabled X-ray analysis and confirmation of the **DIA** structure [34]. Schleyer further improved the synthesis to make it relevant for practical applications [35]. The modification included the use of a more efficient precursor, **Binor-S** [36], that, after hydrogenation and treatment with AlBr$_3$, gave **DIA** in a yield of over 60% for the last step (Scheme 1.2) [35, 37].

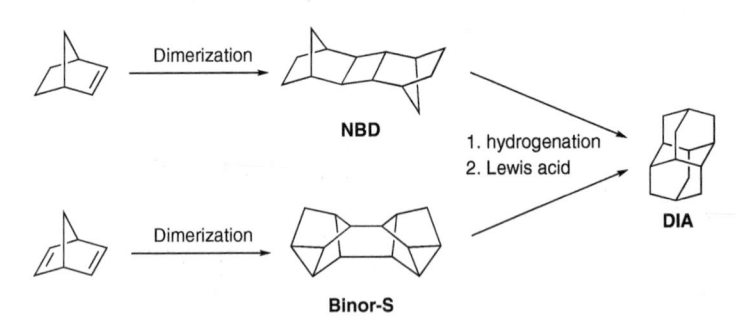

Scheme 1.2 First synthesis of **DIA** from the norbornene dimer (**NBD**) and multi-gram synthesis using **Binor-S** as a precursor.

Schleyer also prepared **TRIA** in a similar way from bis-cyclopropanated polycyclic dimer of cyclooctatetraene (**DCOT**) using the AlBr$_3$/tert-BuBr sludge catalyst [38]. Reductive cleavage of the cyclopropane rings followed by isomerization resulted in **TRIA**, albeit in a very low yield (under 5%) for the final step. McKervey made a slight improvement in **TRIA** synthesis by using **1COOHDIA** as a starting molecule that underwent cage rearrangement as the last step [39], and he could further improve the synthesis by using **Binor-S** heated on a platinum catalyst that gave cyclic olefins [40]. Diels–Alder reaction of these olefins with butadiene and Lewis acid-catalyzed rearrangement afforded the target hydrocarbon with satisfactory preparative yield (Scheme 1.3).

The attempt to expand the stabilomeric approach to the preparation of diamondoids higher than **TRIA** failed. In particular, the isomerization of C$_{22}$H$_{28}$ hydrocarbons derived from the hydrogenation of the Diels–Alder adduct of 1,3-cyclohexadiene with the 38.5 °C melting cyclooctatetraene dimer (**DCOT**) gave unwanted ethano-bridged isomer "bastardane" (**BAST**) rather than **121TET**

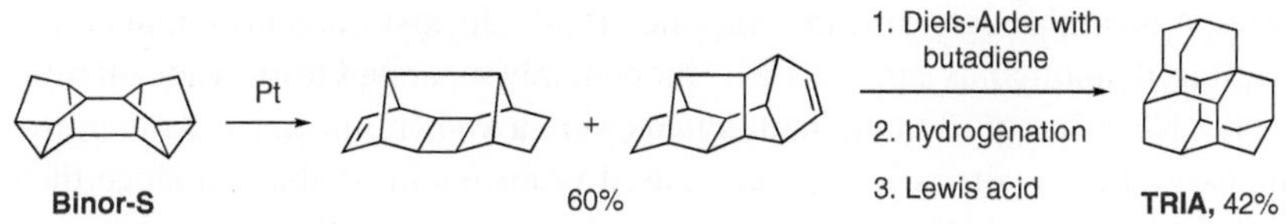

Scheme 1.3 Efficient synthesis of **TRIA** from **Binor-S** by McKervey.

(Scheme 1.4) [41]. Despite **BAST** being thermodynamically less stable than **121TET**, it is unreactive toward Lewis acids because further rearrangement is likely to involve unstable quaternary carbon intermediates.

Of the three possible **TET** isomers, only **121TET** has been successfully synthesized to date [42]. McKervey applied a ring expansion reaction on **DIA** 1,6-dicarboxylic acid (**16COOHDIA**) that resulted in cyclization to the isomeric polycyclic dienes that upon isomerization provided **121TET** (Scheme 1.4). The yield was, however, quite low (10%) but gave enough material to determine the X-ray crystal structure of the target hydrocarbon [43]. Later, it was noted that the final step of the reaction is highly sensitive to the variations in the temperature of the catalyst bed [44].

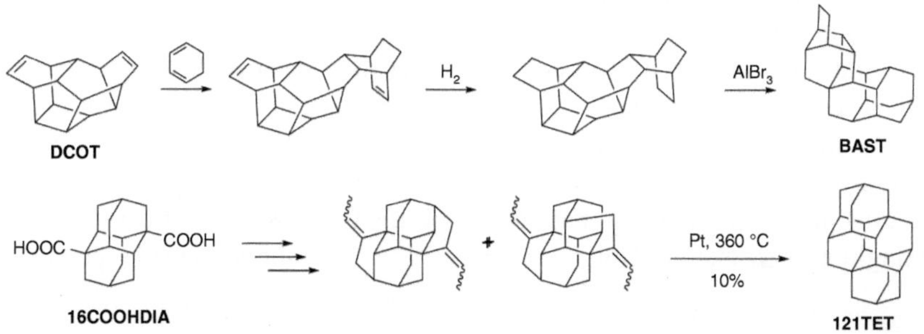

Scheme 1.4 An unsuccessful attempt to prepare **121TET** by stabilomeric synthesis (top) and multi-step synthesis from diamantane 1,6-dicarboxylic acid (**16COOHDIA**) by McKervey (bottom).

To date, no other diamondoid hydrocarbons aside from these four are available synthetically, although many attempts have been made to achieve this goal [45]. The reason is either in the presence of the "dead end" intermediates (like **BAST**) [41], or the increase in the number of isomers with similar thermodynamic stability as the cage size increases [45b, 46], meaning that the Lewis acid rearrangement procedure gives a complex mixture of products that cannot be easily separated.

However, there is a silver lining: diamondoids are readily available from nature! Since diamondoids are naturally occurring molecules that are formed by catagenesis in oil deposits, they can be isolated from mature crude oil, natural gas, and hydrocarbon-rich sediments, as described in Chapter 2. A historical curiosity is that **AD** was first isolated from an oil reservoir near Hodonin, Czechoslovakia, in 1933 [47] and was only later prepared synthetically [24, 28]. Other diamondoids are also present in oil [48] and may have been formed by Lewis acid-catalyzed rearrangement promoted by acidic minerals. The challenge is to harness them from complex oil mixtures. A breakthrough came in 2003 when Dahl, Liu, and Carlson demonstrated that various diamondoids can indeed be individually separated from crude oil mixtures [49]. Note that until then diamondoids were a known unwanted issue in the oil industry: these hydrocarbons often caused problems in production since they clogged pipes and machinery during oil refinement as they had the tendency to condense and solidify on the cool equipment parts. This behavior comes as no surprise since diamondoids have moderate solubility and high melting points (e.g., 268 °C for **AD**, 237 °C for **DIA**, and 221 °C for **TRIA**), while at the same time they are very volatile and resistant to oxidation. These issues arising in the industrial processing of oil actually provide a clue for the isolation and purification of individual diamondoids. Dahl et al. took advantage of these solidification-inducing properties and separated 21 different diamondoids by using a combination of various purification methods and multiple high-performance liquid chromatography (HPLC) runs. In the end, derivatives ranging from **TET**s up to undecamantane were identified, and X-ray analyses were used to unambiguously confirm the structures of some of them (Figure 1.7) [49, 50]. Even though it has been known for a long time that lower diamondoids are present in oil [47, 51], the occurrence of higher diamondoids in oil deposits was finally unequivocally proven by this isolation endeavor.

As was already mentioned, the synthetic preparation of higher diamondoids by carbocationic rearrangements was not feasible due to many possible isomeric structures that are often energetically comparable, thus leading to reaction pathways where the intermediates become trapped in local energy minima, resulting in the formation of many unwanted side products. However, we saw that higher diamondoids exist in oil, and they are thermodynamically the most stable hydrocarbons, meaning that their chemical synthesis in the laboratory simply must be possible.

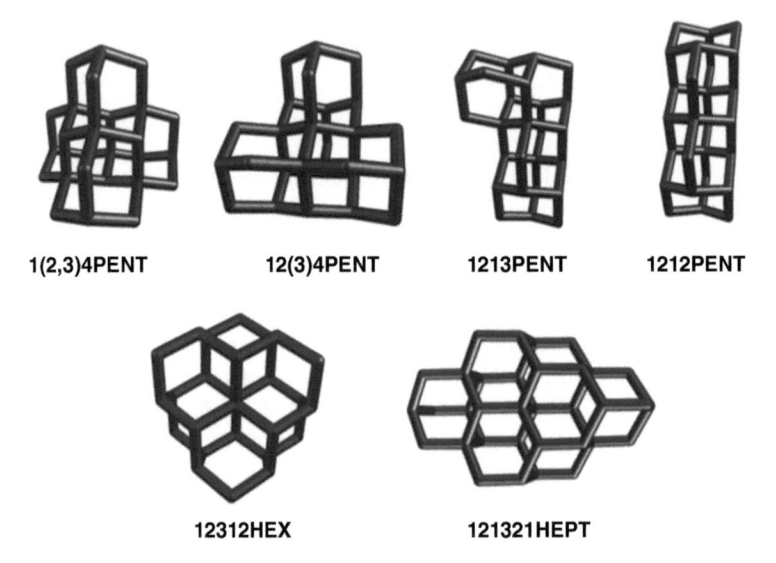

Figure 1.7 Structures of higher diamondoids obtained from X-ray crystal analyses [49, 50]. Hydrogens were omitted for clarity.

1.3 Preparation of Diamondoids

Inspired by this rationale, a groundbreaking preparative procedure has been developed that provides synthetic access to the higher homologs [52]. Since diamondoids form in oil that has undergone thermal cracking, the suggestion was that free-radical cracking reactions are an alternative to the proposed acid-catalyzed carbocationic rearrangements. Astonishingly, upon mimicking the reaction conditions that are present during cracking processes, **TET** and **PENT** isomers form when **TRIA** is subjected to sealed tube pyrolysis experiments. The obtained amounts were small, and alkylated **TRIA** derivatives were the predominant products; however, this experiment provided an excellent proof-of-principle that radical pathways must not be disregarded for diamondoid generation. Similarly, **PENT** isomers could be produced from **TET**s using the same procedure, and the most favored stereoisomers were those with the least steric crowding (Figure 1.8).

In addition, **PENT** isomers that require breaking and reconstruction of the parent **TET** cage were disfavored. This finding strongly implies that the parent diamondoid cage actively directs further cage growth and that only surface hydrogens are involved in the bond construction process. The addition of isobutene

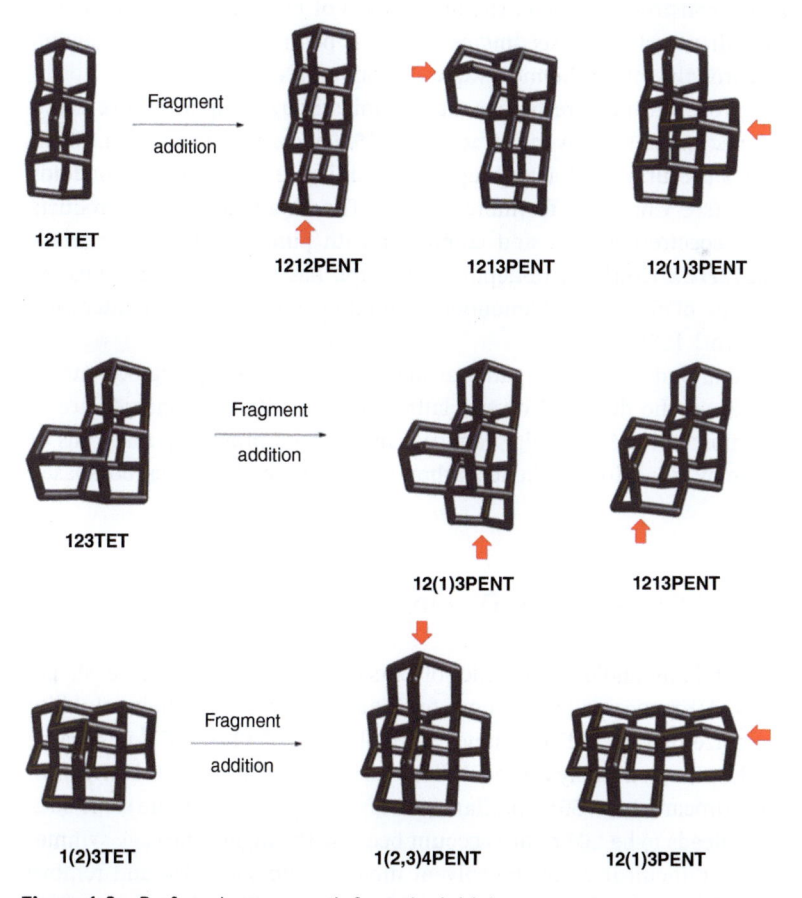

Figure 1.8 Preferred cage growth from the initial tetramantane isomers from the least crowded apical positions of the cage. Hydrogens were omitted for clarity.

Figure 1.9 Cutout from the crystal lattice of diamond with the superimposed structure of the **AD** cage (red) and SEM image of a **12312HEX** crystal magnified 150 times. Hydrogens were omitted for clarity. Source: Reproduced from Ref. [50], Wiley-VCH, 2003.

greatly improved the reaction yields since the isobutyl moiety is essentially all that is needed for the construction of the next cage. What is still unclear is the exact mechanistic pathway for this ring closure because it is possible that introduced isobutyl sources form either free radicals that add one carbon atom at a time or form isobutyl radicals that add directly to the cage. The merit of the described sequential cage construction lies in the preparation of microcrystalline diamond: by using large diamondoids as seeding centers, it is possible to trigger nucleation and diamond growth akin to chemical vapor deposition methods [52].

Another approach to prepare higher diamondoids is by using a plasma and a supercritical medium as the solvent, either xenon [53] or carbon dioxide [54]. When **AD** is used as a precursor and a seeding center, a mixture of higher diamondoids typically forms (see Chapter 6 for more details). The obtained reaction products were analyzed spectroscopically and compared with pure samples where available, e.g., **12312HEX** [53a]. An in-depth monograph has recently been published describing the use of plasma for diamondoid generation, and we refer the interested reader to that work [55].

Since nanodiamonds have the spatial arrangement of their cage carbon atoms superimposable on the diamond crystal lattice (Figure 1.9), diamondoids are at the core of nanoscience [56]. Their properties make them promising materials of the future, side by side with graphene, carbon nanotubes, and other carbon-based entities.

1.4 Physical Properties of Diamondoids

The solubility of diamondoids in organic solvents correlates with their size, shape, and symmetry. A practical rule of thumb for diamondoids is that their solubility drops as their size increases. For instance, the solubility of **12312HEX** in organic solvents is so low that it initially caused difficulties in measuring the ^{13}C NMR spectrum of this hydrocarbon [5, 50]. For diamondoids of approximately the same size, symmetry also needs to be taken into account because the higher the cage symmetry is, the more difficult it is for the solvent molecules to solubilize and remove the molecules from the solid state. For example, the solubility of apically substituted **DIA** derivatives with axial symmetry is lower than that of the analogously substituted medial compounds. As expected, the parent hydrocarbons have higher solubility in nonpolar solvents (Figure 1.10) [57], but it is lower than that of other alicyclic hydrocarbons. As previously mentioned, poor diamondoid solubility and high volatility often cause technical difficulties in oil refinement since they are prone to clogging pipes when present in substantial amounts due to continuous processing.

Nanotechnological diamondoid applications require highly pure starting materials. Commonly used procedures for diamondoid purification include multistage distillation or HPLC [49]. Despite being effective, both methods have some drawbacks from ecological and economical points of view: the former consumes large quantities of energy, while the latter uses large amounts of organic solvents. In addition, HPLC purification suffers from the problem of having to use a suitable detector because diamondoid solutions are not easily detectable with common HPLC analyzers. This leaves the differential refractive index as the only option for detection, but it is three to four orders of magnitude less sensitive than other common detectors. These purification issues were addressed and circumvented by using vapor transport techniques to grow large diamondoid single crystals [58]. The process enables efficient growth of crystals up to around 1 cm^3 (Figure 1.11) under mild conditions without using organic solvents because diamondoids readily form ordered crystals at room temperature, a feature not common for other saturated hydrocarbons. Moreover, with their relatively high vapor pressures, diamondoids can easily sublime. Thus, single crystal growth of diamond molecules through the vapor phase is a viable and effective method for diamondoid purification, a technique that can readily be controlled by temperature manipulation.

Other physical and chemical properties of diamondoid cages that are a direct consequence of their diamond-like structure include high lipophilicity, structural rigidity, thermodynamic stability, high melting points, and facile crystallization [32]. Their often encountered high symmetry can also aid in their structural characterization, as Raman and IR spectroscopies are often a method of choice for analyzing saturated hydrocarbons [59]. In addition to assigning the obtained spectral peaks to their vibrational modes, vibrational spectroscopy is also useful for in-depth analysis of bulk materials and for studying phase changes of diamondoid self-assemblies. As diamondoids are typically available as micrometer-sized powders, knowing their optical properties, like the refractive index, is a useful tool for optoelectronic device applications. For molecular diamondoids, the refractive index increases with decreasing wavelength and increasing polymantane order [60]. Still, the refractive index of the largest diamondoid, **12312HEX** (1.68 at 589 nm), is much lower than that of microscopic diamond (2.42 at 589 nm, Figure 1.12).

Lower diamondoids have low dielectric constants in the range of 2.46 to 2.65, which is significantly lower than that of bulk diamond ($\kappa = 5.66$) or SiO_2 ($\kappa = 4.5$) [61]. This property is postulated to be a consequence of the lower density of diamondoid crystals when compared to, e.g., bulk diamond. Diamondoids therefore have the potential to be used for producing insulating interlayers in nanoelectronic devices.

One of the most fascinating mechanical properties of diamondoid crystals is their resilience in high-pressure environments. This observation is not surprising considering how diamond forms in nature, but it still deserves mentioning in

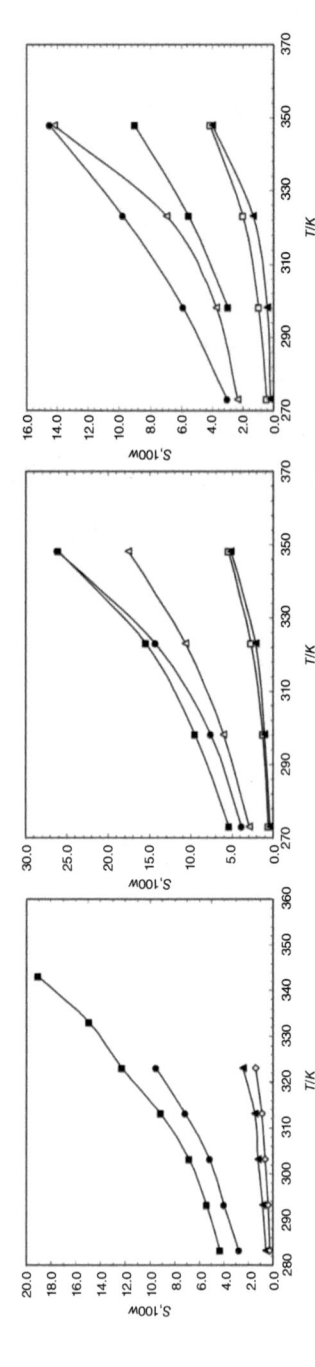

Figure 1.10 Mass fraction solubility (S) of **DIA** (left), **TRIA** (middle), and **121TET** (right) in acetone (▲), ethyl acetate (▲), toluene (●), and cyclohexane (♦) at various temperatures. Source: Reproduced from Ref. [57] with permission from the American Chemical Society, 2008.

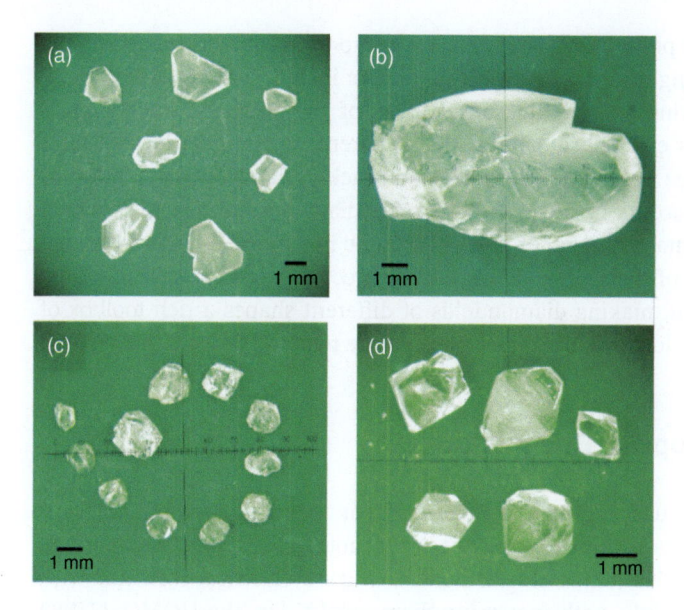

Figure 1.11 Single crystals of **AD** (a,b), **DIA** (c), and **121TET** (d). Source: Reproduced from Ref. [58] with permission from the American Chemical Society, 2010.

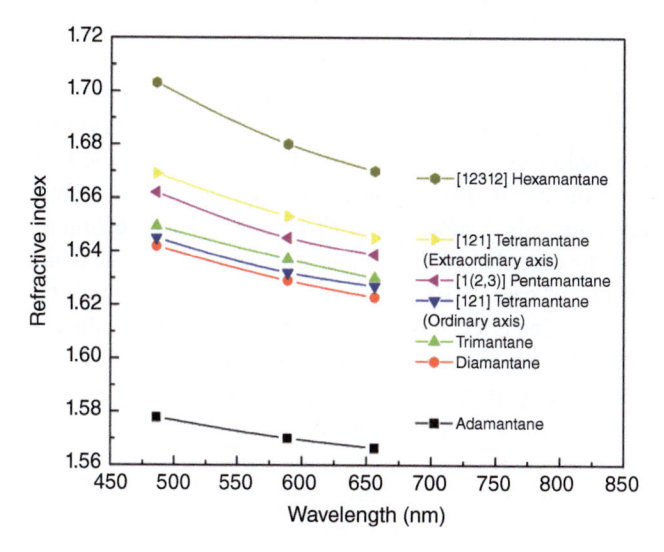

Figure 1.12 Wavelength dependence of refractive index (n) for diamondoids. Source: Reproduced from Ref. [60] with permission from Elsevier, 2008.

the context of their usefulness. Diamondoids at high-pressure conditions readily undergo solid–solid phase transitions without polymerization or decomposition [62]. Such phase transitions can be stimulated by both changes in temperature and/or pressure and often have low energy barriers. For example, **121TET** undergoes a phase transition from the monoclinic $P2_1/n$ structure to a triclinic $P1$ phase starting at pressures of about 13 GPa, and when the pressure is subsequently released, a new solid phase having the *Pc* space group forms at ambient conditions [63]. Both the starting and the ending solid phases in this process have comparable stability, pointing toward the conservation of the structural motifs upon decompression. This example illustrates that different polymorphs formed after pressurization change only their intermolecular packing, and no degradation of the cage molecules occurs because of their stable diamond-like structure, which is a significant advantage over other types of organic materials. What is more, the molecular geometry of the individual cages also affects the compressibility of the corresponding crystal, making diamondoids of different shapes a rich toolbox of nano-sized building blocks for organic, high-pressure material design.

1.5 Spectroscopy of Diamondoids

The uniqueness of unfunctionalized diamondoids when compared to other hydrocarbons becomes especially apparent when noting their spectroscopic and photophysical properties [61]. Bulk diamond is a good electrical insulator with a large band gap. Luckily, with increasing diamondoid size, the HOMO–LUMO gap decreases and shows a tendency to converge close to the fundamental gap of bulk diamond [64]. For example, larger nanodiamonds with sizes up to 2 nm have computed band gap values of 6.7 eV as compared to 5.5 eV for bulk diamond [65]. Structurally well-defined nanodiamonds display strong quantum confinement effects at particle sizes ranging from 0.5 to at least 2 nm, but it is the size and not the shape of the particles that affect the band gap values. This was demonstrated by the comparison of octahedral vs. tetrahedral nanodiamonds that in the end showed the same trends in band gap narrowing. Since there appears to exist a limit to gap narrowing for nanodiamonds, a proposed method for further reduction of the gap is the introduction of external or internal doping substituents (C—H bond substitution and replacement of cage CH/CH_2 groups, respectively). As will be demonstrated in Chapter 12, doping is a very successful strategy for semiconductor applications of diamondoids. Note that such size-property relationships hold true not only for diamondoids but for parent diamond nanoparticles as well: particle sizes must be around 2 nm to start observing an increase in the optical gap (vide infra) [66]. In other words, reducing the size of bulk diamond to the nanoscale has a pronounced effect on the optical gap of the resulting nanoparticle, with a decrease in the optical gap by increasing nanoparticle size being more rapid for diamond than for, e.g., Si or Ge, where quantum confinement effects persist up to 6–7 nm.

Another computational study on diamondoids found that quantum confinement effects essentially disappear in diamondoid structures larger than 1 nm [67]. However, the applied Monte Carlo computations predict a small exciton binding energy and a negative electron affinity (NEA) for nanodiamonds of that size, which is a consequence of significant LUMO delocalization (Figure 1.13). Note here that the HOMO is predominately localized on the C—C bonds inside the nanoparticle, while the delocalized LUMO has a considerable probability outside of the surface boundaries and is mostly composed of the C—H bonds. With increasing

(a) HOMO   (b) LUMO

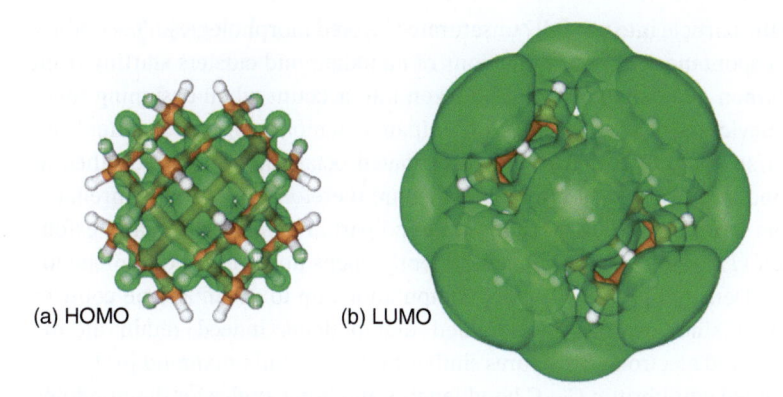

Figure 1.13 Isosurface plots of the square of the (a) HOMO and (b) LUMO of hydrogen-terminated spherical diamond-like nanoparticle $C_{29}H_{36}$. The green isosurfaces include 50% of the charge in each orbital. Source: Reproduced from Ref. [67] with permission from the American Physical Society, 2005.

nanoparticle size, the HOMO ultimately converges to the valence-band maximum of bulk diamond, while the LUMO remains near the surface boundary and does not become a conduction band minimum. Such an almost defect-like nature of the LUMO is the reason why the optical gaps of large diamondoids lie below the gap value of bulk diamond.

As the HOMOs predominately describe the C—C bonding, the vibrational motion of the carbon nuclei has a significant effect on the orbital energies, while the higher-lying orbitals are often Rydberg states that are more sensitive to the vibrational motion of the hydrogen atoms. It follows that in order to reliably compute the optical properties of diamondoids, quantum nuclear dynamics are needed to ensure accurate prediction of their photophysics [68]. Despite such theoretical requirements, recent time-dependent density functional theory (TD-DFT) approaches for vibrationally resolved photoelectron spectra simulations succeeded in coming very close to fully reproducing the experimentally observed vibrational fine structure [69]. What still remains to be tackled in that particular computational approach are errors in computed spectral redshifts upon increasing particle size due to the lack of many-body corrections, a trait that is inherent to the method, as well as the absence of satellites in the high-energy region of the spectra due to electron-nuclear coupling. Further computational studies confirmed that the high diamondoid symmetries lead to forbidden transitions [70]. However, it was also noted that using HOMO–LUMO gaps for the approximation of diamondoid optical gaps was somewhat of an oversimplification, especially since their absorption and emission spectra are vibronically highly structured and often accompanied by broadened and shifted optical bands.

When diamondoids approach the 1 nm range, their crystal morphology itself begins to affect their structural integrity [71]. Theoretical models predicted that while the initial cubic morphology provides relaxation structures comparable to bulk diamond itself, octahedral and cuboctahedral starting materials gradually transform from a pure sp^3 to a mixed sp^3 and sp^2 bonding network, essentially converting the particle into partially unsaturated layered morphologies. This finding implies that spontaneous phase transitions of nanodiamond clusters starting from the outer carbon atom shells need to be taken into account when designing functional nanodevices. However, hydrogen-terminated diamondoids do not suffer from such layer transformation effects. Dehydrogenated octahedral and cuboctahedral nanodiamond particles up to 1 nm in diameter are therefore structurally unreliable and unstable morphologies that readily undergo partial graphitization and exfoliation of their (111) surfaces, which consequently opens up an important niche for applications. Density functional theory computations up to decamantane confirm this reasoning since hydrogen-terminated diamondoids indeed retain atomic arrangements and electronic structures similar to those of bulk diamond [64].

The computed equilibrium C—C bond lengths and bond angles for diamondoids are comparable to those of bulk diamond, providing them with diamond-like properties, especially as the size of the diamondoids increases. Quantum confinement effects were studied using X-ray absorption spectroscopy, and the resulting diamondoid gas phase spectra (Figure 1.14) show that blueshifts in the band edges do not occur with decreasing particle size, as would be expected for typical group IV semiconductors [72]. In other words, diamondoid clusters display some differences from bulk diamond, which can be advantageous for some applications.

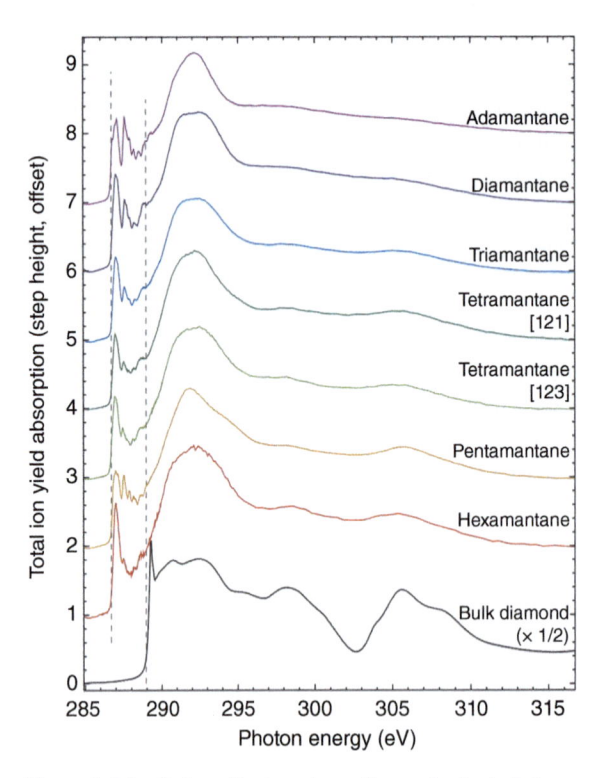

Figure 1.14 Carbon K-edge absorptions of selected diamondoids and bulk diamond. Source: Reproduced from Ref. [72] with permission from the American Physical Society, 2005.

While the information on HOMO–LUMO gaps is fundamentally important, the optical gaps (based on symmetry-allowed transitions) provide more relevant information on the properties of diamondoids. The optical gaps of diamondoids measured in the gas phase [73] are in agreement with the experimental and computed trends in changes of the HOMO–LUMO gaps but display strong shape dependence (Figure 1.15) within different topological families (3D, 2D, and 1D denote tetrahedral, prism, and rod-shaped diamondoids, respectively). This is due to the forbidden HOMO → LUMO transitions for some highly symmetric structures. The optical gaps primarily originate from the single-particle transition from HOMO to LUMO or to LUMO+1 and LUMO+2 depending on the selection rule. For instance, **TET** with C_{2h} symmetry (**121TET**) has the first dipole active transition between HOMO and LUMO+2, while HOMO → LUMO transitions are allowed for C_{3v} and C_2 **TET**s. Such symmetry dependence of optical gaps and excitation energy agreed fully with the experiment (Table 1.3).

While the optical gaps of $C_{26}H_{32}$ **1213PENT** (5.75 eV) and bulk diamond display remarkable similarities, the absolute value of 5.5 eV is still unreachable with diamondoids of this size. This agrees well with computations on larger diamondoids that also predict a rapid decrease of the absorption gap (Figure 1.16, left), which becomes similar to the computation for bulk diamond only for particles of about 1 nm in diameter ($C_{87}H_{78}$) [75].

More detailed TD-DFT computations on small diamondoids with the PBE0 method [74b] reveal that the first 3s-like Rydberg excitation is common for all diamondoids (Figure 1.16, right) except for C_{2h}-symmetric **121TET** and **12312HEX**, where such excitations are dipole-forbidden and, therefore, 3p-like Rydberg transitions occur instead. For tetrahedral **AD** and **1(2,3)4PENT** both $t_2 \to a_1(3s)$ and

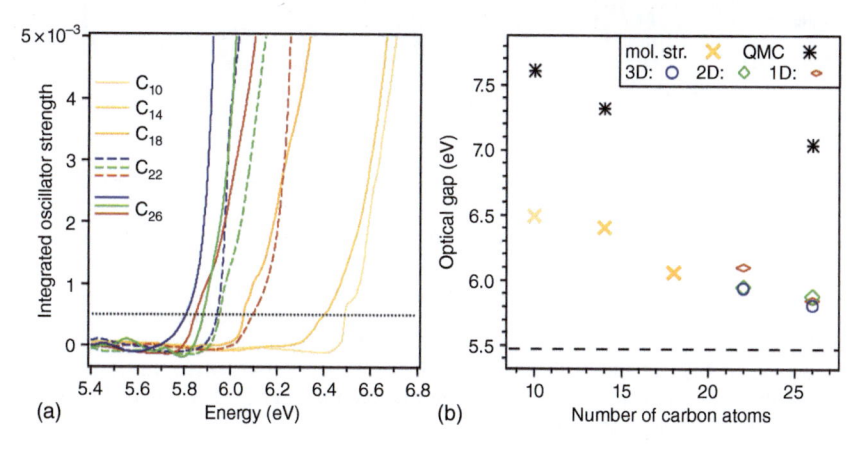

Figure 1.15 (a) Integrated oscillator strength measured for diamondoids **AD** (C_{10}), **DIA** (C_{14}), **TRIA** (C_{18}), as well as for isomeric tetramantanes (C_{22}) and pentamantanes (C_{26}). The threshold defining the optical gap is marked by a dotted line. (b) Experimental optical gaps as a function of size compared to optical gaps derived by quantum Monte Carlo (QMC) calculations. The dashed line marks the energy gap of bulk diamond, and 3D, 2D, and 1D denote tetrahedral, prism, and rod-shaped diamondoids, respectively. Source: Reproduced from Ref. [73] with permission from the American Physical Society, 2009.

1 Description of Diamondoids

Table 1.3 Ground state symmetries, TD-PBE0/aug-cc-pVTZ computed and experimental optical gaps (E_{gap}), and excitation energies of selected diamondoids based on photoluminescence measurements (Source: From [73, 74]).

Diamondoid	Symmetry	Formula	E_{gap} (eV) Comp.	E_{gap} (eV) Exp.	Excitation energy (eV) Exp.
AD	T_d	$C_{10}H_{16}$	6.66	6.49	6.49
DIA	D_{3d}	$C_{14}H_{20}$	6.75	6.40	6.60
TRIA	C_{2v}	$C_{18}H_{24}$	6.12	6.06	6.78
121TET	C_{2h}	$C_{22}H_{28}$	6.25	**6.10**	6.88
1(2)3TET	C_{3v}	$C_{22}H_{28}$	6.04	**5.94**	5.99
123TET	C_2	$C_{22}H_{28}$	6.01	**5.95**	5.98
1(2,3)4PENT	T_d	$C_{26}H_{32}$	5.99	**5.81**	5.93
12(1)3PENT	C_1	$C_{26}H_{32}$	–	5.83	6.20
1212PENT	C_{2v}	$C_{26}H_{32}$	5.86	5.85	–
1213PENT	C_1	$C_{26}H_{32}$	–	5.75	6.31

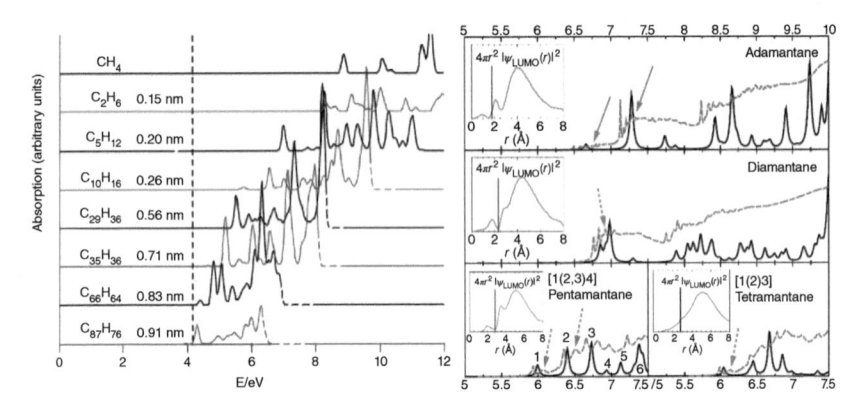

Figure 1.16 *Left*: TD-DFT-computed absorption spectra of diamondoids up to 1 nm in diameter. The vertical dashed line corresponds to the computed absorption gap of bulk diamond. (Source: Reproduced from Ref. [75] with permission from Elsevier, 2005.) *Right*: Computed and experimental absorption spectrum for selected diamondoids. The green (dashed) line shows the experimental spectrum, while the black (continuous) line is the computed spectrum. Inset: radial distribution of 3s-like Rydberg state, where vertical line represents the radius of diamondoid. Source: Reproduced from Ref. [74b] with permission from the American Physical Society, 2009.

$t_2 \rightarrow t_2(3p)$ transitions are allowed, but the latter has a higher density of states that leads to a larger absorption for the 3p-Rydberg excitation [75].

Diamondoids display intrinsic photoluminescence in the ultraviolet spectral region [76]. For example, **AD** exhibits a quite broad UV luminescence band when photoexcited in the gas phase above its principal optical gap of 6.49 eV (Table 1.3)

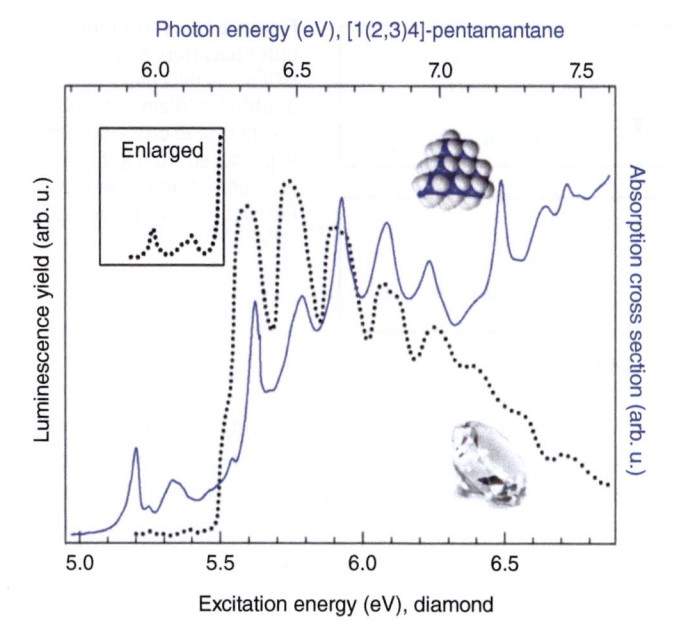

Figure 1.17 Comparison of the optical absorption of **1(2,3)4PENT** (blue line) with the absorption of high-purity type IIa diamond (dotted line). The spectral inset shows the enlarged onset of the bulk diamond spectra. Source: Reproduced from Ref. [73] with permission from the American Physical Society, 2009.

[73]. The optical gap characteristic for diamondoids thus lies in the UV spectral region [77], and the luminescence behavior is attributed to a transition from the delocalized first excited state into different vibrational modes of the electronic ground state [74a]. The majority of such transitions originate from vibrational modes associated with CH wagging and CH_2 twisting of the diamondoid surface atoms (vide infra) [78]. Difficulties in computationally predicting exact lineshapes of diamondoid photoemission spectra arise from electron-vibration coupling that cannot be corrected by applying simple vibration-broadening corrections to the corresponding electronic states [79]. Importantly, the overall optical absorption of diamondoids noticeably changes as a function of their size and shape, and in the case of **1(2,3)4PENT**, the spectrum begins to resemble that of bulk diamond (Figure 1.17). In other words, optical properties change as larger diamondoids are used, demonstrating that diamondoids could form valuable semiconductor nanocrystals for application in light-emitting devices in the deep UV spectral region.

1.6 Ionization Potentials

Since spectral data suggest that diamondoids are also present in space [80], matrix isolation techniques were used to mimic astrophysical conditions. After the deposition of diamondoids in a neon matrix, they were irradiated with high-energy photons, and UV absorption spectra were recorded. This revealed the presence of

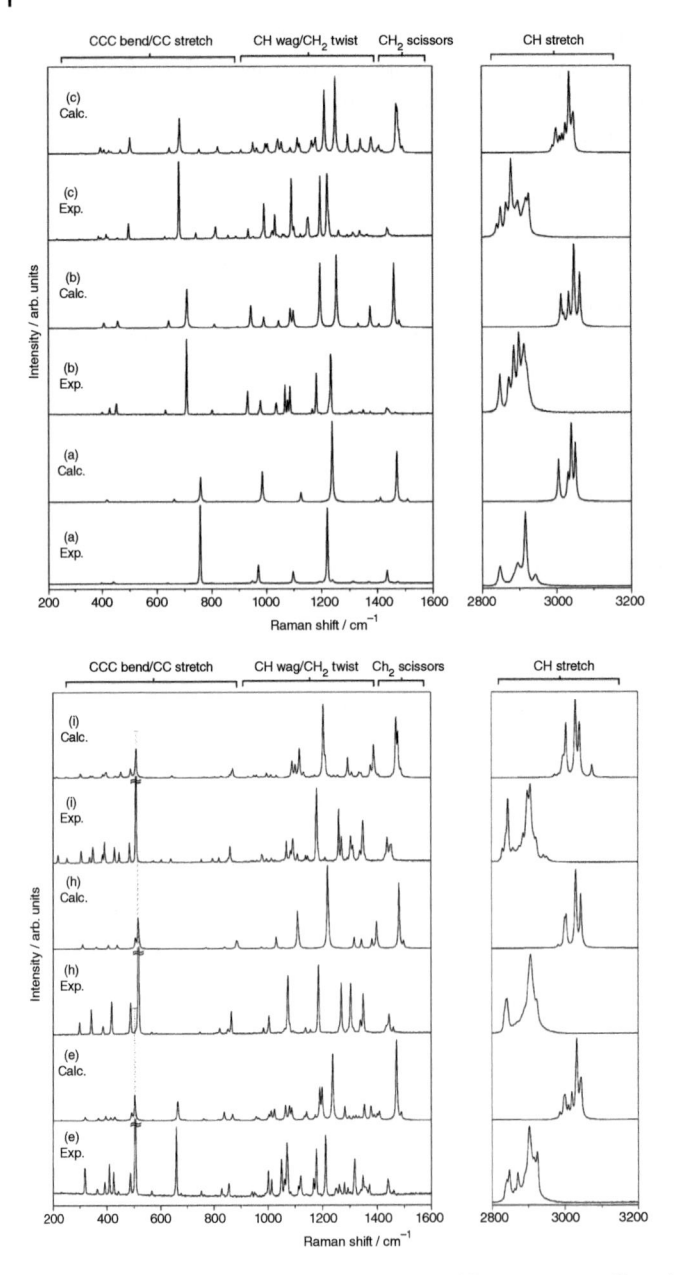

Figure 1.19 *Top*: Experimental and computed Raman spectra (from bottom) for (a) **AD**, (b) **DIA**, (c) and **TRIA**. *Bottom*: Experimental and computed spectra (from bottom) for (e) **1(2)3TRIA**, (h) **1(2,3)4PENT**, and (i) 3-methyl-**1(2,3)4PENT**. Source: Reproduced from Ref. [95] with permission from Elsevier, 2006.

Figure 1.20 Molecular IR absorption of **121TET** (top) and **123TET** (bottom) deposited on Au(111) surface. The blue line shows the unresolved broad peak from STM inelastic electron tunneling spectroscopy (IETS) measurements. Source: Reproduced from Ref. [97] with permission from the American Physical Society, 2013.

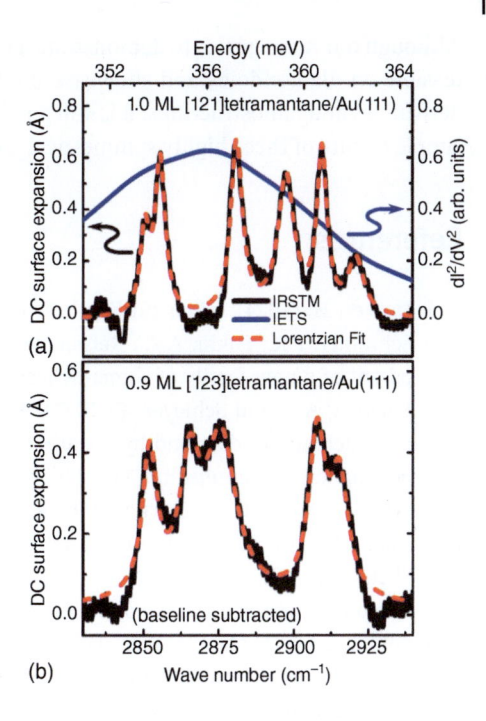

group introduction (so-called "doping" that will be discussed in detail in Chapter 12) can be performed using readily available Raman spectroscopy. In a subsequent study, diamondoid dimers and trimers connected with either a single or a double C—C bond and consisting of combinations of **AD** and **DIA** cages were characterized with valence photoelectron spectroscopy [100]. When a double bond is present in the molecule, it significantly impacts the electronic structure of the HOMO, in contrast to molecules with connecting single bonds, where the impact is only small. Moreover, the orbital superposition of both cages in the singly bonded particles determines the overall electronic structure of such systems. This combination of orbitals directly influences the ionization potentials, as homo-dimers have IP values below those of the corresponding monomers, and the measured IPs for hetero-dimers strongly depend on particle composition. Composite diamondoids like **DIA** dimers can also form van der Waals crystals, and it was confirmed that the central double bond predominately influences optical properties both in the gas phase, where single molecules are observed, and in the crystalline solid state [101]. Note that optical band gaps in such systems are significantly lowered (by 0.5–1.0 eV) when compared to their saturated analogues.

As we have seen, diamondoids underwent much turbulence in the course of their history. They evolved from being structural curiosities isolated from oil to inspiring target molecules that challenged the limits of synthetic organic chemistry, to becoming multi-gram resources available from nature and useful for medicine, and finally, to affording modern-day applicability as unique building blocks in nanomaterial design. It is satisfying to see that this fascinating class of compounds went full circle: it was first found in and today is efficiently extracted from oil deposits.

Although our main goal is to demonstrate a wide range of chemical transformations feasible on diamondoids and showcase the fascinating properties of the prepared derivatives and nanostructures, it is still good to pause for a moment and admire the simple beauty of these highly symmetrical jewels of nature.

References

1 Decker, H. (1924). Wege zur Synthese des Diamanten. *Z. Angew. Chem.* 37: 795.
2 Schreiner, P.R., Fokin, A.A., Reisenauer, H.P. et al. (2009). [123]Tetramantane: parent of a new family of sigma-helicenes. *J. Am. Chem. Soc.* 131: 11292–11293.
3 Balaban, A.T. and Schleyer, P.v.R. (1978). Systematic classification and nomenclature of diamond hydrocarbons-I: graph-theoretical enumeration of polymantanes. *Tetrahedron* 34: 3599–3609.
4 (a) Shenderova, O.A., Zhirnov, V.V., and Brenner, D.W. (2002). Carbon nanostructures. *Crit. Rev. Solid State Mater. Sci.* 27: 227–356. (b) Mochalin, V.N., Shenderova, O., Ho, D., and Gogotsi, Y. (2012). The properties and applications of nanodiamonds. *Nat. Nanotechnol.* 7: 11–23.
5 Fokin, A.A., Tkachenko, B.A., Fokina, N.A. et al. (2009). Reactivities of the prism-shaped diamondoids [1(2)3]tetramantane and [12312]hexamantane (Cyclohexamantane). *Chem. Eur. J.* 15: 3851–3862.
6 Eckroth, D.R. (1967). A method for manual generation of correct von Baeyer names of polycyclic hydrocarbons. *J. Org. Chem.* 32: 3362–3365.
7 (a) Balaban, A.T. (2012). Partitioned-formula periodic tables for diamond hydrocarbons (diamondoids). *J. Chem. Inf. Model.* 52: 2856–2863. (b) Balaban, A.T. (2013). Diamond hydrocarbons and related structures. In: *Diamond and Related Nanostructures* (ed. M.V. Diudea and C.L. Nagy), 1–27. Dordrecht: Springer Netherlands. (c) Balaban, A.T. and Rucker, C. (2013). How to specify the structure of substituted blade-like zigzag diamondoids. *Cent. Eur. J. Chem.* 11: 1423–1430.
8 Cupas, C., Schleyer, P.v.R., and Trecker, D.J. (1965). Congressane. *J. Am. Chem. Soc.* 87: 917–918.
9 Balaban, A.T. (2013). Diamond hydrocarbons revisited: partitioned formula tables of diamondoids. *J. Math. Chem.* 51: 1043–1055.
10 Wang, Y.T., Zhao, Y.J., Liao, J.H., and Yang, X.B. (2018). Theoretical investigations on diamondoids (C_nH_m, n=10–41): nomenclature, structural stabilities, and gap distributions. *J. Chem. Phys.* 148: 014306.
11 Willey, T.M., Lee, J.R.I., Brehmer, D. et al. (2021). X-ray spectroscopic identification of strain and structure-based resonances in a series of saturated carbon-cage molecules: Adamantane, twistane, octahedrane, and cubane. *J. Vac. Sci. Technol. A* 39: 053208.
12 Engler, E.M., Andose, J.D., and Schleyer, P.v.R. (1973). Critical evaluation of molecular mechanics. *J. Am. Chem. Soc.* 95: 8005–8025.
13 Schleyer, P.v.R., Williams, J.E., and Blanchard, K.R. (1970). Evaluation of strain in hydrocarbons. The strain in adamantane and its origin. *J. Am. Chem. Soc.* 92: 2377–2386.

14 (a) Pitzer, K.S. (1955). London force contributions to bond energies. *J. Chem. Phys.* 23: 1735–1735. (b) Pitzer, K.S. and Catalano, E. (1956). Electronic correlation in molecules III. The paraffin hydrocarbons. *J. Am. Chem. Soc.* 78: 4844–4846.

15 Wodrich, M.D., Wannere, C.S., Mo, Y. et al. (2007). The concept of protobranching and its many paradigm shifting implications for energy evaluations. *Chem. Eur. J.* 13: 7731–7744.

16 Joyce, J.P., Shores, M.P., and Rappe, A.K. (2020). Protobranching as repulsion-induced attraction: a prototype for geminal stabilization. *Phys. Chem. Chem. Phys.* 22: 16998–17006.

17 Dorofeeva, O.V. and Ryzhova, O.N. (2019). Enthalpies of formation of diamantanes in the gas and crystalline phase: comparison of theory and experiment. *Struct. Chem.* 30: 615–621.

18 Rablen, P.R. (2020). A procedure for computing hydrocarbon strain energies using computational group equivalents, with application to 66 molecules. *Chemistry* 2: 347–360.

19 Wheeler, S.E., Houk, K.N., Schleyer, P.v.R., and Allen, W.D. (2009). A hierarchy of homodesmotic reactions for thermochemistry. *J. Am. Chem. Soc.* 131: 2547–2560.

20 (a) Gronert, S. (2009). The folly of protobranching: turning repulsive interactions into attractive ones and rewriting the strain/stabilization energies of organic chemistry. *Chem. Eur. J.* 15: 5372–5382. (b) Kemnitz, C.R. (2013). Electron delocalization explains much of the branching and protobranching stability. *Chem. Eur. J.* 19: 11093–11095.

21 Fokin, A.A., Tkachenko, B.A., Gunchenko, P.A. et al. (2005). Functionalized nanodiamonds Part I. An experimental assessment of diamantane and computational predictions for higher diamondoids. *Chem. Eur. J.* 11: 7091–7101.

22 Bondi, A. (1964). van der Waals volumes and radii. *J. Phys. Chem.* 68: 441–451.

23 Böttger, O. (1937). Über einige organische Verbindungen "diamantoider" Struktur. *Ber. Dtsch. Chem. Ges.* 70: 314–325.

24 Prelog, V. and Seiwerth, R. (1941). Über die Synthese des Adamantans. *Chem. Ber.* 1644–1648.

25 Meerwein, H., Kiel, F., Klösgen, G., and Schoch, E. (1922). Über bicyclische und polycyclische Verbindungen mit Brückenbindung. Über das Bicyclo-[1,3,3]-nonan und seine Abkömmlinge. *J. Prakt. Chem.* 104: 161–206.

26 Seiwerth, R. (1996). Prelog's Zagreb school of organic chemistry (1935–1945). *Croat. Chem. Acta* 69: 379–397.

27 Meerwein, H. and Schürmann, W. (1913). Über eine Synthese von Abkömmlingen des Bicyclo-[1,3,3]-nonans. *Liebigs Ann.* 398: 196–242.

28 Prelog, V. and Seiwerth, R. (1941). Über eine neue, ergiebigere Darstellung des Adamantans. *Ber. Bunsen. Phys. Chem* 1769–1772.

29 Stetter, H., Bänder, O.E., and Neumann, W. (1956). Über Verbindungen mit Urotropin-Struktur, VIII. Mitteil: Neue Wege der Adamantan-Synthese. *Chem. Ber.* 89: 1922–1926.

30 Schleyer, P.v.R. (1957). A simple preparation of adamantane. *J. Am. Chem. Soc.* 79: 3292.

31 Schleyer, P.v.R., Donaldson, M.M., Nicholas, R.D., and Cupas, C. (1962). Adamantane (tricyclo[3.3.1.1^{3,7}]decane). *Org. Synth.* 42: 8.

32 Fort, R.C. and Schleyer, P.v.R. (1964). Adamantane: consequences of the diamondoid structure. *Chem. Rev.* 64: 277–300.

33 Schleyer, P.v.R. and Streitwieser, A. (2015). *From the Ivy League to the Honey Pot, The Foundations of Physical Organic Chemistry: Fifty Years of the James Flack Norris Award*, 169–198. Washington, DC: ACS Symposium Series, American Chemical Society.

34 Karle, I.L. and Karle, J. (1965). Crystal and molecular structure of congressane $C_{14}H_{20}$ by X-ray diffraction. *J. Am. Chem. Soc.* 87: 918–920.

35 Gund, T.M., Williams, V.Z., Osawa, E., and Schleyer, P.v.R. (1970). A convenient, high-yield preparation of diamantane (congressane). *Tetrahedron Lett.* 3877–3880.

36 Gund, T.M., Osawa, E., Williams, V.Z. et al. (1974). Preparation of diamantane. Physical and spectral properties. *J. Org. Chem.* 39: 2979–2987.

37 Gund, T.M., Thielecke, W., and Schleyer, P.v.R. (1973). Diamantane - pentacyclo[7.3.1.1^{4,12}.0^{2,7}.0^{6,11}]tetradecane (butanetetraylnaphthalene, 3,5,1,7-1,2,3,4- decahydro). *Org. Synth.* 53: 30–33.

38 Williams, V.Z., Schleyer, P.v.R., Gleicher, G.J., and Rodewald, L.B. (1966). Triamantane. *J. Am. Chem. Soc.* 88: 3862–3863.

39 Burns, W., McKervey, M.A., and Rooney, J.J. (1975). New synthesis of triamantane involving a novel rearrangement of a polycyclic olefin in the gas phase on platinum. *J. Chem. Soc. Chem. Commun.* 965–966.

40 (a) Hamilton, R., McKervey, M.A., Rooney, J.J. et al. (1976). A short synthesis of triamantane. *J. Chem. Soc. Chem. Commun.* 1027–1028. (b) Hollowood, F.S., McKervey, M.A., Hamilton, R., and Rooney, J.J. (1980). Synthesis of triamantane. *J. Org. Chem.* 45: 4954–4958.

41 Osawa, E., Furusaki, A., Hashiba, N. et al. (1980). Thermodynamic rearrangements of larger polycyclic hydrocarbons derived from the 38.5 and 41.5 °C melting dimers of cyclooctatetraene. Crystal and molecular structures of 5-bromoheptacyclo[8.6.0.0^{2,8}.0^{3,13}.0^{4,11}.0^{5,9}.0^{12,16}]hexadecane (5-bromo-[C_2]-bisethanobisnordiamantane), 6,12-dibromoheptacyclo[7.7.0.0^{2,6}.0^{3,15}.0^{4,12}.0^{5,10}.0^{11,16}]hexadecane, and nonacyclo[11.7.1.1^{2,18}.0^{3,16}.0^{4,13}.0^{5,10}.0^{6,14}.0^{7,11}.0^{15,20}]docosane (bastardane). *J. Org. Chem.* 45: 2985–2995.

42 Burns, W., Mitchell, T.R.B., McKervey, M.A. et al. (1976). Synthesis and crystal structure of anti- tetramantane, a large diamondoid fragment. *J. Chem. Soc. Chem. Commun.* 893–895.

43 Roberts, P.J. and Ferguson, G. (1977). *anti*-Tetramantane, a large diamondoid fragment. *Acta Crystallogr. B* 33: 2335–2337.

44 Burns, W., McKervey, M.A., Mitchell, T.R.B., and Rooney, J.J. (1978). A new approach to the construction of diamondoid hydrocarbons. Synthesis of *anti*-tetramantane. *J. Am. Chem. Soc.* 100: 906–911.

45 (a) Schwertfeger, H., Fokin, A.A., and Schreiner, P.R. (2008). Diamonds are a chemist's best friend: Diamondoid chemistry beyond adamantane. *Angew. Chem. Int. Ed.* 47: 1022–1036. (b) McKervey, M.A. (1980). Synthetic approaches to large diamondoid hydrocarbons. *Tetrahedron* 36: 971–992.

46 Hopf, H. (2003). Diamonds from crude oil? *Angew. Chem. Int. Ed.* 42: 2000–2002.

47 Landa, S. and Macháček, V. (1933). Sur l'adamantane, nouvel hydrocarbure extrait du naphte *Collect. Czech. Chem. Commun.* 5: 1–5.

48 Dahl, J.E., Moldowan, J.M., Peters, K.E. et al. (1999). Diamondoid hydrocarbons as indicators of natural oil cracking. *Nature* 399: 54–57.

49 Dahl, J.E., Liu, S.G., and Carlson, R.M.K. (2003). Isolation and structure of higher diamondoids, nanometer-sized diamond molecules. *Science* 299: 96–99.

50 Dahl, J.E.P., Moldowan, J.M., Peakman, T.M. et al. (2003). Isolation and structural proof of the large diamond molecule, cyclohexamantane ($C_{26}H_{30}$). *Angew. Chem. Int. Ed.* 42: 2040–2044.

51 Hala, S., Landa, S., and Hanus, V. (1966). Isolation of tetracyclo[6.3.1.0^{2,6}.0^{5,10}]dodecane and pentacyclo[7.3.1.1^{4,12}.0^{2,7}.0^{6,11}]tetradecane (diamantane) from petroleum. *Angew. Chem. Int. Ed.* 5: 1045–1046.

52 Dahl, J.E.P., Moldowan, J.M., Wei, Z. et al. (2010). Synthesis of higher diamondoids and implications for their formation in petroleum. *Angew. Chem. Int. Ed.* 49: 9881–9885.

53 (a) Stauss, S., Miyazoe, H., Shizuno, T. et al. (2010). Synthesis of the higher-order diamondoid hexamantane using low-temperature plasmas generated in supercritical xenon. *Jpn. J. Appl. Phys.* 49: 070213. (b) Shizuno, T., Miyazoe, H., Saito, K. et al. (2011). Synthesis of diamondoids by supercritical xenon discharge plasma. *Jpn. J. Appl. Phys.* 50: 030207. (c) Oshima, F., Stauss, S., Ishii, C. et al. (2012). Plasma microreactor in supercritical xenon and its application to diamondoid synthesis. *J. Phys. D* 45: 402003. (d) Oshima, F., Stauss, S., Inose, Y., and Terashima, K. (2014). Synthesis and investigation of reaction mechanisms of diamondoids produced using plasmas generated inside microcapillaries in supercritical xenon. *Jpn. J. Appl. Phys.* 53: 010214. (e) Ishii, C., Stauss, S., Kuribara, K. et al. (2015). Atmospheric pressure synthesis of diamondoids by plasmas generated inside a microfluidic reactor. *Diamond Relat. Mater.* 59: 40–46.

54 Nakahara, S., Stauss, S., Kato, T. et al. (2011). Synthesis of higher diamondoids by pulsed laser ablation plasmas in supercritical CO_2. *J. Appl. Phys.* 109: 123304.

55 Stauss, S. and Terashima, K. (2016). *Diamondoids: Synthesis, Properties, and Applications*. Jenny Stanford Publishing p. 242.

56 Bamberg, M., Bursch, M., Hansen, A. et al. (2021). [Cl@Si$_{20}$H$_{20}$]$^-$: Parent siladodecahedrane with endohedral chloride ion. *J. Am. Chem. Soc.* 143: 10865–10871.

57 Chan, Y.C., Choy, K.K.H., Chan, A.H.C. et al. (2008). Solubility of diamantane, trimantane, tetramantane, and their derivatives in organic solvents. *J. Chem. Eng. Data* 53: 1767–1771.

58 Iwasa, A., Clay, W.A., Dahl, J.E. et al. (2010). Environmentally friendly refining of diamond-molecules via the growth of large single crystals. *Cryst. Growth Des.* 10: 870–873.

59 Filik, J. (2010). Diamondoid hydrocarbons. In: *Carbon Based Nanomaterials*, vol. 65–66 (ed. N. Ali, A. Ochsner, and W. Ahmed), 1–26.

60 Choi, J.H., Eichele, C., Lin, Y.C. et al. (2008). Determination of effective refractive index of molecular diamondoids by Becke line method. *Scr. Mater.* 58: 413–416.

61 Clay, W.A., Sasagawa, T., Kelly, M. et al. (2008). Diamondoids as low-kappa dielectric materials. *Appl. Phys. Lett.* 93: 172901.

62 Yang, F., Lin, Y., Baldini, M. et al. (2016). Effects of molecular geometry on the properties of compressed diamondoid crystals. *J. Phys. Chem. Lett.* 7: 4641–4647.

63 Yang, F., Lin, Y., Dahl, J.E.P. et al. (2014). High pressure Raman and X-ray diffraction study of [121]tetramantane. *J. Phys. Chem. C* 118: 7683–7689.

64 McIntosh, G.C., Yoon, M., Berber, S., and Tomanek, D. (2004). Diamond fragments as building blocks of functional nanostructures. *Phys. Rev. B* 70: 045401.

65 Fokin, A.A. and Schreiner, P.R. (2009). Band gap tuning in nanodiamonds: first principle computational studies. *Mol. Phys.* 107: 823–830.

66 Raty, J.Y., Galli, G., Bostedt, C. et al. (2003). Quantum confinement and fullerenelike surface reconstructions in nanodiamonds. *Phys. Rev. Lett.* 90: 037401.

67 Drummond, N.D., Williamson, A.J., Needs, R.J., and Galli, G. (2005). Electron emission from diamondoids: a diffusion quantum Monte Carlo study. *Phys. Rev. Lett.* 95: 096801.

68 (a) Patrick, C.E. and Giustino, F. (2013). Quantum nuclear dynamics in the photophysics of diamondoids. *Nat. Commun.* 4: 1–7. (b) Demjan, T., Voros, M., Palummo, M., and Gali, A. (2014). Electronic and optical properties of pure and modified diamondoids studied by many-body perturbation theory and time-dependent density functional theory. *J. Chem. Phys.* 141: 064308.

69 Xiong, T., Wlodarczyk, R., Gallandi, L. et al. (2018). Vibrationally resolved photoelectron spectra of lower diamondoids: a time-dependent approach. *J. Chem. Phys.* 148: 044310.

70 Banerjee, S. and Saalfrank, P. (2014). Vibrationally resolved absorption, emission and resonance Raman spectra of diamondoids: a study based on time-dependent correlation functions. *Phys. Chem. Chem. Phys.* 16: 144–158.

71 Barnard, A.S., Russo, S.P., and Snook, I.K. (2003). Structural relaxation and relative stability of nanodiamond morphologies. *Diamond Relat. Mater.* 12: 1867–1872.

72 Willey, T.M., Bostedt, C., van Buuren, T. et al. (2005). Molecular limits to the quantum confinement model in diamond clusters. *Phys. Rev. Lett.* 95: 113401.

73 Landt, L., Kluender, K., Dahl, J.E. et al. (2009). Optical response of diamond nanocrystals as a function of particle size, shape, and symmetry. *Phys. Rev. Lett.* 103: 047402.

74 (a) Richter, R., Wolter, D., Zimmermann, T. et al. (2014). Size and shape dependent photoluminescence and excited state decay rates of diamondoids. *Phys.*

Chem. Chem. Phys. 16: 3070–3076. (b) Voros, M. and Gali, A. (2009). Optical absorption of diamond nanocrystals from *ab initio* density-functional calculations. *Phys. Rev. B* 80: 161411.

75 Raty, J.Y. and Galli, G. (2005). Optical properties and structure of nanodiamonds. *J. Electroanal. Chem.* 584: 9–12.

76 Landt, L., Kielich, W., Wolter, D. et al. (2009). Intrinsic photoluminescence of adamantane in the ultraviolet spectral region. *Phys. Rev. B* 80: 205323.

77 Voros, M., Demjen, T., Szilvasi, T., and Gali, A. (2012). Tuning the optical gap of nanometer-size diamond cages by sulfurization: a time-dependent density functional study. *Phys. Rev. Lett.* 108: 267401.

78 Richter, R., Rohr, M.I.S., Zimmermann, T. et al. (2015). Laser-induced fluorescence of free diamondoid molecules. *Phys. Chem. Chem. Phys.* 17: 4739–4749.

79 Gali, A., Demjan, T., Voros, M. et al. (2016). Electron-vibration coupling induced renormalization in the photoemission spectrum of diamondoids. *Nat. Commun.* 7: 11327.

80 (a) Bilalbegović, G., Maksimović, A., and Valencic, L.A. (2018). Tetrahedral hydrocarbon nanoparticles in space: X-ray spectra. *Mon. Not. R. Astron. Soc.* 476: 5358–5364. (b) Bauschlicher, C.W., Liu, Y.F., Ricca, A. et al. (2007). Electronic and vibrational spectroscopy of diamondoids and the interstellar infrared bands between 3.35 and 3.55 μm. *Astrophys. J.* 671: 458–469. (c) Pirali, O., Vervloet, M., Dahl, J.E. et al. (2007). Infrared spectroscopy of diamondoid molecules: new insights into the presence of nanodiamonds in the interstellar medium. *Astrophys. J.* 661: 919–925. (d) Bouwman, J., Horst, S., and Oomens, J. (2018). Spectroscopic characterization of the product ions formed by electron ionization of adamantane. *ChemPhysChem* 19: 3211–3218. (e) Greaves, J.S., Scaife, A.M.M., Frayer, D.T. et al. (2018). Anomalous microwave emission from spinning nanodiamonds around stars. *Nat. Astron.* 2: 662–667.

81 Steglich, M., Huisken, F., Dahl, J.E. et al. (2011). Electronic spectroscopy of FUV-irradiated diamondoids: a combined experimental and theoretical study. *Astrophys. J.* 729: 91.

82 Mutschke, H., Andersen, A.C., Jager, C. et al. (2004). Optical data of meteoritic nano-diamonds from far-ultraviolet to far-infrared wavelengths. *Astron. Astrophys.* 423: 983–993.

83 (a) Lu, A.J., Pan, B.C., and Han, J.G. (2005). Electronic and vibrational properties of diamondlike hydrocarbons. *Phys. Rev. B* 72: 035447. (b) Lenzke, K., Landt, L., Hoener, M. et al. (2007). Experimental determination of the ionization potentials of the first five members of the nanodiamond series. *J. Chem. Phys.* 127: 084320.

84 National Institute of Standard, *NIST Chemistry webbook*. http://webbook.nist.gov/chemistry.

85 Meunier, M., Quirke, N., and Binesti, D. (1999). The calculation of the electron affinity of atoms and molecules. *Mol. Simul.* 23: 109–125.

86 de Urquijo, J., Arriaga, C.A., Cisneros, C., and Alvarez, I. (1999). A time-resolved study of ionization, electron attachment and positive-ion drift in methane. *J. Phys. D* 32: 41–45.

87 Bowers, K.W., Greene, F.D., and Nolfi, G.J. (1963). Radical anions of adamantane and hexamethylenetetramine. *J. Am. Chem. Soc.* 85: 3707.

88 Jones, M.T. (1966). The reported adamantane anion radical. Its relationship to benzene anion radical. *J. Am. Chem. Soc.* 88: 174–176.

89 Li, Q.S., Feng, X.J., Xie, Y., and Schaefer, H.F. (2005). Perfluoroadamantane and its negative ion. *J. Phys. Chem. A* 109: 1454–1457.

90 Irikura, K.K. (2008). Sigma stellation: a design strategy for electron boxes. *J. Phys. Chem. A* 112: 983–988.

91 Li, J.R., Niesner, D., and Fauster, T. (2021). Negative electron affinity of adamantane on Cu(111). *J. Phys. Condens. Matter* 33: 135001.

92 Wang, Y., Kioupakis, E., Lu, X. et al. (2008). Spatially resolved electronic and vibronic properties of single diamondoid molecules. *Nat. Mater.* 7: 38–42.

93 Oomens, J., Polfer, N., Pirali, O. et al. (2006). Infrared spectroscopic investigation of higher diamondoids. *J. Mol. Spectrosc.* 238: 158–167.

94 Richardson, S.L., Baruah, T., Mehl, M.J., and Pederson, M.R. (2006). Cyclohexamantane ($C_{26}H_{30}$): first-principles DFT study of a novel diamondoid molecule. *Diamond Relat. Mater.* 15: 707–710.

95 Filik, J., Harvey, J.N., Allan, N.L. et al. (2006). Raman spectroscopy of diamondoids. *Spectrochim. Acta A Mol. Biomol. Spectrosc.* 64: 681–692.

96 Richardson, S.L., Baruah, T., Mehl, M.J., and Pederson, M.R. (2005). Theoretical confirmation of the experimental Raman spectra of the lower-order diamondoid molecule: cyclohexamantane ($C_{26}H_{30}$). *Chem. Phys. Lett.* 403: 83–88.

97 Pechenezhskiy, I.V., Hong, X., Nguyen, G.D. et al. (2013). Infrared spectroscopy of molecular submonolayers on surfaces by infrared scanning tunneling microscopy: Tetramantane on Au(111). *Phys. Rev. Lett.* 111: 126101.

98 Zhuang, C.Q., Jiang, X., Zhao, J.J.J. et al. (2009). Infrared spectra of hydrogenated nanodiamonds by first-principles simulations. *Phys. E: Low-Dimens. Syst. Nanostructures* 41: 1427–1432.

99 Meinke, R., Richter, R., Merli, A. et al. (2014). UV resonance Raman analysis of trishomocubane and diamondoid dimers. *J. Chem. Phys.* 140: 034309.

100 Zimmermann, T., Richter, R., Knecht, A. et al. (2013). Exploring covalently bonded diamondoid particles with valence photoelectron spectroscopy. *J. Chem. Phys.* 139: 084310.

101 Tyborski, C., Meinke, R., Gillen, R. et al. (2017). From isolated diamondoids to a van-der-Waals crystal: a theoretical and experimental analysis of a trishomocubane and a diamantane dimer in the gas and solid phase. *J. Chem. Phys.* 147: 044303.

2

Naturally Occurring Diamondoids

Diamondoids are considered the smallest representatives of nanodiamonds (for a definition of terms see Ref. [1]) and their formation is likely to be tightly connected to other diamond materials. Naturally occurring diamondoids were extensively scrutinized [2] suggesting a possible link in nature with other nanodiamond constituents [3]. While the conversion of graphite to macroscopic diamond is generally associated with high pressures and heating to overcome unfavorable thermodynamics [4], the situation at the nanolevel is clearly different. Due to the fact that the heat of formation of hydrocarbons depends on the H/C-ratio, the hydrogen-terminated tetragonal carbon nanoparticles are always more stable than single-molecule polyaromatics, simply because molecules with single C—C and C—H bonds are so much more stable than ones with C—C double bonds, no matter how well conjugated they are (Figure 2.1) [5].

Note that even an infinitely large graphene sheet is never more stable than a diamond-like hydrocarbon with the same number of carbon atoms. Graphite, which is *not* a single molecule, is slightly more stable than diamond because of the sheet-sheet (π–π) interactions that tip the balance. From this viewpoint, our choice of graphite as a thermodynamic reference state for carbon is quite unfortunate. However, the preparation of hydrogen-terminated nanodiamonds is still cumbersome as they are stable only at temperatures below 900 °C; at higher temperatures the surface functional groups start to detach, forming radical centers (sometimes oddly referred to as "dangling bonds[1]"). Increased heating causes graphitization of the surface, resulting in the formation of onion-like structures (Figure 2.2) [7].

This process may be effectively monitored through Raman spectra and TEM (Figure 2.2, bottom) [7a, 8] and is reversible in the presence of hydrogen sources when carbon onions are transferred back to nanodiamonds [9]. Thus, the formation of hydrogen-terminated nanodiamonds from carbon materials is largely associated with the presence of hydrogen sources typical for chemical vapor deposition (CVD) synthesis [10]. Remarkably, about 5% of interstellar carbon has been suggested to exist in the form of 1–100 nm nanodiamond particles [11]. The formation of diamond materials in carbon-rich stars does not require very high temperatures

---

1 The notion of "dangling bonds" is a rather curious one because it would make such materials paramagnetic, which they hardly ever are. Even on a highly strained diamond-like surface, two neighboring radical centers form a π-bond. A good example is adamantene [6].

The Chemistry of Diamondoids: Building Blocks for Ligands, Catalysts, Materials, and Pharmaceuticals, First Edition. Andrey A. Fokin, Marina Šekutor, and Peter R. Schreiner.
© 2024 WILEY-VCH GmbH. Published 2024 by WILEY-VCH GmbH.

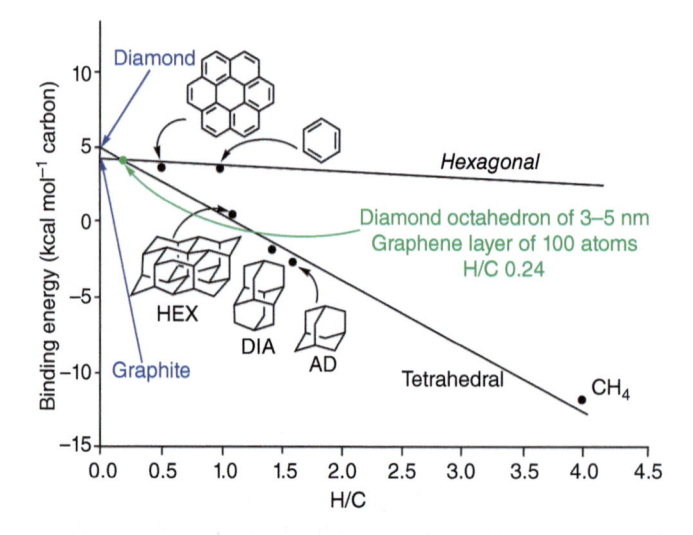

Figure 2.1 Calculated "binding energies" (the H_f per mol of carbon) as a function of the H/C ratio for aliphatic (mostly diamond-like) and (poly)aromatic molecules. The ordinate intercepts the energies of graphite and diamond, whereas the crossing-point (in green) of the two lines occurs at a H/C ratio of ca. 0.24, corresponding to diamond crystallites of 3–5 nm. Source: Graph adopted from Ref. [5b].

and occurs in the gas phase similarly to the low-pressure CVD conditions [10b, 12]. Alternatively, shock-wave diamond formation may be easily achieved in supernovae [11, 13]. Diamonds are common in space [14] and were found in carbon-rich meteorites [12, 15], interstellar dust [16], dense clouds [17], protoplanetary disks [18], suggested to form the cores of the "diamond planets" [19], and have become an increasingly popular object of astrophysical studies [20]. Based on some very peculiar emission bands, the protostellar sources are suggested to contain very small (from 0.75 nm) hydrogen-terminated nanodiamond particles (viewed as higher diamondoids) in neutral or ionized forms [21], initiating numerous recent studies on the charged and excited species generated from diamondoids [21d, 22].

In contrast, Earth utilizes very slow crystallization from inorganic carbonate melts that gives clean macroscopic diamond [3, 23], possibly through H-terminated carbon particles as templates [24]. In the Earth's crust, diamondoids stay at the low end of the hydrogen-terminated diamond material and are present in up to ppm quantities in oils (from the Proterozoic to Cenozoic), condensates [25], coals [26], sediments [27], and metamorphic rocks [28]. The above observation (Figure 2.1) [5b] that hydrogen-terminated diamonds less than 3 nm in size are more stable than graphite particles of the same size may explain the formation of diamondoids in fossil fuels. It is generally believed that diamondoid formation is due to the transformations of diverse organic materials under moderate heating and pressure in the presence of acidic mineral catalysts [29]. Such routes for diamondoid formation in the presence of Lewis acids were confirmed in the lab by many groups working with various organic substrates [30], where even catalytic cracking of methanol gives

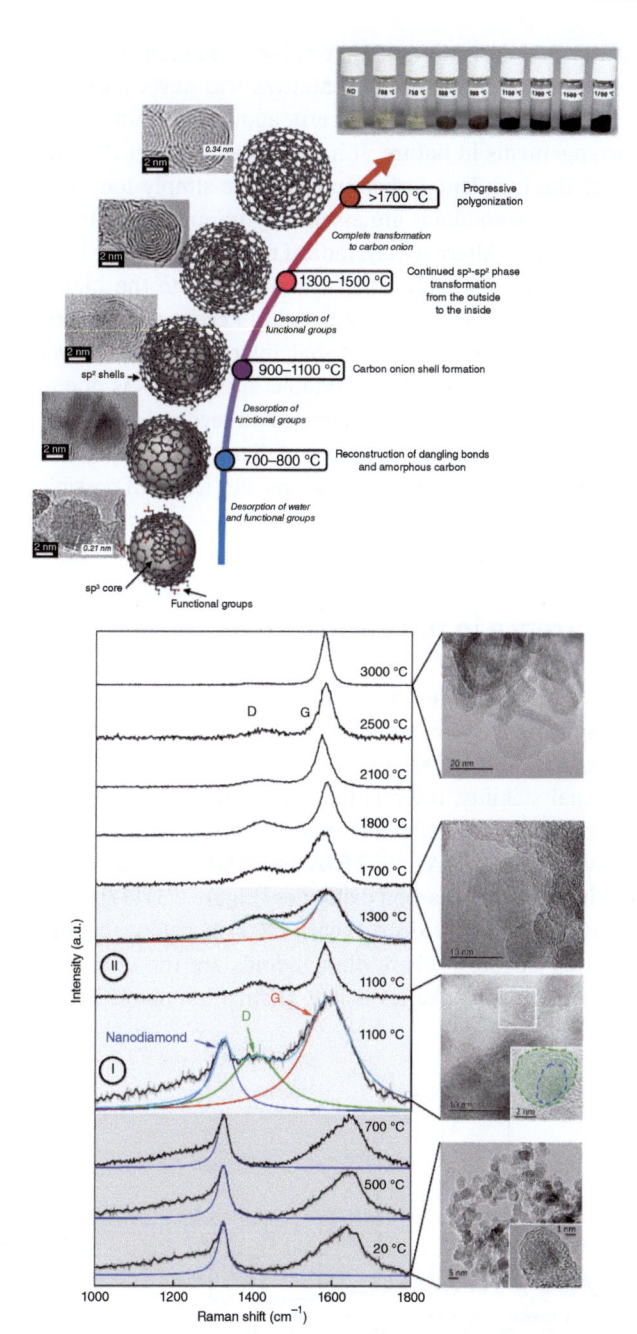

Figure 2.2 *Top*: Transformation from nanodiamond material to carbon onions by annealing based on STM studies. Source: Reproduced from Ref. [7a] with permission from the Royal Society of Chemistry, 2016. *Bottom*: Overview of the Raman spectra of the series of nanodiamond-based samples heat-treated at different temperatures. Black lines represent the smoothed curves, and gray lines show the original spectra. The TEM reveals the onion-like structure of the carbons. Source: Reproduced from Ref. [8] with permission from Elsevier, 2021.

AD together with some of its alkyl derivatives [31]. Yet, the formation of higher diamondoids such as tetramantane and other representatives was never observed under such "stabilomeric" conditions, raising concern about their formation through acid-catalyzed rearrangements in nature. It has also been shown that the barriers for rearrangement of the incipient carbenium ions are simply too high, so that hydrocarbons other than diamondoids are essentially trapped and further rearrangements are blocked [30a, 32]. Alternatively, radical mechanisms have been suggested that are controlled by the C/H ratio that actually leads to the higher diamondoids preferentially [10b] (see Chapter 1 for details). Recent mechanistic studies [33] on the cracking of tetracosane suggest that hydrocarbons of lower molecular weight react with the pre-organized diamondoid lattice. This is due to the higher thermal stability of the cage structure of diamondoid seeds.

In this Chapter, we will not only discuss the abundance of diamondoids in nature and their role as markers in geosciences but also address some key issues of possible mechanisms of their formation.

2.1 Diamondoid Occurrence in the Earth's Crust

Diamondoids and their alkyl derivatives are present in virtually all oils [34], but their concentrations vary significantly [35], commonly ranging from approximately 40–500 µg g^{-1} for adamantane (**AD**) and 5–200 µg g^{-1} for diamantane (**DIA**) derivatives. Due to their high thermal stability, the relative diamondoid abundance in petroleum increases during cracking [36]. After the removal of unsaturated compounds from mature and heavy oils, **AD**, **DIA**, and **TRIA** and their alkyl derivatives together with n-alkanes dominate in hydrocarbon distillates (Figure 2.3) [37].

Distillation of the condensate from very deep (nearly 7000 m below the surface) petroleum reservoirs gives fractions where diamondoids are the dominant components and where the content of n-alkanes is low. Distillation above 200 °C

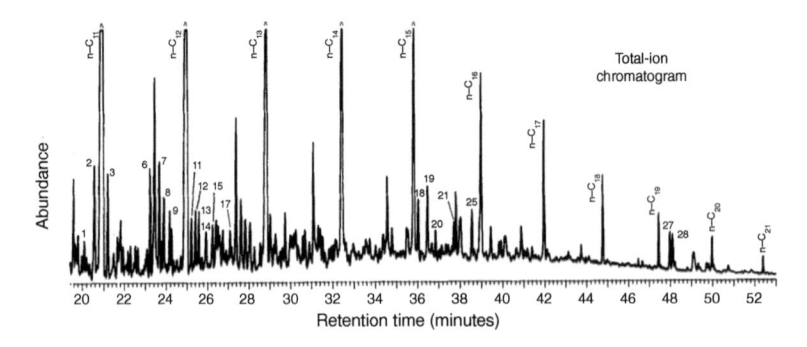

Figure 2.3 A historical (1992) partial total-ion chromatogram of the saturated hydrocarbon fraction from oil distillation, where the peaks from 1 to 17 are adamantanes, from 18 to 25 – diamantanes, and 27 and 28 minutes are triamantanes. Source: Reproduced from Ref. [37] with permission from Elsevier, 1992.

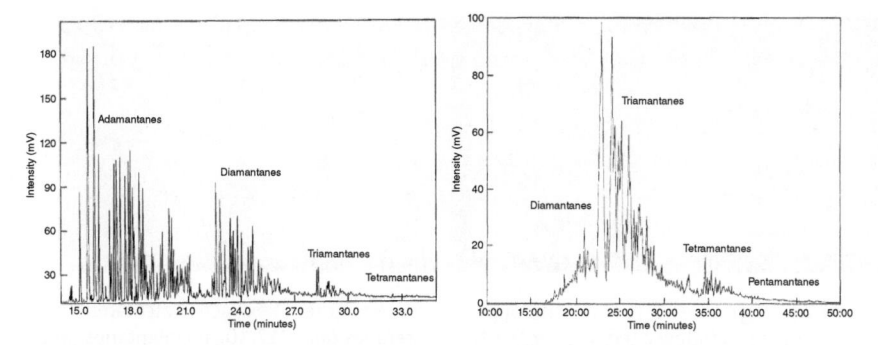

Figure 2.4 *Left:* Gas chromatogram of the very deep diamondoid-rich condensate sample showing clusters of peaks representing diamondoids up to tetramantanes. *Right:* High-temperature distillation fraction (343 °C) enriched with higher diamondoids. Source: Reproduced from Ref. [38] with permission from Elsevier, 1995.

gives fractions enriched with **AD** and **DIA** and their alkyl derivatives. Higher distillation temperatures give fractions with higher diamondoids up to hexamantane (Figure 2.4) [38]. The isolation of unsubstituted lower diamondoids from such mixtures is possible because of their crystallinity and high melting points (see below). Note that most alkyladamantanes are either low-melting-point solids or liquids.

In oils, the typical relative ratios of adamantanes:diamantanes:triamantanes:tetramantanes:pentamantanes:hexamantanes is 100 : 50 : 15 : 5 : 1 : 0.1 [39]. Refined products are usually enriched [40] with the more volatile lower diamondoids: the concentrations of **AD** and alkylated **AD**s in crude oil are up to 500 µg g^{-1}; weathered diesel fuel may contain up to 600 µg g^{-1} of **DIA**s [35]. The presence of diamondoids substantially increases the efficiency of standard US missile fuel JP-10 from 39.6 K to 44.7 K MJ m^{-3} [41]. The abundance of diamondoids may also be used for chemical fingerprinting of petroleum products [42] and non-aqueous phase liquids (NAPL) [43]. Diamondoids are especially useful in source identifications as **AD** and **DIA** derivatives are more stable than other biomarkers such as polycyclic terpanes and steranes [44]. Quantitative analysis through gas-chromatography-tandem mass spectrometry (GC–MS/MS) allows quantification of diamondoids in oil directly without special sample preparation [45] and serves as an effective method for determining the origins of condensates and natural gases [25g]. However, the most effective and representative analytical tool is the combination of 2D gas chromatography coupled with time-of-flight mass spectrometry (GC–GC-TOFMS) developed for diamondoid analysis independently by two groups [46]. GC–GC-TOFMS allows distinguishing not only between diamondoids and other components but also determining the relative abundance of diamondoids in oil extracts (Figure 2.5, left). With this method, certain hydrocarbon domains can be analyzed separately to gain information about the ratios of isomers and homologues (Figure 2.5, right). The GC–GC-TOFMS method is becoming increasingly popular for the analysis of highly complex oil samples [47].

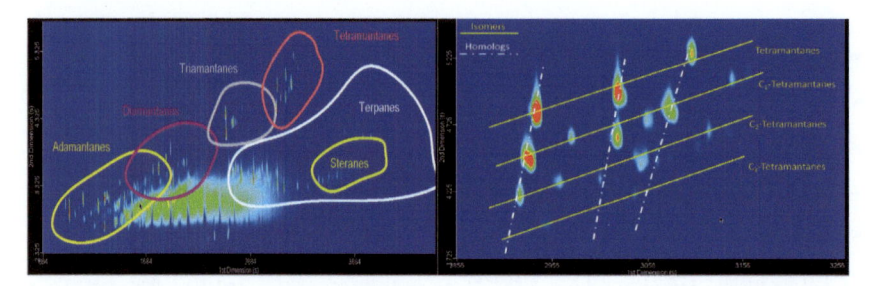

Figure 2.5 *Left*: GC–GC-TOFMS contour plot of crude mature oil extracts. Calculated intensities are as follows: terpanes (m/z 191.11), steranes (m/z 217.10), adamantanes (m/z 135), diamantanes (m/z 187), triamantanes (m/z 239.5), tetramantanes (m/z 291.60); *Right*: Chromatographic surface of tetramantanes. Intensities were calculated from the sum of m/z 291, 305, 319, and 333. The diagonal solid lines represent isomers of tetramantane (m/z 291 and 292), C_1-tetramantanes (m/z 291 and 306), C_2-tetramantanes (m/z 305 and 320), and C_3-tetramantanes (m/z 319 and 334). Source: Reproduced from Ref. [46a] with permission from Elsevier, 2013.

2.2 Diamondoids in Geochemical Studies

Diamondoids are widely used in studies of maturity [48], of cracking and mixing [30b, 36], biodegradation [37, 49], thermochemical sulfate reduction (TSR) [50], and migration [25a] of oils, gas washing, and phase fractionation [51], as well as for differentiating organic facies [52]. The role of diamondoids in geochemical studies has been extensively reviewed [39]. For a long time, it was believed that natural oil could exist only at moderately hot temperatures and decompose at depths lower than 5 km ("oil deadline") and at temperatures higher than 180 °C [53]. The discovery of hot oil reservoirs and experiments on oil cracking clearly demonstrated that oil can survive [54] temperatures above 200 °C and depths of 8500 ± 500 m forming ultralarge oil fields [55]. Moreover, in deep reservoirs, uncracked oil is not unusual if expelled from a source rock [56].

The diamondoid content and alkyl diamondoid isomer ratios are very useful in estimating the extent of oil cracking, and many such "diamondoid indices" were suggested to validate the degree and type of biodegradation of oil in reservoirs; this is very useful in understanding its origin and aspects of biological as well as geological evolution [51]. The progressive abundance of diamondoids after cracking allows the determination of oil maturity based on the relative abundance of lower and higher diamondoids [36]. As the stability of isomeric alkyl diamondoid derivatives depends on the position of the substituent, e.g., 2-methyl **AD** is less stable than 1-methyl **AD** and apical 4-methyl **DIA** is more stable than other methyl diamantanes [57], the "methyladamantane" and "methyldiamantane" indices based on their relative abundance were suggested [48b]. It was shown that the abundance of the more stable isomers of methyl **AD** and **DIA** increases with the age of the oil, but this parameter is not always reliable [34c, 58]. Very recently, higher ethanodiamantanes and ethanotriamantanes were found in very mature condensate samples [34c, 59]. The presence of pristine and alkylated ethanoadamantane and ethanodiamantane was confirmed

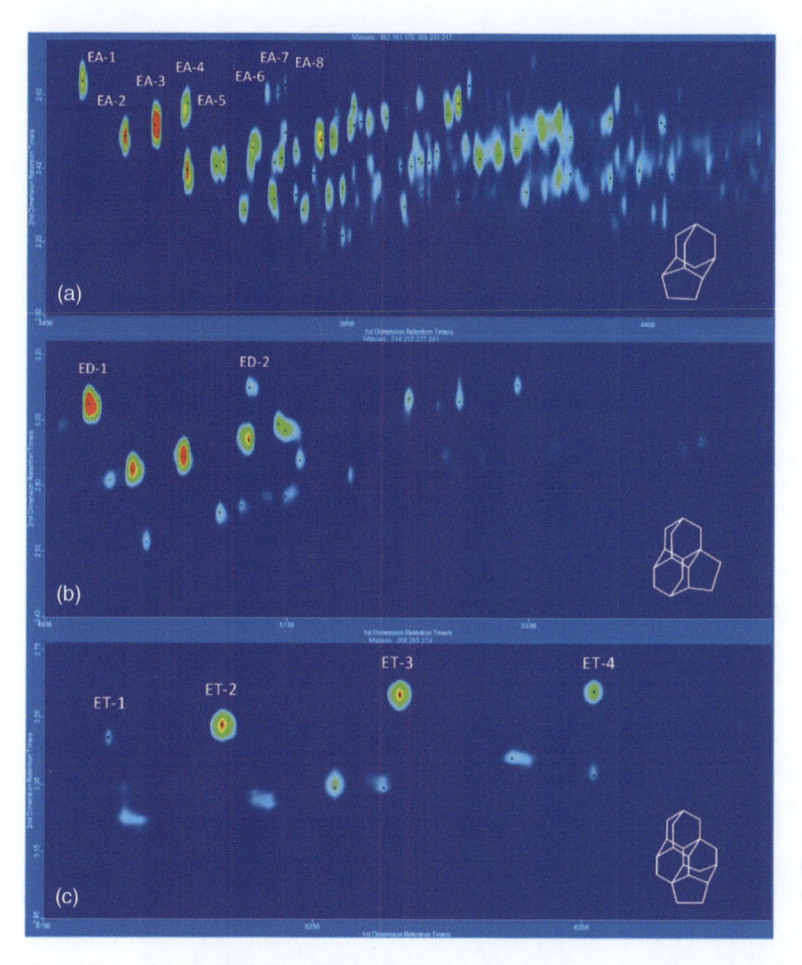

Figure 2.6 GC–GC-TOFMS analyses of ethanoadamantanes in oil condensates display the presence of (a) pristine (**EA-1**) and alkylated ethanoadamantanes (**EA-2–EA-8**), (b) isomeric ethanodiamantanes (**ED-1** and **ED-2**), and (c) isomeric ethanotriamantanes (**ET-1–ET-4**). Source: Reproduced from Ref. [59a] with permission from the American Chemical Society, 2018.

(Figure 2.6) by comparison with standard samples, while yet unknown ethanotriamantanes were identified based on mass-spectral data only [59a]. The presence of large amounts and the high robustness of ethanodiamondoids in oils are in full agreement with the fact that they appear to be the most stable among other isomeric diamondoid derivatives [60], and ethanodiamondoid content was suggested as an indicator for oil cracking instead of methyldiamondoid indices [34c, 59a].

Diamondoids were also used recently as tracers of gas invasion into oil reservoirs that alter the properties of oil [61]. A strong correlation exists between the diamondoid content and the extent of the gas invasion. Black oil reservoirs with a relatively low gas/oil ratio (GOR) have oil-dissolved gases likely co-generated with the oil and are characterized by a large variety and abundance of diamondoids (Figure 2.7). In contrast, reservoirs with high GOR (i.e., high gas invasion) lack diamondoids.

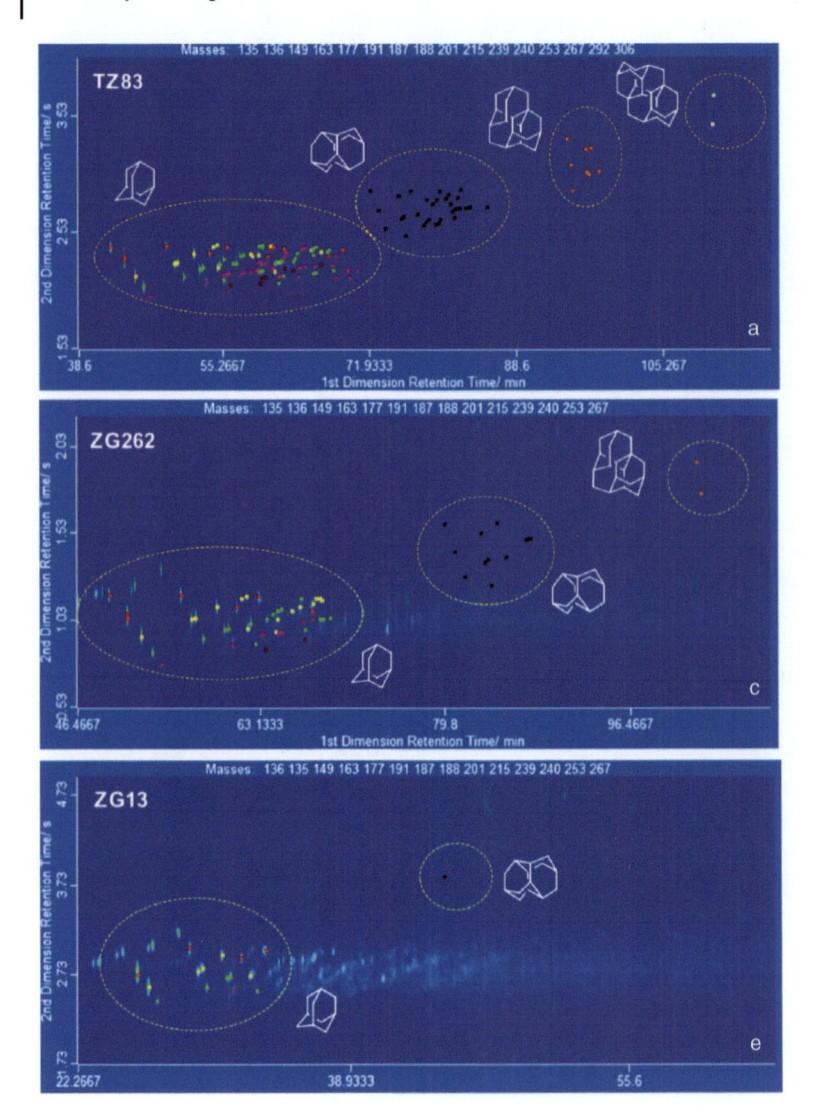

Figure 2.7 GC–GC-MS diamondoid distribution in petroleum liquids altered by gas invasion. Sample a is severely altered, sample c is weakly altered, and sample e is essentially unaltered. Source: Reproduced from Ref. [61] with permission from Elsevier, 2019.

2.3 Diamondoid Formation in the Earth's Crust

For a long time, it was assumed that the geological formation of diamondoids is associated with thermodynamically controlled carbocationic rearrangements [62] of isomeric hydrocarbons for which diamondoids such as **AD** and **DIA** are stabilomers. This lab synthesis was pioneered by Schleyer and co-workers who discovered a simple preparation of **AD** from tetrahydrodicyclopentadiene [32a] and of **DIA** from the norbornene dimer [32b] in the presence of Lewis acids (see Chapter 1 for details). The concept of "stabilomeric" synthesis arose also from the success of the Lewis-acid-catalyzed laboratory synthesis of cage molecules from diverse organic precursors, even including very dissimilar structures such as steroids, nujol, or abietic acid [30a], and this procedure is applicable for the preparation of high-density diamondoid fuels [63]. The isomer ratios of thus obtained diamondoid alkyl derivatives are close to those found in crude oils [34d, 64]. Lower diamondoids also form from organic materials through simple pyrolysis [30b].

The situation with higher diamondoids starting from tetramantane is not as simple. The main argument against the ionic mechanism is based on the fact that all attempts to prepare higher diamondoids starting from tetramantane from isomeric precursors through Lewis-acid-catalyzed carbocationic rearrangements failed in the lab. For instance, the acid-catalyzed isomerization of $C_{22}H_{28}$ hydrocarbons gave a structure with fused **DIA** and 2,8-ethanonoradamantane units rather than with the anticipated tetramantane core [65]. These experimental results challenged the hypothesis about the carbocationic pathways for higher diamondoid formation in nature. The existence of various bifurcating reaction paths, fragmentations, and the presence of many energetically close-lying carbocation intermediates make the geological acid-catalyzed formation of higher diamondoids highly unlikely. Another argument against the carbocationic mechanism is that the isomeric diamondoid distributions in the lithosphere do not correlate with their thermodynamic stability. The fact that diamondoids are present in oils formed in non-acidic carbonate strata also provides evidence against the carbocationic paths for their formation [25h]. Strong support against a cationic mechanism arose from cracking experiments where **AD** and **DIA** form from n-alkanes at only 450 °C [66]. The formation of diamondoids by cracking the components of crude oil was demonstrated independently [66, 67].

A mechanistic reconciliation was achieved after breakthrough experiments when higher diamondoids were shown to form from lower ones at 500 °C. In these experiments, the yields are greatly increased in the presence of isobutane, which serves as a diamondoid homologation unit [10b]. Careful analysis of isomeric hydrocarbon ratios from these experiments agrees with the notion that diamondoid homologation with isobutane is primarily sterically controlled. For instance, only less sterically crowded **1(2,3)4PENT** and **12(1)3PENT** were found in the high-temperature homologation of **1(2)3TET** (Figure 2.8) [10b].

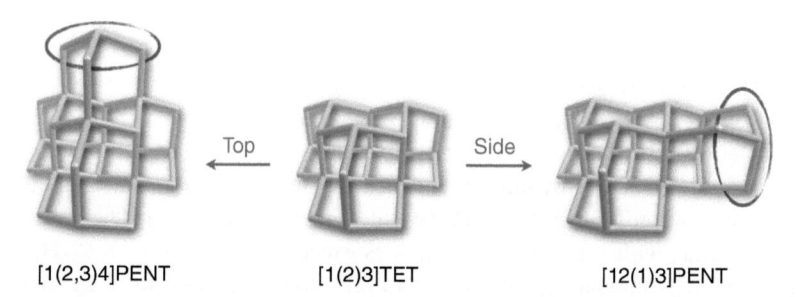

Figure 2.8 The least sterically crowded pentamantanes **[1(2,3)4]PENT** and **[12(1)3]PENT** dominate in the homologation of [1(2)3]tetramantane (**[1(2)3]TET**) in the presence of isobutane at 500 °C. Source: Reproduced from Ref. [10b], Wiley-VCH, 2010.

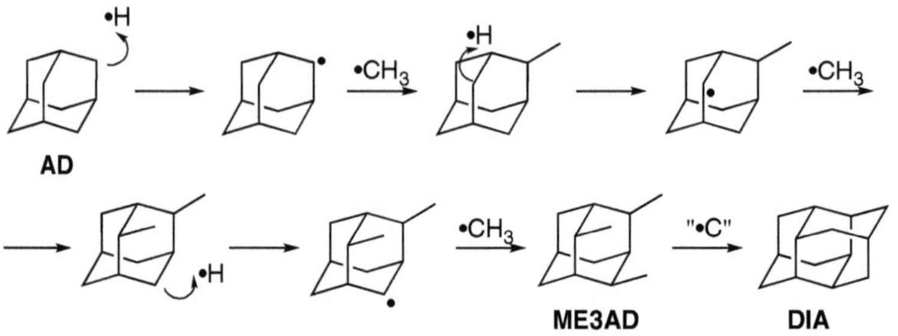

Figure 2.9 Suggested radical path for the formation of diamantane (**DIA**) from adamantane (**AD**) and CH$_4$ under microplasma conditions through subsequent hydrogen abstraction/recombination reactions and 2,4,9-trimethyladamantane (**ME3AD**) reaction with carbon-centered radicals at the terminal step. Source: Adapted from Ref. [73].

These experimental studies confirm that radical processes under conditions that may be present in the Earth's lithosphere may be responsible for the formation of higher diamondoids. The radical mechanism is additionally supported by the formation of nanodiamonds in CVD experiments in which **AD**s [68] and higher diamondoids [69] seed diamond nucleation. Remarkably, **TRIA** requires lower temperatures and pressures for diamond formation than **AD** and **DIA** [70]. Presumably, this is due to the presence of quaternary carbon atoms in **TRIA** that facilitate the formation of cubic diamond cages without breaking carbon bonds. Higher diamondoid derivatives may have certain advantages as seeding materials, as was shown with the example of 7-dichlorophosphoryl[1(2,3)4]pentamantane chemically bound to a vertically oriented silicon wafer [71]. The nanodiamond could be easily recognized through the characteristic bulk optical vibrational mode at 1332 cm^{-1} (see Chapter 6 for more details).

Higher diamondoids also form in low-temperature plasmas generated in supercritical xenon, utilizing lower diamondoids as precursors, again through the radical mechanism [72]. A possible reaction mechanism for the formation of higher diamondoids from **AD** involves repeated hydrogen abstraction and the addition of carbon-containing fragments, as is illustrated in Figure 2.9, for the formation of **DIA** from **AD** in the presence of methane [73]. The terminal step, i.e., the formation of **DIA** from 2,4,6-trimethyl adamantane (**ME3AD**) requires the addition of a carbon-centered radical. Notably, alkyl diamondoids form only in the presence of H$_2$/CH$_4$ mixtures [73].

2.4 Large-scale Isolation of Natural Diamondoids

The first isolation of **AD** from a very heavy crude oil was achieved by Landa and Macháček in 1933 [74] by careful rectification followed by steam distillation and crystallization [75]. It was pointed out that **AD** separates as "beautiful octahedral crystals" with a melting point that is exceptionally high for a C$_{10}$H$_{16}$ hydrocarbon [75]. The identity of this "natural" **AD** was confirmed by the independent synthesis by Prelog in 1941 (Figure 2.10) [76].

Based on the balance between the heavy oil distillation products and the very low **AD** content in oil, Landa and Macháček assumed the presence of **AD** homologues

Figure 2.10 The first photograph of synthetic **AD** crystals was taken by Prelog and Seiwerth in 1941. Source: Reproduced from Ref. [76] John Wiley & Sons, Inc. 1941

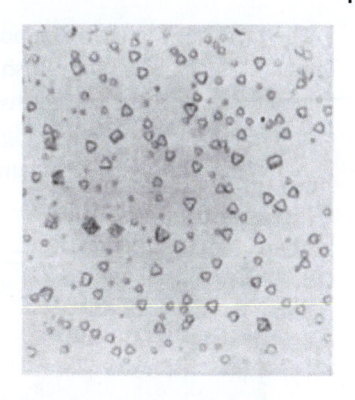

and developed a method for separation through the formation of crystal clathrates with thiourea. This allowed the isolation of many alkyl **AD**s from petroleum; even dimeric **ADAD**s were identified [75]. A very simple procedure for the isolation of **AD** from raw oil was developed in 1959 utilizing the cycloparaffin fraction with a boiling point of ca. 190 °C being subjected to azeotropic distillation in the presence of $(C_4F_9)_3N$. The **AD** crystals can then simply be collected in the condenser and resublimed. The amount of **AD** in the original petroleum was estimated to be 0.0004% [77]. As diamondoids usually co-elute together with *cyclo-* and *iso-*alkanes, their purely chromatographic separation requires a combination of techniques utilizing customized molecular sieves with well-defined pore sizes [78] or the formation of inclusion complexes with β-cyclodextrin [79].

DIA was isolated in 1964 from a petroleum fraction boiling between 200 and 300 °C through adduct formation with thiourea followed by extractive crystallization [80]. Purification through preparative gas chromatography, subsequent sublimation, and crystallization gave a pure sample identical to the material available synthetically [32b].

The isolation of higher diamondoids represents an increasingly difficult task as their content in raw oil is an order of magnitude lower than that of **AD** and **DIA**. Quantitative diamondoid analysis [36] utilizing deuterium-labeled diamondoids shows that relative diamondoid content is higher in mature oils (Figure 2.11) [81].

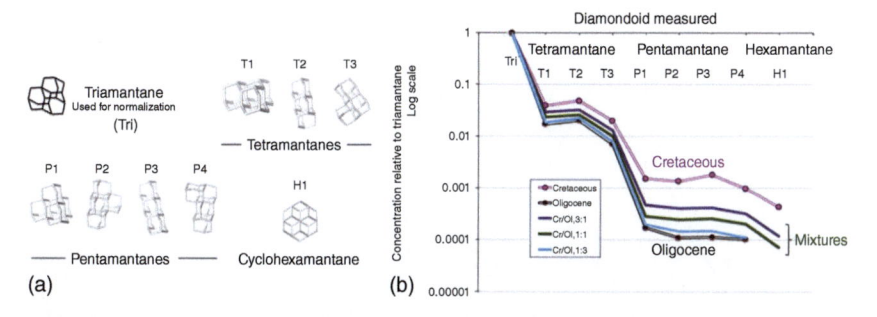

Figure 2.11 (a) Structures of higher diamondoids measured along with internally isotopically labeled standards. (b) Relative (to triamantane) content of higher diamondoids in Oligocene (immature) and Cretaceous (mature) oil samples and their mixtures mimicking the diamondoid content in black oils. Source: Reproduced from Ref. [81] with permission from Elsevier, 2015.

2 Naturally Occurring Diamondoids

The fact that higher diamondoids accumulate in mature oils where traditional biomarkers, e.g., steranes vanish (Figure 2.12, top) [82] is useful for the determination of thermal maturity. Selected samples of Mississippian and Woodford oils were examined in order to analyze higher diamondoid distributions (tetramantanes, pentamantanes, and cyclohexamantane (Figure 2.12, center), and the general trends

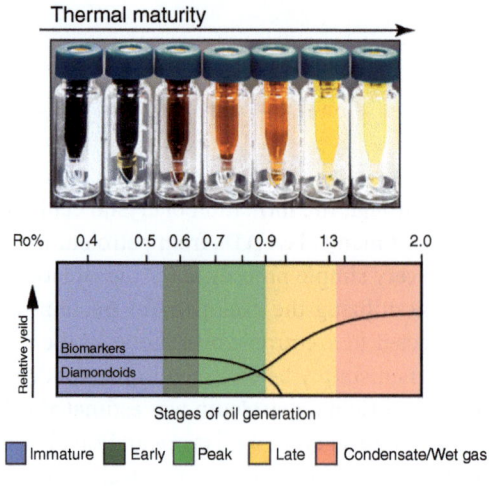

Figure 2.12 *Top*: Petroleum at different thermal maturity stages where biomarkers start to disappear and diamondoids start to accumulate at the late oil window. *Center*: Quantitative extended diamondoid analysis of Mississippian and Woodford-sourced oils. *Bottom*: Ternary diagram comparing the relative abundance of different pentamantane isomers from a collection of unknown hydrocarbons and known endmember crude oil. Source: Reproduced from Ref. [82] with permission from Elsevier, 2021.

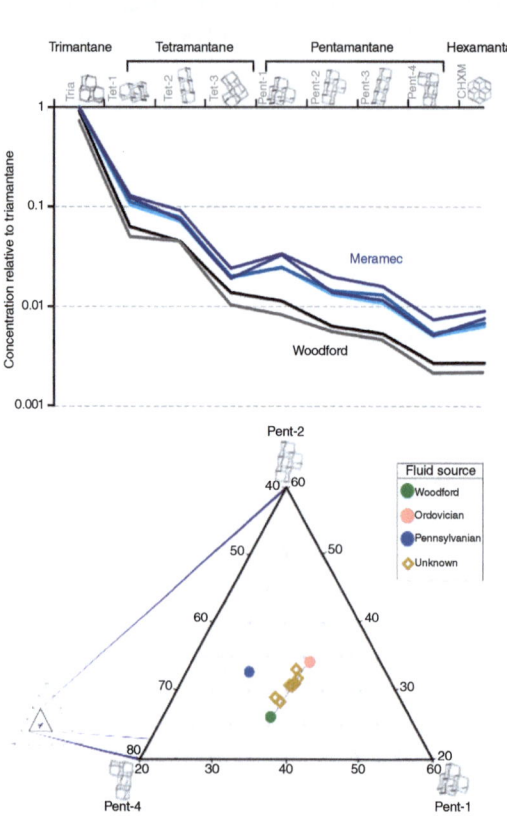

are close to those previously determined for Oligocene and Cretaceous oils (vide supra) [81]. The distributions of higher diamondoids are thus useful to determine oil families and to associate oils and condensates with their source rocks. The ratio of isomers is also valuable for geochemical analysis. For instance, the distribution of isomeric pentamantanes is distinctly different for Devonian and Ordovician oils (Figure 2.12, bottom).

The first large-scale isolation of higher diamondoids from natural gas condensates, including those from the Norphlet Formation, Gulf of Mexico, was developed as late as 2003 [83]. The fractions obtained by vacuum distillation above 345 °C (atmospherically equivalent boiling point) were pyrolyzed at 400–450 °C to remove non-diamondoid admixtures. Further removal of polar and unsaturated compounds was achieved by liquid chromatography on silver-impregnated silica gel. Subsequent reverse-phase HPLC and crystallization from acetone gave pure samples of higher diamondoids. As a result, all four tetramantanes, nine pentamantanes, one hexamantane, two heptamantanes, two octamantanes, one nonamantane, one decamantane, and one undecamantane were isolated and characterized (Figure 2.13). Additional efforts led to the isolation and partial

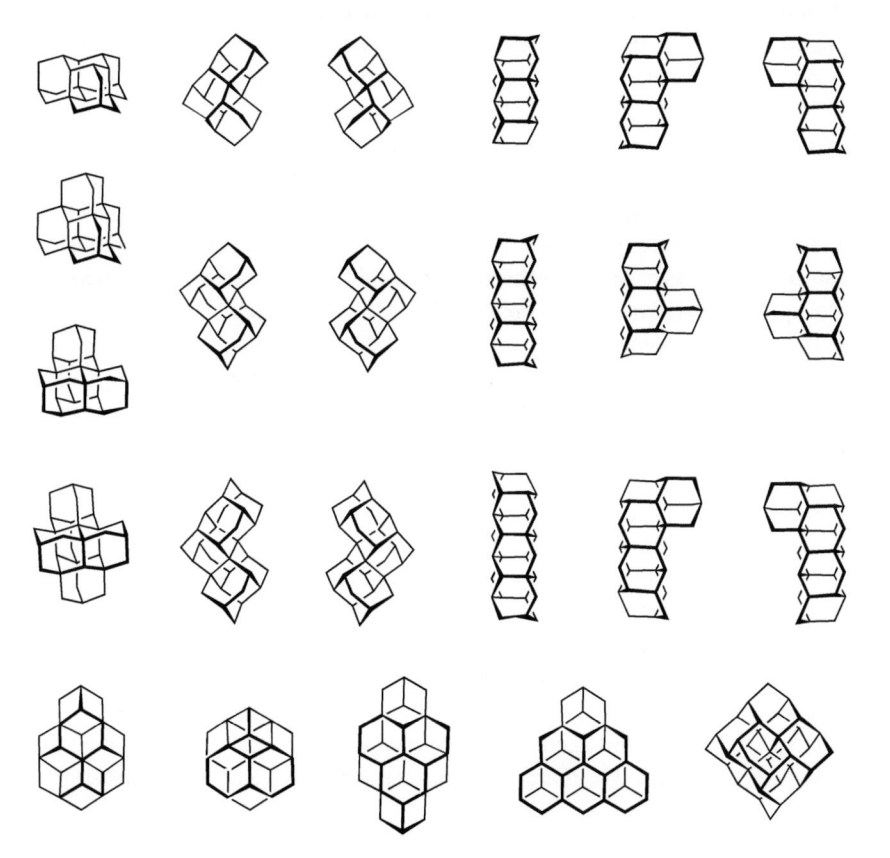

Figure 2.13 Higher diamondoids isolated from natural gas condensates and characterized Source: Adapted from Ref. [83].

NMR-characterization of extremely poorly soluble D_{3d}-symmetric cyclohexamantane ([12 312]hexamantane) [84]. Due to the high relaxation times of spatially crowded quaternary carbon atoms, the observation of their ^{13}C NMR resonances required a significant increase of the time delay between pulses [85].

The combined distillation/crystallization technique developed by MolecularDiamond Technologies allowed the isolation of higher diamondoids in multigram quantities: **TRIA** and **DIA** were isolated up to 100 kg. This accelerated the development of diamondoid chemistry in the mid-2000s.

2.5 Alternative Natural Sources of Diamondoids

Besides being present in virtually all oils diamondoids are also abundant in coals and organic-compound-rich sedimentary rocks. The presence of diamondoids in coals was first postulated in 1959 [86] and then confirmed experimentally [87]; there are a number of diamondoids that survive even coal liquefaction conditions [88]. The diamondoid concentration in coal depends on its maturity, which is usually estimated based on the vitrinite reflectance (R_o). In immature coals, the diamondoid concentrations are low (0.02 to 0.41 ppm) [89]. The abundance reaches maximum values of 10 ppm in moderately mature coals and again drops in higher-maturity samples [25c, 25g, 90] (Figure 2.14), possibly due to thermal destruction [89]. The latter was shown by separate experiments on **DIA** thermolysis resulting in the formation of a number of polyaromatic compounds. This is a result of entropy-driven decomposition similar to that observed for the thermolysis of hydrogen-terminated nanodiamonds upon lack of hydrogen sources (see Figure 2.2). The kinetic instability of diamondoids vs. their aromatic counterparts under thermolytic conditions may originate from lower BDEs of sp^3C—H relative sp^2C—H bonds in hydrocarbons [91].

The diamondoid content in coal is so distinct that it can be used as a maturity marker. Organogeochemical investigations of coals representing a maturity range R_o from 0.74% to 3.55% were presented for comparison, and **DIA**-based parameters proved to be excellent for the characterization of high-rank samples, whereas the use of biomarkers such as bicyclic sesquiterpanes and tetracyclic diterpanes had considerable limitations [26b].

As the diamondoid content is much higher in raw petroleum, only a few examples of their isolation from coal are documented. **AD** was isolated from coal in 1973 [26a], utilizing extraction, vacuum distillation, and steam distillation. Separation from other saturated hydrocarbons [92] through clathrates with thiourea and further recrystallization gave **AD** in very low yield.

The general trends in diamondoid distribution in sedimentary source rocks (silicates or carbonates that accumulate hydrocarbons) follow those observed for raw oils [28a, 93]. Recently, diamondoids were isolated from petroleum source

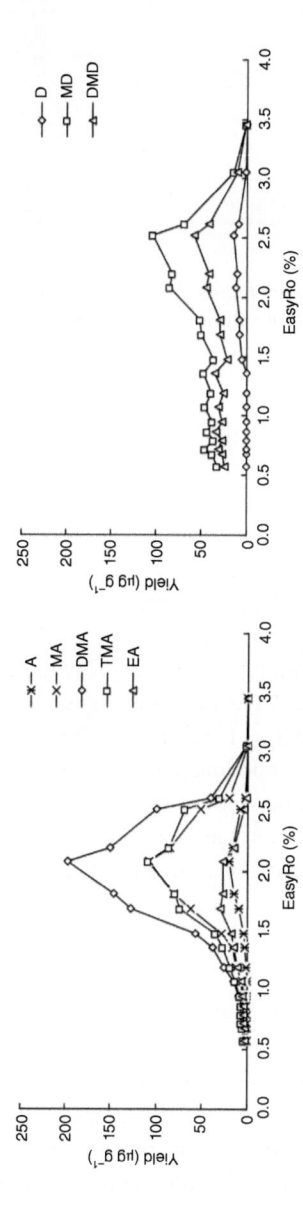

Figure 2.14 Variation in the yields (µg g^{-1} oil) of different types of diamondoids in the oil-cracking experiments with vitrinite reflectance equivalence calculated using the "EasyRo" method: *left*) adamantanes, *right*) diamantanes. A = adamantane, MA = methyladamantanes, EA = ethyladamantanes, DMA = dimethyladamantanes, TMA = trimethyladamantanes, D = diamantane, MD = methyldiamantanes, DMD = dimethyldiamantanes. Source: Reproduced from Ref. [25c] with permission from Elsevier, 2012.

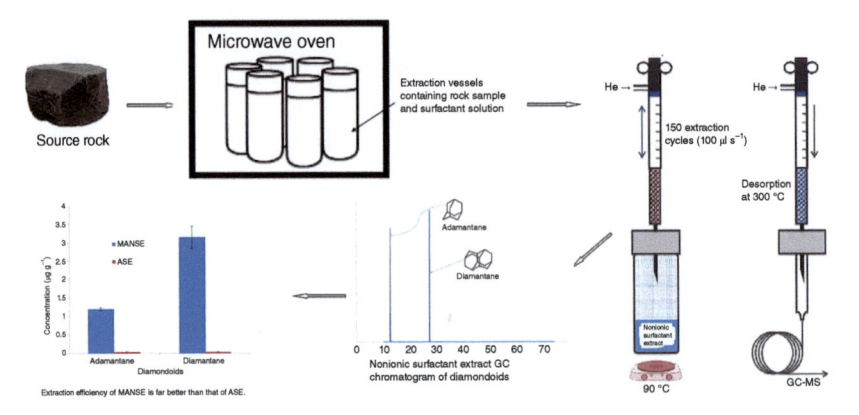

Figure 2.15 Microwave-assisted in-tube nonionic extraction (MANSE) of diamondoids from petroleum source rock with surfactant solution. Source: Reproduced from Ref. [94] with permission from Elsevier, 2019.

rocks by direct microwave-assisted nonionic surfactant extraction (MANSE) (Figure 2.15) [94], and this method is more efficient than solvent extraction. The MANSE extracts were analyzed by GC–MS and display alkyl diamondoid isomeric ratio distributions similar to those obtained from solvent extractions.

The fact that diamondoids are abundant in source rocks provides evidence that they form prior to oil generation [89]. A diamondoid index can be used as a maturity parameter, e.g., to estimate the degree of cracking of oil-cracking gas [95] and even allows to distinguish between different types of rocks [52, 96]. Analogously to crude oil [97], the concentration of **ADs** and **DIAs** in rock extracts and kerogen depends on maturation: it reaches a maximum at $R_o = 1.7\%$ and vanishes at $R_o > 3\%$ [28b]. If only **DIAs** are analyzed, the maximum diamondoid abundance increases to $R_o = 4\%$ [89]. This clearly shows that studies based on accounting for lower diamondoids and their alkyl derivatives only [25f, 28b, 89, 96, 98] leave questions about the abundance of higher diamondoids in mature source rocks, especially in ultradeep reservoirs, unanswered [99].

2.6 Other Diamondoid Derivatives in Nature

Besides pristine diamondoid hydrocarbons, many of their functional derivatives were found in nature. In addition to diamondoid acids [100] and diacids [101], possibly formed from oxidative biodegradation of alkyl diamondoid derivatives, crude oils contain thiadiamondoids and diamondoid thiols [102]. Thiaadamantane (**S_AD**) was isolated for the first time in 1952 by the Anglo-Iranian Oil Company from the kerosene fraction of Iranian oil [103]. It was collected as a solid after distillation and purification through sublimation and crystallization. The resulting pure **S_AD** was characterized by elementary analysis and through the desulfurization to bicyclo[3.3.1]nonane [103]; later on, it was compared to the material

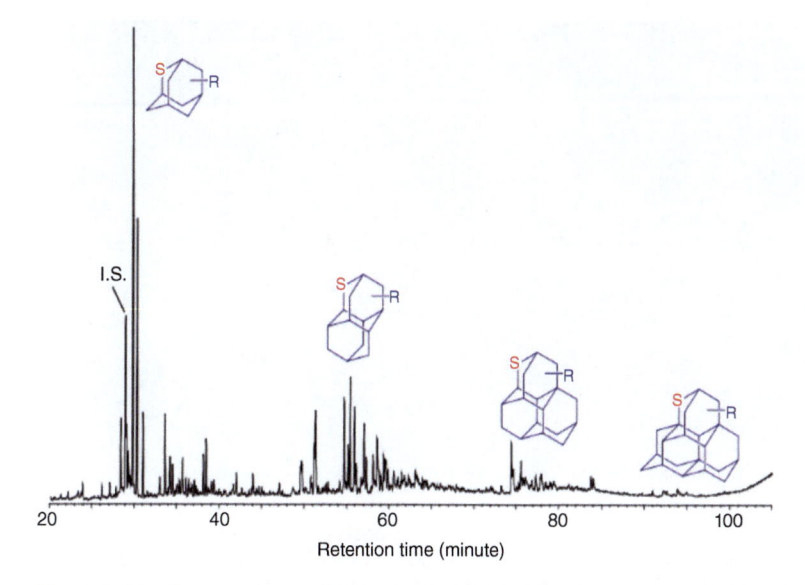

Figure 2.16 Chromatogram of the organosulfur fraction showing the distribution of thiadiamondoids, including thiaadamantanes, thiadiamantanes, thiatriamantanes, and thiatetramantanes (I.S. = internal standard, D$_3$-1-methyl-2-thiaadamantane). Source: Reproduced from Ref. [102b] with permission from Elsevier, 2012.

prepared synthetically by Stetter [104]. This essentially repeats the story of **AD**, which was also first isolated from natural sources and only then prepared in the lab (vide supra). Further studies demonstrate that thiadiamondoids form from various precursors upon oxidation of organic compounds with inorganic sulfates (thermochemical sulfate reduction, TSR) under geological conditions [105]. The existence of thiadiamondoids in low-temperature shallow reservoirs, where TSR conditions cannot be reached, is associated with vertical migration [106] from deeper strata [56, 107]. Various thiadiamondoids, including higher ones, form in sulfur-rich kerogen and have been identified as the main components of organosulfur oil distillation fractions. The separation from highly TSR-altered oils was achieved through two-layer Ag-impregnated silica gel column chromatography [102b], where thiadiamondoids as heavy as thiatetramantanes were identified (Figure 2.16).

The naturally occurring thiadiamondoid palette is not limited to the mono-sulfur derivatives. Many isomeric oligothiadiamondoids and their alkyl derivatives were identified through GC–MS [108] and GC–GC-TOFMS [109] analyses of the condensates from deeply buried Cambrian and Ordovician carbonate reservoirs in the Tarim Basin, China (Figure 2.17). The condensate sample series contained 267 thiadiamondoids with a total of 28 mg g^{-1}, which is about seven times lower than diamondoid hydrocarbon content in the same condensate [109b]. It was suggested that thiadiamondoids do not preferentially concentrate relative to diamondoid hydrocarbons during TSR [110].

Recently, thiadiamondoids and diamondoid thiols were identified in lower Cambrian oil, where five isomeric thiapentamantanes were found among higher

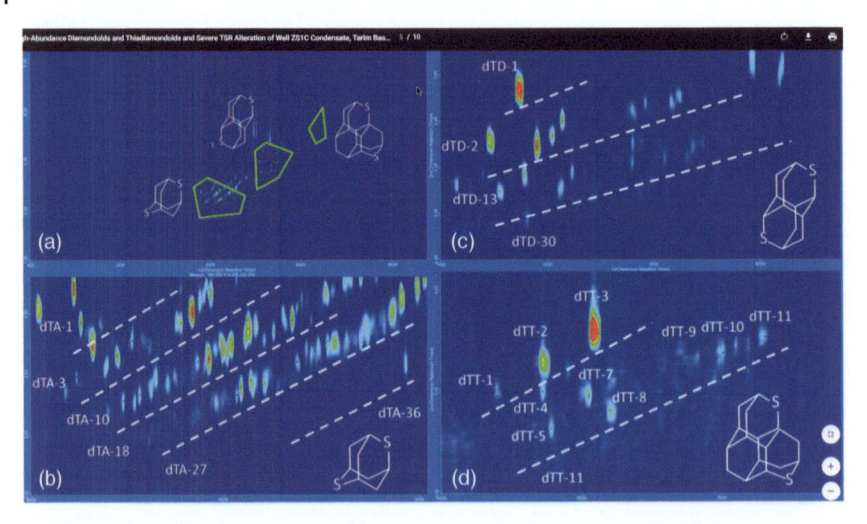

Figure 2.17 Dithiadiamondoids from the GC–GC-MS analysis of condensates from Cambrian strata of the Tarim Basin: a m/z 186 + 200 + 214 + 228 + 224 + 238 + 252 + 266 + 280 + 276 + 290 + 304 chromatogram highlighting thiadiamondoids with one to three cages, b m/z 186 + 200 + 214 + 228 chromatogram highlighting dithiaadamantanes, c m/z 224 + 238 + 252 + 266 + 280 – dithiadiamantanes, and d m/z 276 + 290 + 304 - dithiatriamantanes. Source: Reproduced from Ref. [109b] with permission from the American Chemical Society, 2018.

thiadiamondoids [111]. The formation of diamondoid thiols was attributed to C—H-bond oxidations under TSR conditions [102a]. Modeling the TSR conditions in the lab clearly showed that both thiadiamondoids and diamondoid thiols may form from non-sulfur-containing diamondoids, and it seems likely they are precursors of thiadiamondoids and diamondoid thiols in nature. The latter, though, display much lower thermal stability than parent hydrocarbons [112]. A possible mechanism of sulfurization involves hydrogen abstraction from C—H bonds from diamondoids with sulphanyl radicals (HS•) generated from H_2S under TSR conditions [113]. Thus, the generation of diamondoids during TSR may again be attributed to radical reactions [114], as suggested for diamondoid growth in the presence of alkanes [10b] or their formation from alkane cracking [66]. The hydrogen abstraction from alkanes with sulphanyl radicals is endothermic but associated with low activation barriers [115]. Thus formed alkyl radicals may be efficiently oxidized under TSR conditions and further trapped with sulfur-containing nucleophiles. The incorporation of sulfur into the structure of a diamondoid changes the electronic properties of the molecules, and the practical potential of such "doping" is discussed in Chapters 5 and 12.

References

1 Schwertfeger, H., Fokin, A.A., and Schreiner, P.R. (2008). Diamonds are a chemist's best friend: diamondoid chemistry beyond adamantane. *Angew. Chem. Int. Ed.* 47: 1022–1036.

2 (a) Schoell, M. and Carlson, R.M.K. (1999). Diamondoids and oil are not forever. *Nature* 399: 15–16. (b) Nekhaev, A.I., Bagrii, E.I., and Maksimov, A.L. (2011). Petroleum nanodiamonds: new in diamondoid naphthenes. *Pet. Chem.* 51: 86–95. (c) Gordadze, G.N. (2008). Geochemistry of cage hydrocarbons. *Pet. Chem.* 48: 241–253. (d) Nekhaev, A.I. and Maksimov, A.L. (2019). Diamondoids in oil and gas condensates (review). *Pet. Chem.* 59: 1108–1117.

3 Simakov, S.K. (2018). Nano- and micron-sized diamond genesis in nature: an overview. *Geosci. Front.* 9: 1849–1858.

4 Schrand, A.M., Hens, S.A.C., and Shenderova, O.A. (2009). Nanodiamond particles: properties and perspectives for bioapplications. *Crit. Rev. Solid State* 34: 18–74.

5 (a) Langenhorst, F. and Campione, M. (2019). Ideal and real structures of different forms of carbon, with some remarks on their geological significance. *J. Geol. Soc. London* 176: 337–347. (b) Badziag, P., Verwoerd, W.S., Ellis, W.P., and Greiner, N.R. (1990). Nanometre-sized diamonds are more stable than graphite. *Nature* 343: 244–245.

6 Alberts, A.H., Strating, J., and Wynberg, H. (1973). Adamantene. *Tetrahedron Lett.* 3047–3050.

7 (a) Zeiger, M., Jackel, N., Mochalin, V.N., and Presser, V. (2016). Review: carbon onions for electrochemical energy storage. *J. Mater. Chem. A* 4: 3172–3196. (b) Duan, X.G., Tian, W.J., Zhang, H.Y. et al. (2019). Sp(2)/sp(3) framework from diamond nanocrystals: a key bridge of carbonaceous structure to carbocatalysis. *ACS Catalysis* 9: 7494–7519.

8 Schupfer, D.B., Badaczewski, F., Peilstocker, J. et al. (2021). Monitoring the thermally induced transition from sp(3)-hybridized into sp(2)-hybridized carbons. *Carbon* 172: 214–227.

9 Xiao, J., Ouyang, G., Liu, P. et al. (2014). Reversible nanodiamond-carbon onion phase transformations. *Nano Lett.* 14: 3645–3652.

10 (a) Angus, J.C., Sunkara, M., Sahaida, S.R., and Glass, J.T. (1992). Twinning and faceting in early stages of diamond growth by chemical vapor deposition. *J. Mater. Res.* 7: 3001–3009. (b) Dahl, J.E.P., Moldowan, J.M., Wei, Z. et al. (2010). Synthesis of higher diamondoids and implications for their formation in petroleum. *Angew. Chem. Int. Ed.* 49: 9881–9885.

11 Tielens, A., Seab, C.G., Hollenbach, D.J., and McKee, C.F. (1987). Shock processing of interstellar dust: diamonds in the sky. *Astrophys. J.* 319: L109–L113.

12 Lewis, R.S., Ming, T., Wacker, J.F. et al. (1987). Interstellar diamonds in meteorites. *Nature* 326: 160–162.

13 Haggerty, S.E. (1999). Earth and planetary sciences – a diamond trilogy: superplumes, supercontinents, and supernovae. *Science* 285: 851–860.

14 Guillois, O., Ledoux, G., and Reynaud, C. (1999). Diamond infrared emission bands in circumstellar media. *Astrophys. J.* 521: L133–L136.

15 (a) Alexander, C.M.O., Cody, G.D., De Gregorio, B.T. et al. (2017). The nature, origin and modification of insoluble organic matter in chondrites, the major source of Earth's C and N. *Geochem.* 77: 227–256. (b) Alexander, C.M.O.,

Russell, S.S., Arden, J.W. et al. (1998). The origin of chondritic macromolecular organic matter: a carbon and nitrogen isotope study. *Meteorit. Planet. Sci.* 33: 603–622. (c) Sephton, M.A. (2002). Organic compounds in carbonaceous meteorites. *Nat. Prod. Rep.* 19: 292–311.

16 (a) Xie, Y.X., Ho, L.C., Li, A.G., and Shangguan, J.Y. (2018). The widespread presence of nanometer-size dust grains in the interstellar medium of galaxies. *Astrophys. J.* 867: 91. (b) Dartois, E. (2019). Interstellar carbon dust. *C-J. Carbon Res.* 5: 80.

17 Allamandola, L.J., Sandford, S.A., Tielens, A., and Herbst, T.M. (1993). Diamonds in dense molecular clouds: a challenge to the standard interstellar medium paradigm. *Science* 260: 64–66.

18 Goto, M., Henning, T., Kouchi, A. et al. (2009). Spatially resolved 3 μm spectroscopy of elias 1: origin of diamonds in protoplanetary disks. *Astrophys. J.* 693: 610–616.

19 Becher, C. (2014). Fluorescent nanoparticles: diamonds from outer space. *Nat. Nanotechnol.* 9: 16–17.

20 Marks, N.A., Lattemann, M., and McKenzie, D.R. (2012). Nonequilibrium route to nanodiamond with astrophysical implications. *Phys. Rev. Lett.* 108: 075503.

21 (a) Bilalbegović, G., Maksimović, A., and Valencic, L.A. (2018). Tetrahedral hydrocarbon nanoparticles in space: X-ray spectra. *Mon. Not. R. Astron. Soc.* 476: 5358–5364. (b) Bauschlicher, C.W., Liu, Y.F., Ricca, A. et al. (2007). Electronic and vibrational spectroscopy of diamondoids and the interstellar infrared bands between 3.35 and 3.55 mm. *Astrophys. J.* 671: 458–469. (c) Pirali, O., Vervloet, M., Dahl, J.E. et al. (2007). Infrared spectroscopy of diamondoid molecules: new insights into the presence of nanodiamonds in the interstellar medium. *Astrophys. J.* 661: 919–925. (d) Bouwman, J., Horst, S., and Oomens, J. (2018). Spectroscopic characterization of the product ions formed by electron ionization of adamantane. *ChemPhysChem* 19: 3211–3218. (e) Greaves, J.S., Scaife, A.M.M., Frayer, D.T. et al. (2018). Anomalous microwave emission from spinning nanodiamonds around stars. *Nat. Astron.* 2: 662–667.

22 (a) Patzer, A., Schütz, M., Möller, T., and Dopfer, O. (2012). Infrared spectrum and structure of the adamantane cation: direct evidence for Jahn-teller distortion. *Angew. Chem. Int. Ed.* 51: 4925–4929. (b) Candian, A., Bouwman, J., Hemberger, P. et al. (2018). Dissociative ionisation of adamantane: a combined theoretical and experimental study. *Phys. Chem. Chem. Phys.* 20: 5399–5406. (c) Crandall, P.B., Müller, D., Leroux, J. et al. (2020). Optical spectrum of the adamantane radical cation. *Astrophys. J. Lett.* 900: L20. (d) Hernandez-Rojas, J. and Calvo, F. (2019). The structure of adamantane clusters: atomistic vs. coarse-grained predictions from global optimization. *Front. Chem.* 7: 573. (e) Demjan, T., Voros, M., Palummo, M., and Gali, A. (2014). Electronic and optical properties of pure and modified diamondoids studied by many-body perturbation theory and time-dependent density functional theory. *J. Chem. Phys.* 141: 064308. (f) Gali, A., Demjan, T., Voros, M. et al. (2016). Electron-vibration

coupling induced renormalization in the photoemission spectrum of diamondoids. *Nat. Commun.* 7: 11327.

23 Shirey, S.B., Cartigny, P., Frost, D.J. et al. (2013). Diamonds and the geology of mantle carbon. In: *Carbon in Earth*, vol. 75 (ed. R.M. Hazen, A.P. Jones, and J.A. Baross), 355–421.

24 Frezzotti, M.L. (2019). Diamond growth from organic compounds in hydrous fluids deep within the Earth. *Nat. Commun.* 10: 4952.

25 (a) Sassen, R. and Post, P. (2008). Enrichment of diamondoids and C-13 in condensate from Hudson canyon, US Atlantic. *Org. Geochem.* 39: 147–151. (b) Giruts, M.V., Stroeva, A.R., Gadzhiev, G.A. et al. (2014). Adamantanes C-11-C-13 in biodegraded and nonbiodegraded condensates. *Pet. Chem.* 54: 10–15. (c) Fang, C.C., Xiong, Y.Q., Liang, Q.Y., and Li, Y. (2012). Variation in abundance and distribution of diamondoids during oil cracking. *Org. Geochem.* 47: 1–8. (d) Mei, M., Bissada, K.K., Malloy, T.B. et al. (2018). Improved method for simultaneous determination of saturated and aromatic biomarkers, organosulfur compounds and diamondoids in crude oils by GC-MS/MS. *Org. Geochem.* 116: 35–50. (e) Cheng, X., Hou, D.J., and Xu, C.G. (2018). The effect of biodegradation on adamantanes in reservoired crude oils from the Bohai Bay Basin, China. *Org. Geochem.* 123: 38–43. (f) Gentzis, T. and Carvajal-Ortiz, H. (2018). Comparative study of conventional maturity proxies with the methyldiamondoid ratio: examples from West Texas, the Middle East, and northern South America. *Int. J. Coal Geol.* 197: 115–125. (g) Mei, M., Bissada, K.K., Malloy, T.B. et al. (2018). Origin of condensates and natural gases in the almond formation reservoirs in southwestern Wyoming, USA. *Org. Geochem.* 124: 164–179. (h) Gordadze, G.N. and Aref'yev, O.A. (1997). Adamantanes of genetically different crude oils. *Pet. Chem.* 37: 381–389.

26 (a) Imuta, K. and Ouchi, K. (1973). Isolation of adamantane from coal extract. *Fuel* 52: 301–302. (b) Bocker, J., Littke, R., Hartkopf-Froder, C. et al. (2013). Organic geochemistry of Duckmantian (Pennsylvanian) coals from the Ruhr Basin, western Germany. *Int. J. Coal Geol.* 107: 112–126.

27 (a) Shimoyama, A. and Yabuta, H. (2002). Mono- and bicyclic alkanes and diamondoid hydrocarbons in the cretaceous/tertiary boundary sediments at Kawaruppu, Hokkaido, Japan. *Geochem. J.* 36: 173–189. (b) Berwick, L., Alexander, R., and Pierce, K. (2011). Formation and reactions of alkyl adamantanes in sediments: carbon surface reactions. *Org. Geochem.* 42: 752–761.

28 (a) Fang, C.C., Xiong, Y.Q., Li, Y. et al. (2015). Generation and evolution of diamondoids in source rock. *Mar. Pet. Geol.* 67: 197–203. (b) Li, Y., Chen, Y., Xiong, Y.Q. et al. (2015). Origin of adamantanes and diamantanes in marine source rock. *Energy Fuel* 29: 8188–8194. (c) Forkner, R., Fildani, A., Ochoa, J., and Moldowan, J.M. (2021). Linking source rock to expelled hydrocarbons using diamondoids: an integrated approach from the northern Gulf of Mexico. *J. Petrol. Sci. Eng.* 196: 108015.

29 Wei, Z.B., Moldowan, J.M., and Paytan, A. (2006). Diamondoids and molecular biomarkers generated from modern sediments in the absence and presence of minerals during hydrous pyrolysis. *Org. Geochem.* 37: 891–911.

30 (a) Nomura, M., Schleyer, P.v.R., and Arz, A.A. (1967). Alkyladamantanes by rearrangement from diverse starting materials. *J. Am. Chem. Soc.* 89: 3657–3659. (b) Fang, C.C., Xiong, Y.Q., Li, Y. et al. (2013). The origin and evolution of adamantanes and diamantanes in petroleum. *Geochim. Cosmochim. Acta* 120: 109–120.

31 Wei, Y.X., Li, J.Z., Yuan, C.Y. et al. (2012). Generation of diamondoid hydrocarbons as confined compounds in SAPO-34 catalyst in the conversion of methanol. *Chem. Commun.* 48: 3082–3084.

32 (a) Schleyer, P.v.R. (1957). A simple preparation of adamantane. *J. Am. Chem. Soc.* 79: 3292. (b) Cupas, C.; Schleyer, P.v.R.; Trecker, D. J. Congressane. *J. Am. Chem. Soc.* 1965, 87, 917–918; (c) Williams, V. Z.; Schleyer, P.v.R.; Gleicher, G. J.; Rodewald, L. B. Triamantane. *J. Am. Chem. Soc.* 1966, 88, 3862–3863.

33 Ender, C.P., Liang, J.X., Friebel, J. et al. (2022). Mechanistic insights of seeded diamond growth from molecular precursors. *Diamond Relat. Mater.* 122.

34 (a) Giruts, M.V., Derbetova, N.B., Erdnieva, O.G. et al. (2013). Identification of tetramantanes in crude oils. *Pet. Chem.* 53: 285–287. (b) Giruts, M.V., Badmaev, C.M., Erdnieva, O.G. et al. (2012). Identification of triamantanes in crude oils. *Pet. Chem.* 52: 65–67. (c) Pytlak, L., Kowalski, A., Gross, D., and Sachsenhofer, R.F. (2017). Composition of diamondoids in oil samples from the alpine foreland basin, Austria: potential as indices of source rock facies, maturity and biodegradation. *J. Petroleum Geol.* 40: 153–171. (d) Musayev, I.A., Bagrii, Y.I., Kurashova, E.K. et al. (1983). Diamantane and 4-methyldiamantane in naphthalane and Russian crude oils. *Pet. Chem.* 23: 182–185. (e) Goodwin, N.R.J., Abdullayev, N., Javadova, A. et al. (2020). Diamondoids and basin modelling reveal one of the world's deepest petroleum systems, South Caspian basin, Azerbaijan. *J. Pet. Geol.* 43: 133–149.

35 Yang, C., Wang, Z.D., Hollebone, B.P. et al. (2006). GC/MS quantitation of diamondoid compounds in crude oils and petroleum products. *Environ. Forensics* 7: 377–390.

36 Dahl, J.E., Moldowan, J.M., Peters, K.E. et al. (1999). Diamondoid hydrocarbons as indicators of natural oil cracking. *Nature* 399: 54–57.

37 Wingert, W.S. (1992). G.C.-m.s. analysis of diamondoid hydrocarbons in Smackover petroleums. *Fuel* 71: 37–43.

38 Lin, R. and Wilk, Z.A. (1995). Natural occurrence of tetramantane ($C_{22}H_{28}$), pentamantane ($C_{26}H_{32}$) and hexamantane ($C_{30}H_{36}$) in a deep petroleum reservoir. *Fuel* 74: 1512–1521.

39 Ma, A. (2016). Advancement in application of diamondoids on organic geochemistry. *J. Nat. Gas Geosci.* 1: 257–265.

40 Li, Y., Xiong, Y.Q., Chen, Y., and Tang, Y.J. (2014). The effect of evaporation on the concentration and distribution of diamondoids in oils. *Org. Geochem.* 69: 88–97.

41 Chung, H.S., Chen, C.S.H., Kremer, R.A. et al. (1999). Recent developments in high-energy density liquid hydrocarbon fuels. *Energy Fuel* 13: 641–649.

42 Stout, S.A., Douglas, G.S., and Uhler, A.D. (2016). Chemical fingerprinting of gasoline and distillate fuels. In: *Standard Handbook Oil Spill Environmental Forensics: Fingerprinting and Source Identification*, 2nd ed. (Eds. S.A. Stout and Z. Wang), 509–564. Elsevier.

43 Stout, S.A. and Douglas, G.S. (2004). Diamondoid hydrocarbons – application in the chemical fingerprinting of natural gas condensate and gasoline. *Environ. Forensics* 5: 225–235.

44 Wang, Z.D., Yang, C., Hollebone, B., and Fingas, M. (2006). Forensic fingerprinting of diamondoids for correlation and differentiation of spilled oil and petroleum products. *Environ. Sci. Technol.* 40: 5636–5646.

45 Liang, Q.Y., Xiong, Y.Q., Fang, C.C., and Li, Y. (2012). Quantitative analysis of diamondoids in crude oils using gas chromatography-triple quadrupole mass spectrometry. *Org. Geochem.* 43: 83–91.

46 (a) Silva, R.C., Silva, R.S.F., de Castro, E.V.R. et al. (2013). Extended diamondoid assessment in crude oil using comprehensive two-dimensional gas chromatography coupled to time-of-flight mass spectrometry. *Fuel* 112: 125–133. (b) Wang, G.L., Shi, S.B., Wang, P.R., and Wang, T.G. (2013). Analysis of diamondoids in crude oils using comprehensive two-dimensional gas chromatography/time-of-flight mass spectrometry. *Fuel* 107: 706–714.

47 (a) Wang, Y.C., Ma, W.Y., Zhou, N. et al. (2017). Analyzing crude oils from the Junggar Basin (NW China) using comprehensive two-dimensional gas chromatography coupled with time-of-flight mass spectrometry (GCxGC-TOFMS). *Acta Geochim.* 36: 66–73. (b) Wang, G.L., Simoneit, B.R.T., Shi, S.B. et al. (2018). A GCxGC-TOFMS investigation of the unresolved complex mixture and associated biomarkers in biodegraded petroleum. *Acta Geol. Sin.-Eng.* 92: 1959–1972. (c) Scarlett, A.G., Despaigne-Diaz, A.I., Wilde, S.A., and Grice, K. (2019). An examination by GC x GC-TOFMS of organic molecules present in highly degraded oils emerging from Caribbean terrestrial seeps of Cretaceous age. *Geosci. Front.* 10: 5–15.

48 (a) Jiang, W.M., Li, Y., Fang, C.C. et al. (2021). Diamondoids in petroleum: their potential as source and maturity indicators. *Org. Geochem.* 160: 104298. (b) Chen, J.H., Fu, J.M., Sheng, G.Y. et al. (1996). Diamondoid hydrocarbon ratios: novel maturity indices for highly mature crude oils. *Org. Geochem.* 25: 179–190.

49 (a) Williams, J.A., Bjoroy, M., Dolcater, D.L., and Winters, J.C. (1986). Biodegradation in South Texas Eocene oils – effects on aromatics and biomarkers. *Org. Geochem.* 10: 451–461. (b) Grice, K., Alexander, R., and Kagi, R.I. (2000). Diamondoid hydrocarbon ratios as indicators of biodegradation in Australian crude oils. *Org. Geochem.* 31: 67–73.

50 Machel, H.G., Krouse, H.R., and Sassen, R. (1995). Products and distinguishing criteria of bacterial and thermochemical sulfate reduction. *Appl. Geochem.* 10: 373–389.

51 Jiang, W.M., Li, Y., and Xiong, Y.Q. (2020). Reservoir alteration of crude oils in the Junggar Basin, Northwest China: insights from diamondoid indices. *Mar. Petr. Geol.* 119: 104451.

52 Schulz, L.K., Wilhelms, A., Rein, E., and Steen, A.S. (2001). Application of diamondoids to distinguish source rock facies. *Org. Geochem.* 32: 365–375.

53 Landes, K.K. (1967). Eometamorphism, and oil and gas in time and space. *AAPG Bull.* 51: 828–841.

54 (a) Price, L.C. (1993). Thermal stability of hydrocarbons in nature: limits, evidence, characteristics, and possible controls. *Geochim. Cosmochim. Acta* 57: 3261–3280. (b) Hill, R.J., Tang, Y.C., Kaplan, I.R., and Jenden, P.D. (1996). The influence of pressure on the thermal cracking of oil. *Energy Fuel* 10: 873–882. (c) Hill, R.J., Tang, Y.C., and Kaplan, I.R. (2003). Insights into oil cracking based on laboratory experiments. *Org. Geochem.* 34: 1651–1672.

55 Zhu, G., Li, J., Zhang, Z. et al. (2020). Stability and cracking threshold depth of crude oil in 8000 m ultra-deep reservoir in the Tarim Basin. *Fuel* 282: 118777.

56 Zhu, G.Y., Milkov, A.V., Chen, F.R. et al. (2018). Non-cracked oil in ultra-deep high-temperature reservoirs in the Tarim basin, China. *Mar. Petr. Geol.* 89: 252–262.

57 (a) Clark, T., Knox, T.M., Mackle, H. et al. (1975). Calorimetric evaluation of enthalpies of formation of some bridged-ring hydrocarbons. Comparison with data from empirical force field calculations. *J. Am. Chem. Soc.* 97: 3835–3836. (b) Dorofeeva, O.V. and Ryzhova, O.N. (2019). Enthalpies of formation of diamantanes in the gas and crystalline phase: comparison of theory and experiment. *Struct. Chem.* 30: 615–621.

58 Li, J.G., Philp, P., and Cui, M.Z. (2000). Methyl diamantane index (MDI) as a maturity parameter for lower Palaeozoic carbonate rocks at high maturity and overmaturity. *Org. Geochem.* 31: 267–272.

59 (a) Zhu, G.Y., Wang, M., Zhang, Y., and Zhang, Z.Y. (2018). Higher ethanodiamondoids in petroleum. *Energy Fuel* 32: 4996–5000. (b) Wang, M., Zhu, G.Y., Milkov, A.V., and Chi, L.X. (2020). Comprehensive molecular compositions and origins of DB301 crude oil from deep strata, Tarim Basin, China. *Energy Fuel* 34: 6799–6810.

60 Farcasiu, D., Wiskott, E., Osawa, E. et al. (1974). Ethanoadamantane – most stable C12H18 isomer. *J. Am. Chem. Soc.* 96: 4669–4671.

61 Zhu, G.Y., Zhang, Z.Y., Milkov, A.V. et al. (2019). Diamondoids as tracers of late gas charge in oil reservoirs: example from the Tazhong area, Tarim Basin, China. *Fuel* 253: 998–1017.

62 McKervey, M.A. (1974). Adamantane rearrangements. *Chem. Soc. Rev.* 3: 479–512.

63 (a) Xie, J.W., Jia, T.H., Gong, S. et al. (2020). Synthesis and thermal stability of dimethyl adamantanes as high-density and high-thermal-stability fuels. *Fuel* 260: 116424. (b) Xie, J.W., Zhang, X.W., Xie, J.J. et al. (2019). Acid-catalyzed rearrangement of tetrahydrotricyclopentadiene for synthesis of high density alkyl-diamondoid fuel. *Fuel* 239: 652–658.

References

64 Bagrii, Y.I., Sanin, P.I., Vorobeva, N.S., and Petrov, A.A. (1967). Hydrocarbon composition of extracts obtained from petroleum fractions by extractive crystallization with thiourea. *Pet. Chem.* 7: 159–163.

65 Osawa, E., Engler, E.M., Godleski, S.A. et al. (1980). Application of force-field calculations to organic chemistry. 10. Bridgehead reactivities of ethanoadamantane – bromination and solvolysis of bromides. *J. Org. Chem.* 45: 984–991.

66 Gordadze, G.N. and Giruts, M.V. (2008). Synthesis of adamantane and diamantane hydrocarbons by high-temperature cracking of higher n-alkanes. *Pet. Chem.* 48: 414–419.

67 Giruts, M.V. and Gordadze, G.N. (2007). Generation of adamantanes and diamantanes by thermal cracking of polar components of crude oils of different genotypes. *Pet. Chem.* 47: 12–22.

68 (a) Tiwari, R.N. and Chang, L. (2010). Growth, microstructure, and field-emission properties of synthesized diamond film on adamantane-coated silicon substrate by microwave plasma chemical vapor deposition. *J. Appl. Phys.* 107: 103305. (b) Mandal, S., Thomas, E.L.H., Jenny, T.A., and Williams, O.K. (2016). Chemical nucleation of diamond films. *ACS Appl. Mater. Interfaces* 8: 26220–26225. (c) Sangphet, S., Siriroj, S., Sriplai, N. et al. (2018). Enhanced ferromagnetism in mechanically exfoliated CVD-carbon films prepared by using adamantane as precursor. *Appl. Phys. Lett.* 112: 242406. (d) Liang, J.X., Ender, C.P., Zapata, T. et al. (2020). Germanium iodide mediated synthesis of nanodiamonds from adamantane "seeds" under moderate high-pressure high-temperature conditions. *Diamond Relat. Mater.* 108: 108000. (e) Ekimov, E.A., Kudryavtsev, O.S., Mordvinova, N.E. et al. (2018). High-pressure synthesis of nanodiamonds from adamantane: myth or reality? *ChemNanoMat* 4: 269–273.

69 (a) Zhang, J., Ishiwata, H., Babinec, T.M. et al. (2016). Hybrid group IV nanophotonic structures incorporating diamond silicon-vacancy color centers. *Nano Lett.* 16: 212–217. (b) Gebbie, M.A., Ishiwata, H., McQuade, P.J. et al. (2018). Experimental measurement of the diamond nucleation landscape reveals classical and nonclassical features. *Proc. Natl. Acad. Sci. U. S. A.* 115: 8284–8289.

70 Park, S., Abate, I.I., Liu, J. et al. (2020). Facile diamond synthesis from lower diamondoids. *Sci. Adv.* 6: eaay9405.

71 Tzeng, Y.K., Zhang, J., Lu, H.Y. et al. (2017). Vertical-substrate MPCVD epitaxial nanodiamond growth. *Nano Lett.* 17: 1489–1495.

72 (a) Stauss, S., Miyazoe, H., Shizuno, T. et al. (2010). Synthesis of the higher-order diamondoid hexamantane using low-temperature plasmas generated in supercritical xenon. *Jpn. J. Appl. Phys.* 49: 070213. (b) Shizuno, T., Miyazoe, H., Saito, K. et al. (2011). Synthesis of diamondoids by supercritical xenon discharge plasma. *Jpn. J. Appl. Phys.* 50: 030207.

73 Stauss, S., Ishii, C., Pai, D.Z. et al. (2014). Diamondoid synthesis in atmospheric pressure adamantane-argon-methane-hydrogen mixtures using a continuous flow plasma microreactor. *Plasma Sources Sci. Technol.* 23: 035016.

74 Landa, S. and Macháček, V. (1933). Sur l'adamantane, nouvel hydrocarbure extrait du naphte. *Collect Czech. Chem. Commun.* 5: 1–5.
75 Landa, S. (1963). Adamantane and its homologues. *Curr. Sci.* 32: 485–489.
76 Prelog, V. and Seiwerth, R. (1941). Über die Synthese des Adamantans. *Chem. Ber.* 1644–1648.
77 Mair, B.J., Shamaiengar, M., Krouskop, N.C., and Rossini, F.D. (1959). Isolation of adamantane from petroleum. *Anal. Chem.* 31: 2082–2083.
78 (a) He, M., Moldowan, J.M., Nemchenko-Rovenskaya, A., and Peters, K.E. (2012). Oil families and their inferred source rocks in the Barents Sea and northern Timan-Pechora Basin, Russia. *AAPG Bull.* 96: 1121–1146.
(b) Nguyen, T.X. and Philp, R.P. (2016). Separation of diamondoids in crude oils using molecular sieving techniques to allow compound-specific isotope analysis. *Org. Geochem.* 95: 1–12.
79 Huang, L., Zhang, S.C., Wang, H.T. et al. (2011). A novel method for isolation of diamondoids from crude oils for compound-specific isotope analysis. *Org. Geochem.* 42: 566–571.
80 Hala, S., Landa, S., and Hanus, V. (1966). Isolation of tetracyclo[6.3.1.0^{2,6}.0^{5,10}]dodecane and pentacyclo[7.3.1.1^{4,12}.0^{2,7}.0^{6,11}]tetradecane (diamantane) from petroleum. *Angew. Chem. Int. Ed.* 5: 1045–1046.
81 Moldowan, J.M., Dahl, J., Zinniker, D., and Barbanti, S.M. (2015). Underutilized advanced geochemical technologies for oil and gas exploration and production-1. The diamondoids. *J. Petrol. Sci. Eng.* 126: 87–96.
82 Atwah, I., Moldowan, J.M., Koskella, D., and Dahl, J. (2021). Application of higher diamondoids in hydrocarbon mudrock systems. *Fuel* 284: 118994.
83 Dahl, J.E., Liu, S.G., and Carlson, R.M.K. (2003). Isolation and structure of higher diamondoids, nanometer-sized diamond molecules. *Science* 299: 96–99.
84 Dahl, J.E.P., Moldowan, J.M., Peakman, T.M. et al. (2003). Isolation and structural proof of the large diamond molecule, cyclohexamantane ($C_{26}H_{30}$). *Angew. Chem. Int. Ed.* 42: 2040–2044.
85 Fokin, A.A., Tkachenko, B.A., Fokina, N.A. et al. (2009). Reactivities of the prism-shaped diamondoids [1(2)3]tetramantane and [12312]hexamantane (Cyclohexamantane). *Chem. Eur. J.* 15: 3851–3862.
86 Friedel, R.A. and Queiser, J.A. (1959). Ultraviolet-visible spectrum and the aromaticity of coal. *Fuel* 38: 369–380.
87 (a) Chakrabartty, S.K. and Berkowitz, N. (1974). Studies on the structure of coals. 3. Some inferences about skeletal structures. *Fuel* 53: 240–245.
(b) Chakrabartty, S.K. and Berkowitz, N. (1976). Aromatic structures in coal. *Nature* 261: 76–77.
88 Aczel, T., Gorbaty, M.L., Maa, P.S., and Schlosberg, R.H. (1979). Stability of adamantane and its derivatives to coal-liquefaction conditions, and its implications toward the organic structure of coal. *Fuel* 58: 228–230.
89 Wei, Z., Moldowan, J.M., Jarvie, D.M., and Hill, R. (2006). The fate of diamondoids in coals and sedimentary rocks. *Geology* 34: 1013–1016.

90 Fang, C., Wu, W., Liu, D., and Liu, J. (2016). Evolution characteristics and application of diamondoids in coal measures. *J. Nat. Gas Geosci.* 1: 93–99.

91 McMillen, D.F. and Golden, D.M. (1982). Hydrocarbon bond dissociation energies. *Annu. Rev. Phys. Chem.* 33: 493–532.

92 Chakrabartty, S.K. and Kretschmer, H.O. (1972). Studies on the structure of coals: Part 1. The nature of aliphatic groups. *Fuel* 51: 160–163.

93 (a) Akinlua, A., Ibeachusim, B.I., Adekola, S.A. et al. (2020). Diamondoid geochemistry of Niger Delta source rocks: implication for petroleum exploration. *Energy Sources A: Recovery Util. Environ. Eff.* 1–11. (b) Esegbue, O., Jones, D.M., van Bergen, P.F., and Kolonic, S. (2020). Quantitative diamondoid analysis indicates oil cosourcing from a deep petroleum system onshore Niger Delta Basin. *AAPG Bull.* 104: 1231–1259.

94 Akinlua, A., Jochmann, M.A., Lorenzo-Parodi, N. et al. (2019). A green approach for the extraction of diamondoids from petroleum source rock. *Anal. Chim. Acta* 1091: 23–29.

95 Shen, W.B., Chen, J.F., Wang, Y.Y. et al. (2019). The origin, migration and accumulation of the Ordovician gas in the Tazhong III region, Tarim Basin, NW China. *Mar. Petr. Geol.* 101: 55–77.

96 Jiang, W.M., Li, Y., and Xiong, Y.Q. (2018). The effect of organic matter type on formation and evolution of diamondoids. *Mar. Pet. Geol.* 89: 714–720.

97 Jiang, W.M., Li, Y., and Xiong, Y.Q. (2019). Source and thermal maturity of crude oils in the Junggar Basin in Northwest China determined from the concentration and distribution of diamondoids. *Org. Geochem.* 128: 148–160.

98 Gordadze, G.N., Kerimov, V.Y., Gaiduk, A.V. et al. (2017). Hydrocarbon biomarkers and diamondoid hydrocarbons from late Precambrian and lower Cambrian rocks of the Katanga saddle (Siberian Platform). *Geochem. Int.* 55: 360–366.

99 Zhu, G.Y., Zhang, Z.Y., Zhou, X.X. et al. (2018). Preservation of ultradeep liquid oil and its exploration limit. *Energy Fuel* 32: 11165–11176.

100 Bowman, D.T., Slater, G.F., Warren, L.A., and McCarry, B.E. (2014). Identification of individual thiophene-, indane-, tetralin-, cyclohexane-, and adamantane-type carboxylic acids in composite tailings pore water from Alberta oil sands. *Rapid Commun. Mass Spectrom.* 28: 2075–2083.

101 Lengger, S.K., Scarlett, A.G., West, C.E., and Rowland, S.J. (2013). Diamondoid diacids (' O-4 ' species) in oil sands process-affected water. *Rapid Commun. Mass Spectrom.* 27: 2648–2654.

102 (a) Wei, Z., Moldowan, J.M., Fago, F. et al. (2007). Origins of thiadiamondoids and diamondoidthiols in petroleum. *Energy Fuel* 21: 3431–3436. (b) Wei, Z.B., Walters, C.C., Moldowan, J.M. et al. (2012). Thiadiamondoids as proxies for the extent of thermochemical sulfate reduction. *Org. Geochem.* 44: 53–70.

103 Birch, S.F., Cullum, T.V., Dean, R.A., and Denyer, R.L. (1952). Thiaadamantane. *Nature* 170: 629–630.

104 Stetter, H., Held, H., and Schulte-Oestrich, A. (1962). Über Verbindungen mit Urotropin-Struktur, XXIII. Synthese des 2-Thia-adamantans. *Chem. Ber.* 95: 1687–1681.

105 Ma, A.L., Jin, Z.J., Zhu, C.S. et al. (2023). Formation mechanism of thiadiamondoids: evidence from thermochemical sulfate reduction simulation experiments with model compounds. *Org. Geochem.* 177: 104554.

106 Chakhmakhchev, A., Sanderson, J., Pearson, C., and Davidson, N. (2017). Compositional changes of diamondoid distributions caused by simulated evaporative fractionation. *Org. Geochem.* 113: 224–228.

107 Zhu, G.Y., Zhang, Y., Zhou, X.X. et al. (2019). TSR, deep oil cracking and exploration potential in the Hetianhe gas field, Tarim Basin, China. *Fuel* 236: 1078–1092.

108 Cai, C.F., Xiao, Q.L., Fang, C.C. et al. (2016). The effect of thermochemical sulfate reduction on formation and isomerization of thiadiamondoids and diamondoids in the lower Paleozoic petroleum pools of the Tarim Basin, NW China. *Org. Geochem.* 101: 49–62.

109 (a) Zhu, G.Y., Wang, H.T., and Weng, N. (2016). TSR-altered oil with high-abundance thiaadamantanes of a deep-buried Cambrian gas condensate reservoir in Tarim Basin. *Mar. Pet. Geol.* 69: 1–12. (b) Zhu, G.Y., Zhang, Y., Wang, M., and Zhang, Z.Y. (2018). Discovery of high-abundance diamondoids and thiadiamondoids and severe TSR alteration of well ZS1C condensate, Tarim Basin, China. *Energy Fuel* 32: 7383–7392.

110 Gvirtzman, Z., Said-Ahmad, W., Ellis, G.S. et al. (2015). Compound-specific sulfur isotope analysis of thiadiamondoids of oils from the Smackover formation, USA. *Geochim. Cosmochim. Acta* 167: 144–161.

111 Ma, A.L., Jin, Z.J., Zhu, C.S., and Gu, Y. (2018). Detection and significance of higher thiadiamondoids and diamondoidthiols in oil from the Zhongshen 1C well of the Tarim Basin, NW China. *Sci. China Earth Sci.* 61: 1440–1450.

112 Xiao, Q.L., Sun, Y.G., He, S. et al. (2019). Thermal stability of 2-thiadiamondoids determined by pyrolysis experiments in a closed system and its geochemical implications. *Org. Geochem.* 130: 14–21.

113 Zhu, G.Y., Wang, P., Wang, M. et al. (2019). Occurrence and origins of thiols in deep strata crude oils, Tarim Basin, China. *ACS Earth Space Chem.* 3: 2499–2509.

114 Peng, Y.Y., Cai, C.F., Fang, C.C. et al. (2022). Diamondoids and thiadiamondoids generated from hydrothermal pyrolysis of crude oil and TSR experiments. *Sci. Rep.* 12: 196.

115 Zeng, Z., Altarawneh, M., Oluwoye, I. et al. (2016). Inhibition and promotion of pyrolysis by hydrogen sulfide (H_2S) and sulfanyl radical (SH). *J. Phys. Chem. A* 120: 8941–8948.

3

Diamondoids as Alkane CH Activation Models

Hydrocarbon chemistry is critical for many areas of the modern world ranging from fuel technology to the production of fine chemicals. The direct conversion of methane to methanol [1] is a notable example where thousands of research projects [2] still have not resulted in a full-fledged industrial application: alternative steam reforming of methane to syngas still is technically superior for methanol production through the simple sequence $CH_4 + H_2O \rightarrow CO + H_2 \rightarrow CH_3OH$ [3]. Hope is associated with ionic methane conversion in the presence of SO_3 that has recently been scaled up to 200 kg of pure methanesulfonic acid produced per week [4] where, however, high temperatures and pressures are still required. The intense current debates around this reaction illustrate the difficulties in mechanistic interpretations of alkane reactivity in electrophilic media [5].

Current approaches to ionic methane transformations are still associated with low conversions in order to avoid over-functionalizations [3]. On the other hand, while the C—H-substitutions of small alkanes occur with ease and have been known since 1840 as Dumas' chlorination [6], radical reactions are not very useful for higher hydrocarbons as it is difficult to discriminate between the inequivalent C—H-sites. In summary, we are still far from solving key problems in the chemistry of unactivated alkanes, namely overreactions and low selectivities [7].

Remarkably, diamondoids, also belonging to saturated hydrocarbons, are a clear exception as their C—H substitutions are quite selective and overreactions are avoided in many cases. This makes diamondoids useful C—H activation models both in mechanistic studies and for the development of new alkane functionalization methods [8]. It was a great surprise when Landa found that **AD** reacts with neat bromine under very mild conditions and forms 1-bromoadamantane as the sole product almost quantitatively [9]. Since then, many other electrophilic and radical reagents have been employed for diamondoids often with high selectivities. This is due to the combination of several factors, namely high symmetries and cage polarizabilities as well as the fact that elimination reactions of the cage intermediates are hampered because anti-Bredt olefins would form. As the cage frameworks are able to transfer hyperconjugative and polarization effects over long distances [10], mono-substitution deactivates the cage often avoiding over-functionalizations. Additionally, Lewis acid catalysis or utilization of more active reagents makes

---

The Chemistry of Diamondoids: Building Blocks for Ligands, Catalysts, Materials, and Pharmaceuticals,
First Edition. Andrey A. Fokin, Marina Šekutor, and Peter R. Schreiner.
© 2024 WILEY-VCH GmbH. Published 2024 by WILEY-VCH GmbH.

polyfunctionalizations still viable. As a result, preparative diamondoid chemistry is exceedingly rich and provides universal models for mechanistic studies. Still, diamondoid C—H-substitutions are closely connected to general alkane activation scenarios (Figure 3.1). Alkane activation with electron-deficient species **E** is formally almost always oxidative in nature but proceeds in a wide range from purely electrophilic (**TS1**) to solely radical (**TS2**) pathways while also including oxidative single-electron transfer (SET) oxidations. Less common is molecule-induced homolytic cleavage (for the most recent review see Ref. [12]), where two radicals form from two closed-shell species, which is characteristic for the reactions of alkanes with metal-oxo species [13], with substantial charge transfer observed in transition structures **TS3**. Indeed, charge transfer dominates over the unpaired spin effects in hydrogen abstractions from hydrocarbons by electrophilic reagents [11].

Among these transformations, the reactions with charged electrophiles **E⁺** (**TS1**, Figure 3.1) are the most intriguing, as they are often termed as "hydride transfer" even though this contradicts the C—H-bond polarization where hydrogen carries a partial positive charge. Consequently, H-abstraction/substitution reactions with electrophiles should be better viewed as a proton (or hydrogen) transfer [14]. Olah suggested that the attack of electrophiles on alkanes occurs due to "the electron donor ability of the σ bonds via two-electron, three-center bond formation" [1c]. This suggestion was based on the existence of the two-electron, three-centered (2e-3c) bonding [15] in prototypical methonium cation CH_5^+, which is stable

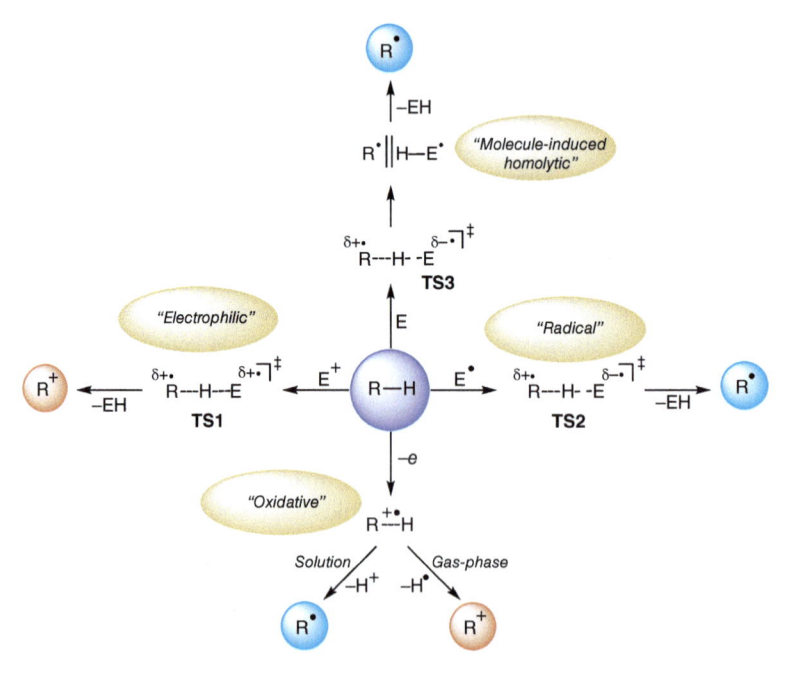

Figure 3.1 General pathways for the C—H-bond activation in alkanes with various electron-deficient reagents **E** also including single-electron transfer oxidations. Source: Reproduced from Ref. [11b], Wiley-VCH, 2004.

both in the gas [16] and condensed [17] phases [18]. However, all attempts to find the 2e-3c transition structures for model reactions of methane with electrophiles failed, and the direct attack on the carbon atom was found to be operative instead [19]. Nonetheless, the activation of the C—H-bonds in methane is not a good model for the reactions of diamondoids with electrophiles, as the reactivity of 3° C—H-bonds is rather different, thus perhaps suggesting isobutane as a more appropriate model. In contrast to methane, isobutane upon protonation forms a weakly bound cluster of the *tert*-butyl cation with dihydrogen [20] rather than 2e-3c bonded structures that are so characteristic of methane. All attempts to model the transition structures for the reaction of isobutane with other charged electrophiles such as water oxide H_2OO^+ [21], protonated hydrogen peroxide $H_3O_2^+$ [22], Cl_3^+, and HCl_2^+ [23] gave linear or almost linear [*t*Bu•••H•••E]$^+$ transition structures **TS1**, structurally similar to those found for the hydrogen abstractions with radicals [8b]. Thus, from a mechanistic viewpoint, activation of the 3° C—H cage bonds with charged electrophiles and radicals **E·** (**TS2**, Figure 3.1) proceeds in a similar fashion, i.e., through the linear transition structures [23]. When charged electrophiles are involved, the activation typically occurs as H-coupled electron transfer (HCET), which comprises the transfer of an electron and a hydrogen atom, such as the total transfer of two electrons mirrors consecutive one-electron oxidations. The HCET mechanism suggested by two of us [23, 24] was found to be operative in the C—H-activations with a variety of electrophiles [25]. In this instance, electrophiles and radicals occupy opposite ends of the same mechanistic spectrum [26], with the only difference being the degree of charge transfer from the hydrocarbon to the electrophile, as exemplified with **AD** (Figure 3.2). Even though both closed-shell electrophiles and radicals form linear transition structures *en route* to the C—H-abstraction, the exact nature of the reaction can be assessed by the degree of charge transfer from the hydrocarbon to the electrophile. Hence, the leftmost part of the scale in Figure 3.2 is reserved for radicals of low electrophilicity (the charge transfer in the transition structures is below 0.1*e* as for alkyl radicals Alk·),

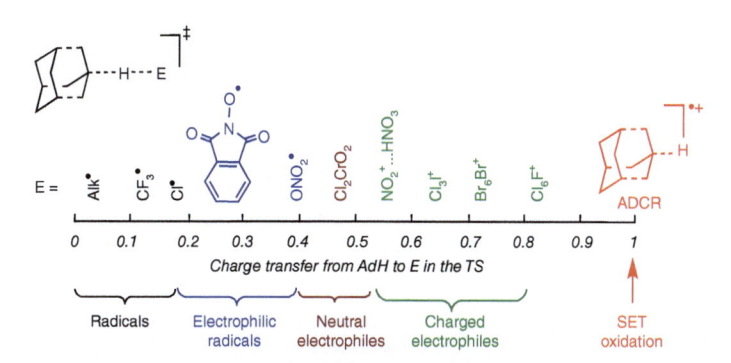

Figure 3.2 Hydrogen abstraction from the bridgehead position of **AD** through the H-coupled electron transfer mechanism. Source: Adopted from Ref. [26]. The degree of charge transfer is expressed for transition structures with various electron-deficient species and ends with the outer-sphere SET pathway with the formation of adamantane radical cation (**ADCR**).

while the middle area (0.3–0.4e) is occupied by electrophilic radicals such as nitroxyl- (NO$_2$O·) and PINO (phthalimide-N-oxyl). Going further right on the scale leads to increasingly pronounced electron transfer from the hydrocarbon to the reagent with neutral and charged electrophiles and ends with the outer-sphere SET pathway, i.e., formation of a free adamantane radical cation (**ADCR**) with an electron-depleted C—H-bond.

Note that the product distributions are different in the reaction of diamondoids in electrophilic media when compared to exclusive radical reaction conditions. For example, **AD** reacts with neat bromine to give 1-bromoadamantane (**1BRAD**) with 100% regioselectivity, whereas free radical halogenations give a mixture of secondary and tertiary C—H-substitution products [27]. The difference in the energetic stability of 1- and 2-adamantyl cations (possibly generated in the presence of electrophiles) cannot be responsible for such a high preference for **1BRAD** vs. 2-bromo-derivative (**2BRAD**) since bromination is not a thermodynamically controlled substitution. Furthermore, the corresponding adamantyl radicals have comparable heats of formation [28], meaning that 2° and 3° C—H-bonds are of comparable strength, and the reaction selectivity depends solely on steric and polar effects in the transition state [8b]. All of these observations can be rationalized by hydrogen abstractions being accompanied by substantial charge transfer from the hydrocarbon to the electrophile. As a result, the hydrocarbon moiety resembles the structure of **ADCR** (Figure 3.2). Due to its high symmetry, such selective HCET from bridgehead positions is in line with cage polarization and therefore highly favorable, explaining the exclusive formation of **1BRAD** in the reaction with bromine. The underlying mechanism for C—H-brominations with neat bromine involves polybromonium cations (Br$_n$$^+$) that are weakly electrophilic, and the process is best viewed as an HCET reaction [23] as described above (Figure 3.2). Polyhalogen cations are effective H-abstractors and HCET through linear transition structures (Figure 3.3) that explain the observed regioselectivity because electron

Figure 3.3 Hydrogen-coupled electron transfer (HCET) transition structures for the reaction of **AD** with model polyhalogen cations BrCl$_4$$^+$ (left) and Hal$_7$$^+$ (right). Critical bond lengths in Å and reaction barriers at B3LYP/6-31G(d,p) for H, C, Cl, and Br (3-21G for I). Source: Reproduced from Ref. [23]. with permission from the American Chemical Society, 2002.

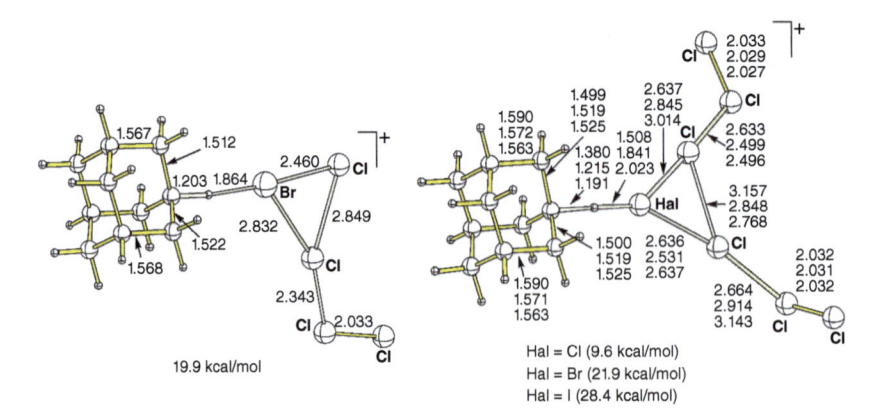

transfer involving secondary C—H bonds is energetically unfavorable. The relative reaction barriers are the lowest for electronegative Cl_7^+ and the highest for ICl_6^+ heptahalogen $(Hal_7)^+$ electrophiles (Figure 3.3).

The found regioselectivites are therefore a direct consequence of the largely cationic nature of the corresponding transition structures. What is more, the suggested HCET reaction mechanism was also substantiated by measuring high reaction orders (ca. 7.5 for **AD** bromination) and by kinetic isotope effect (KIE) studies. The experimentally observed large KIE ($k_H/k_D = 3.9 \pm 0.2$) values are characteristic of linear transition structures and also point toward the partial radical character of the involved species, once more emphasizing the comparability of these electrophilic reactions with radical reactions [23]. The mechanism of adamantane halogenation with polyhalogen electrophiles was recently termed as "cluster halogenation." [29] The new kinetic study [29] not only confirmed the previous parameters for the bromination but was also extended to the chlorination of **AD** with ICl, and it was concluded that the mechanisms of cluster bromination with Br_2 and chlorination with ICl are similar. These findings illustrate the general importance of charge transfer in alkane C—H-halogenation reactions.

Extraordinarily high 3°/2° C—H-bond selectivities are also typical for the reactions of diamondoids with 100% nitric acid [30], where electrophilic nitronium nitrate $NO_2^+NO_3^-$ is responsible for the activation step. Recent kinetic studies demonstrate that 1,3,5,7-tetramethyladamantane, which lacks 3° C—H-bonds, reacts with nitric acid ca. 17000 more slowly than 1,3,5-trimethyladamantane [31], clearly showing that polarization of the adamantane cage by the electrophile through the secondary CH_2 positions is much less favorable. From a mechanistic perspective, the transition structures for $HNO_3 \bullet\bullet\bullet NO_2^+$ (which is also viewed as protonated $NO_2^+NO_3^-$, a solvated model electrophile) attacking the C—H-fragment display (Figure 3.4) a linear geometry of the C$\bullet\bullet\bullet$H$\bullet\bullet\bullet$E fragment as shown for **1(2)3TET** (HCET mechanism, see Figure 3.2). Since only moderate overall charge transfer occurs in such transition structures as **1(2)3TET7** and **1(2)3TET2**, steric factors play a much more important role in hydrogen abstractions with nitronium reagents than in the case of halogenations. As a result, nitroxylations display inverse reaction selectivities (apical) when compared to bromination, where medial derivatives dominate [33]. Note that nitroxylations do not produce any bridge C—H substitution products, as is expected for diamondoids reacting with oxidizing electrophiles (vide supra).

The HCET mechanism was recently suggested [34] for the CH-substitution in adamantane and diamantane derivatives with the quinuclidine radical cation **QUINOBS** (OBs = OSO_2Ph, Figure 3.5). The high observed *tert*-selectivities of the C—H-substitution of the adamantane cage were attributed to "increased charge-transfer character in the HAT transition state **TS-1**" [34]. This suggestion is additionally supported by the fact that the electron-withdrawing substituent (OBs) on quinuclidine increases the reaction rate. This additional driving force of hydrogen transfer is determined by strengthening the N—H bond.

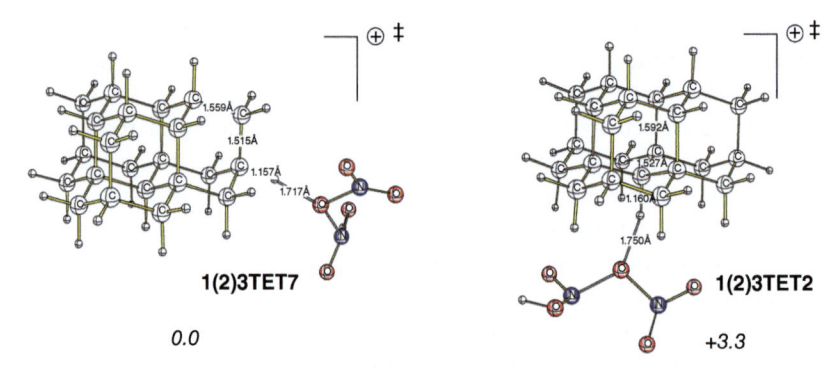

Figure 3.4 Energetically lowest (apical, **1(2)3TET7**) and highest (medial, **1(2)3TET2**) lying hydrogen-coupled electron transfer transition structures for the reaction of **1(2)3TET** with protonated nitronium nitrate (HNO$_2$•••NO$_3$)$^+$ as the model electrophile. Critical bond lengths in Å, reaction barriers at B3PW96/cc-pVDZ in kcal mol^{-1}. Source: Reproduced from Ref. [32], Wiley-VCH, 2009.

Figure 3.5 The hydrogen abstraction from **AD** with the quinuclidine radical cation derivative **QUINOBS** and the HCET transition structure **TS-1** for hydrogen transfer. Source: Reproduced from ref. [34] with permission from the American Chemical Society, 2019.

As a consequence, the most widely used and most useful reagents in diamondoid chemistry are charged electrophiles. Examples of practical diamondoid bridgehead substitutions include oxidations and nitroxylations [31a, 35], with concentrated sulfuric [36] and fuming nitric [37] acid, respectively. These reactions often give the bridgehead carbenium ions as a result of kinetically controlled H-abstractions followed by the recombination with the nucleophilic part of the reagent. A historically eminent method for diamondoid cation generation with superacids, as pioneered by Olah [38] and independently by Hogeveen [39], is very useful for mechanistic and structural studies [40].

Among the diamondoid transformations with radical species, the most selective substitutions are observed for highly electrophilic radicals that occupy the middle part of the reactivity scale shown in Figure 3.2. The PINO- and nitroxyl-radicals display almost exclusive 3° C—H-bond substitutions for **AD** and **DIA**. For radicals with low electrophilicity like Hal$_3$C·, steric factors play a decisive role in positional selectivity. The larger the radical, the more selective the reaction will be for the tertiary positions since the attack of a bulkier radical on the cage bridge is disfavored owing to repulsive interactions with the hydrogens of the neighboring methylene groups.

This has been demonstrated by applying phase-transfer catalytic (PTC) conditions for radical halogenations of diamondoids using bulky trihalomethyl radicals [41].

The above approaches for selective C—H-bond activations combined with the use of a variety of reagents led to many practical strategies toward a plethora of functionalized diamondoids. Over many years, preparative diamondoid chemistry has been well presented, and the reader is encouraged to refer to these comprehensive works, including reviews focusing on cage synthesis and functionalizations [42], strategies for the introduction of functional groups [26, 43], applications of designed scaffolds [44], medicinal applications [45], and examples in other fields of interest [46]. The following chapter presents an overview of the typically used functionalization methods when designing diamondoid scaffolds and is a reference guide to the preparative diamondoid chemistry available to date. Adamantane chemistry is covered only to the extent that is needed for comparison with higher diamondoids. The chosen arbitrary division of the subchapters is therefore not based on the mechanistic aspects of the reactions [26] but rather on the synthetic availability of the target product classes.

References

1 (a) Gesser, H.D., Hunter, N.R., and Prakash, C.B. (1985). The direct conversion of methane to methanol by controlled oxidation. *Chem. Rev.* 85: 235–244. (b) Munz, D. and Strassner, T. (2015). Alkane C-H functionalization and oxidation with molecular oxygen. *Inorg. Chem.* 54: 5043–5052. (c) Olah, G.A. and Molnar, A. (2003). *Hydrocarbon Chemistry*, 2nd ed. Hoboken, NJ: John Wiley & Sons. (d) Tinberg, C.E. and Lippard, S.J. (2011). Dioxygen activation in soluble methane monooxygenase. *Acc. Chem. Res.* 44: 280–288. (e) Otsuka, K. and Wang, Y. (2001). Direct conversion of methane into oxygenates. *Appl. Catal. A – Gen.* 222: 145–161. (f) Tomkins, P., Ranocchiari, M., and van Bokhoven, J.A. (2017). Direct conversion of methane to methanol under mild conditions over Cu-zeolites and beyond. *Acc. Chem. Res.* 50: 418–425. (g) da Silva, M.J. (2016). Synthesis of methanol from methane: challenges and advances on the multi-step (syngas) and one-step routes (DMTM). *Fuel Process. Technol.* 145: 42–61.

2 (a) Zakaria, Z. and Kamarudin, S.K. (2016). Direct conversion technologies of methane to methanol: an overview. *Renew. Sust. Energ. Rev.* 65: 250–261. (b) Ravi, M., Ranocchiari, M., and van Bokhoven, J.A. (2017). The direct catalytic oxidation of methane to methanol – a critical assessment. *Angew. Chem. Int. Ed.* 56: 16464–16483.

3 Sun, L.L., Wang, Y., Guan, N.J., and Li, L.D. (2020). Methane activation and utilization: current status and future challenges. *Energy Technol.* 8: 1900826.

4 Diaz-Urrutia, C. and Ott, T. (2019). Activation of methane to CH_3^+: a selective industrial route to methanesulfonic acid. *Science* 363: 1326–1329.

5 (a) Roytman, V.A. and Singleton, D.A. (2019). Comment on "Activation of methane to CH_3^+: a selective industrial route to methanesulfonic acid". *Science* 364: aax7083. (b) Diaz-Urrutia, C. and Ott, T. (2020). Response to comment

on "Activation of methane to CH_3^+: a selective industrial route to methanesulfonic acid". *Science* 369: aax9966. (c) Blankenship, A.N., Ravi, M., Newton, M.A., and van Bokhoven, J.A. (2021). Heterogeneously catalyzed aerobic oxidation of methane to a methyl derivative. *Angew. Chem. Int. Ed.* 60: 18138–18143. (d) Li, H.D., Hu, Y.C., Li, L.F. et al. (2021). Synthesis of methanesulfonic acid directly from methane: the cation mechanism or the radical mechanism? *J. Phys. Chem. Lett.* 12: 6486–6491.

6 Dumas, J. (1840). Ueber die Einwirkung des Chlors auf den aus essigsauren Salzen entstehenden Kohlenwasserstoff. *Ann. Chem. Pharm.* 33: 187–189.

7 (a) Liao, K., Negretti, S., Musaev, D.G. et al. (2016). Site-selective and stereoselective functionalization of unactivated C–H bonds. *Nature* 533: 230–234. (b) Goldberg, K.I. and Goldman, A.S. (2017). Large-scale selective functionalization of alkanes. *Acc. Chem. Res.* 50: 620–626. (c) Strong, P.J., Xie, S., and Clarke, W.P. (2015). Methane as a resource: can the methanotrophs add value? *Environ. Sci. Technol.* 49: 4001–4018. (d) Hashiguchi, B.G., Bischof, S.M., Konnick, M.M., and Periana, R.A. (2012). Designing catalysts for functionalization of unactivated C–H bonds based on the CH activation reaction. *Acc. Chem. Res.* 45: 885–898.

8 (a) Schwartz, N.A., Boaz, N.C., Kalman, S.E. et al. (2018). Mechanism of hydrocarbon functionalization by an iodate/chloride system: the role of ester protection. *ACS Catal.* 8: 3138–3149. (b) Fokin, A.A. and Schreiner, P.R. (2002). Selective alkane transformations via radicals and radical cations: insights into the activation step from experiment and theory. *Chem. Rev.* 102: 1551–1593.

9 Landa, S., Kriebel, S., and Knobloch, E. (1954). O adamantanu a jeho derivatech I. *Chem. List.* 48: 61–64.

10 (a) Adcock, W., Coope, J., Shiner, V.J., and Trout, N.A. (1990). Evidence for 2-fold hyperconjugation in the solvolysis of 5-(trimethylsilyl) and 5-(trimethylstannyl)-2-adamantyl sulfonates. *J. Organomet. Chem.* 55: 1411–1412. (b) Adcock, W. and Trout, N.A. (1991). Transmission of polar substituent effects in the adamantane ring system as monitored by ^{19}F NMR: hyperconjugation as a stereoinductive factor. *J. Organomet. Chem.* 56: 3229–3238.

11 (a) Saouma, C.T. and Mayer, J.M. (2014). Do spin state and spin density affect hydrogen atom transfer reactivity? *Chem. Sci.* 5: 21–31. (b) Schreiner, P.R. and Fokin, A.A. (2004). Selective alkane C–H-bond functionalizations utilizing oxidative single-electron transfer and organocatalysis. *Chem. Rec.* 3: 247–257.

12 Sandhiya, L., Jangra, H., and Zipse, H. (2020). Molecule-induced radical formation (MIRF) reactions – a reappraisal. *Angew. Chem. Int. Ed.* 59: 6318–6329.

13 (a) Tkachenko, B.A., Shubina, T.E., Gusev, D.V. et al. (2003). Mechanisms of CH activation in alkanes by chromium-oxo reagents. *Theor. Exp. Chem.* 39: 90–95. (b) Rappe, A.K. and Jaworska, M. (2003). Mechanism of chromyl chloride alkane oxidation. *J. Am. Chem. Soc.* 125: 13956–13957.

14 Mestres, J., Duran, M., and Bertran, J. (1994). Ab initio electronic analysis of the hydride transfer in the $[CH_3\text{-}H\text{-}CH_3]^+$ system. *Theor. Chim. Acta* 88: 325–338.

15 Schreiner, P.R., Kim, S.J., Schaefer, H.F. et al. (1993). CH_5^+: the never-ending story or the final word? *J. Chem. Phys.* 99: 3716–3720.

16 Talrose, V.L. and Lyubimova, A.K. (1952). Secondary processes in the ion source of the mass spectrometer. *Dokl. Akad. Nauk SSSR* 86: 909–912.

17 White, E.T., Tang, J., and Oka, T. (1999). CH_5^+: the infrared spectrum observed. *Science* 284: 135–137.

18 Schreiner, P.R. (2000). Does CH_5^+ have (a) "structure?" – a tough test for experiment and theory. *Angew. Chem. Int. Ed.* 39: 3239–3241.

19 (a) Schreiner, P.R., Schleyer, P.v.R., and Schaefer, H.F. (1993). Mechanisms of electrophilic substitutions of aliphatic hydrocarbons. $CH_4 + NO^+$. *J. Am. Chem. Soc.* 115: 9659–9666. (b) Olah, G.A., Hartz, N., Rasul, G., and Prakash, G.K.S. (1995). Electrophilic substitution of methane revisited. *J. Am. Chem. Soc.* 117: 1336–1343.

20 Hiraoka, K. and Kebarle, P. (1976). Stabilities and energetics of pentacoordinated carbonium ions. The isomeric $C_2H_7^+$ ions and some higher analogues: $C_3H_9^+$ and $C_4H_{11}^+$. *J. Am. Chem. Soc.* 98: 6119–6125.

21 Bach, R.D., Su, M.D., Aldabbagh, E. et al. (1993). A theoretical model for the orientation of carbene insertion into saturated hydrocarbons and the origin of the activation barrier. *J. Am. Chem. Soc.* 115: 10237–10246.

22 Bach, R.D. and Su, M.D. (1994). The transition state for the hydroxylation of saturated hydrocarbons with hydroperoxonium ion. *J. Am. Chem. Soc.* 116: 10103–10109.

23 Fokin, A.A., Shubina, T.E., Gunchenko, P.A. et al. (2002). H-coupled electron transfer in alkane C-H activations with halogen electrophiles. *J. Am. Chem. Soc.* 124: 10718–10727.

24 Gunchenko, P.A. and Fokin, A.A. (2012). Mechanisms of activation of C–H bonds in framework compounds: theory and experiment. *Theor. Exp. Chem.* 47: 343–360.

25 (a) de Petris, G., Rosi, M., Ursini, O., and Troiani, A. (2013). The oxidative mechanism in electrophilic C-H activation: the case of CH_2F_2 and CH_2Cl_2. *Chem. Asian J.* 8: 588–595. (b) Wang, Z.C., Wu, X.N., Zhao, Y.X. et al. (2010). Room-temperature methane activation by a bimetallic oxide cluster $AlVO_4^+$. *Chem. Phys. Lett.* 489: 25–29. (c) de Petris, G., Cartoni, A., Troiani, A. et al. (2010). Double C-H activation of ethane by metal-free $SO_2^{·+}$ radical cations. *Chem. Eur. J.* 16: 6234–6242. (d) de Petris, G., Troiani, A., Rosi, M. et al. (2009). Methane activation by metal-free radical cations: experimental insight into the reaction intermediate. *Chem. Eur. J.* 15: 4248–4252. (e) Shchapin, I.Y., Vasil'eva, V.V., Nekhaev, A.I., and Bagrii, E.I. (2006). On a possible radical-cation mechanism of the biomimetic oxidation of the saturated hydrocarbon 1,3-dimethyladamantane in a Gif-type system containing a Fe_2^+ salt, picolinic acid, and pyridine. *Kinet. Catal.* 47: 624–637. (f) Kirillova, M.V., Kuznetsov, M.L., Reis, P.M. et al. (2007). Direct and remarkably efficient conversion of methane into acetic acid catalyzed by amavadine and related vanadium complexes. A synthetic and a theoretical DFT mechanistic study. *J. Am. Chem. Soc.* 129: 10531–10545.

26 Fokin, A.A. and Schreiner, P.R. (2012). Selective alkane CH bond substitutions: strategies for the preparation of functionalized diamondoids (nanodiamonds).

In: *Strategies and Tactics in Organic Synthesis*, vol. 8 (ed. M. Harmata), 317–350. Academic Press, Cambridge, Massachusetts.
27 Fokin, A.A. and Schreiner, P.R. (2003). Metal-free, selective alkane functionalizations. *Adv. Synth. Catal.* 345: 1035–1052.
28 Kruppa, G.H. and Beauchamp, J.L. (1986). Energetics and structure of the 1-and 2-adamantyl radicals and their corresponding carbonium ions by photoelectron spectroscopy. *J. Am. Chem. Soc.* 108: 2162–2169.
29 Shernyukov, A.V., Salnikov, G.E., Krasnov, V.I., and Genaev, A.M. (2022). Cluster halogenation of adamantane and its derivatives with bromine and iodine monochloride. *Org. Biomol. Chem.* 20: 8515–8527.
30 (a) Moiseev, I.K., Bagrii, E.I., Klimochkin, Y.N. et al. (1985). Adamantanol nitrates in nucleophilic substitution reactions. *Bull. Acad. Sci. USSR, Div. Chem. Sci.* 34: 1983–1985. (b) Moiseev, I.K., Bagrii, E.I., Klimochkin, Y.N. et al. (1985). Synthesis of alkyladamantanol nitrates. *Bull. Acad. Sci. USSR, Div. Chem. Sci.* 34: 1980–1982.
31 (a) Klimochkin, Y.N. and Moiseev, I.K. (1988). Kinetics of the reaction of adamantane and its derivatives with nitric acid. *Zh. Org. Khim.* 24: 557–560. (b) Klimochkin, Y.N. and Ivleva, E.A. (2021). Reaction of 1,3,5,7-tetramethyladamantane with nitric acid. *Russ. J. Org. Chem.* 57: 845–848.
32 Fokin, A.A., Tkachenko, B.A., Fokina, N.A. et al. (2009). Reactivities of the prism-shaped diamondoids [1(2)3]tetramantane and [12312]hexamantane (cyclohexamantane). *Chem. Eur. J.* 15: 3851–3862.
33 Fokin, A.A., Schreiner, P.R., Fokina, N.A. et al. (2006). Reactivity of [1(2,3)4]pentamantane (T_d-pentamantane): a nanoscale model of diamond. *J. Org. Chem.* 71: 8532–8540.
34 Yang, H.B., Feceu, A., and Martin, D.B.C. (2019). Catalyst-controlled C-H functionalization of adamantanes using selective H-atom transfer. *ACS Catal.* 9: 5708–5715.
35 Klimochkin, Y.N., Abramov, O.V., Moiseev, I.K. et al. (2000). Reactivity of cage hydrocarbons in the nitroxylation reaction. *Pet. Chem.* 40: 415–418.
36 (a) Geluk, H.W. and Schlatmann, J.L. (1971). Hydride transfer reactions of the adamantyl cation (IV): synthesis of 1,4- and 2,6-substituted adamantanes by oxidation with sulfuric acid. *Rec. Trav. Chim. Pays Bas* 90: 516–520. (b) Geluk, H.W. and Schlatmann, J.L. (1967). A convenient synthesis of adamantanone. *Chem. Commun.* 426.
37 Fokina, N.A., Tkachenko, B.A., Merz, A. et al. (2007). Hydroxy derivatives of diamantane, triamantane, and 121 tetramantane: selective preparation of bis-apical derivatives. *Eur. J. Org. Chem.* 4738–4745.
38 (a) Olah, G.A., Lukas, J., and Stable carbonium ions. XLVII. (1967). Alkylcarbonium ion formation from alkanes via hydride (alkide) ion abstraction in fluorosulfonic acid-antimony pentafluoride-sulfuryl chlorofluoride solution. *J. Am. Chem. Soc.* 89: 4739–4744. (b) Olah, G.A. and Schlosberg, R.H. (1968). Chemistry in super acids. I. Hydrogen exchange and polycondensation of methane and alkanes in FSO_3H-SbF_5 ("magic acid") solution. Protonation of alkanes and the intermediacy of CH_5^+ and related hydrocarbon ions. The high

chemical reactivity of "paraffins" in ionic solution reactions. *J. Am. Chem. Soc.* 90: 2726–2627.

39 (a) Hogeveen, H. and Bickel, A.F. (1967). Chemistry and spectroscopy in strongly acidic solutions: electrophilic substitution at alkane-carbon by protons. *Chem. Commun.* 635–636. (b) Hogeveen, H. and Bickel, A.F. (1967). Formation of trimethylcarbonium ions from isobutane and protons. Basicity of isobutene. *Rec. Trav. Chim. Pays Bas* 86: 1313–1315.

40 Olah, G.A., Prakash, G.K.S., Shih, J.G. et al. (1985). Stable carbocations. 258. Bridgehead adamantyl, diamantyl, and related cations and dications. *J. Am. Chem. Soc.* 107: 2764–2772.

41 (a) Schreiner, P.R., Lauenstein, O., Butova, E.D. et al. (2001). Selective radical reactions in multiphase systems: phase-transfer halogenations of alkanes. *Chem. Eur. J.* 7: 4996–5003. (b) Lauenstein, O., Fokin, A.A., and Schreiner, P.R. (2000). Kinetic isotope effects for the C-H activation step in phase-transfer halogenations of alkanes. *Org. Lett.* 2: 2201–2204.

42 (a) Fort, R.C. and Schleyer, P.v.R. (1964). Adamantane: consequences of the diamondoid structure. *Chem. Rev.* 64: 277–300. (b) Moiseev, I.K., Makarova, N.V., and Zemtsova, M.N. (1999). Reactions of adamantane in electrophilic media. *Russ. Chem. Bull.* 68: 1102–1121.

43 (a) Bagrii, Y.I. and Karaulova, Y.N. (1993). Activation of C-H bonds and functionalization of hydrocarbons of the adamantane series – review. *Pet. Chem.* 33: 183–201. (b) Bagrii, E.I., Safir, R.E., and Arinicheva, Y.A. (2010). Methods of the functionalization of hydrocarbons with a diamond-like structure. *Pet. Chem.* 50: 1–16. (c) Bagrii, E.I., Nekhaev, A.I., and Maksimov, A.L. (2017). Oxidative functionalization of adamantanes (review). *Pet. Chem.* 57: 183–197. (d) Hrdina, R. (2019). Directed C-H functionalization of the adamantane framework. *Synthesis* 51: 629–642. (e) Weigel, W.K., Dang, H.T., Feceu, A., and Martin, D.B.C. (2021). Direct radical functionalization methods to access substituted adamantanes and diamondoids. *Org. Biomol. Chem.* 20: 10–36. (f) Grover, N. and Senge, M.O. (2020). Synthetic advances in the C-H activation of rigid scaffold molecules. *Synthesis-Stuttgart* 52: 3295–3325.

44 (a) de Lozanne, A. (2008). A sludge-to-diamond story. *Nat. Mater.* 7: 10–12. (b) Marchand, A.P. (2003). Diamondoid hydrocarbons – delving into nature's bounty. *Science* 299: 52–53. (c) Marchand, A.P. (1995). Polycyclic cage compounds: reagents, substrates, and materials for the 21st century. *Aldrichim. Acta* 28: 95–104. (d) Schwertfeger, H. and Schreiner, P.R. (2010). Future of diamondoids. *Chem. Unserer Zeit* 44: 248–253. (e) Hohman, J.N., Claridge, S.A., Kim, M., and Weiss, P.S. (2010). Cage molecules for self-assembly. *Mater. Sci. Eng. Rep.* 70: 188–208. (f) Gunawan, M.A., Hierso, J.-C., Poinsot, D. et al. (2014). Diamondoids: functionalization and subsequent applications of perfectly defined molecular cage hydrocarbons. *New J. Chem.* 38: 28–41. (g) Agnew-Francis, K.A. and Williams, C.M. (2016). Catalysts containing the adamantane scaffold. *Adv. Synth. Catal.* 358: 675–700. (h) Stauss, S. and Terashima, K. (2016). *Diamondoids: Synthesis, Properties, and Applications*, 242. Jenny Stanford Publishing. (i) Seidel, S.R. and Stang, P.J. (2002). High-symmetry coordination cages via self-assembly.

Acc. Chem. Res. 35: 972–983. (j) Cook, T.R., Zheng, Y.R., and Stang, P.J. (2013). Metal-organic frameworks and self-assembled supramolecular coordination complexes: comparing and contrasting the design, synthesis, and functionality of metal-organic materials. *Chem. Rev.* 113: 734–777.

45 (a) Shokova, E.A. and Kovalev, V.V. (2016). Biological activity of adamantane-containing mono- and polycyclic pyrimidine derivatives (a review). *Pharm. Chem. J.* 50: 63–75. (b) Wanka, L., Iqbal, K., and Schreiner, P.R. (2013). The lipophilic bullet hits the targets: medicinal chemistry of adamantane derivatives. *Chem. Rev.* 113: 3516–3604. (c) Stockdale, T.P. and Williams, C.M. (2015). Pharmaceuticals that contain polycyclic hydrocarbon scaffolds. *Chem. Soc. Rev.* 44: 7737–7763. (d) Spilovska, K., Zemek, F., Korabecny, J. et al. (2016). Adamantane - a lead structure for drugs in clinical practice. *Curr. Med. Chem.* 23: 3245–3266. (e) Klimochkin, Y.N., Shiryaev, V.A., and Leonova, M.V. (2015). Antiviral properties of cage compounds. New prospects. *Russ. Chem. Bull.* 64: 1473–1496. (f) Lamoureux, G. and Artavia, G. (2010). Use of the adamantane structure in medicinal chemistry. *Curr. Med. Chem.* 17: 2967–2978.

46 (a) Baranov, N.I., Bagrii, E.I., Safir, R.E. et al. (2022). Advances in the chemistry of unsaturated adamantane derivatives (a review). *Pet. Chem.* 62: 352–375. (b) Bagrii, E.I. and Saginaev, A.T. (1983). Unsaturated adamantane derivatives. *Usp. Khim.* 52: 1538–1567. (c) Shvekheimer, M.G.A. (1996). Adamantane derivatives containing heterocyclic substituents in bridgehead positions. Synthesis and properties. *Usp. Khim.* 65: 603–647. (d) Khardin, A.P. and Radchenko, S.S. (1982). Adamantane polymer derivatives. *Usp. Khim.* 51: 480–506. (e) Vakili-Nezhaad, G.R. (2007). Thermodynamic properties of diamondoids. In: *Molecular Building Blocks for Nanotechnology: From Diamondoids to Nanoscale Materials and Applications*, vol. 109 (ed. G.A. Mansoori, T.F. George, L. Assoufid, and G. Zhang), 7–28. (f) Zhou, Y.J., Brittain, A.D., Kong, D.Y. et al. (2015). Derivatization of diamondoids for functional applications. *J. Mater. Chem. C* 3: 6947–6961.

4

Preparative Diamondoid Functionalizations

The introduction of a variety of functional groups into diamondoid scaffolds is crucial for applications of the resulting derivatives. Functionalization not only introduces polarity in the structure but also creates surface attachment points with further applications in synthetic, material, and medicinal chemistry. For precise tailoring of material properties, the functional groups need to be introduced in the hydrocarbon cage on well-defined positions, which is quite a challenging task as the number of inequivalent positions increases with a decrease in the cage symmetry (Figure 4.1) [1].

Unlike **AD**, which has only one type of 3° C—H-bond, the higher homologues possess several nonequivalent sites. For example, **DIA** and **PENT** have two different 3° C—H-bonds, **TRIA** four, and C_2-symmetric **123TET** six. Especially attractive among diamondoid building blocks are apical derivatives, as they possess higher surface affinities due to steric considerations [2]. Owing to the ability to transfer the electronic effects through the cage, selective polysubstitutions of diamondoids are possible for adamantane and diamantane; however, this is more difficult for higher diamondoids. This is associated with the exponential growth of the number of possible isomers, as already six positional isomers exist for *tert*-disubstituted **DIA** (Figure 4.1, bottom) and 16 for *tert*-disubstituted **TRIA**. Since these bridgehead bonds are different in reactivity, they can be both a blessing and a curse, depending on the ease of their activation under particular reaction conditions. If sufficiently diverse in reactivity, tertiary cage C—H-bonds can give rise to selective and scalable functionalizations. Some of these are outlined in the following.

4.1 Halogenations

The chemistry of diamondoids got a fulminant start when Landa discovered that **AD** readily and nearly quantitatively reacts with neat bromine, giving only **1BRAD** [3, 4]. Before, it was assumed that alkanes in general were unreactive toward electrophiles under standard conditions and were only considered through the lens of radical chemistry. Soon it became apparent that electrophilic halogenations were characteristic of many other diamondoids that also readily react with neat bromine, exclusively giving tertiary bromides. In particular, the bromination of **DIA**

The Chemistry of Diamondoids: Building Blocks for Ligands, Catalysts, Materials, and Pharmaceuticals,
First Edition. Andrey A. Fokin, Marina Šekutor, and Peter R. Schreiner.
© 2024 WILEY-VCH GmbH. Published 2024 by WILEY-VCH GmbH.

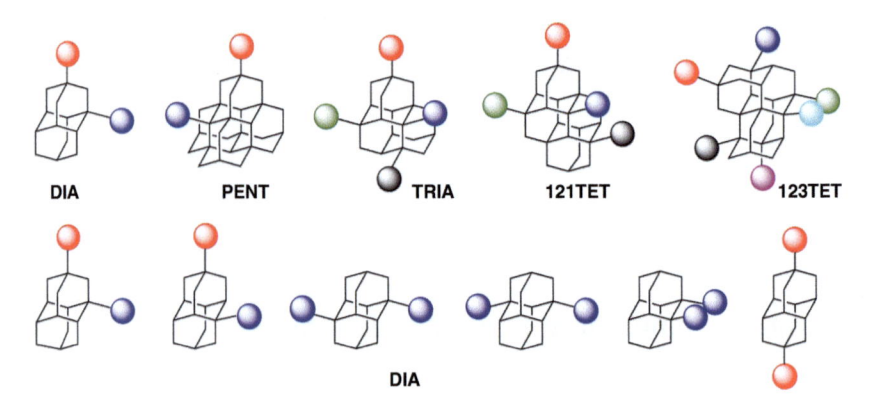

4 Preparative Diamondoid Functionalizations

Figure 4.1 Top: Selected diamondoids with highlighted apical (red spheres) and medial substituted bridgehead tertiary carbons. Bottom: All possible *tert*-disubstituted diamantanes with apical and medial positions of the cage. Different colors are attributed to non-equivalent C–H positions.

gives **1BRDIA** [5], and a mixture of *tert*-bromides forms [6] in the case of **TRIA** (Scheme 4.1). However, the reaction is very different from aromatic electrophilic substitutions as the nucleophile is incorporated into the cage structure upon halogenation [7] with I^+Cl^- (exclusive formation of **1CLAD**) and with anhydrous nitric acid [8], where nitronium nitrate $NO_2{}^+NO_3{}^-$ is the active reagent giving rise to nitroxy-derivatives. In this regard, **AD** is very different from the reactions of benzene with electrophiles and might be considered as "antibenzene."

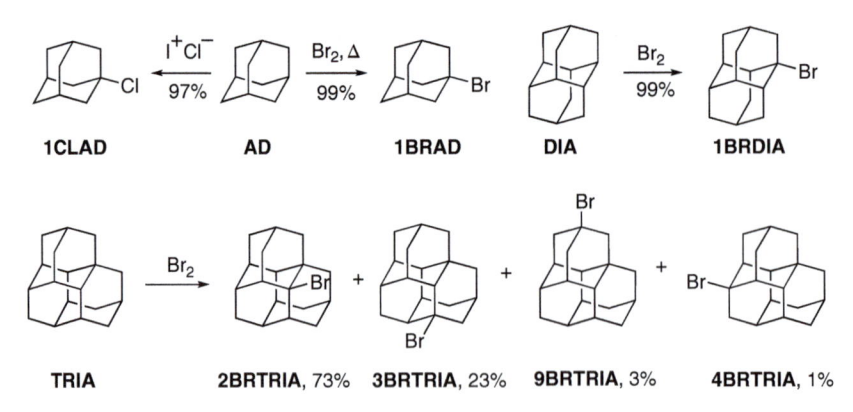

Scheme 4.1 Monohalogenations of **AD**, **DIA**, and **TRIA** lead to tertiary bromides exclusively.

Monobrominations of higher diamondoids occur at room temperature or lower, and sometimes dilution with inert solvents is desirable. The statistical probability of the substitution needs to be taken into consideration as well, for example, in **DIA** the medial positions are at a clear advantage over the apical ones, and the medial bromide **1BRAD** forms almost quantitatively. For **TRIA**, the selectivity drops since many different medial positions are similarly reactive, affording various monobromides, but the medial bromides **2BRTRIA** and **3BRTRIA** still dominate (Scheme 4.1) [6b]. For higher diamondoids, bridgehead substitution is determined by favorable polarization of the cage through charge transfer from the hydrocarbon to the electrophile, and this correlates well with the stability of the corresponding carbocations. As a rule of thumb, medial carbocations are usually more stable than apical ones, the reason being that cationic carbons located closer to the geometrical center of the molecule facilitate more effective delocalization of the positive charge through the entire cage. This may be illustrated with the electrostatic potentials of the medial (**DIA1+** and **121TET2+**) and apical (**DIA4+** and **121TET6+**) carbocations derived from **DIA** and **121TET** (Figure 4.2).

Higher diamondoids like **1(2)3TET**, **121TET**, **123TET**, **1(2,3)4PENT**, and **12312HEX** [6b, 10–12], often selectively react with bromine because high cage symmetry translates into a lower number of equivalent bridgehead positions, affording the corresponding medial bromide for tetrahedral **PENT** almost selectively (Scheme 4.2). Note that even if the reaction is not fully selective, purification of the bromides can be accomplished with simple recrystallizations owing to the high symmetry of the products. Even though a complex mixture of bromides forms upon bromination of **123TET**, the medial bromide **7BR123TET** was isolated in satisfactory yield through double recrystallization from *n*-hexane [12]. In general, the positional selectivities for the bromination can be confidently predicted based on the stabilities of the corresponding carbocations [9], as bromine preferentially substitutes the C—H-bond for which the more stable carbocation forms, as stated above.

As mentioned previously, apical monobromides usually form only in trace amounts in neat bromine, which is unfortunate since they are very attractive building blocks for nanotechnology applications [13]. Apical 4-bromodiamantane (**4BRDIA**) forms when adding trace amounts of AlBr$_3$ to the bromination reaction mixture, but the selectivity and reproducibility of this procedure is not satisfactory for routine applications [14]. The apical derivatives can be obtained in somewhat higher yields by using radical brominations under phase-transfer catalytic (PTC)

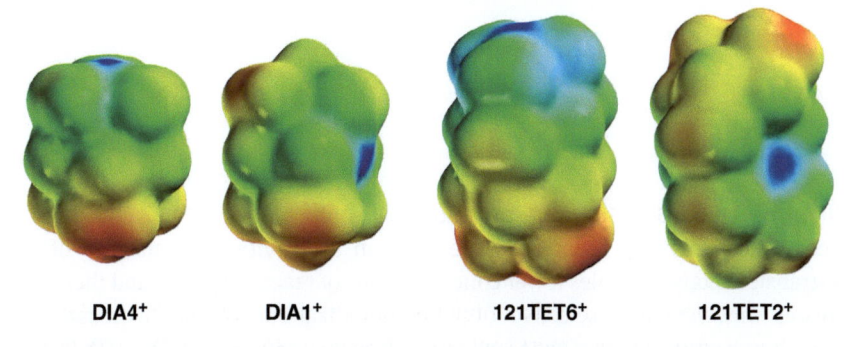

Figure 4.2 Left to right, the distribution of the electrostatic potentials in 4- and 1-diamantyl, and 6- and 2- [121]tetramantyl cations. The positive charge is more effectively delocalized if located close to the center of the cage. Source: Reproduced from Ref. [9] Wiley-VCH, 2005.

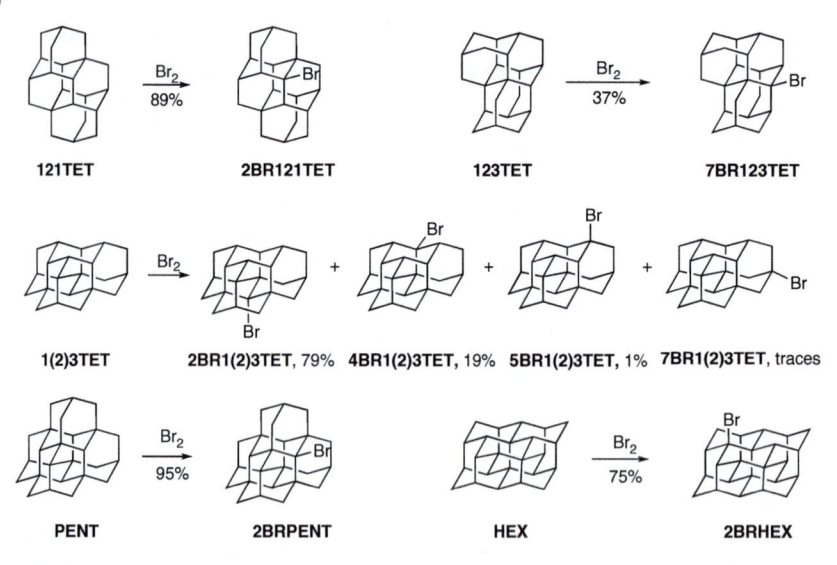

4 Preparative Diamondoid Functionalizations

Scheme 4.2 Monobrominations of higher diamondoids **121TET**, **123TET**, **1(2)3TET**, **PENT**, and **HEX** give mostly medially substituted bromides.

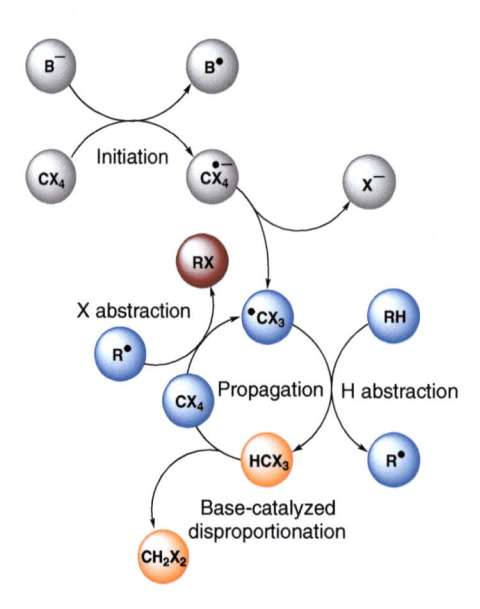

Figure 4.3 Radical C−H-halogenations (X=Br, I) under phase-transfer catalytic conditions. Source: Adopted from Ref. [1].

conditions (Figure 4.3) [15a]. In contrast to unselective free radical halogenations, PTC conditions allow for control of the reaction selectivity since the use of a phase-transfer catalyst enables lower concentrations of reactive species and thereby suppresses unwelcome unselective over-functionalizations of the hydrocarbon skeleton. It was suggested that the reaction involves the reduction of CX_4 with OH^- with the formation of $CX_4^{·-}$ (gray part of Figure 4.3) that generates the CX_3-radical. The latter carries the propagation cycle (blue color), thus avoiding the formation of halogen radicals that are known to react with diamondoids unselectively.

PTC-bromination of **AD** in the CBr$_4$/NaOH system afforded **1BRAD** as the dominant C—H-substitution product (Scheme 4.3) [16]. Under homogenous conditions in CH$_2$Cl$_2$ solution, the bromination of **AD** with CBr$_4$ in the presence of *t*BuOK is also possible and gives a mixture of tertiary and secondary bromides in a 3.3 : 1 ratio [17]. Using CBr$_4$ in the NaOH/CH$_2$Cl$_2$/*n*-Bu$_4$NBr system with **DIA** gives a mixture containing all possible monobromination products (Scheme 4.3) [9]. The PTC protocol is also applicable for direct CH-iodination (vide infra) of **AD** [18] and **DIA** [9] with iodoform HCI$_3$, where CI$_3$-radicals are responsible for the CH-activation step. Note once again that, unlike carbocations, secondary and tertiary diamondoid radicals are of similar stability, and the reaction selectivities with radicals depend solely on steric and polar effects in the corresponding transition structures [19b]. Despite marked improvements in this radical approach when compared to conventional homogeneous free radical protocols, the direct monohalogenation of diamondoids in apical positions remains only partially solved, with better alternatives presenting themselves by introducing different functional groups, namely alcohols (vide infra).

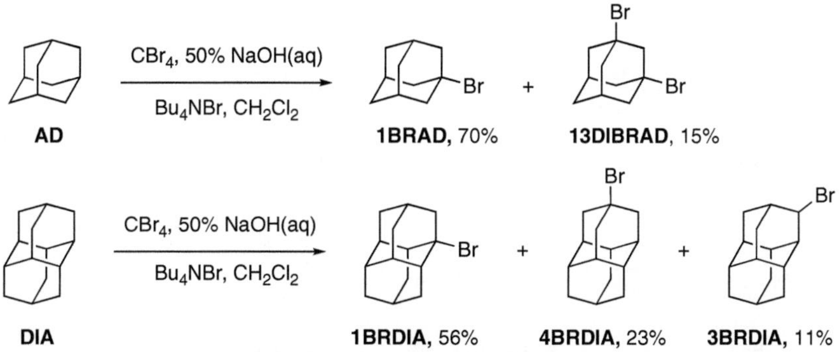

Scheme 4.3 Radical bromination using phase-transfer catalytic (PTC) conditions.

Next to consider is the selective introduction of more than one bromine into diamondoid scaffolds. As already stated, monobromination of **AD** requires mild reaction conditions, but the introduction of the second bromine cannot proceed without a Lewis acid catalyst. As more bromines are introduced [20], the conditions need to be rather harsh (Scheme 4.4) [21b, 22], and higher catalyst loadings facilitate the polybrominations. Through using stronger Lewis acids, even 1,3,5-tribromoadamantane (**135TRIBRAD**) [20] and, under harsh reaction conditions with Br$_2$/AlBr$_3$ in a sealed tube, 1,3,5,7-tetrabromoadamantane (**1357TETRABRAD**) can be obtained in good yields.

Starting from **DIA**, the issues with reaction selectivity become prominent and, what is more, the resulting mixtures of polybromo-derivatives are quite tedious to purify, either by crystallization or chromatographically [14]. The product distributions of diamondoid dibromo-derivatives are, in addition to medial/apical selectivity, also governed by the position and inductive effect of the first substituent and the presence of Lewis acids that change the regioselectivity in favor of the apical substitution. Attempts to control product distributions by applying strictly defined

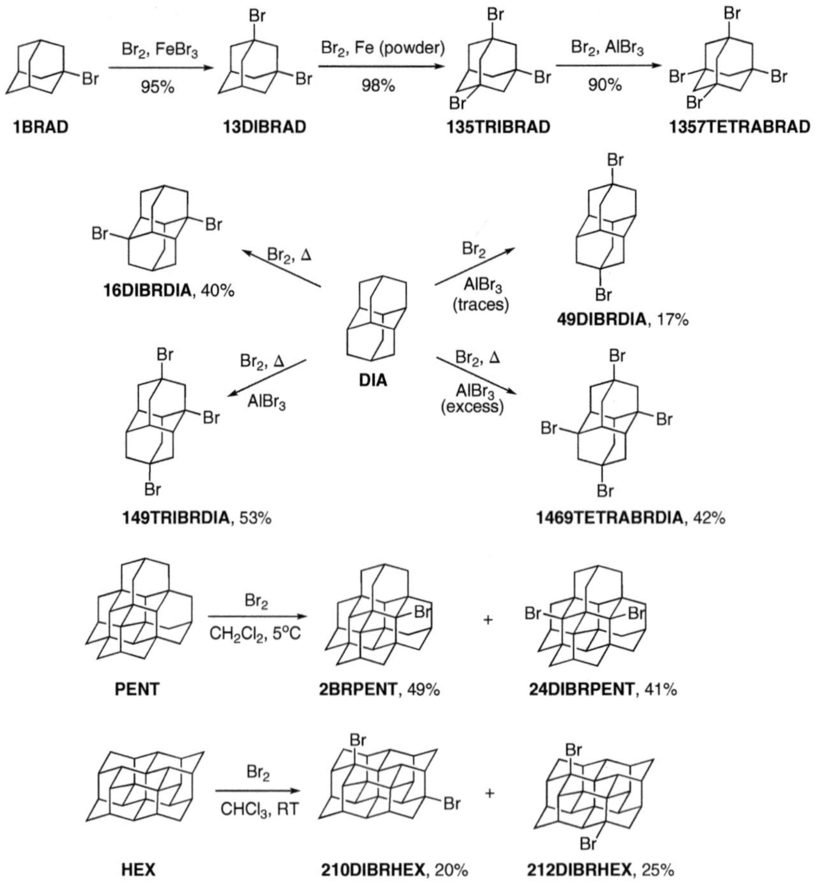

Scheme 4.4 Polybrominations of **AD**, **DIA**, **PENT**, and **HEX**.

reaction conditions, e.g., the use of Fe with Freon-113 as the solvent [23], are not necessarily always successful, and Lewis acid-catalyzed reactions often remain uncontrollable. One exception to these difficulties is the preparation of **16DIBRDIA**, which can be readily obtained from **DIA** in refluxing neat bromine without the use of Lewis acid catalysts, and the product can be easily purified by crystallization because of its high symmetry [24]. In general, since the most stable diamondoid cations have positive charges close to the geometrical center of the molecule, bromination can readily afford mono- and bis-medial bromides that do not require the use of Lewis acid catalysts, thereby limiting this approach. Some dibrominations of larger cages occur with ease, such as in the case of **PENT** and **HEX,** which in the reaction with Br_2 in the CH_2Cl_2 or $CHCl_3$ solutions, give respective dibromides **24DIBRPENT** [10] and **212DIBRHEX** in satisfactory yields [11].

Although diamondoid brominations have been the most widely explored halogenation, the introduction of other halogens in the cage structures has also been accomplished over the years. In this context, we must mention the curious case of polyhalogenated **FCLBRIAD**, the first pseudotetrahedral chiral compound incorporating all stable halogens. It was prepared [25] through sequential radical halogenations without halogen exchange by using previously mentioned PTC conditions (Scheme 4.5). The low overall yield comes as no surprise since every incorporated halogen increasingly deactivates the cage and hampers C—H-substitution. Thus, the reaction conditions have to be increasingly harsher with every step [25] so that highly reactive perfluoroalkyl radicals [26] were employed for the iodination of trihalo-derivative **FCLBRAD** in the terminal step.

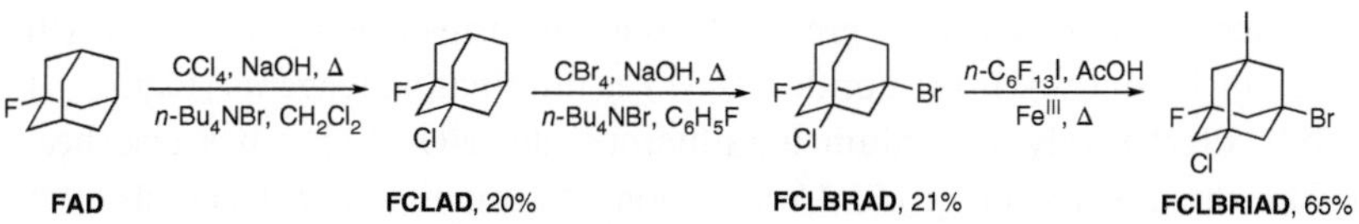

Scheme 4.5 Preparation of pseudotetrahedral tetrahalogenated adamantane **FCLBRIAD** under PTC conditions followed by iodination with n-perfluorohexyliodide.

As for halogen derivatives of other diamondoids, mono- and dichlorides of **DIA** were prepared [9, 27], including the apical **49DICLDIA** that was obtained with chlorosulfonic acid (Scheme 4.6) [28]. The chlorination of **DIA** with chloromethanes in the presence of zeolites [29] or rhodium complexes [30] gives a mixture of medial (**1CLDIA**) and apical (**4CLDIA**) chlorodiamantanes. The brominations with bromomethanes in the presence of zeolites also afforded a mixture of medial and apical **DIA** derivatives [31]. Chromyl chloride oxidation is not selective enough, and a mixture of **4CLDIA** and **1CLDIA** along with **1OHDIA** ensues [9]. Iodo-derivatives of **AD** and **DIA** can again be obtained by a radical reaction under PTC conditions

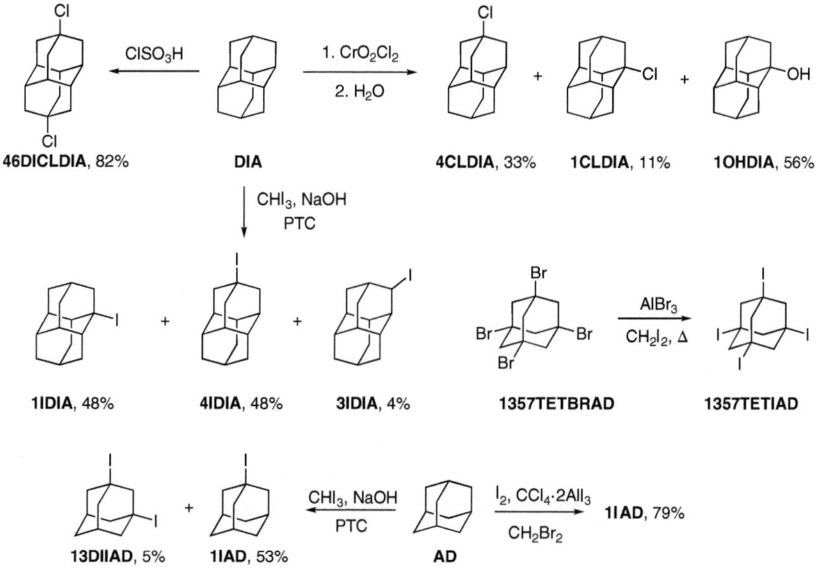

Scheme 4.6 Preparation of selected diamondoid chloro- and iodo-derivatives.

4 Preparative Diamondoid Functionalizations

with HCl$_3$ and NaOH in dichloromethane [9, 18] or by halogen exchange reactions [21b, 32]. Due to the minimal deactivation of the cage after the incorporation of the first iodine substituent, substantial amounts of 1,3-diiodoadamantane **13DIIAD** form under the PTC-iodination conditions [18]. The increase in selectivity was achieved in the presence of β-cyclodextrin, where bis-iodination is suppressed due to the encapsulation of 1-iodoadamantane (**1IAD**) [33]. Alternatively, **1IAD** was obtained selectively in 79% yield through the electrophilic iodination of **AD** with I$_2$ in the presence of the 2AlI$_3$•••CCl$_4$ complex (Scheme 4.6) [34].

Fluoro-derivatives can also be prepared from the corresponding bromides with AgF or diethylaminosulfur trifluoride (DAST) as fluorinating reagents (Figure 4.4) [21b, 35]. Alternatively, nitronium tetrafluoroborate (NO$_2$$^+BF_4$$^-$) has also been applied for the preparation of **1FDIA** as an ionic fluorinating agent that abstracts halides [36]. The bromide precursors do not necessarily need to be without any other functional group bound to the diamondoid scaffold, as the mild conditions for fluorine exchange can tolerate other substituents. In particular, acetate and chloroacetamide functional groups remain unchanged in the presence of

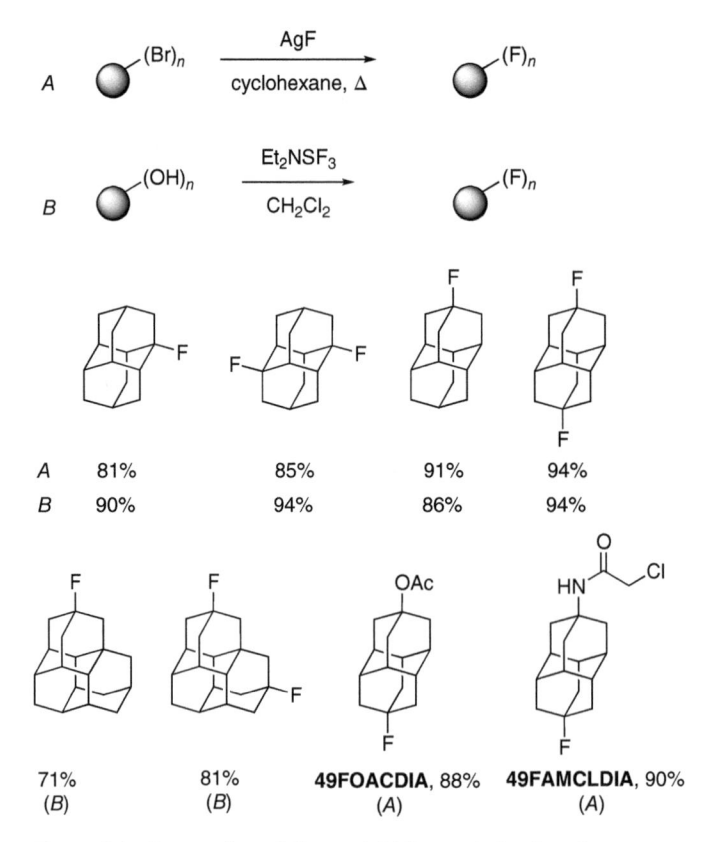

Figure 4.4 Preparation of diamondoid fluoro-derivatives from corresponding bromides with AgF (method A) or from alcohols with diethylaminosulfur trifluoride (DAST) (method B).

AgF, giving rise to fluorohydroxy- (**49FOACDIA**) and fluoroaminodiamantane- (**49FAMCLDIA**) 4,9-derivatives from the corresponding bromides [35].

In general, the combination of C—H-bromination and halogen exchange is a powerful tool for the preparation of various halodiamondoids. However, bromination is not always viable as the product is inevitably contaminated with bromine. From this point, the alternative, "bromine-free" methods should be used for primary diamondoid functionalizations [37]. A viable option is C—H-bond nitroxylation [38a, 39, 40] followed by acid-catalyzed exchange [9, 41] that gives rise to halogen-free diamondoid derivatives.

4.2 Diamondoid Alcohols and Ketones

Oxidation of the cage hydrocarbon skeleton that ultimately affords alcohols and ketones is arguably one of the most useful transformations in diamondoid chemistry. Hydroxy-diamondoids are the key precursors for the preparation of various derivatives such as amines, acetamides, carboxylic acids, and many others, even at large scale [42]. Unlike halogenation reactions that often prove to be irreproducible and result in product mixtures that are difficult to separate, cage alcohols and ketones are, in principle, easily obtainable, and their isolation is straightforward. While radical oxidations of **AD** usually produce a mixture of 1- and 2-derivatives, transformations in electrophilic media in most cases [21b, 43b, c, 44d, 45] allow selective preparation of C—H-bond oxidation products. Upon heating, CH_2-oxidations occur, e.g., with concentrated sulfuric acid **AD**, which is converted mostly to adamantanone (**O=AD**) together with tar products (see below). With oleum, the active electrophilic species is believed to be protonated sulfur trioxide (HSO_3^+), and due to the high oxidation power, deeper oxidations take place. For example, the reaction of **AD** with fuming sulfuric acid with 20% SO_3 gives a mixture of hydroxyl-derivatives whose additional oxidation with chromium oxide affords a mixture of **13OHAD** and ketones **O=AD**, **26DIO=AD**, and **1OH4O=AD** that can be readily separated using a combination of chromatography and crystallization (Scheme 4.7) [46a].

The oxidation of **AD** with anhydrous nitric acid [40] occurs via hydrogen abstraction with electrophilic nitronium species, i.e., nitronium nitrate $NO_2^+NO_3^-$ and gives substantial amounts of apical tertiary derivatives in the case of higher diamondoids [6b, 38a, 39, 47]. Note that the reactivity of diamondoids toward electrophiles increases with the size of the substrate: **DIA** reacts 14.8 times faster than **AD** in the CH_2Cl_2/HNO_3 system [39]. The ability of other nitronium reagents, such as $NO_2^+BF_4^-$ and $NO_2^+OAc_f^-$ to activate C—H-bonds of saturated systems was shown for many examples, including cycloalkanes, that are generally unreactive toward electrophiles [48]. For these charged electrophiles, the H-abstraction leads to the corresponding carbocations [49] that are trapped by the nucleophilic species present in the reaction with 100% HNO_3 (nitrates form in CH_2Cl_2 [39, 50] and acetamides in CH_3CN after hydrolysis), i.e., formally nucleophilic substitution takes place [8a]. Diamondoid nitrates can be quantitatively hydrolyzed to alcohols by simple water

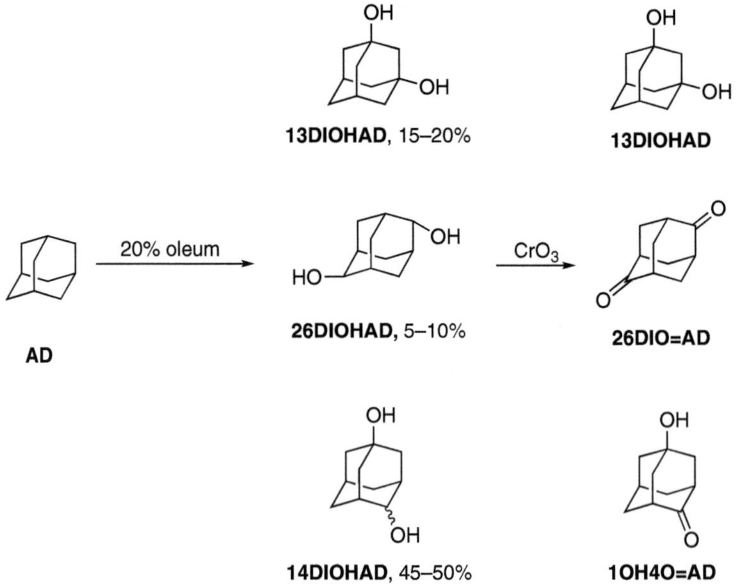

Scheme 4.7 Oxidation of **AD** with 20% oleum followed by Jones oxidation simplifies the separation of resulting mixtures.

treatment or hydrogenation in the case of substrates, which are prone to rearrange [1, 50, 51]. In non-nucleophilic solvents (CH$_2$Cl$_2$, CCl$_4$, CH$_3$NO$_2$), substantial amounts of nitro derivatives may form upon reaction with nitronium reagents NO$_2^+$X$^-$. Such ambiguous results prompted a number of experimental mechanistic studies on the C—H-bond substitution reactions of diamondoids with nitrogen-containing electrophiles [39, 49, 52]. The oxidation of diamondoids with fuming nitric acid occurs only at the bridgehead positions, and the product ratio is kinetically controlled. For example, oxidation of **DIA** affords **4OHDIA** as the major product, but longer reaction times lead to **14DIOHDIA** (Scheme 4.8) [47]. If left to equilibrate in sulfuric acid, apical **49DIOHDIA** forms almost exclusively. Since **49DIOHDIA** is a highly

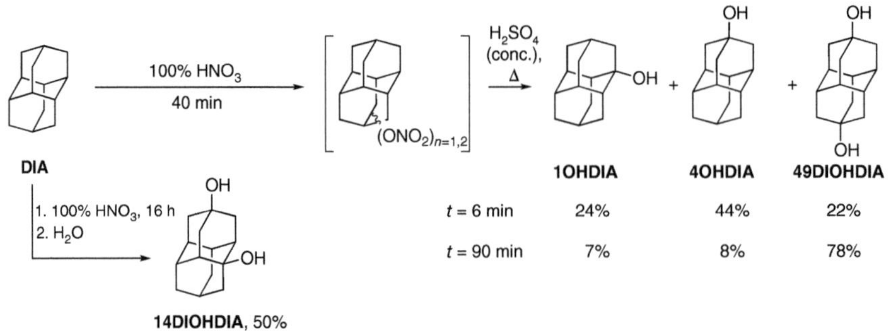

Scheme 4.8 Oxidations of **DIA** with fuming nitric acid and further isomerizations with sulfuric acid to give bis-apical dialcohol **49DIOHDIA**. Source: From Ref. [47].

symmetric molecule and has low solubility, it can be isolated from the reaction mixture by simple filtration of the precipitate formed after pouring the reaction mixture onto ice. Note that both nitroxy- and hydroxy-diamondoids can be isomerized with sulfuric acid [53], but the nitroxy-derivatives are much more soluble. Good miscibility and homogeneous reaction conditions are very important for the functionalization of diamondoids in electrophilic media since higher diamondoids are far more reactive than the lower ones and often require shorter reaction times, low temperatures, and/or higher dilution.

When oxidizing **TRIA**, one has to take into account that this diamondoid has four different bridgehead C—H-bond positions and consequently can form many more alcohols (Scheme 4.9). Although nitroxylation is not very selective, the main product in the mixture after hydrolysis is the medial derivative **3OHTRIA**. Upon

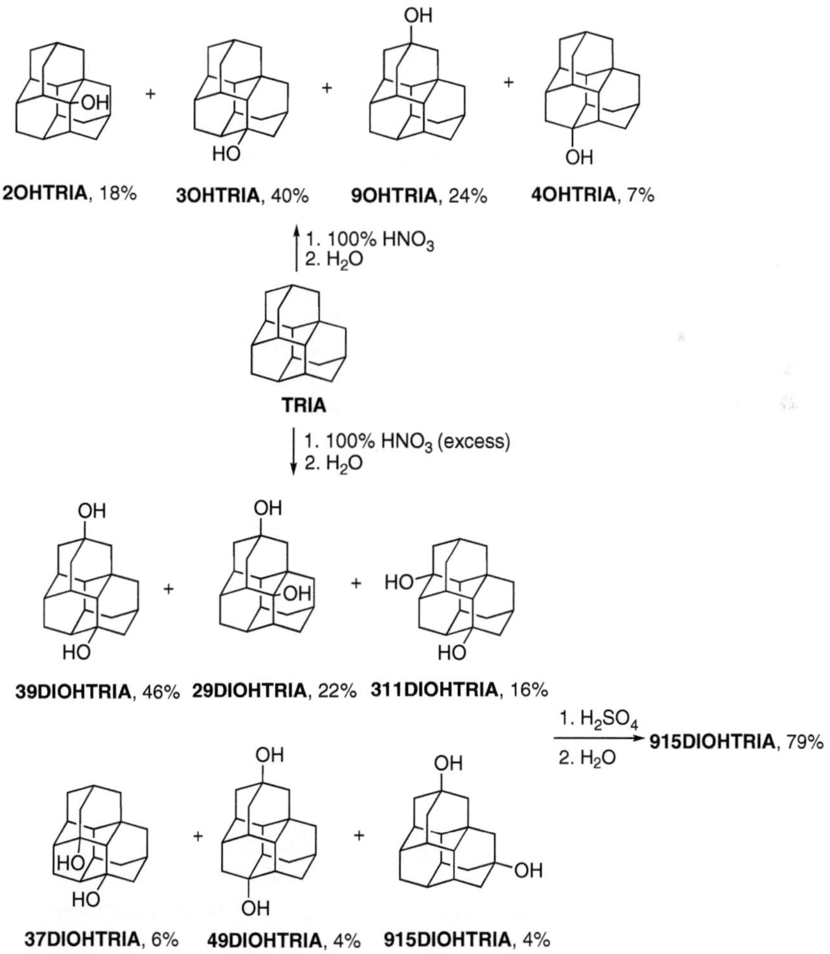

2OHTRIA, 18%   **3OHTRIA**, 40%   **9OHTRIA**, 24%   **4OHTRIA**, 7%

1. 100% HNO$_3$
2. H$_2$O

TRIA

1. 100% HNO$_3$ (excess)
2. H$_2$O

39DIOHTRIA, 46%   **29DIOHTRIA**, 22%   **311DIOHTRIA**, 16%

1. H$_2$SO$_4$
2. H$_2$O → **915DIOHTRIA**, 79%

37DIOHTRIA, 6%   **49DIOHTRIA**, 4%   **915DIOHTRIA**, 4%

Scheme 4.9 Oxidation of **TRIA** with fuming nitric acid and isomerization to bis-apical dihydoxy-derivative **915DIOHTRIA** with concentrated sulfuric acid. Source: From Ref. [47].

prolonging the reaction time and providing an excess of nitric acid, a mixture of diols forms after hydrolysis with the kinetically controlled product **39DIOHTRIA** being predominant [47]. Apical derivative **915DIOHTRIA** is accessible through thermodynamically controlled isomerization of diol mixtures after sufficient time in sulfuric acid. Additionally, since symmetric **915DIOHTRIA** has lower solubility in organic solvents, it can be isolated from the reaction mixture by crystallization. To summarize, in the case of **TRIA** the oxidation with nitric acid, various medially substituted alcohols form under kinetic control, but under thermodynamic conditions, bis-apical **915DIOHTRIA** is the major reaction product obtained with a 79% yield (Scheme 4.9).

As for **121TET** oxidation, mononitroxylation selectivity is expected to be even lower than for **TRIA**, but again, due to the higher thermodynamic stability, the apical bis-diol **613DIOH121TET** is the major product after equilibration in sulfuric acid (Scheme 4.10) [47]. Surprisingly, nitroxylation of *iso*-tetramantane **1(2)3TET** shows higher selectivity, with the apical alcohol dominating the product mixture. This finding can be explained by steric shielding of the medial positions since these C—H bonds are located on the "planar" surface of the molecule and reagents cannot easily approach them. This is in full agreement with the relative energies of the HCET for the reaction of **1(2)3TET** with model electrophile protonated nitronium nitrate $[HNO_2NO_3]^+$ (see Chapter 3, Figure 3.4). The incorporation of the first substituent

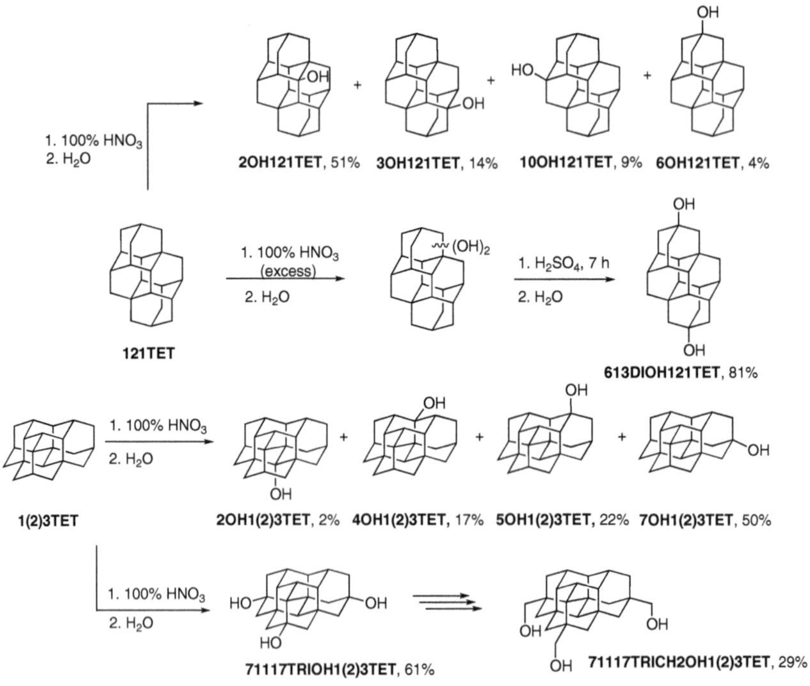

Scheme 4.10 Oxidations of tetramantanes with fuming nitric acid, isomerization to bis-apical dihydoxy-derivative **613DIOH121TET** with concentrated sulfuric acid, and preparation of tripodal triol **71117TRICH2OH1(2)3TET** based on the **1(2)3TET** core.

into the apical position of **1(2)3TET** deactivates the neighboring medial positions, such that the incorporation of three groups is possible with high selectivity. Thus obtained tris-hydroxy derivative **71117TRIOH1(2)3TET** was used to prepare the tripodal rigid surface anchor **71117TRICH2OH1(2)3TET** [11].

Larger, but highly symmetric, **PENT** reacts even more regioselectively in nitroxylation reactions and, after hydrolysis, forms almost exclusively the apical hydroxy derivative **7OHPENT** (Figure 4.5) [10]. The incorporation of the first group has only little influence on the reactivity of the distant C—H bonds, so that the bis-derivative **711DIOHPENT** may be obtained in high yields just by increasing the time for the nitroxylation. The reactivity of **PENT** is a clear example of how varying the nature of the electrophile can switch the regioselectivity of the *tert*C—H-bond substitution due to steric hindrance. While nitroxylation gives apical, bromination exclusively provides access to the medially substituted cages in mono- and di-substitution reactions (Scheme 4.4).

Note that the medial surfaces of **1(2)3TET** and **PENT** reproduce the surface of [13]graphane, which was recently suggested as a measure for the steric size of the groups X (Figure 4.5) [54]. The six adjacent axial hydrogen atoms in such systems provide a nearly circular constriction on the substituent close to its point of attachment with constant steric features. Accordingly, the steric hindrance is maximized at the medial positions of **PENT**, preventing this position from undergoing nitroxylation.

Hence, the preparation of various diamondoid alcohols via a combination of nitroxylation in fuming (100%) HNO$_3$ and isomerization with H$_2$SO$_4$ followed by hydrolysis affords kinetically or thermodynamically controlled C—H-bond mono- and di-substitutions [47]. Other advantages of these combined mineral acid oxidations/isomerizations are high yields, reproducibility and scalability, low reagent and reaction work-up costs, and, something that is especially important, the ease of subsequent functional group transformation of the ensuing alcohols to

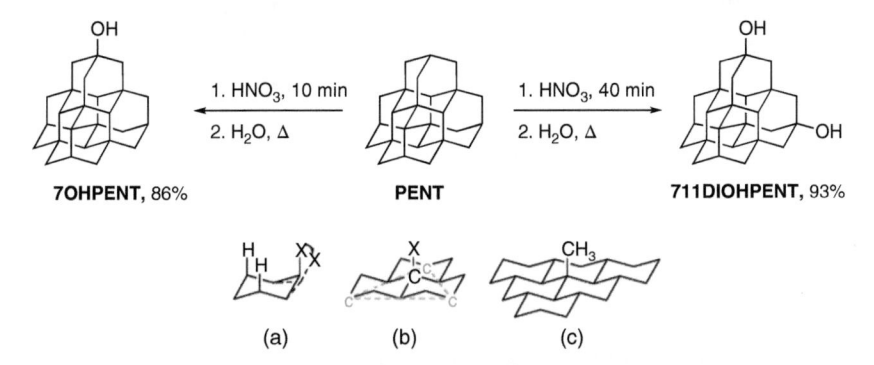

Figure 4.5 *Top*: Oxidation of **PENT** with fuming nitric acid gives apical derivatives **7OHPENT** and **711DIOHPENT** exclusively because the medial positions are highly sterically hindered. *Bottom*: Model to measure the steric size of group X. (a) Flexing of axial-substituted cyclohexane to minimize steric strain, (b) plane defined by the three outer carbon atoms of X-[13]graphane (in green), (c) Me-[22]graphane. Source: Reproduced from Ref. [54] with permission from the American Chemical Society, 2021.

4 Preparative Diamondoid Functionalizations

a large variety of other functional derivatives. Additionally, this approach provides a way toward "bromine-free" diamondoid derivatives that are important for drug syntheses. It is a good alternative to the direct Ritter reaction of diamondoids with H_2SO_4/CH_3CN [55].

Nitroxylations and brominations of diamondoids followed by acid-catalyzed isomerization often lead to symmetric diols, whose further transformations first require desymmetrization. 2,2,2-Trifluoroethanol as a nucleophile for the selective monoprotection of diamondoid diols has proven exceptionally efficient for the desymmetrization of 1,3-dihydroxyadamantane and higher diamondoids (Figure 4.6). The corresponding monoesters were isolated in >50% yield after column chromatography of the reaction mixtures [56].

Alternative selective monoprotection of diamondoid diols with silyl reagents [57] gives rise to unequally substituted cages omitting chromatography. More specifically, the *tert*-butyldimethylsilyl (TBDMS) protecting group can selectively be introduced only on one diamondoid alcohol group if the reaction is performed in 1-methylimidazole as the solvent in the presence of I_2 (Scheme 4.11). The basis for the selectivity of this reaction is the relatively low solubility of the monoprotected alcohol in the reaction media, thereby escaping the second silylation step.

Most importantly, the separation of mono- and di-silylated derivatives is possible without column chromatography and can be performed at a multigram scale. An example of the synthetic applicability of these monoprotected alcohols is the preparation of apical **4OHDIA** through a very simple chromatography-free sequence (Scheme 4.12) [57]. This approach is also useful for the preparation of apical alcohols of **TRIA** and **121TET** from their apical bis-dialcohols.

The oxidation with fuming acids is, of course, not the only way to obtain hydroxy-diamondoids. These products can also be prepared through the hydrolysis

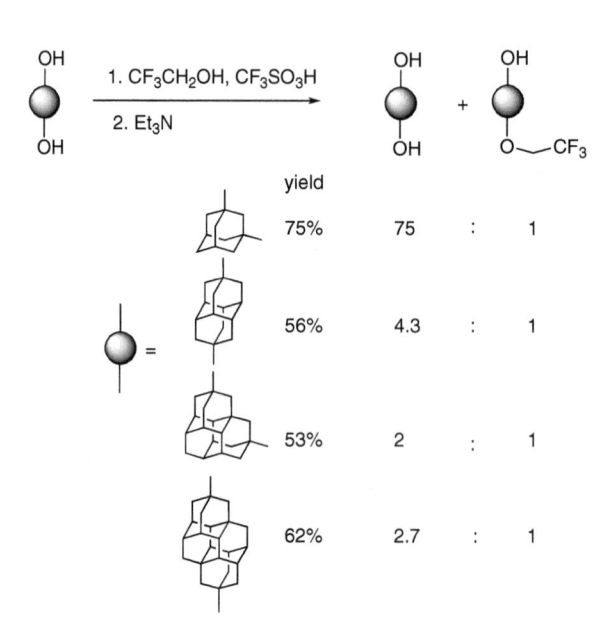

Figure 4.6 Desymmetrization of apical diamondoid diols with 2,2,2-trifluoroethanol in acidic media and yields of monoesters after column chromatography. Source: From Ref. [56].

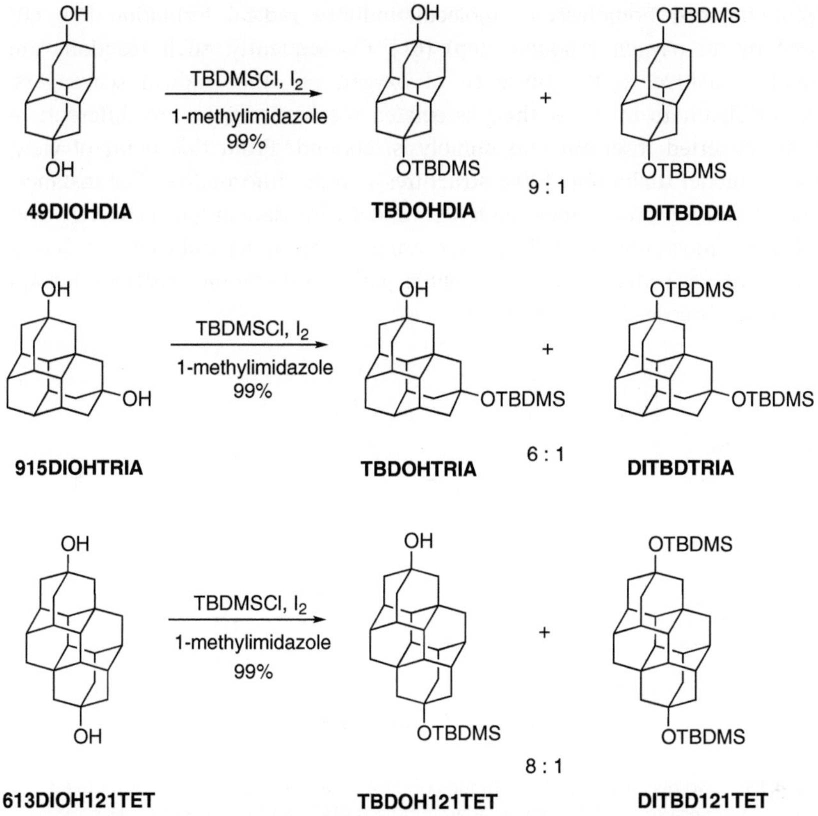

Scheme 4.11 Monoprotection of diamondoid diols with *tert*-butyldimethylsilyl chloride (TBDMSCl) allows for omitted chromatography separation. Source: From Ref. [57].

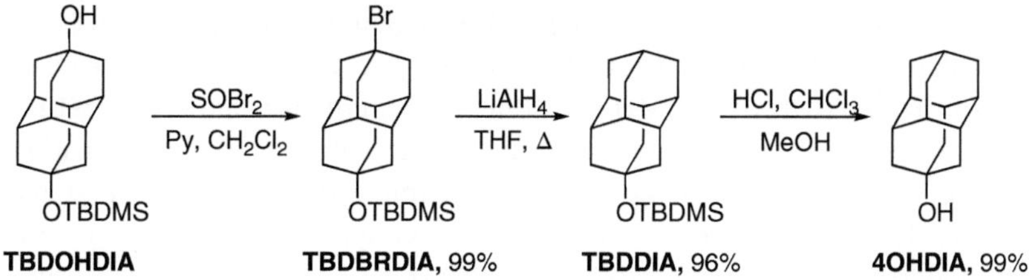

Scheme 4.12 Chromatography-free synthesis of the apical hydroxyl-derivative **4OHDIA** from easily assessable mono-protected diamantane 4,9-diol. Source: From Ref. [57].

of their halogen bromo- or chloro-precursors [28a, 58]. Direct diamondoid hydroxylation with neutral electrophiles can be accomplished also with *m*-chloroperbenzoic acid (MCPBA), Pb(OAc)$_4$, dimethyldioxirane (DMD), etc., albeit often with somewhat lower yields or selectivities; hence, these have only limited application in diamondoid chemistry. The mechanisms of these transformations are often poorly understood since they either proceed via concerted C—H-bond insertions or form radicals from neutral closed-shell reactants during the transformation (molecule-induced homolysis or molecule-induced radical formation [59, 60] followed by an oxygen rebound step) [61]. Consequently, such reactions are inherently sensitive to the presence of oxygen or other radical scavengers, and it is difficult to interpret their associated mechanisms, i.e., to differentiate between concerted insertion and homolysis/rebound. From this point of view, the use of model diamondoid-like structures is quite informative. For instance, the reactivities of propellanes such as 1,3-dehydroadamantane (**DHAD**) and 3,6-dehydrohomoadamantane (**DHHAD**) are quite revealing and clearly different [62]. This allows concerted and homolytic paths in dioxirane reactions toward σ-bonds to be discerned (Scheme 4.13).

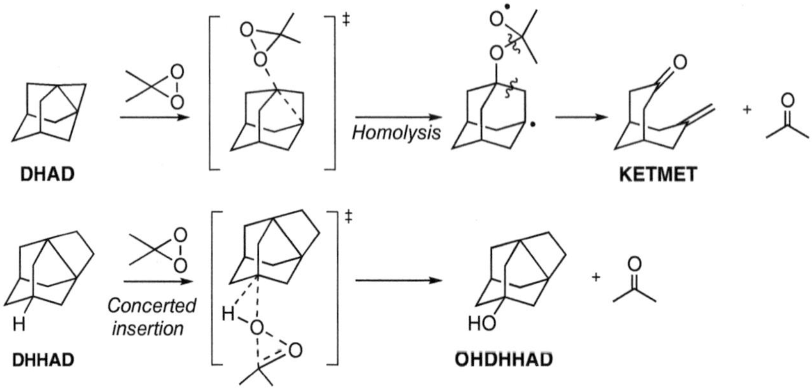

Scheme 4.13 Two distinctly different modes of dioxirane reactivity: Molecule-induced homolytic fragmentation of 1,3-dehydroadamantane (**DHAD**) and concerted C−H-bond insertion into CH bond of 3,6-dehydrohomoadamantane (**DHHAD**). Source: From Ref. [62].

While **DHAD**, which contains inverted carbon atoms, is very reactive toward radicals and electrophiles [63], propellane **DHHAD** contains a "normal" central C−C bond and demonstrates typical alkane behavior [64]. As a result, the reactivity of **DHAD** and **DHHAD** toward DMD is distinctly different (Scheme 4.13). **DHAD** adds DMD to the central C−C bond through molecule-induced biradical formation followed by the fragmentation to the unsaturated ketone **KETMET**. Yet, more stable propellane **DHHAD** favors *tert*C−H-bond insertion, typical for saturated hydrocarbons, and forms hydroxy-derivative **OHDHHAD** [62]. All these reactions occur under mild reaction conditions ruling out homolytic cleavage of DMD, thus excluding radical attack of the DMD biradical and clearly showing that only very reactive σ-bonds (like the central C−C bond in **DHAD**) can be homolyzed by ground-state singlet DMD.

As C−H-bond insertions with dioxiranes are not as sensitive to polar effects, mono-hydroxylations of diamondoids higher than **AD** are not very selective. Still, while DMD typically preferentially attacks bridgehead C−H bonds, the degree of substitution determines the number of OH groups incorporated into the cage. For example, DMD readily reacts with **DIA**, giving predominately **1OHDIA** and some **4OHDIA,** but diols also form (Scheme 4.14) [1, 9]. This illustrates the challenge

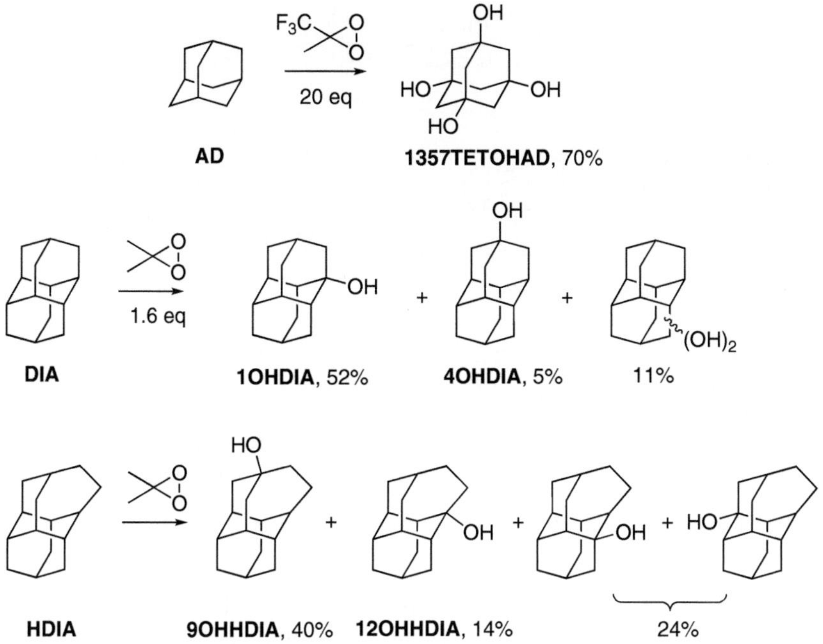

Scheme 4.14 Diamondoid hydroxylations using dimethyldioxirane.

when employing dioxiranes for diamondoid functionalization: overoxidation can occur readily. The reason for this problem lies in the nature of the reaction itself, meaning that it is difficult to stop it after only one C—H bond reacted since there are other tertiary groups still available and the EWD groups in the cage have only a little deactivation effect on the rate of the insertion of DMD. Dioxirane oxidations are therefore primarily used for polyhydroxylations where even the direct conversion of **AD** to 1,3,5,7-tetrahydroxyadamantane (**1357TETRAOHAD**) is possible [65]. Additionally, if commonly used electrophiles cause the rearrangement of the cage, as is the case for the hydroxylation of homodiamantane (**HDIA**), dioxiranes are also useful (Scheme 4.14). While bromination or nitroxylation causes the rearrangement of the homodiamantane cage to methyldiamantane derivatives, oxidation with dioxirane occurs with conservation of the cage and predominantly gives apical 9-hydroxyhomodiamantane (**9OHHDIA**) [52c].

Many alternative approaches were tested for the functionalization of **DIA** including the reaction with dichlorocarbene, which gives a 1.7 : 1 mixture of 1- and 4-insertion products [66]. High-valence transition metal acetates predominantly give medial derivatives [67], as do aerobic oxidations with ruthenium-substituted polyoxometalates [68] and phosphorylations in the presence of Lewis acids [69]. Lewis acid-catalyzed bis-apical diphenylation with benzene [70], direct amidation in the MeCN/CBrCl$_3$/Mo(CO)$_6$ system [71], hydroxylations with peracids [72], as well as metal-oxo reagents, are also useful, but there are some exceptions, like the previously mentioned chromyl chloride CrO$_2$Cl$_2$, which causes diamantane chlorination [9].

4 Preparative Diamondoid Functionalizations

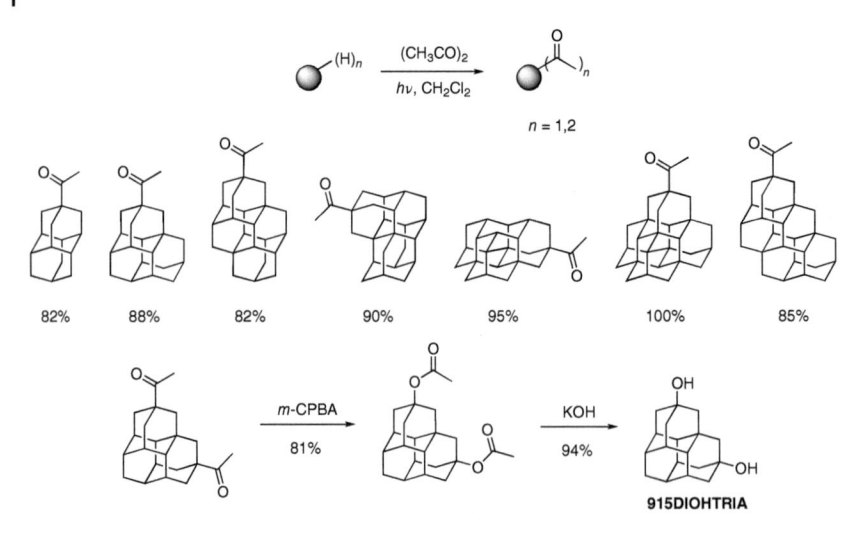

Figure 4.7 Photoacetylations of diamondoids with triplet diacetyl.

Another option for cage oxidation is photoacetylation with butadione ("diacetyl," [CH$_3$CO]$_2$), which is one of the most selective and useful radical reactions for diamondoid functionalization since it readily affords apically substituted derivatives (Figure 4.7) due to the steric bulk of the H-abstracting species [6b, 11, 73].

This reaction is initiated photochemically (for the most recent review on photochemically induced hydrogen transfer from hydrocarbons, see Ref. [74]), and hydrogen abstraction with triplet diacetyl generates the apical products. A systematic study reveals that for photoacetylations of diamondoids, solvent and steric effects play a bigger role than electron transfer effects, which are typically the driving force for medial substitutions in electrophilic media [73a]. Furthermore, it appears that the preference toward the apical positions (ratio exceeded 4 : 1 for all studied cases) is governed by the higher cage polarizabilities in the apical direction [73a]. An important advantage of the acetylated products is their facile conversion to alcohols through Baeyer–Villiger oxidation followed by simple hydrolysis. This is especially useful in the preparation of isomeric pentamantane alcohols, where monohydroxylations are cumbersome, and it remains the only way to make apical derivatives of [1212]pentamantane.

On the other hand, the medial acetyl derivatives are easily accessible from their respective carboxylic acids. In contrast to 1-adamantylmethyl ketone, which is commercially available and displays typical ketone reactivity, highly sterically hindered 1-diamantylmethyl ketone (**1ACDIA**) does not react with many reducing or Grignard reagents. Under typical Corey–Chaykovsky methylenation reaction conditions [75], the reaction proceeds anomalously, where not only isopropylidene diamantane (**1IPDDIA**) forms but, at higher temperatures, diolefination to **1DIENDIA** occurs [76]. Moreover, with a large base excess, **1ACDIA** undergoes dimethylenation to cyclopropane derivative **1ACCPDIA** (Scheme 4.15) [78]. A possible explanation is that due to steric hindrance toward the attack of the Corey

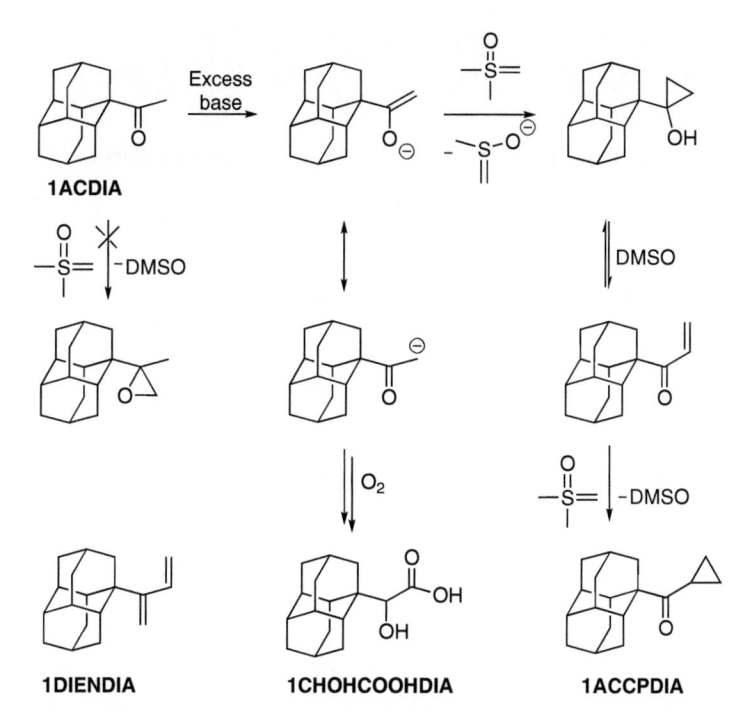

Scheme 4.15 Unusual reactivity of sterically congested 1-diamantylmethyl ketone (**1ACDIA**) in the presence of base excess. While the oxidation of intermediate enolate gives acid **1CHOHCOOHDIA** selectively [77] in the presence of dimethylsulfoxonium methylide, dimethylenation to cyclopropane derivative **1ACCPDIA** takes place [78] instead of expected diolefination of **1ACDIA** to **1DIENDIA**.

reagent on the carbonyl group of **1ACDIA** base-catalyzed enolization takes place, which is followed by double methylenation (Scheme 4.15, right). Another example of the anomalous behavior of ketone **1ACDIA** is its aerobic oxidation in the presence of NaOH, where the intermediate enolate is oxidized (Scheme 4.15, center) to 2-hydroxy-2-(1-diamantyl)acetic acid (**1CHOHCOOHDIA**) [77].

A fascinating recent example of vicinally substituted diamantane derivatives illustrates that there is still much to explore in the field of diamondoid functionalizations. Hydroxylation approaches considered until now rely on the simple rule that incorporation of the first functional group deactivates the neighboring sites of the cage, both electronically and sterically. Consequently, no vicinal derivatives are accessible via direct C—H-poly functionalizations of the cage, even though they would be very attractive bidentate diamondoid building blocks. Recently it was revealed, however, that readily available 1-hydroxydiamantane **1OHDIA** reacts with a bromine/potassium carbonate mixture in chloroform to give **12BROHDIA** in good yield (Scheme 4.16) [79]. The proposed mechanism involves a retro-Barbier-type fragmentation of the corresponding hypobromide to **BRODIA** that is characteristic of tertiary alcohols in the presence of bromine [80], but, in this case, the follow-up reaction reconstructs the diamantane cage through subsequent

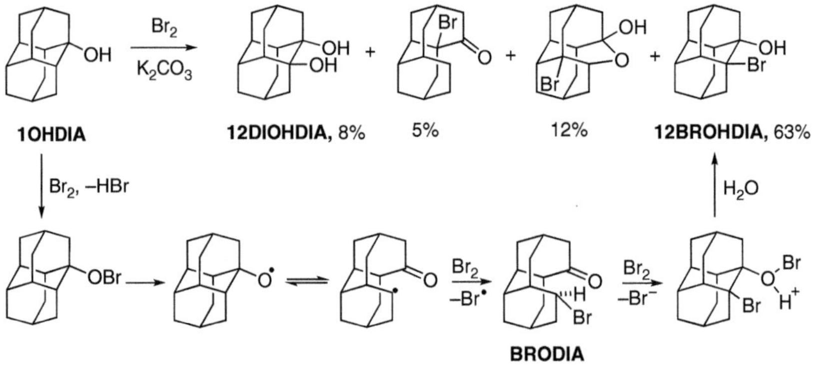

Scheme 4.16 Proposed mechanism of the retro-Barbier type fragmentation of 1-hydroxydiamantane (**1OHDIA**), followed by the cage reconstruction of intermediate ketone **BRODIA** to 1-bromo-2-hydroxydiamantane (**12BROHDIA**) [79].

bromination/dehydrobromination steps. The structures of the minor side products are in full agreement with the fragmentation/cyclization reaction scheme.

Diol **12DIOHDIA**, obtained quantitatively through the hydrolysis of **12BROHDIA**, is a useful building block for the preparation of various 1,2-derivatives of diamantane. For example, carbonate **12CO3DIA** was prepared with N,N-carbonyldiimidazole, sulfite **12SO3DIA** through the reaction with SOCl$_2$, phosphate **12PO3DIA** with POCl$_3$, and diamine **12DINH2DIA** through the reaction with chloroacetonitrile followed by hydrolysis (Scheme 4.17). Diamine **12DINH2DIA** is seen as a promising conformationally rigid bidentate ligand and was effectively enantioseparated through its diastereomeric salts with (+)-tartaric acid. The Pt-complexes of **12DINH2DIA** display pronounced anti-proliferative activity [81].

Another method for accessing 1,2-disubstituted **AD** and **DIA** derivatives is based on the imine clip-and-cleave concept mediated by copper(I) complexes (Figure 4.8) [82]. The aldehyde **CHOAD** first reacts with N,N-diethylethylenediamine giving the corresponding imine, which serves as an effective bidentate ligand (directing scaffold) for complexation with copper from [Cu(CH$_3$CN)$_4$][CF$_3$SO$_3$]. The formed intermediate complex is exposed to dioxygen and forms a bis(μ-oxido)-dicopper complex, which undergoes oxygen insertion (Figure 4.8). Subsequent cleavage under acidic conditions affords hydroxylated aldehyde **CHOOHAD** selectively. This hydroxylation protocol leads to the oxidation exclusively at the bridge position and therefore excludes any free radical aerobic oxidation that would, as we have seen, give bridgehead products. Furthermore, due to the rigid nature of the adamantyl dicopper complex, only hydroxylation at the β-C—H position is feasible, meaning that the stereochemical arrangement of the available C—H bonds dictates the reaction selectivity. As for **DIA**, both bridge and bridgehead neighboring positions are sterically available for oxidation, but the reaction at the tertiary bonds is both kinetically and thermodynamically favored: DFT computations [82] utilizing model spectator and reactive ligands (Figure 4.8) revealed that the barrier for oxygen insertion is only

Scheme 4.17 Preparation of 1,2-substituted **DIA** derivatives through diol **12DIOHDIA**.

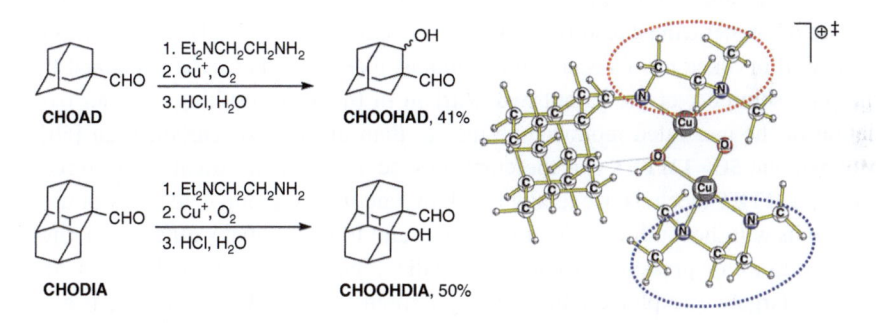

Figure 4.8 Aerobic copper(I)-mediated β-hydroxylation of diamondoid aldehydes and (right) the RI-BLYP-D3/def2-TZVP(SDD)/COSMO(CH$_2$Cl$_2$)//RI-PBE-D3/def2-TZVP computed transition structure for the oxygen insertion into the CH bond of diamantane 1-carbaldehyde (**CHODIA**) through the bis(µ-oxido)-dicopper complex (left). The CF$_3$SO$_3^-$ counterion is not shown for clarity; the model spectator and reactive ligands are marked in blue and red, respectively. Source: From Ref. [82].

4.7 kcal mol^{-1}, that is ca. 5 and 9 kcal mol^{-1} lower than the barrier for the analogous insertion into β-CH$_2$ groups of **DIA** and **AD**, respectively.

The keto-group is another good starting point for the selective preparation of diamondoid derivatives. Methylene group oxidations of diamondoid scaffolds can be achieved using concentrated sulfuric acid at elevated temperatures (Scheme 4.18). As the intramolecular hydrogen migration is sterically hampered [83], the reactions occur through intermolecular hydrogen exchange between tertiary diamondoidyl cations and the methylene positions of the neutral cages.

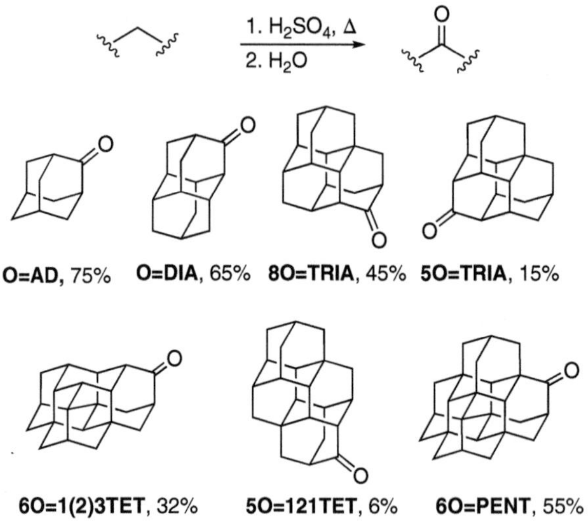

Scheme 4.18 Preparation of diamondoid ketones.

Reagents: 1. H₂SO₄, Δ; 2. H₂O

O=AD, 75%   O=DIA, 65%   8O=TRIA, 45%   5O=TRIA, 15%

6O=1(2)3TET, 32%   5O=121TET, 6%   6O=PENT, 55%

The oxidation of **AD** in hot sulfuric acid is still the most selective method for the preparation of the corresponding monoketone **O=AD**, and the reaction proceeds with the least amount of side products [84]. Similarly, **O=DIA** and many other diamondoid ketones can be readily obtained by this method [6b, 85]. Lower yields and formation of diketones for some diamondoids, like **121TET**, can be explained by the poor solubility of the parent hydrocarbon in the acid, resulting in faster oxidation of the generated monoketone moiety than of the hydrocarbon itself [6b]. Monoketone **5O=121TET** can therefore only be accessed via oxidation from secondary **121TET** alcohols. Generally speaking, methylene group oxidations of diamondoids with hot sulfuric acid are highly useful if not too many different methylene positions are present. The chemistry of diamondoidyl ketones higher than **AD** remains largely unexplored and is mostly limited to **O=DIA**, which gives rise to many 3-diamantane derivatives [85a, 86]. Direct C—H-functionalization of **O=DIA** is possible with many oxidants, but always with the formation of mixtures of medial and apical substitution products [87].

Incorporation of electronegative heteroatoms into the cage ("internal doping") markedly reduces the reactivity of diamondoids, as found for the bromination and nitroxylation of oxaadamantane **OAD**, for which the remote CH position "5" is the most reactive [88]. The bromination of the C—H-bond of **OAD**, however, is possible under radical conditions employing the PTC protocol of the CBr₄/NaOH system [16, 89]. Higher oxadiamondoids can be functionalized using strong electrophiles, and bromination and nitroxylation/hydrolysis were performed on **ODIA** and **5OTRIA** (Figure 4.9) [85b]. Unlike their pure hydrocarbon counterparts, the oxa-derivatives react quite selectively with electrophiles; for example, **ODIA** and **5OTRIA** give upon bromination only **6BRODIA** and **2BR5OTRIA**, respectively. Similarly, the reaction of **ODIA** with 100% nitric acid and subsequent hydrolysis

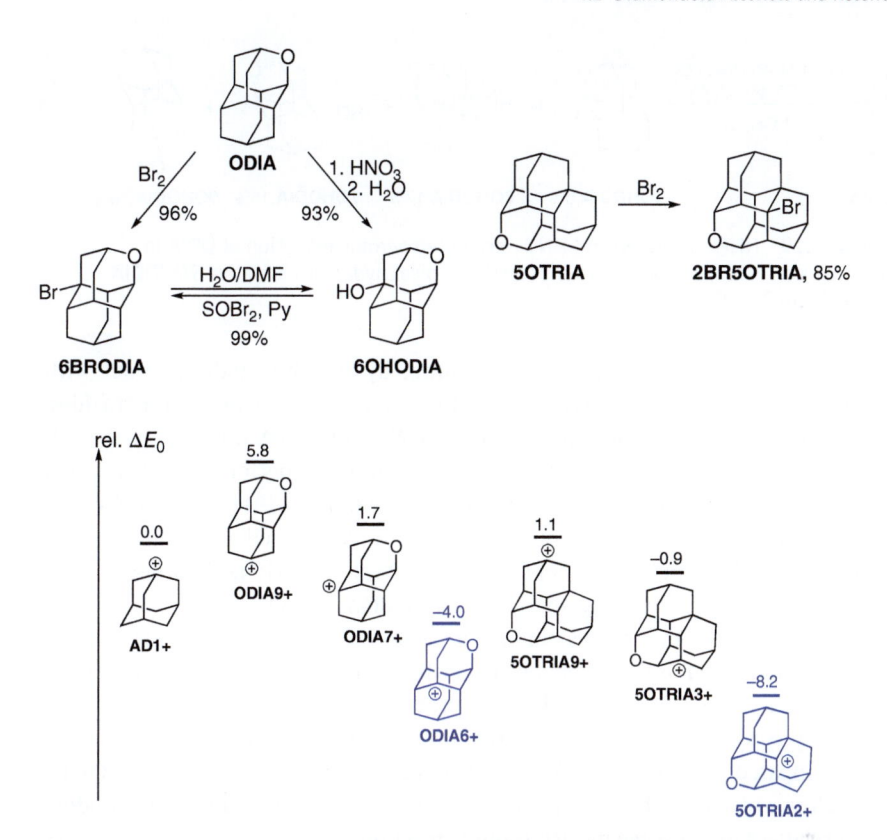

Figure 4.9 Functionalizations of oxadiamondoids and the stabilities of selected oxadiamantyl (**ODIA+**) and 5-oxatriamantyl (**5OTRIA+**) cations relative to 1-adamantyl (**AD1+**) cations (based on respective homodesmotic equations at MP2/cc-pVDZ level of theory, kcal mol^{-1}). Source: Reproduced from Ref. [85b] with permission from the American Chemical Society, 2009.

only gives **6OHODIA**. It follows that the position of the dopant in the diamondoid cage significantly influences the selectivity of the substitution reaction. For brominations, the observed selectivity correlates with the stability of carbocations in the corresponding positions (Figure 4.9) [85b]. Note that the 1-diamantyl cation is ca. 5 kcal mol^{-1} more stable than the 1-adamantyl cation (**AD1+**), indicating that oxygen doping generally destabilizes the cage cations. Among isomeric γ-cations, oxadiamant-6-yl (**ODIA6+**) and 5-oxatriamant-2-yl (**5OTRIA2+**) are much less destabilized, thus determining the selectivity of **ODIA** and **5OTRIA** reactions with electrophiles toward medial derivatives.

The low reactivity of oxadiamondoids toward electrophiles and the unavailability of apical substitution derivatives prompted the development of radical C—H bond functionalizations. The metal-free aerobic N-hydroxyphthalimide (NHPI)-catalyzed oxidation of **ODIA** in highly polar trifluoroacetic acid (TFA) promoted by HNO$_3$ (Scheme 4.19) indeed demonstrates that other substitution patterns are also accessible [90]. More specifically, the electrophilic PINO radical generated from NHPI

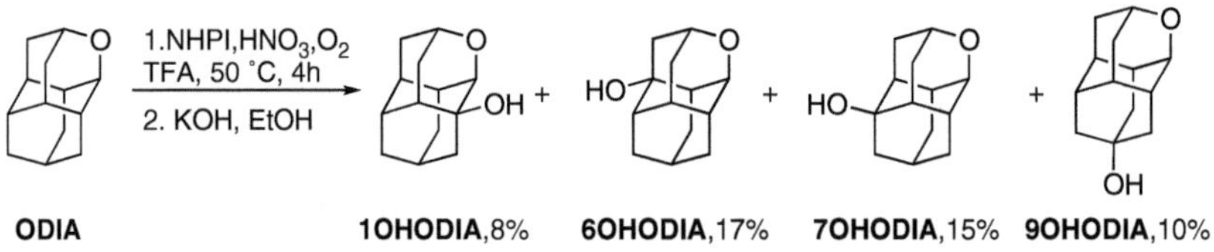

Scheme 4.19 NHPI-catalyzed and HNO$_3$-promoted aerobic oxidation of **ODIA** in trifluoroacetic acid gives otherwise inaccessible apical hydroxyl-derivative **9OHODIA**. Source: From Ref. [90].

performs C—H-abstractions, and the ensuing hydrocarbon radicals react with molecular oxygen, giving peroxyl radicals first, then intermediate hydroperoxides, and finally oxidation products after hydrolysis. Note that TFA was used as the solvent because it was demonstrated that the transition structures for the H-abstraction with PINO are characterized by substantial charge transfer [91] and therefore the reaction benefits from a highly polar medium [90]. While medial substitution still dominates, a substantial amount of the apical derivative **9OHODIA** forms, thereby providing access to apical derivatives of O-doped diamondoids.

The evidence that this process indeed involves radicals despite the presence of nitric acid that could in theory enable an electrophilic reaction as well, comes from the observation that the reaction does not proceed when NHPI is not present. This is not surprising since we saw that **ODIA** has lower reactivity toward electrophiles when compared to its parent **DIA**. The PINO-radical-promoted oxidation of **ODIA** is therefore a viable method for the preparation of apically substituted diamondoid oxa-derivatives inaccessible by electrophilic reactions.

To summarize, once initial diamondoid cage substitution is achieved, further functional group transformations give rise to a plethora of derivatives. In continuation of this chapter, examples for the preparation of various functionalized diamondoid derivatives will be presented.

4.3 Carboxylic Acids and Their Derivatives

The preparation of diamondoid carboxylic acids typically involves a Koch–Haaf reaction in HCOOH/H$_2$SO$_4$ starting from the hydrocarbon or the corresponding bromides or alcohols [58]. When the parent hydrocarbon is oxidized, the presence of *tert*-butanol under acidic conditions is needed because the in situ-generated *tert*-butyl cation performs an H-abstraction from the diamondoid cage giving rise to diamondoidyl cations. While the preparation of **1COOHAD** from either **1BRAD** or **1OHAD** is straightforward [92], introduction of the acid functionality becomes more challenging for the **DIA** cage because Koch–Haaf conditions applied to **DIA** exclusively give rise to **1COOHDIA** [58], while apical **4COOHDIA** is accessible from the parent apical bromide **4BRDIA** under high dilution to prevent intermolecular hydrogen exchange (Scheme 4.20) [58, 93] that would scramble the product ratio in favor of the medial derivative [94].

Scheme 4.20 Preparation of apical **4COOHDIA** under high dilution conditions. Source: From Ref. [58].

Performing the Koch–Haaf reaction on a mixture of diamantane polybromides obtained from the bromination reaction with trace amounts of AlBr$_3$ readily yields polyacid derivatives. These polyacids are easier to separate than the parent bromides because of substantial differences in the solubility of their ammonium salts [94]. However, this approach is not as practical as the stepwise bromination of readily available alcohols using thionyl bromide followed by a Koch–Haaf reaction (Figure 4.10) [95]. The advantage of using alcohols is the precise control of the product distribution and excellent overall yields, even for larger cages like **TRIA**, **TET**, and **PENT**.

In addition to compounds where the carboxylic functional group is bound directly to the diamondoid scaffold, alcohols and bromides can also be used as a starting point to prepare diamondoid acetic acids via the Bott reaction in H$_2$SO$_4$/Cl$_2$C=CH$_2$ [96]. For example, the preparation of 1-adamantyl acetic acid (**1CH2COOHAD**)

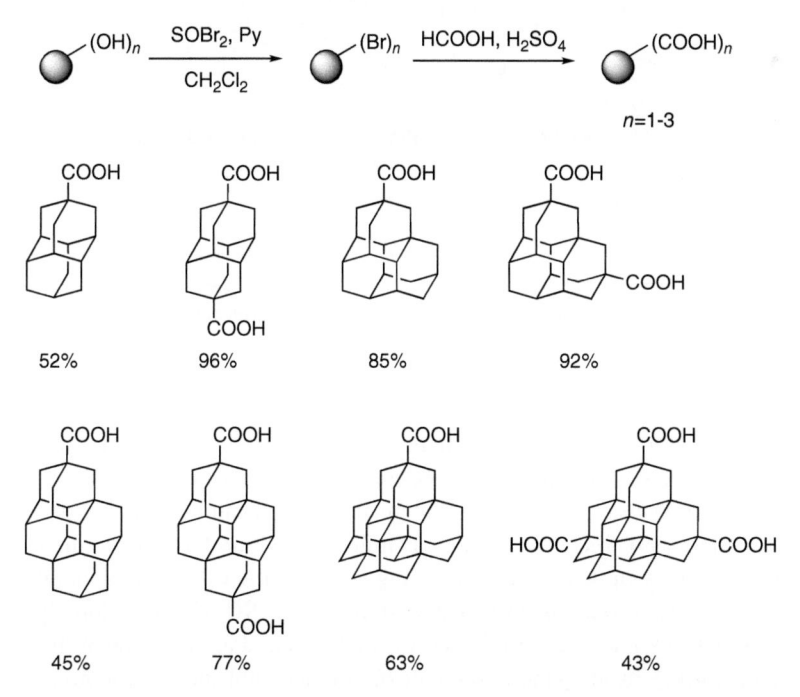

Figure 4.10 Stepwise preparation of diamondoid carboxylic acids through their corresponding mono- and dibromides.

proceeds from **1BRAD** using 1,1-dichloroethylene; the diacid **13DICH2COOHAD** can be obtained with a slightly modified protocol (Scheme 4.21) [96]. Note, however, that the Bott reaction often suffers from significant drawbacks like rapid reagent decomposition, byproduct formation, and low yields, and it is usually more suitable for compounds with bridgehead carbon atoms. Since this reaction is sensitive to steric hindrance around the reaction center where the positive charge is generated, the higher thermodynamic stabilities of diapical **DIA** and **TRIA** derivatives enable the preparation of diamondoid diacetic acids in one step. Introduction of longer chains onto the sterically hindered medial cage positions can be accomplished through radical reactions, e.g., by using methyl acrylate, azobis(isobutyronitrile) (AIBN) as the radical initiator, and $n\text{Bu}_3\text{SnH}$ as the terminal reductant [95]. The reaction in refluxing toluene gives the corresponding diamantane and triamantane propionic acids **1CH23COOHDIA** and **2CH23COOHTRIA** with high preparative yields (Scheme 4.21).

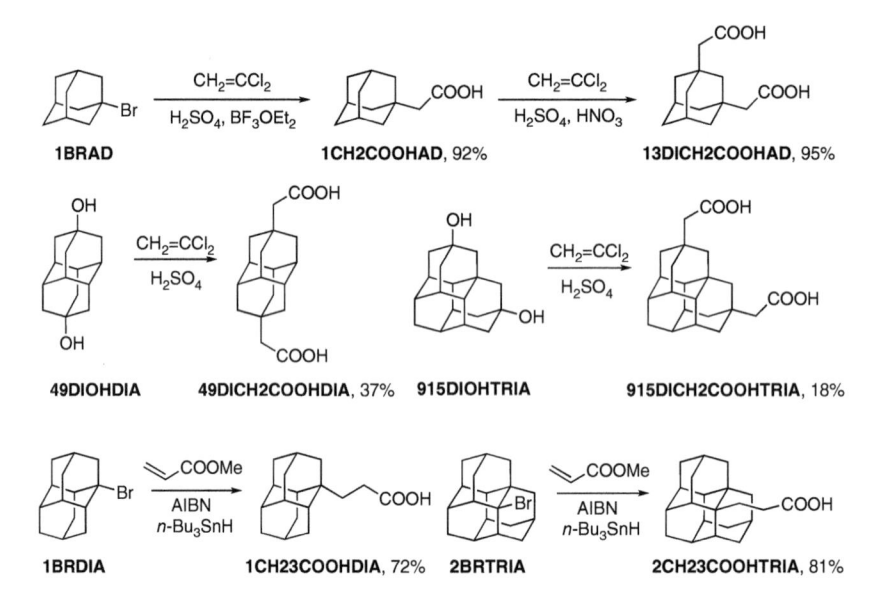

Scheme 4.21 Preparation of diamondoid acetic and propionic acids. Source: From Ref. [95].

Other approaches for the preparation of diamondoid carboxylic acids include photoacetylation [73b] as well as hydrolysis of the corresponding nitriles [97], although the reverse reaction is also possible [98]. The nitriles in turn can be obtained from diamondoid halides [99]. Recently, a relatively mild method for diamondoid direct C—H-cyanation has been described [100] using NHPI as the catalyst to generate diamondoidyl radicals through hydrogen atom transfer to PINO. The resulting cage radical was trapped by *p*-toluenesulfonyl cyanide (TsCN) and afforded the desired cyanides (Figure 4.11). In situ generation of PINO was accomplished by combining NHPI with cerium (IV) nitrate (CAN) in the presence of a base. When applied to **DIA**, the direct C—H-cyanation unfortunately produces a mixture of **1CNDIA** and

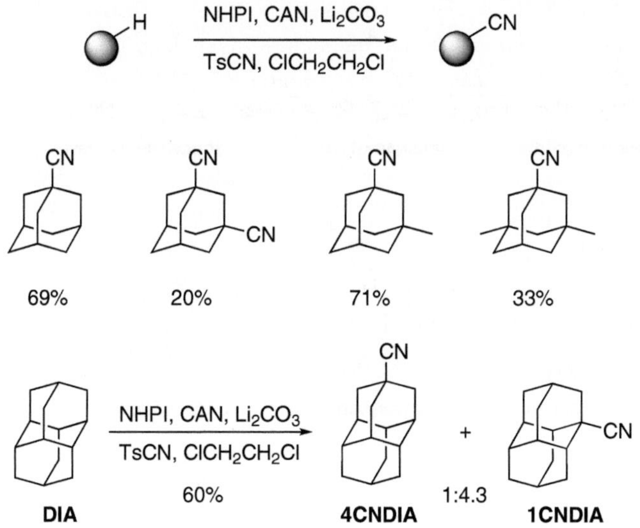

Figure 4.11 Preparation of diamondoid nitriles using NHPI-catalyzed cyanation. Source: From Ref. [100].

4CNDIA with a ratio of 4.3 : 1 in favor of the medial derivative. Nevertheless, this method not only allows the direct functionalization of the diamondoid skeleton but also tolerates a broad range of functional groups like triple bonds, acids, acetamides, azides, etc., and thereby provides access to orthogonally functionalized diamondoid building blocks.

A special class of carboxylic acid derivatives that needs to be mentioned here are amino acids, which are important building blocks for artificial peptide synthesis [101]. The amine and the acid functional groups can be introduced directly into the diamondoid cage either on the methylene [102] or the bridgehead positions. The introduction of these two groups demands a somewhat more elaborate synthesis consisting of several steps. For example, the preparation of **3COOH1NH3AD** starts with the parent monoacid **COOHAD** and undergoes bromination and subsequent acetamidation to afford the target adamantane amino acid upon hydrolysis (Scheme 4.22). The preparation of apical **4COOH9NH3DIA** was even more challenging: first, apical **49DIOHDIA** needed to be monoprotected to afford the unequally disubstituted derivative **4OETF9OHDIA** that undergoes a Ritter reaction with chloroacetonitrile to give **4OETF9NACDIA** (Scheme 4.22) [56]. The etheral group was then converted to a trifluoroacetoxy group in **4OACF9NACDIA**, and subsequent cleavage of both functional groups afforded **4OH9NH2DIA**. After Koch–Haaf reaction and introducing the methyl ester group for easier purification (**4COOH9NH2DIA** is poorly soluble in organic solvents), the final hydrolysis gave the target amino acid in the form of its hydrochloride salt **4COOH9NH3DIA**.

An alternative approach for the preparation of γ-aminoadamantane carboxylic acids uses a direct C—H-amidation reaction [103]. Straightforward amidation without prior halogenation of the tertiary position allows for large-scale preparation

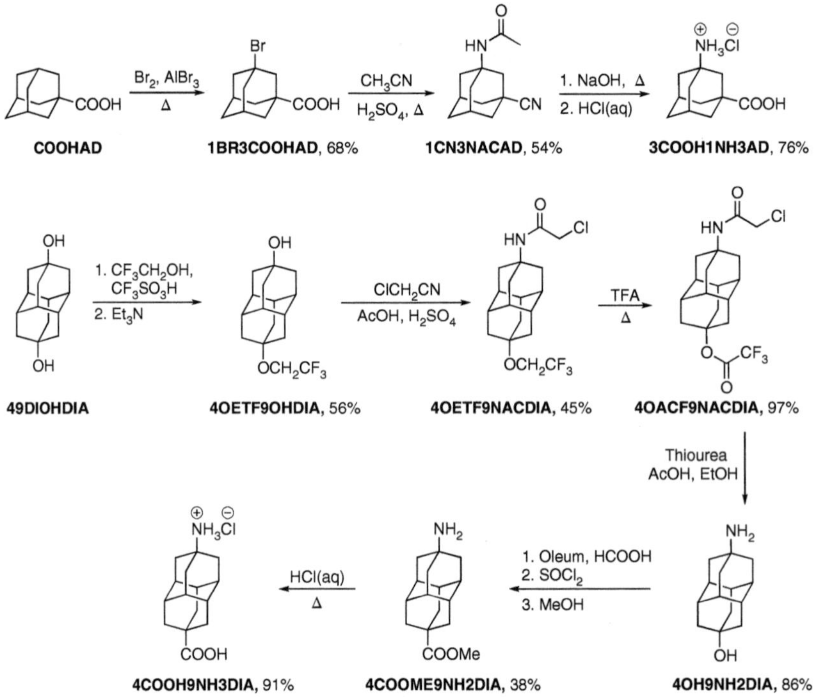

Scheme 4.22 Preparation of adamantane and diamantane amino acids.

of **AD** amides that are subsequently easily converted to amino acids. The key feature of this Ritter-type protocol is the use of a nitrating mixture (HNO₃/H₂SO₄) that activates the tertiary C—H bonds with the in situ-generated NO₂⁺ species. Quenching with either acetonitrile or amides affords the corresponding **AD** acetamide (Figure 4.12).

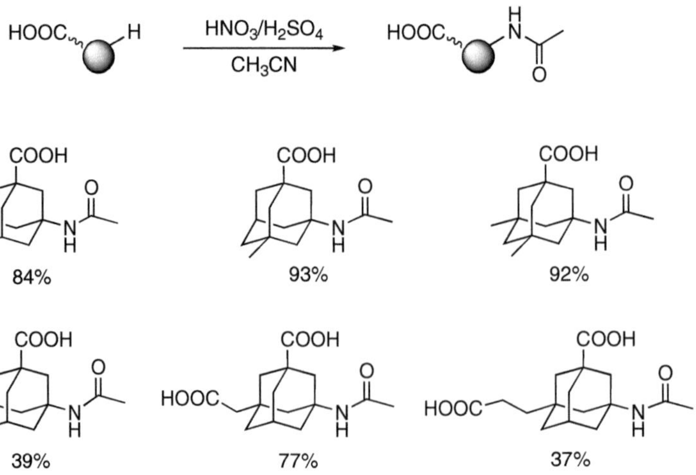

Figure 4.12 Preparation of **AD** γ-amino acid derivatives by direct C—H-amidation.

4.4 Nitrogen-Containing Compounds

Unnatural amino acids built upon diamondoid cages are of high importance since they can be incorporated into various peptide structures, thus making them potentially bioactive [104] and also useful for organocatalysis [105].

4.4 Nitrogen-Containing Compounds

As mentioned previously, the hydrolysis of the acetamido group readily affords the corresponding amine, and this reaction was applied for the first preparation of hydrochloride of 1-aminoadamantane (**1NH2AD**) [106], used as a commercial drug Amantadine®. Acetamides can be hydrolyzed either under acidic or basic reaction conditions. However, they first need to be introduced into the cage scaffold [58, 107], typically by applying Ritter reaction conditions (acetonitrile and sulfuric acid) on diamondoid bromides or alcohols. Many of the above methods for amine preparation unfortunately do not allow for facile scale-up, and a solution for this is the optimized two-step procedure starting with diamondoid alcohols [108]. The treatment of alcohols in the presence of chloroacetonitrile [109] leads to the acid-catalyzed exchange of the hydroxyl group for the chloroacetamide moiety. The resulting product can be cleaved under slight heating in the presence of thiourea to afford the desired amine (Figure 4.13). This procedure is applicable for the preparation of diamondoid mono- and diamines [108].

Another way of preparing tertiary diamondoid amines is the hydrogenation of the corresponding azides (Scheme 4.23) [28b, 110]. The highest-yielding azidation method found so far is the combination of trimethylsilyl azide with a tin tetrachloride catalyst that gives good yields and offers broad applicability [110a]. Other procedures for obtaining diamondoid amines include the direct amination with $NCl_3/AlCl_3$, which, however, suffers from low yields and often results in product mixtures [112]. The reduction of diamondoid oximes [113] that are readily available from the parent ketones or one-pot reductive amination of diamantanone (**O=DIA**) [111] is also useful but can only provide amines at the methylene positions, such as 3-aminodiamantane **3NH2DIA** (Scheme 4.23).

Diamondoid amine synthesis can also start with the corresponding acid derivatives. In that case, the carboxylic acid group reacts with diphenylphosphoryl

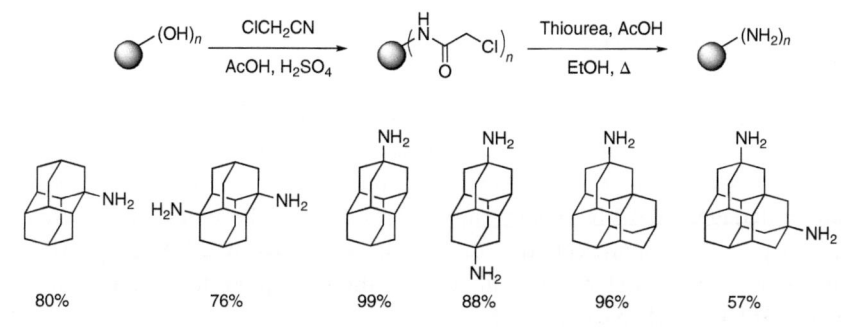

Figure 4.13 Two-step procedure for obtaining diamondoid amines via a Ritter-type reaction with chloroacetonitrile followed by hydrolysis. Source: From Ref. [108].

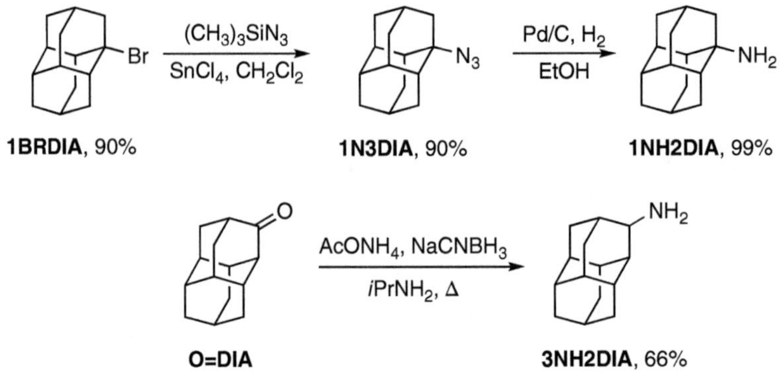

Scheme 4.23 Preparation of tertiary and secondary diamantane amines. Source: From Refs. [28b, 110b, 111].

azide in *t*BuOH followed by hydrolysis of the formed intermediate to the target amine [108].

The exchange of the hydroxyl group for nitrogen nucleophiles is usually achieved under acidic conditions to generate respective cage carbocations [114]. If acid-free conditions are needed, the generation of carbocations is possible through the oxidation of phosphinites obtained from the corresponding alcohols [114]. The oxidative decomposition of phosphinites is achieved with diisopropyl azodicarboxylate, followed by the addition of various nucleophiles (Scheme 4.24).

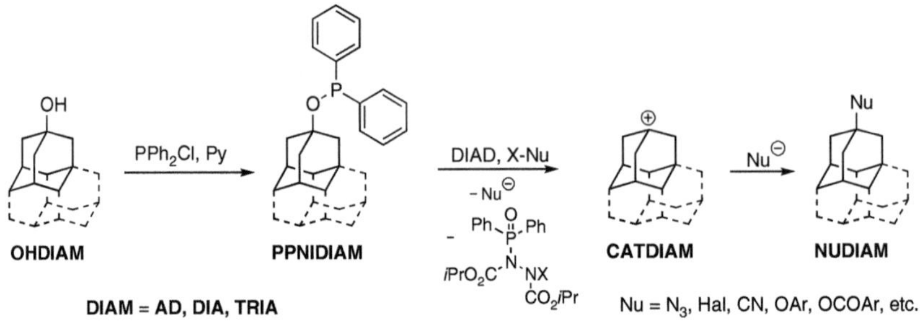

Scheme 4.24 Generation of cage carbocations from respective alcohols is achieved under acid-free conditions through the reaction of phosphinites with azodicarboxylate in the presence of nucleophiles. Source: From Ref. [114].

Molecules with two different functionalities present at the hydrocarbon cage are very useful diamondoid building blocks, as mentioned for the preparation of cage amino acids (see Scheme 4.22). Following this reasoning, in addition to obtaining the amine **4OH9NH2DIA** by concurrent hydrolysis of both the amide and the ester group, the preceding chloroacetamide diamantane derivative **4OETF9NACCLDIA** can be immediately hydrolyzed with thiourea to afford the protected amino alcohol **4OETF9NH2DIA** (Scheme 4.25) [56].

Although it is intuitive that simple reduction of diamondoid nitro-derivatives could readily afford amines, the practical reaction sequence actually proceeds

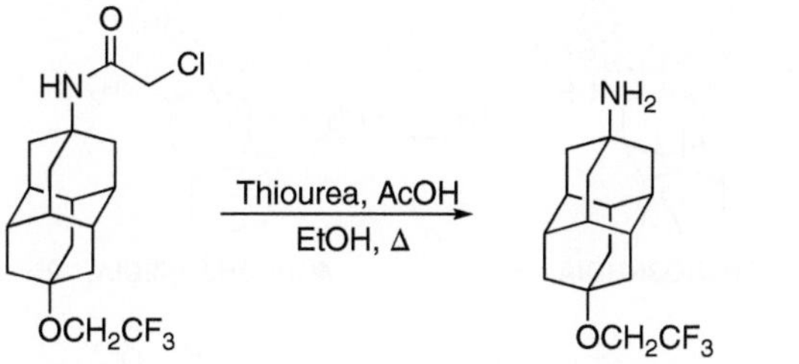

4OETF9NACCIDIA **4OETF9NH2DIA**, 93%

Scheme 4.25 Preparation of amino alcohol derivative **4OETF9NH2DIA**. Source: From Ref. [56].

the other way around. Namely, it is much easier to selectively introduce the amine functional group into the cage in a series of high-yielding steps starting from bromides or alcohols than it would be to achieve selective nitration of the parent hydrocarbon. In the case of diamondoids, methods for introducing a nitro group often give product mixtures that are low-yielding or need special reagents [52b]. Consequently, diamondoid amines are the precursors of choice as they readily react with MCPBA in 1,2-dichloroethane to yield nitro-functionalized products (Figure 4.14) [115]. Diamines of diamantane are also precursors for the preparation of respective diisocyanates through the reactions of **49DINH2DIA** and **16DINH2DIA** with triphosgene [116].

Lastly, a very modern synthetic approach for intramolecular C—H-bond amination enables the direct introduction of nitrogen functionalities into the cage framework and easily affords vicinal cage disubstitution products [117]. The starting sulfamate ester **4CH2SONH2DIA** undergoes an Rh(II)-catalyzed intramolecular nitrene C—H-bond insertion to give the heterocyclic product **4CH2SO3NHDIA**

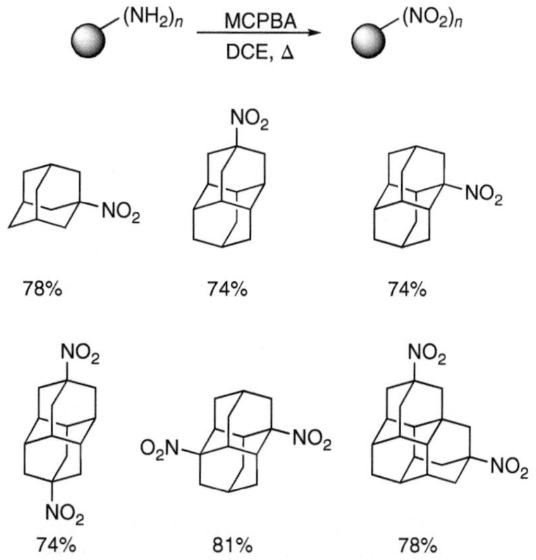

Figure 4.14 Oxidations of diamondoid amines to nitro-derivatives. Source: From Ref. [115].

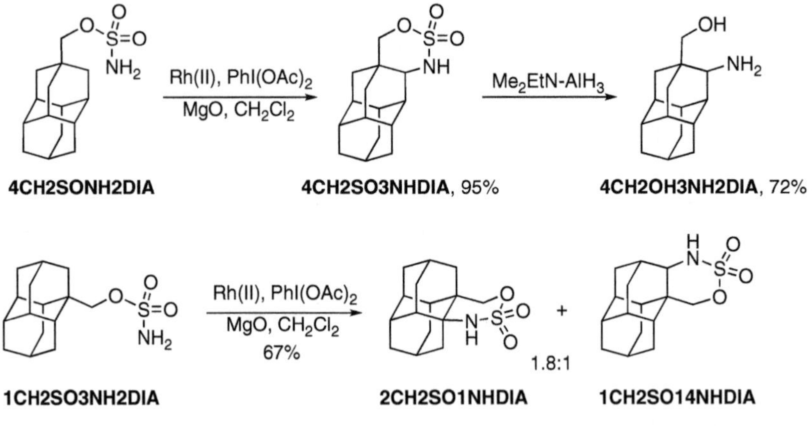

Scheme 4.26 Preparation of vicinally disubstituted diamantane amines. Source: From Ref. [117].

that can be cleaved into the corresponding amino alcohol **4CH2OH3NH2DIA** (Scheme 4.26).

Several analogs of vicinally substituted **DIA** derivatives were prepared as well, including the highly elusive bis-medial 1,2-insertion product **2CH2SO1NHDIA** [117]. Note that this approach also provides the means to prepare a wide variety of neighboring bridge-bridgehead substituted diamondoids that present an exciting new class of synthetically valuable chiral building blocks.

A similar intramolecular strategy was used in the directed acetoxylation of diamondoids, utilizing picolylamide as a bidentate directing group (Scheme 4.27) [118]. This reaction requires Pd(OAc)$_2$ as the catalyst and diacetoxyiodobenzene

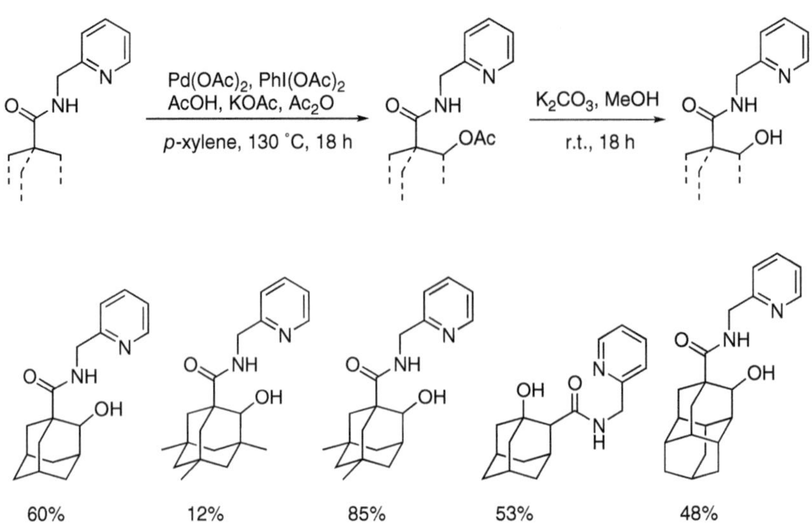

Scheme 4.27 Acetoxylation of diamondoids is followed by hydrolysis to β-hydroxy carboxylic acid derivatives. Source: From Ref. [118].

(PhI(OAc)$_2$) as the oxidant and provides β-hydroxy carboxylic acid derivatives as the main products. Of course, the corresponding diols and triols are common side products of this approach, but the mixtures can be separated by column chromatography. Polyacetoxylation occurs predominantly when an excess of oxidant is used. Note that cleavage of the directing group in strongly acidic conditions can readily give the corresponding β-substituted carboxylic acid that is also an attractive building block for further cage functionalizations and other synthetic post-functionalization protocols [118].

Effective intramolecular CH-amination of diamondoid carbamates is achieved in the presence of Rh-acetate, which is accompanied by nitrene insertion to give corresponding diamondoid vicinal derivatives (Scheme 4.28) [119]. Among others is the possible application of cyclic carbamates through their N-methylation followed by thermally induced ring contraction in triflic acid that gives corresponding nor-diamondoid carbaldehydes. The cascade reaction consists of a triflic acid-promoted decarboxylation and 1,2-alkyl shift in thus formed carbocationic intermediate, followed by hydrolysis. Even double-ring contraction is possible, which gives a new diamondoid building block bisnordiamantane bis-carbaldehyde (**CHO2NORDIA**).

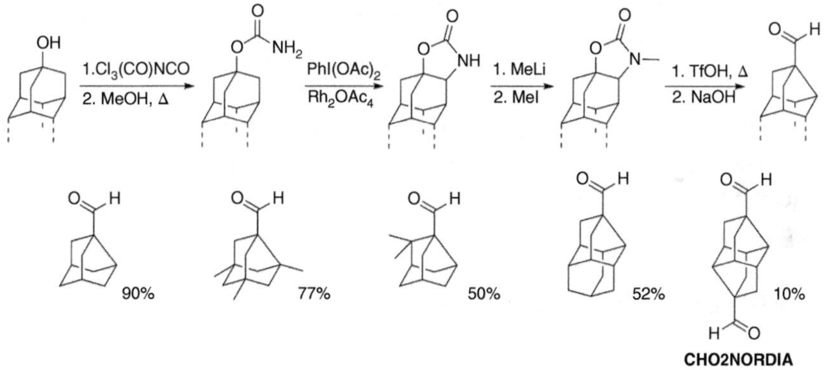

Scheme 4.28 The intramolecular vicinal CH-amination of diamondoids through Rh(II)-promoted nitrene insertion and the preparation of respective carbaldehydes through acid-catalyzed thermolysis of intermediate N-methyl carbamates (the yields for the triflic acid-promoted decarboxylation/rearrangement are shown). Source: From Ref. [119].

Sterically crowded adamantane and diamantane amides [120] were obtained through the Schotten–Baumann reactions of the corresponding amine with **ADCOCL** in the presence of a base. Reduction of amides with BH$_3$·THF gave the corresponding amines (Figure 4.15). Among the most crowded **DIANHCOAD** and **N_ADCOAD** only the latter displays substantial deviation from the CNC plane, i.e., the C—O bond is 16.0° twisted out-of-plane below the azaadamantane unit. Such out-of-planarity is much higher than observed for other amides whose distortion originates from steric repulsions (usually less than 4°), but still much lower than observed in sterically restricted lactams [121], where the amide moiety is locked within the cycle(s), such as in Kirby's most twisted amide **ADAMID** [122]. Note,

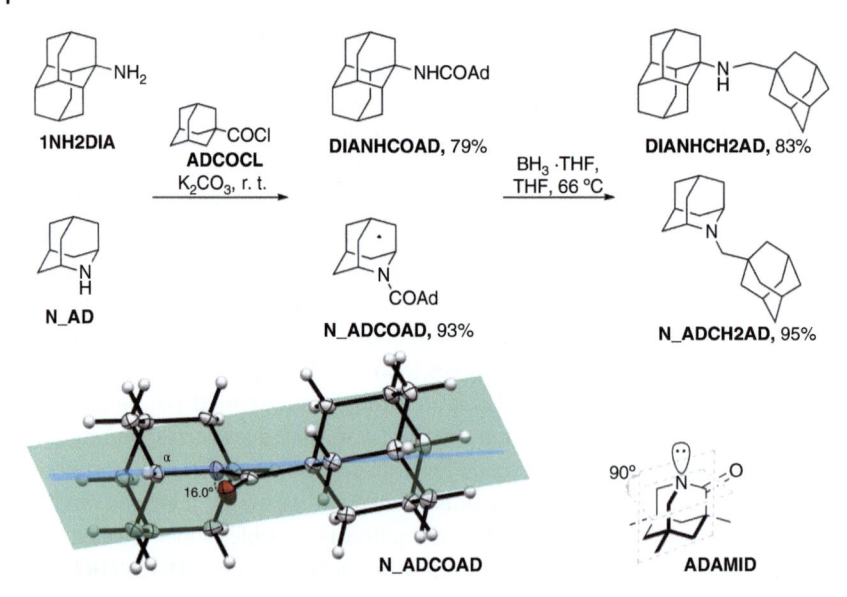

Figure 4.15 The preparation of sterically encumbered diamondoid amines and amides. The most twisted sterically congested (**N_ADCOAD**) and sterically restricted (Kirby's amide, **ADAMID**) amides. Amid **N_ADCOAD** displays a 16.0° out-of-plane twist of the amide functional group. Source: Adopted from Ref. [120], Wiley-VCH, 2022.

however, that in contrast to **N_ADCOAD** the "twist" in **ADAMID** is not a simple torsion angle but rather mirrors the out-of-plane bending on the nitrogen atom as defined by Winkler and Dunitz [123].

4.5 Phosphorous- and Sulfur-Containing Compounds

Another useful venue of diamondoid functionalizations is the introduction of phosphorous-containing functional groups that typically proceed either from the corresponding alcohols or parent hydrocarbons. Organophosphorous compounds often find applications as ligands for transition metals, and diamondoid derivatives are no exception. The use of the Lewis acid-based $AlCl_3/PCl_3/CH_2Cl_2$ system known as the Clay–Kinnear–Perren phosphorylation [124] for diamondoids [69] is somewhat tedious and often difficult to reproduce [125]. Additionally, the known problems of this reaction include isomerization of *n*-alkyl chlorides to branched phosphonic acid dichlorides, elimination of methylene groups (e.g., 2-methyl-2-butyl chloride affording *tert*-butylphosphonic dichloride), instances of low reactivity, formation of mixtures of alkyl phosphonic acid dichlorides and dialkyl phosphonic acid chlorides, etc. [126] All of this makes the utilization of Brønsted instead of Lewis acids far more attractive. Phosphorylations in H_2SO_4 or TFA were first performed on adamantane derivatives [127], and these were expanded to higher cages, additionally allowing the use of hydroxy derivatives instead of bromides. The phosphorylations of diamondoid alcohols with PCl_3 under

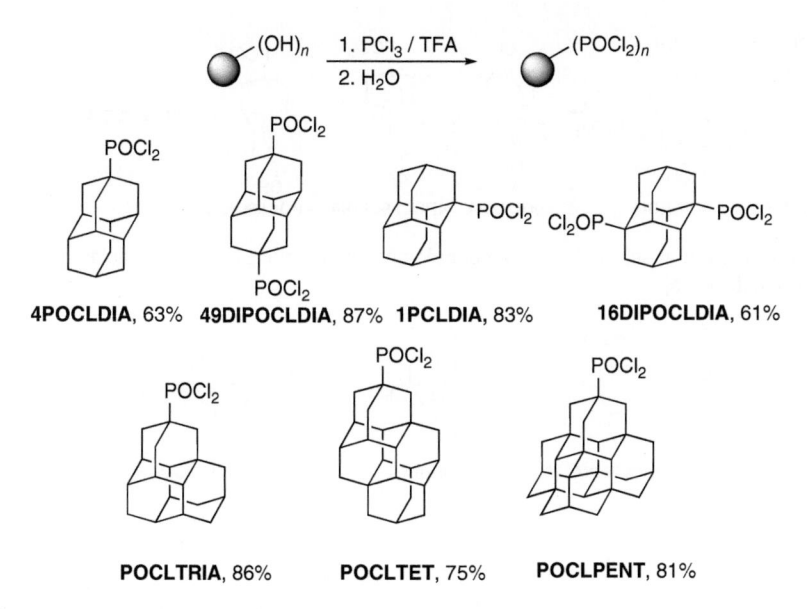

Figure 4.16 Preparation of diamondoid phosphonic acid dichlorides.

acidic conditions give phosphonic acid dichlorides in excellent yields (Figure 4.16) [128]. Since diamondoid cations are generated in the presence of a strong acid, they can be trapped with nucleophilic PCl$_3$, thus affording the phosphonic acid dichlorides in one step. Some of these compounds, such as dichlorophosphonate **16POCL2DIA**, display exceptionally high thermal stabilities and melt without decomposition at >360 °C.

Note that varying the Brønsted acid used for phosphorylation can have an effect on the reaction selectivity. For example, in trifluoroacetic acid, the phosphorylation of **49DIOHDIA** affords exclusively the bis-apical phosphonic acid dichloride **49DIPOCLDIA**, while sulfuric acid gives a separable product mixture whose exact composition depends on the reaction conditions (Scheme 4.29) [128b]. The proposed reason for the different reactivity of **49DIOHDIA** with those two acids is its good solubility in TFA when compared to its poor solubility in H$_2$SO$_4$. In sulfuric acid, the reaction therefore proceeds quite slowly, and the substituted diamantyl cation can be attacked by nucleophiles other than PCl$_3$, resulting in a variety of products giving rise to unsymmetrically substituted derivatives (**4POCL9CLDIA** and **4POCL9OHDIA**). The latter was successfully employed as a nonporous material for gas sensors [129].

The preparation of diamondoid phosphonic acid derivatives is not limited solely to dichlorides; dialkyl- or alkylarylphosphinic acid chlorides (Scheme 4.30) are obtained when using phosphorylations in protic acid media [128b] as well under Lewis acid catalysis [125]. In the latter case the bulky bis-diamondoidyl chlorophosphonates **POCLBISDIA** and **POCLBISTRIA** were obtained in good yields directly from corresponding hydrocarbons.

Subsequent C—H-bond functionalizations of diamondoid phosphonic acid dichlorides are also feasible and proceed selectively at the apical position of

110 | *4 Preparative Diamondoid Functionalizations*

Scheme 4.29 Control of selectivity in diamondoid phosphorylations in Brønsted acids. Source: From Ref. [128b].

Scheme 4.30 Preparation of dialkyl or alkylaryl phosphinic acid chlorides. Source: From Refs. [125, 128b].

4POCLDIA because the electron-withdrawing group deactivates the medial positions (Scheme 4.31) [128b]. As for the medially substituted derivative **1POCLDIA**, the C—H-bond functionalizations occur at positions most distant from the already present group in the cage, and a mixture of 1,4-(**1POCL4OHDIA**) and 1,6-(**1POCL6OHDIA**) derivatives arise. Remarkably, the POCl$_2$-group tolerates 100% nitric acid and follow-up hydrolysis of the nitrate intermediates.

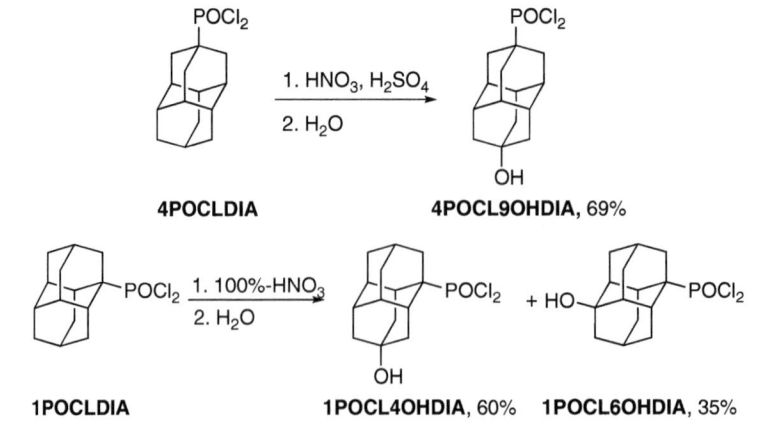

Scheme 4.31 Functionalizations of diamondoid phosphonic acid dichlorides. Source: From Ref. [128b].

4.5 Phosphorous- and Sulfur-Containing Compounds | 111

Phosphonic acid derivatives are excellent precursors for the preparation of the corresponding diamondoid phosphines. Unlike typical primary phosphines that are often pyrophoric, diamondoid phosphines show remarkable air stability, and the increase in bulkiness around the phosphorus atom provided by the diamondoid cage induces resistance toward oxidation [128b]. Primary or secondary diamondoid phosphines are accessible through the reduction of the corresponding phosphorylated derivatives either with LiAlH$_4$ or HSiCl$_3$ (Figure 4.17) [125, 128b].

Alternatively, the phosphorylation [130] of the diamantane cage is achieved through the reactions of diamantyl triflate with secondary phosphines and is applicable even for the preparation of sterically congested derivatives (Scheme 4.32).

Tertiary diamondoid phosphines are also known, either as neutrals or as their corresponding salts [125], and they find application as ligands in catalysis. The phosphorous atom can be substituted with alkyl or a combination of alkyl and aryl subunits. One curious example of a very bulky trialkylphosphine is the recently prepared

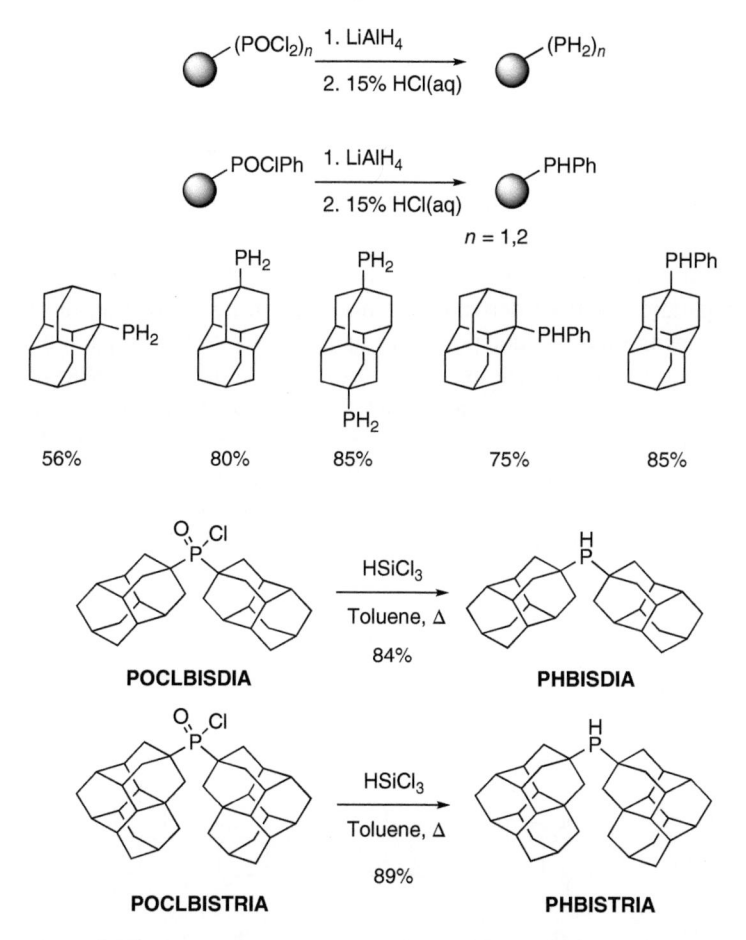

Figure 4.17 Preparation of diamondoid phosphines through reduction of the corresponding phosphorylated derivatives. Source: From Refs. [125, 128b].

Scheme 4.32 The preparation of phosphorylated diamantane derivatives through the reactions of secondary phosphines with 1-diamantyl triflate. Source: From Ref. [130].

triadamantylphosphine [131]. Despite the challenge of introducing a third sterically demanding adamantyl substituent on the central phosphorous, synthesis of **AD3P** was achieved by condensation of **AD2PH** with adamantyl acetate **1OACAD** in good overall yield (Scheme 4.33). The intermediate protonated form is neutralized with Et$_3$N that afforded **AD3P**, a stable alkylphosphine quite resistant to air oxidation.

Scheme 4.33 Preparation of triadamantyl-1-phosphine **AD3P**. Source: From Ref. [131].

The introduction of a sulfur atom into diamondoids is especially convenient through the preparation of thiol derivatives. Since the synthesis of bridgehead derivatives precludes the use of thiolation reagents applicable to the preparation of primary and secondary thiols, the use of thiourea as a sulfur transfer reagent proved to be highly advantageous. First applied to diamondoids in 1990 [112], condensation of thiourea with the corresponding bromides or alcohols readily affords the corresponding tertiary diamandoidyl thiols (Figure 4.18) [132]. Note that the use of a CH$_3$COOH/HBr mixture prevents side reactions stemming from diamondoid carbocation condensation with solvents like ethanol. The reaction of

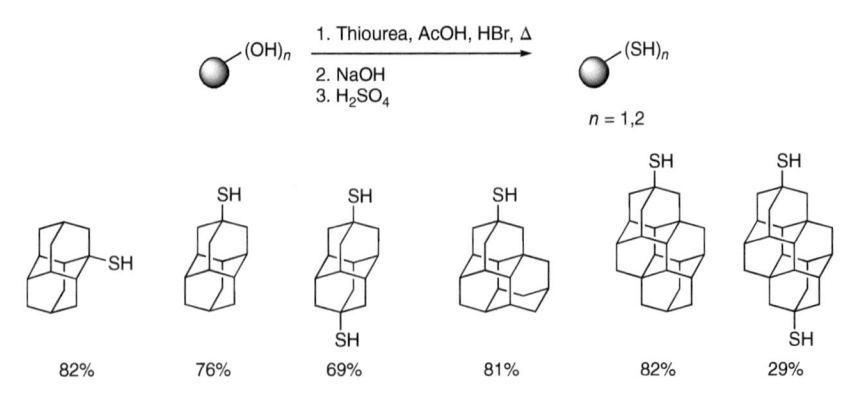

Figure 4.18 Preparation of diamondoid thiols. Source: From Ref. [132].

diamondoid alcohols with thiourea includes the formation of a thiouronium salt intermediate that is hydrolyzed under basic conditions and results in the formation of the target thiol in one step. Various applications of thiolated diamondoids are described in Chapter 5 and in a recent review article [13].

4.6 Single Electron Oxidations of Diamondoids

Another important and versatile tool for diamondoid activation is the use of single-electron transfer (SET) oxidants. Since saturated hydrocarbons possess relatively high oxidation potentials, reactions producing their radical cations (usually termed as σ-radical cations) can only proceed when strong oxidants are used. Cage hydrocarbons possess highly delocalized HOMOs, so it follows that the produced diamondoid radical cations display significant spin/charge delocalization within the cages [9] throughout the overlapping system of elongated *syn*-periplanar C—C and C—H bonds. Due to their triply degenerate HOMOs T_d-symmetric **AD** and **PENT** exhibit classic Jahn–Teller (JT) distortions [133] to C_{3v}-symmetric structures upon ionization (Figure 4.19). Such behavior is very different from the parent methane radical cation, which undergoes secondary distortions to a C_{2v}-structure [134] due to a combination of JT active e and t_2 vibrations [135] and fast tunneling [136]. In contrast, the adamantane radical cation (**ADCR**) elongates only one *tert*C—H bond [137], and this determines the direction of subsequent proton loss. As a result, very selective functionalizations of **AD** under SET-oxidations are observed in solution (vide infra) [138]. The ionization of diamondoids with nondegenerate HOMOs is accompanied by secondary distortions that lower their symmetry. The only exception is the diamantane radical cation (**DIARC**), which retains the D_{3d}-symmetry [139] of the neutral. Tetramantanes **121TET** and **123TET** belong to the nondegenerate point groups (C_{2h} and C_2, respectively) and thus retain their symmetry upon ionization [9]. In all cases, the distortion of the cages occurs in the apical directions since more bonds are involved in charge delocalization as compared to medial orientation, i.e., the apical tertiary C—H bonds undergo elongations (Figure 4.19) that ultimately lead to proton [139] or hydrogen [140] loss (depending on the media). Note that the ionization potentials become lower with increasing cage size, meaning that oxidation of higher diamondoids is increasingly more facile (vide infra).

Due to short lifetimes, the experimental structure assignments of σ-hydrocarbon radical cations are rather difficult [141]. Under EI, XUV or VUV ionization **AD** and **DIA** undergo fast fragmentation and produce various hydrocarbon shards [142]. Recent [143] time-resolved XUV ionization studies on **AD** and **DIA** show that the fragmentation of diamondoid radical cation starts with the opening of the carbon cage. While being perfectly stable in the gas phase [137c] (the barrier toward the cage fragmentation is ca. 40 kcal mol^{-1} for **ADRC**) [144] due to the weakening of one *tert*C—H bond, in the condensed state **ADRC** undergoes fast paramagnetic relaxation [145] and behaves as strong Brønsted acid [138]. The 1,3-dimethyl-analogue of **ADRC** is extremely short-lived and loses a proton even

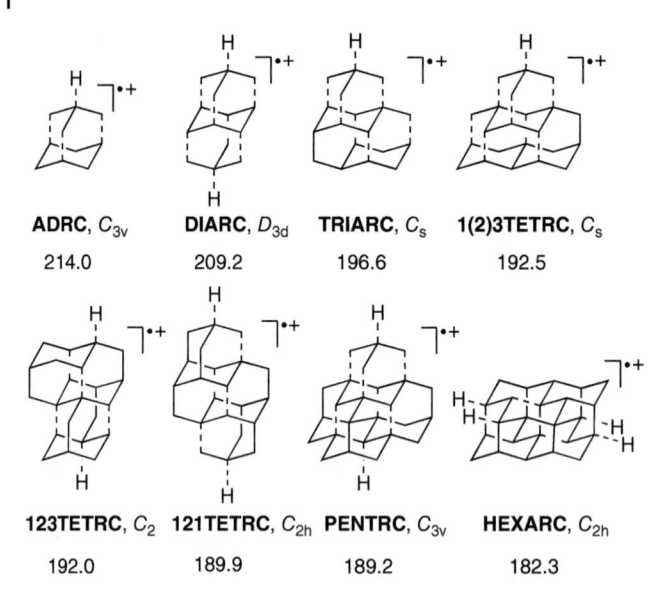

ADRC, C_{3v}	DIARC, D_{3d}	TRIARC, C_s	1(2)3TETRC, C_s
214.0	209.2	196.6	192.5

123TETRC, C_2	121TETRC, C_{2h}	PENTRC, C_{3v}	HEXARC, C_{2h}
192.0	189.9	189.2	182.3

Figure 4.19 Structures and symmetries of diamondoid radical cations and vertical ionization potentials (IP_v, kcal mol^{-1}) of the corresponding diamondoids computed at the UB3LYP/6-31G(d) level of theory. Source: Adopted from Ref. [9].

at 77 K in a freon matrix [146e]. Photoelectron [147] and chemical [148] ionization spectroscopic studies on **ADRC** generated in the gas phase display strong line broadening due to JT-distortions and vibrational coupling [149], making the experimental structure of **ADRC** elusive. Only recently, the first true experimental evidence for the e $T \otimes (e+t_2)$ type of JT-distortion in **ADRC** was provided, and the elongation of the *tert*C—H bond and formation of theoretically predicted C_{3v}-structure were confirmed: He-tagging IR photodissociation (IRPD) spectroscopy allowed to identify a weakly bound cluster of **ADRC** with helium (**ADRCHE2**) [150] and with water (**ADRCH2O**), clearly confirming the C_{3v}-symmetry of **ADRC** (Figure 4.20) [151]. This result not only provided the structure of **ADRC** but also verified the relevance of inexpensive DFT methods such as B3LYP for studying the ionized species derived from diamondoids.

An alternative way to generate **ADRC** is the ionization of **AD** by charged ("He$^+$ doped") helium droplets with [152] or without [153] extra water molecules. Recently measured absorption spectrum of **ADRC** in helium droplets [154] and computed with equation of motion (EOM) CCSD/aug-cc-pVDZ method reveal one additional to low energy optically allowed transition previously observed by Dopfer et al. [151]. Still, no vibrational substructure of the **ADRC** absorption spectrum was observed, possibly due to lifetime broadening.

In contrast to **AD**, despite also belonging to the degenerate point group, **DIA** is not JT-active as its a_{1g}-HOMO is none-degenerate and transforms into SOMO of the same topology upon ionization [9, 139]. Very recently, the structure of **DIARC** generated by electron ionization was probed experimentally [155]. The optical spectrum was obtained by electronic photodissociation (EPD) of mass-selected ions with

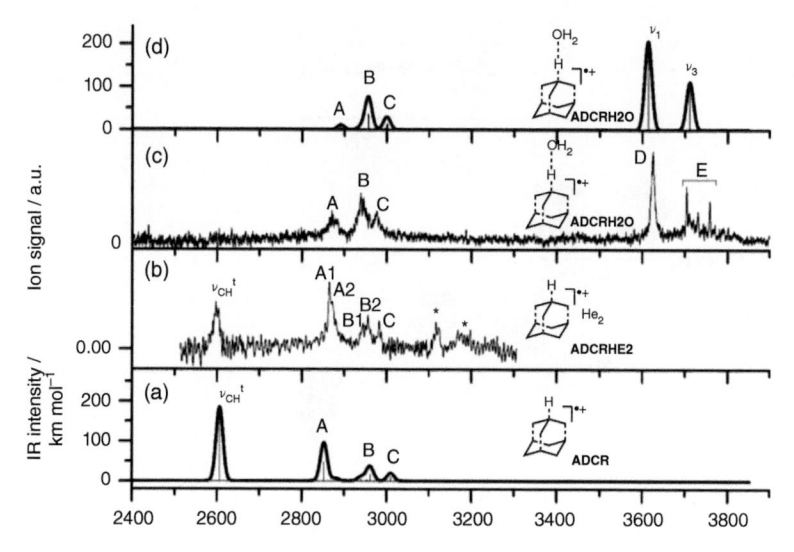

Figure 4.20 (a) B3LYP-D3/cc-pVTZ computed IR absorption spectra of the free adamantane radical cation (**ADRC**); (b) experimental IR photodissociation spectrum of a weakly bound cluster of **ADCR** with helium (**ADRCHE2**); experimental IRPD (c) and computed IR (d) spectra of the complex of adamantane radical cation with water (**ADRCH2O**). Source: Adopted from Ref. [151], Wiley-VCH, 2020.

Scheme 4.34 Reaction steps in SET oxidation of **AD**.

trapping at 5 K (Figure 4.21). The TD-DFT calculations of the optical absorption spectrum of **DIARC** were limited to the first six vertically excited electronic doublet states. The experimental EPD spectrum in the range of 400–1000 nm is in full agreement with the transitions computed for the D_{3d}-symmetric structure of **DIARC**, first suggested by us [9, 139] and later confirmed [156] by others. It should be noted, however, that even at very low temperatures, the optical spectrum of **DIARC** is poorly resolved due to the short lifetime of the excited states and vibronic coupling.

The elongation of the C—H bond in **ADCR** directly translates into its reactivity: proton loss in solution usually occurs on the position that has the most elongated bridgehead C—H bond and thereby leads to pronounced C—H activation regioselectivity (Scheme 4.34) [141]. Accordingly, the product distribution varies depending on the use of one (SET) or two-electron oxidation reactions, where, in fact, SET oxidation followed by proton loss from the radical cation is the most selective way to generate the 1-adamantyl radical **ADRAD**.

From a practical synthetic perspective, diamondoid radical cations can also be generated using photochemical oxidation or electrochemical methods. In line with the

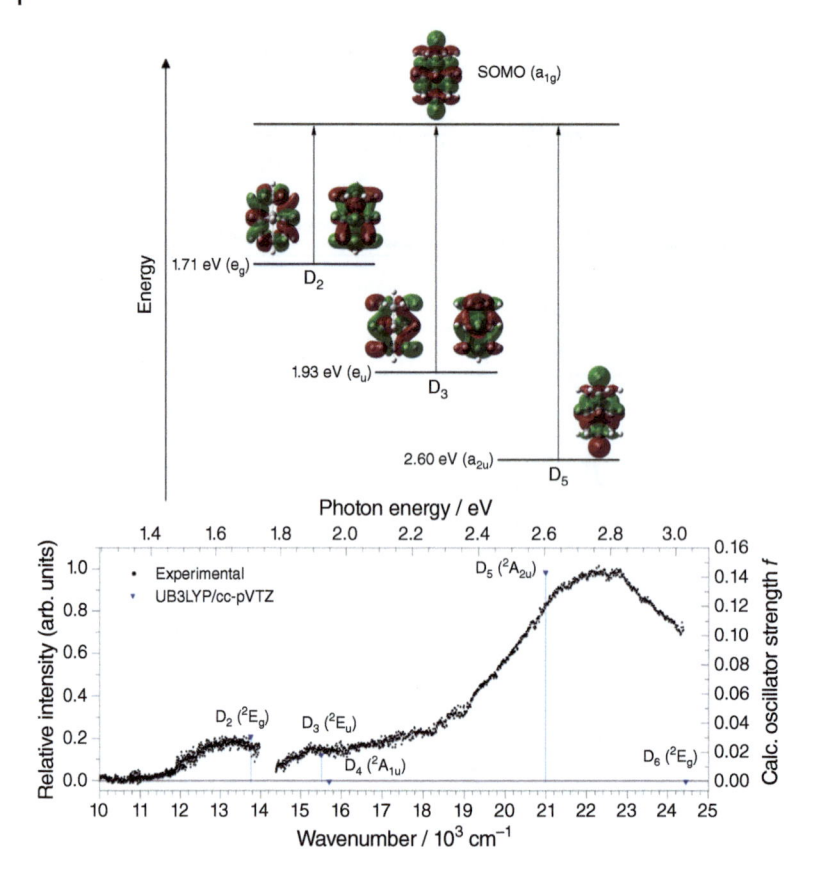

116 | 4 Preparative Diamondoid Functionalizations

Figure 4.21 Top: The DFT orbitals involved in the electron transitions in diamantane radical cation (**DIARC**), where SOMO closely corresponds to the none-degenerate a_{1g}-HOMO of the neutral **DIA**. Bottom: The experimental electronic photodissociation spectrum of **DIARC** in the range 400–1000 nm compared to the vertical transitions (in blue) computed at TD-DFT UB3LYP/cc-pVTZ level of theory. Source: Reproduced from Ref. [155] with permission from IOP Publishing, 2022.

generation and reactivity of radical cations described above, the photooxidation of diamondoids with 1,2,4,5-tetracyanobenzene (TCB) gives only apical substitution products (Figure 4.22) [141]. The first step of these reactions is a SET (Equation 1, Figure 4.22) from the hydrocarbon to TCB in the singlet excited state, which is among the most powerful organic oxidants (+3.44 V vs. SCE) [157]. Subsequent proton loss from the diamondoid radical cation makes the diamondoid radical capable of coupling with the aromatic reagent.

The substitution regioselectivities under the photoinduced SET-oxidations with TCB are in full accord with the gas-phase computed structures of the diamondoid radical cations, where the proton loss occurs from the apical cage positions (Figure 4.22).

Upon anodic oxidation, **AD** and derivatives readily react in acetonitrile at a Pt anode to give the acetamides in excellent yield (Figure 4.23) [158]. Just like in

4.6 Single Electron Oxidations of Diamondoids

Figure 4.22 SET oxidation of diamondoids in the presence of photoexcited TCB. Source: From Ref. [141].

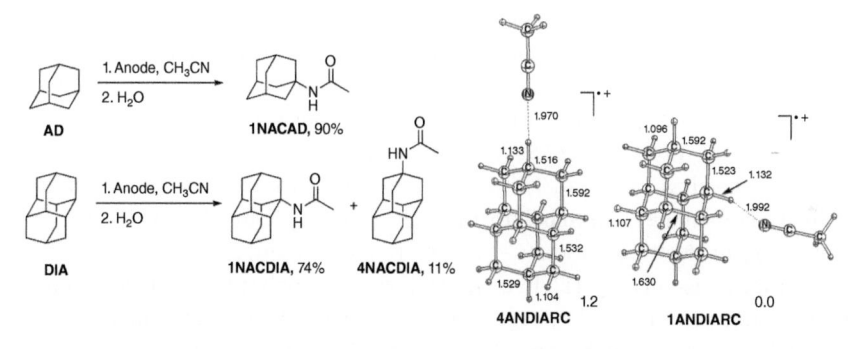

Figure 4.23 Anodic oxidation of **AD** and **DIA** and the B3LYP-D3/6–31(d,p) computed relative energies (ΔH^{298}, kcal mol^{-1}) of the complexes of the diamantane radical cation with acetonitrile at the apical (**4ANDIARC**) and medial (**1ANDIARC**) positions. Source: Adopted from Ref. [139].

the photochemical reaction, electron transfer and subsequent proton loss afford the cage radical, but then further oxidation to the adamantyl cation follows; the latter reacts with acetonitrile, i.e., a double SET process (ECEC) takes place. Note that both photoinduced SET and anodic oxidation of **AD** give only bridgehead substitution products, but in the case of higher diamondoids mixtures of apical and medial derivatives form in the anodic oxidation, as was the case for **DIA,** where both medial (**1NACDIA**) and apical (**4NACDIA**) acetamides were found [159]. This result contradicts the gas-phase computations where only one diamantane radical cation **DIARC** structure with elongated apical C—H bonds was localized (Figure 4.23). A probable reason for the predominant formation of the medial product **1NACDIA** in acetonitrile is the higher stability of the corresponding medial complex of **DIA** radical cation (**1ANDIARC**) [139] with the solvent. The apical complex **4ANDIARC** is 1.2 kcal mol^{-1} less stable at the B3LYP-D3/6–31(d,p) level.

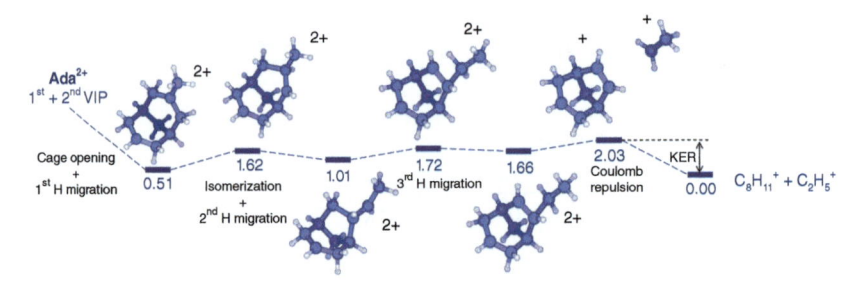

Figure 4.24 B3LYP/6-31G(d) computed decomposition of adamantane dication (**Ada**$^{2+}$) to small hydrocarbons (relative energies in eV). Source: Reproduced from Ref. [162] with permission from the Royal Society of Chemistry, 2022.

These results demonstrate the necessity of solvation in predicting the reactivity of hydrocarbons with oxidants in solution through radical cationic intermediates. The most stable gas-phase structures may not necessarily parallel the relative stabilities of the solvated species. In the case of alkyladamantanes, similar bond elongation also takes place, but the C—C bond between the alkyl substituent and the adamantane cage also becomes longer and is increasingly more prone to dissociate both in the gas phase and in condensed media as the alkyl fragment becomes larger [141, 160]. For example, in alkyladamantane radical cations, the length of the Ad–Alk bond increases in the order methyl < ethyl < isopropyl < tert-butyl adamantane (1.646, 1.720, 1.885, and 2.552 Å, respectively) [160], leading to increasingly easier loss of the alkyl fragment. This is also reflected in the experimental findings since alkyl adamantanes indeed behave differently under photoinduced SET (C—H substitution) and electrooxidation (C—C bond fragmentation) conditions, but in all cases, the cage is sustained [141]. In contrast, if electron-rich substituent is present, the ionization is accompanied by cage fragmentation as was shown for amantadine (1-aminoadamantane), which opens the cage upon ionization [161].

Very recently, the fate of adamantane dication (**Ada**$^{2+}$, Figure 4.24) produced from double-core ionization of **AD** was studied experimentally by energy-resolved electron-ion multi-coincidence spectroscopy, where base edge and on Auger decay charged products of fragmentation were identified [162]. **Ada**$^{2+}$ exothermically opens the cage without a barrier and thus forms an open dication that undergoes a set of H-shifts and fragmentations to smaller hydrocarbon cations C$_2$H$_5^+$ and C$_8$H$_{11}^+$. This reaction path differs from the one observed for the fragmentation of **ADRC**, which is more stable in gas phase, and where C$_8$H$_{12}^{·+}$ forms at the initial stages of decomposition as a result of the loss of the neutral ethylene molecule [144]. This study presents a model for the X-ray-prompted dissociation of diamondoids in space through multiple-charged species.

4.7 Other Diamondoid Derivatives

The C—C bond-forming methods are arguably one of the most important reaction classes in organic synthesis since they enable the expansion of the starting molecular

4.7 Other Diamondoid Derivatives | 119

backbone and the buildup of larger scaffolds. One example of a C—C-forming reaction on diamondoids is the straightforward Lewis acid-catalyzed arylation of **DIA** that exclusively affords **49PHDIA** in excellent yield (Scheme 4.35) [70]. The formation of the bis-apical derivative as the sole product is typical for the thermodynamically controlled transformations of **DIA** in electrophilic media, as described previously.

Scheme 4.35 Arylation of **DIA** in the presence of a Lewis acid.

Although **49PHDIA** was designed to be the backbone of a molecular rotor [70], scaffolds like these can be used as convenient linear building blocks, spacers, or polymer elements. Another class of diamondoid derivatives of practical importance in material sciences is terminal diamondoid 1,3-dienes made available by an oxetane ring opening reaction [78, 163]. The previously described diamondoid photoacetylation with diacetyl $(CH_3CO)_2$ (Figure 4.25) readily affords methyl ketones that can be used as starting materials for the preparation of 1,3-dienes. The two-step procedure includes double methylenation [164] (Yurchenko reaction) with dimethylsulfoxonium methylide, followed by acid-catalyzed ring opening of the thus formed oxetanes. Note that the corresponding homoallylic alcohols are key intermediates in going from oxetanes to 1,3-dienes [163].

In addition to the expected 1,3-diene products, cyclopropyl ketone derivatives were also isolated as side products for medial acetyl derivatives [78] since the hydrocarbon cage hinders the attack of the reagent on the carbonyl group.

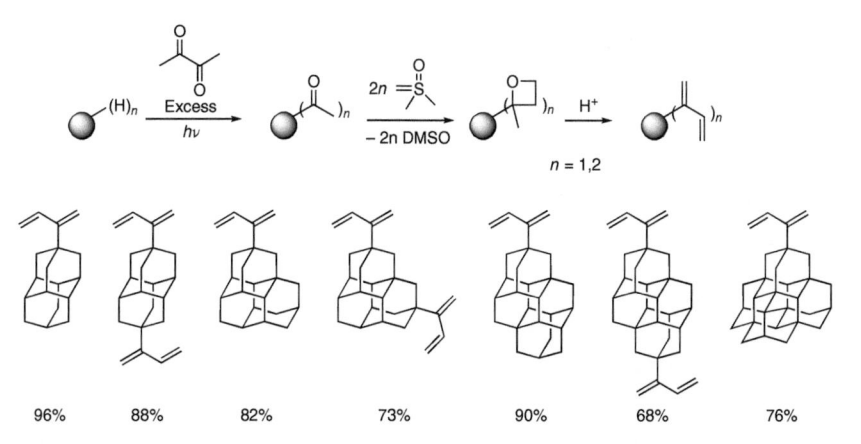

Figure 4.25 Preparation of terminal diamondoid 1,3-dienes through oxetane ring opening. Source: From Ref. [163].

4 Preparative Diamondoid Functionalizations

The synthesis of a related class of unsaturated compounds, vinyl diamondoids, has also been developed and includes the esterification of diamondoid acetic acids, followed by the reduction and conversion of the obtained alcohols to bromides. The last step in the transformation is dehydrobromination with tBuOK in either DMSO or tBuOH, giving the vinyl derivatives (Figure 4.26) [165].

Generally, the construction of the internal C—C bonds in diamondoid cages is difficult due to Bredt rule restrictions (1,2-adamantene was characterized under cryogenic matrix isolation conditions) [166]. However, partially cage-opened dienes

Figure 4.26 Preparation of vinyl diamondoids. Source: From Ref. [165].

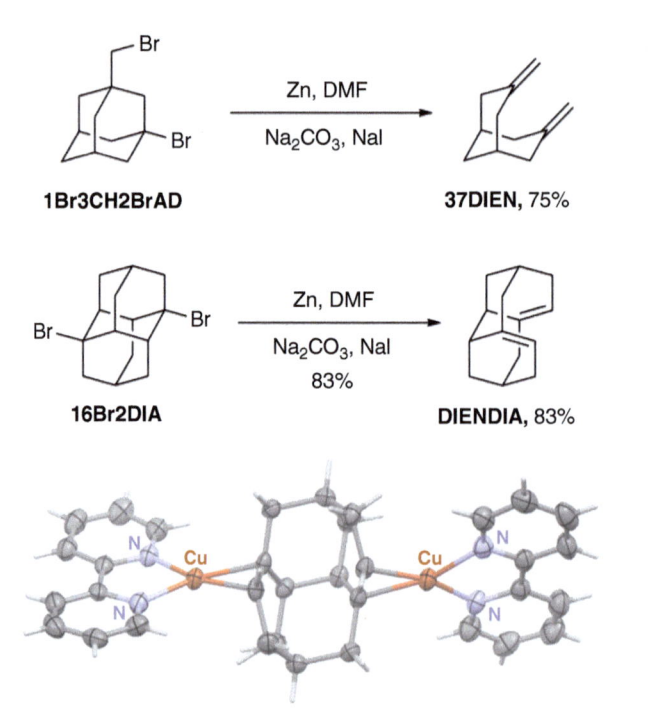

Figure 4.27 Preparation of cage-opened diamondoids and the X-ray crystal structure of [Cu$_2$(bipy)$_2$(**DIENDIA**)]$^{2+}$ complex. Source: From Ref. [169].

are stable and were prepared by reductive cleavage of the 1,4-dibromo derivatives with zinc powder in DMF. The dienes **37DIEN** [167] and **16DIENDIA** [168] have been described (Figure 4.27), and the draw of such compounds is their ability to complex with transition metals [169]. Furthermore, the formed oligomeric **16DIENDIA** structure (a polymeric copper complex) is a fascinating example of a metal/diamondoid molecular scaffold with long, rigid organic bridges between the transition metals. Oligomerization can be suppressed by adding bipyridine, and the resulting single complex formally comprises a p–n–p-conductivity junction (Figure 4.27).

Another typical C—C bond-forming technique for diamondoids is the Grignard reaction, in this case usually as a cross-coupling between a diamondoid bromide and alkyl magnesium bromides [5, 58, 170]. Some metalated diamondoids have also been prepared and isolated, including organomagnesium [171], organolithium [172], and organozinc [173] derivatives (Scheme 4.36).

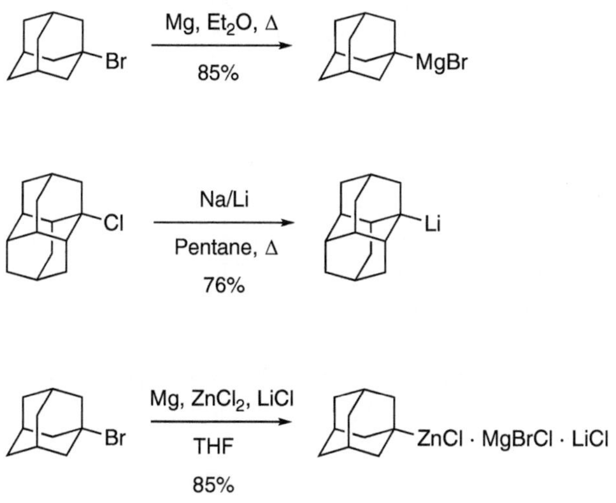

Scheme 4.36 Preparation of organometallic diamondoid derivatives.

The synthetic examples given in this chapter illustrate that the functionalization of diamondoid cages offers a wealth of derivatives applicable to vast areas of research and technology. The uniqueness of the diamondoid structures that are a consequence of the fused sp^3-carbon framework and that translates to structural rigidity, uniformity, and spatial directionality enables the use of diamondoid building blocks in nanomaterial sciences, catalysis, medicinal chemistry, supramolecular chemistry, and beyond.

References

1 Fokin, A.A. and Schreiner, P.R. (2012). Selective alkane CH bond substitutions: strategies for the preparation of functionalized diamondoids (nanodiamonds).

In: *Strategies and Tactics in Organic Synthesis*, vol. 8 (ed. M. Harmata), 317–350. Academic Press, Cambridge, Massachusetts.

2 Willey, T.M., Fabbri, J.D., Lee, J.R.I. et al. (2008). Near-edge X-ray absorption fine structure spectroscopy of diamondoid thiol monolayers on gold. *J. Am. Chem. Soc.* 130: 10536–10544.

3 Landa, S., Kriebel, S., and Knobloch, E. (1954). O adamantanu a jeho derivatech I. *Chem. List.* 48: 61–64.

4 Stetter, H. and Schwartz, E.F. (1968). Compounds with urotropine structure. 40. Ring closure reactions of bicyclo[3.3.1]nonadi-2,6-ene. *Chem. Ber.* 101: 2464–2467.

5 Gund, T.M., Nomura, M., Williams, V.Z. et al. (1970). Functionalization of diamantane (congressane). *Tetrahedron Lett.* 4875–4878.

6 (a) Hollowood, F., Karim, A., McKervey, M.A. et al. (1978). Regioselective functionalization of triamantane. *J. Chem. Soc. Chem. Commun.* 306–308. (b) Schreiner, P.R., Fokina, N.A., Tkachenko, B.A. et al. (2006). Functionalized nanodiamonds: triamantane and [121]tetramantane. *J. Org. Chem.* 71: 6709–6720.

7 Yurchenko, A.G., Kulik, N.I., Kuchar, V.P. et al. (1986). On the mechanism of liquid phase halogenation of adamantane derivatives. *Tetrahedron Lett.* 27: 1399–1402.

8 (a) Moiseev, I.K., Bagrii, E.I., Klimochkin, Y.N. et al. (1985). Adamantanol nitrates in nucleophilic substitution reactions. *Bull. Acad. Sci. USSR, Div. Chem. Sci.* 34: 1983–1985. (b) Moiseev, I.K., Bagrii, E.I., Klimochkin, Y.N. et al. (1985). Synthesis of alkyladamantanol nitrates. *Bull. Acad. Sci. USSR, Div. Chem. Sci.* 34: 1980–1982.

9 Fokin, A.A., Tkachenko, B.A., Gunchenko, P.A. et al. (2005). An experimental assessment of diamantane and computational predictions for higher diamondoids. *Chem. Eur. J.* 11: 7091–7101.

10 Fokin, A.A., Schreiner, P.R., Fokina, N.A. et al. (2006). Reactivity of [1(2,3)4] pentamantane (T_d-pentamantane): a nanoscale model of diamond. *J. Org. Chem.* 71: 8532–8540.

11 Fokin, A.A., Tkachenko, B.A., Fokina, N.A. et al. (2009). Reactivities of the prism-shaped diamondoids [1(2)3]tetramantane and [12312]hexamantane (cyclohexamantane). *Chem. Eur. J.* 15: 3851–3862.

12 Schreiner, P.R., Fokin, A.A., Reisenauer, H.P. et al. (2009). [123]Tetramantane: parent of a new family of sigma-helicenes. *J. Am. Chem. Soc.* 131: 11292–11293.

13 Yeung, K.W., Dong, Y.Q., Chen, L. et al. (2020). Nanotechnology of diamondoids for the fabrication of nanostructured systems. *Nanotechnol. Rev.* 9: 650–669.

14 Gund, T.M., Schleyer, P.v.R., Unruh, G.D., and Gleicher, G.J. (1974). Diamantane. III. Preparation and solvolysis of diamantyl bromides. *J. Org. Chem.* 39: 2995–3003.

15 (a) Schreiner, P.R., Lauenstein, O., Butova, E.D. et al. (2001). Selective radical reactions in multiphase systems: phase-transfer halogenations of alkanes. *Chem. Eur. J.* 7: 4996–5003. (b) Lauenstein, O., Fokin, A.A., and Schreiner,

P.R. (2000). Kinetic isotope effects for the C-H activation step in phase-transfer halogenations of alkanes. *Org. Lett.* 2: 2201–2204.

16 Schreiner, P.R., Lauenstein, O., Kolomitsyn, I.V. et al. (1998). Selective C-H activation of aliphatic hydrocarbons under phase-transfer conditions. *Angew. Chem. Int. Ed.* 37: 1895–1897.

17 Emery, K.J., Young, A., Arokianathar, J.N. et al. (2018). KO^{t}Bu as a single electron donor? Revisiting the halogenation of alkanes with CBr_4 and CCl_4. *Molecules* 23: 1055.

18 Schreiner, P.R., Lauenstein, O., Butova, E.D., and Fokin, A.A. (1999). The first efficient iodination of unactivated aliphatic hydrocarbons. *Angew. Chem. Int. Ed.* 38: 2786–2788.

19 (a) Saouma, C.T. and Mayer, J.M. (2014). Do spin state and spin density affect hydrogen atom transfer reactivity? *Chem. Sci.* 5: 21–31. (b) Schreiner, P.R. and Fokin, A.A. (2004). Selective alkane C-H-bond functionalizations utilizing oxidative single-electron transfer and organocatalysis. *Chem. Rec.* 3: 247–257.

20 Delimarsky, R.E., Rodionov, V.N., and Yurchenko, A.G. (1988). Preparative synthesis of 1,3,5-tribromoadamantane. *Ukr. Khim. Zh.* 54: 437–438.

21 (a) Fort, R.C. and Schleyer, P.v.R. (1964). Adamantane: consequences of the diamondoid structure. *Chem. Rev.* 64: 277–300. (b) Moiseev, I.K., Makarova, N.V., and Zemtsova, M.N. (1999). Reactions of adamantane in electrophilic media. *Russ. Chem. Bull.* 68: 1102–1121.

22 Stetter, H. and Wulff, C. (1960). Über Verbindungen mit Urotropin-Struktur, XVIII. Über die Bromierung des Adamantans. *Chem. Ber.* 93: 1366–1371.

23 Davis, M.C. and Liu, S.G. (2006). Selective apical bromination of diamantane and conversion to the dihydroxy and dicarboxylic acid derivatives. *Synth. Commun.* 36: 3509–3514.

24 Gund, T.M., Schleyer, P.v.R., and Hoogzand, C. (1971). Ionic bromination of diamantane. *Tetrahedron Lett.* 1583–1586.

25 Schreiner, P.R., Fokin, A.A., Lauenstein, O. et al. (2002). Pseudotetrahedral polyhaloadamantanes as chirality probes: synthesis, separation, and absolute configuration. *J. Am. Chem. Soc.* 124: 13348–13349.

26 Liguori, L., Bjorsvik, H.R., Bravo, A. et al. (1997). A new direct homolytic iodination reaction of alkanes by perfluoroalkyl iodides. *Chem. Commun.* 1501–1502.

27 (a) Faulkner, D., Glendinning, R.A., Johnston, D.E., and McKervey, M.A. (1971). Functionalisation reactions of diamantane. *Tetrahedron Lett.* 1671–1674. (b) McKervey, M.A., Rooney, J.J., and Johnston, D.E. (1972). Conformational preference of the chloro-substituent: equilibration of 1- and 4-chlorodiamantane. *Tetrahedron Lett.* 1547–1550.

28 (a) Blaney, F., Johnston, D.E., McKervey, M.A., and Rooney, J.J. (1975). Diamondoid rearrangements in chlorosulphonic acid. A highly regioselective route to apically disubstituted diamantanes. *Tetrahedron Lett.* 99–100. (b) Šekutor, M., Molčanov, K., Cao, L.P. et al. (2014). Design, synthesis, and X-ray structural analyses of diamantane diammonium salts: guests for cucurbit[n]uril (CB[n]) hosts. *Eur. J. Org. Chem.* 2014: 2533–2542.

29 Khusnutdinov, R.I., Shchadneva, N.A., Mayakova, Y.Y. et al. (2018). Halogenation of diamantane by halomethanes under the action of zeolites. *Russ. J. Gen. Chem.* 88: 869–873.

30 Khusnutdinov, R.I., Shchadneva, N.A., and Dzhemilev, U.M. (1991). New efficient method for the synthesis of haloadamantanes involving rhodium complexes. *Bull. Acad. Sci. USSR, Div. Chem. Sci.* 40: 2528–2529.

31 Khusnutdinov, R.I., Shchadneda, N.A., Mayakova, Y.Y. et al. (2018). Amidation of diamantane with organic nitriles and CBr$_4$ under the action of granulated zeolite FeHY. *Russ. J. Gen. Chem.* 88: 658–663.

32 Sollott, G.P. and Gilbert, E.E. (1980). A facile route to 1,3,5,7-tetraaminoadamantane. Synthesis of 1,3,5,7-tetranitroadamantane. *J. Org. Chem.* 45: 5405–5408.

33 Rezaei-Seresht, E., Rahmandoost, M., and Mahdavi, B. (2019). Green and selective iodination of diamondoid adamantane by beta-cyclodextrin as a molecular reactor. *J. Incl. Phenom. Macrocycl.* 95: 51–54.

34 Akhrem, I., Orlinkov, A., Vitt, S., and Chistyakov, A. (2002). First examples of superelectrophile initiated iodination of alkanes and cycloalkanes. *Tetrahedron Lett.* 43: 1333–1335.

35 Schwertfeger, H., Wuertele, C., Hausmann, H. et al. (2009). Selective preparation of diamondoid fluorides. *Adv. Synth. Catal.* 351: 1041–1054.

36 (a) Olah, G.A., Shih, J.G., Krishnamurthy, V.V., and Singh, B.P. (1984). Preparation and ^{13}C NMR spectroscopic study of fluoroadamantanes and fluorodiamantanes: study of ^{13}C-^{19}F NMR coupling constants. *J. Am. Chem. Soc.* 106: 4492–4500. (b) Krishnamurthy, V.V., Shih, J.G., Singh, B.P., and Olah, G.A. (1986). Unexpected formation of 1,4,7,9-tetrafluorodiamantane in the reaction of 1,4,9-tribromodiamantane with NO$_2^+$BF$_4^-$/pyridinium polyhydrogen fluoride (PPHF). *J. Org. Chem.* 51: 1354–1357. (c) Olah, G.A., Shih, J.G., Singh, B.P., and Gupta, B.G.B. (1983). Synthetic methods and reactions; 110. Fluorination of 1-haloadamantanes and -diamantane with nitronium tetrafluoroborate/pyridine polyhydrogen fluoride or sodium nitrate/pyridine polyhydrogen fluoride. *Synthesis* 713–715.

37 (a) Schreiner, P.R.; Fokin, A.A.; Wanka, L.; Wolfe, D.M., Verfahren zur Herstellung von 1-Formamido-3,5-dimethyladamantan, EP1989175A1. 2006; (b) Ivleva, E.A. and Klimochkin, Y.N. (2017). Convenient synthesis of memantine hydrochloride. *Org. Prep. Proced. Int.* 49: 155–162. (c) Vu, B.D., Ba, N.M.H., Pham, V.H., and Phan, D.C. (2020). Simple two-step procedure for the synthesis of memantine hydrochloride from 1,3-dimethyl-adamantane. *ACS Omega* 5: 16085–16088.

38 (a) Klimochkin, Y.N. and Moiseev, I.K. (1988). Kinetics of the reaction of adamantane and its derivatives with nitric acid. *Zh. Org. Khim.* 24: 557–560. (b) Klimochkin, Y.N. and Ivleva, E.A. (2021). Reaction of 1,3,5,7-tetramethyladamantane with nitric acid. *Russ. J. Org. Chem.* 57: 845–848.

39 Klimochkin, Y.N., Abramov, O.V., Moiseev, I.K. et al. (2000). Reactivity of cage hydrocarbons in the nitroxylation reaction. *Pet. Chem.* 40: 415–418.

40 Klimochkin, Y.N., Ivleva, E.A., and Zaborskaya, M.S. (2021). Synthesis of diamantane derivatives in nitric acid media. *Russ. J. Org. Chem.* 57: 186–194.

41 Clark, T., Knox, T.M., McKervey, M.A., and Mackle, H. (1980). Thermochemistry of bridged-ring substances. Enthalpies of formation of diamantan-1-, -3-, and -4-ol and of diamantanone. *J. Chem. Soc. Perkin Trans.* 2: 1686–1689.

42 Dahl, J.E.; Carlson, R.M.; Liu, S. Optical uses of diamondoid-containing materials, WO2004054047A2. 2003

43 (a) Bagrii, Y.I. and Karaulova, Y.N. (1993). Activation of C-H bonds and functionalization of hydrocarbons of the adamantane series – review. *Pet. Chem.* 33: 183–201. (b) Bagrii, E.I., Safir, R.E., and Arinicheva, Y.A. (2010). Methods of the functionalization of hydrocarbons with a diamond-like structure. *Pet. Chem.* 50: 1–16. (c) Bagrii, E.I., Nekhaev, A.I., and Maksimov, A.L. (2017). Oxidative functionalization of adamantanes (review). *Pet. Chem.* 57: 183–197. (d) Hrdina, R. (2019). Directed C-H functionalization of the adamantane framework. *Synthesis* 51: 629–642. (e) Weigel, W.K., Dang, H.T., Feceu, A., and Martin, D.B.C. (2021). Direct radical functionalization methods to access substituted adamantanes and diamondoids. *Org. Biomol. Chem.* 20: 10–36. (f) Grover, N. and Senge, M.O. (2020). Synthetic advances in the C-H activation of rigid scaffold molecules. *Synthesis* 52: 3295–3325.

44 (a) de Lozanne, A. (2008). A sludge-to-diamond story. *Nat. Mater.* 7: 10–12. (b) Marchand, A.P. (2003). Diamondoid hydrocarbons – delving into nature's bounty. *Science* 299: 52–53. (c) Marchand, A.P. (1995). Polycyclic cage compounds: reagents, substrates, and materials for the 21st century. *Aldrichim. Acta* 28: 95–104. (d) Schwertfeger, H. and Schreiner, P.R. (2010). Future of diamondoids. *Chem. Unserer Zeit* 44: 248–253. (e) Hohman, J.N., Claridge, S.A., Kim, M., and Weiss, P.S. (2010). Cage molecules for self-assembly. *Mater. Sci. Eng. Rep.* 70: 188–208. (f) Gunawan, M.A., Hierso, J.-C., Poinsot, D. et al. (2014). Diamondoids: functionalization and subsequent applications of perfectly defined molecular cage hydrocarbons. *New J. Chem.* 38: 28–41. (g) Agnew-Francis, K.A. and Williams, C.M. (2016). Catalysts containing the adamantane scaffold. *Adv. Synth. Catal.* 358: 675–700. (h) Stauss, S. and Terashima, K. (2016). *Diamondoids: Synthesis, Properties, and Applications*, 242. Jenny Stanford Publishing. (i) Seidel, S.R. and Stang, P.J. (2002). High-symmetry coordination cages via self-assembly. *Acc. Chem. Res.* 35: 972–983. (j) Cook, T.R., Zheng, Y.R., and Stang, P.J. (2013). Metal-organic frameworks and self-assembled supramolecular coordination complexes: comparing and contrasting the design, synthesis, and functionality of metal-organic materials. *Chem. Rev.* 113: 734–777.

45 Schwertfeger, H., Fokin, A.A., and Schreiner, P.R. (2008). Diamonds are a chemist's best friend: diamondoid chemistry beyond adamantane. *Angew. Chem. Int. Ed.* 47: 1022–1036.

46 (a) Geluk, H.W. and Schlatmann, J.L. (1971). Hydride transfer reactions of the adamantyl cation (IV): synthesis of 1,4- and 2,6-substituted adamantanes by oxidation with sulfuric acid. *Rec. Trav. Chim. Pays Bas* 90: 516–520. (b) Geluk,

H.W. and Schlatmann, J.L. (1967). A convenient synthesis of adamantanone. *Chem. Commun.* 426.

47 Fokina, N.A., Tkachenko, B.A., Merz, A. et al. (2007). Hydroxy derivatives of diamantane, triamantane, and 121 tetramantane: selective preparation of bis-apical derivatives. *Eur. J. Org. Chem.* 4738–4745.

48 Bach, R.D., Taaffee, T.H., and Holubka, J.W. (1980). Reaction of saturated organic compounds with acetyl and trifluoroacetyl nitrate. *J. Org. Chem.* 45: 3439–3442.

49 Bach, R.D., Holubka, J.W., Badger, R.C., and Rajan, S.J. (1979). Mechanism of the oxidation of alkanes with nitronium tetrafluoroborate in acetonitrile. Evidence for a carbenium ion intermediate. *J. Am. Chem. Soc.* 101: 4416–4417.

50 Moiseev, I.K., Belyaev, P.G., Barabanov, N.V. et al. (1975). Synthesis of adamantane nitrates – new method for preparation of 1-hydroxy adamantane and 1,3-dihydroxyadamantane. *Zh. Org. Khim.* 11: 214–215.

51 Klimochkin, Y.N., Yudashkin, A.V., Zhilkina, E.O. et al. (2017). One-pot synthesis of cage alcohols. *Russ. J. Org. Chem.* 53: 971–976.

52 (a) Vodička, L., Janku, J., and Burkhard, J. (1978). Bromination of adamantane in presence of nitric acid. *Collect. Czechoslov. Chem. Commun.* 43: 1410–1412. (b) Olah, G.A., Ramaiah, P., Rao, C.B. et al. (1993). Electrophilic reactions at single bonds. 25. Nitration of adamantane and diamantane with nitronium tetrafluoroborate. *J. Am. Chem. Soc.* 115: 7246–7249. (c) Fokin, A.A., Zhuk, T.S., Pashenko, A.E. et al. (2014). Functionalization of homodiamantane: oxygen insertion reactions without rearrangement with dimethyldioxirane. *J. Org. Chem.* 79: 1861–1866.

53 (a) Johnston, D.E., Rooney, J.J., and McKervey, M.A. (1972). Equilibration of diamantan-1-ol and diamantan-4-ol. Conformational enthalpy of the hydroxy-group, and an unusual example of how entropy and symmetry factors can influence relative thermodynamic stabilities. *J. Chem. Soc. Chem. Commun.* 29–30. (b) Courtney, T., Johnston, D.E., Rooney, J.J., and McKervey, M.A. (1972). The chemistry of diamantane. Part I. Synthesis and some functionalisation reactions. *J. Chem. Soc. Dalton Trans.* 1: 2691–2696.

54 McFord, A.W., Butts, C.P., Fey, N., and Alder, R.W. (2021). 3× axial vs 3× equatorial: the ΔG_{GA} value is a robust computational measure of substituent steric effects. *J. Am. Chem. Soc.* 143: 13573–13578.

55 Agarwal, N. L.; Mistri, P. P.; Patel, N. M. An improved process for the preparation of memantine hydrochloride, EP2882291B1. 2015.

56 Schwertfeger, H., Würtele, C., Serafin, M. et al. (2008). Monoprotection of diols as a key step for the selective synthesis of unequally disubstituted diamondoids (nanodiamonds). *J. Org. Chem.* 73: 7789–7792.

57 Kahl, P., Tkachenko, B.A., Novikovsky, A.A. et al. (2014). Efficient preparation of apically substituted diamondoid derivatives. *Synthesis* 46: 787–798.

58 Gund, T.M., Nomura, and M., Schleyer, P.v.R. Diamantane. II. (1974). Preparation of derivatives of diamantane. *J. Org. Chem.* 39: 2987–2994.

59 Sandhiya, L., Jangra, H., and Zipse, H. (2020). Molecule-induced radical formation (MIRF) reactions – a reappraisal. *Angew. Chem. Int. Ed.* 59: 6318–6329.

References | 127

60 Ruchardt, C., Gerst, M., and Ebenhoch, J. (1997). Uncatalyzed transfer hydrogenation and transfer hydrogenolysis: two novel types of hydrogen-transfer reactions. *Angew. Chem. Int. Ed.* 36: 1407–1430.

61 Bach, R.D. (2016). The DMDO hydroxylation of hydrocarbons via the oxygen rebound mechanism. *J. Phys. Chem. A* 120: 840–850.

62 Fokin, A.A., Tkachenko, B.A., Korshunov, O.I. et al. (2001). Molecule-induced alkane homolysis with dioxiranes. *J. Am. Chem. Soc.* 123: 11248–11252.

63 (a) Scott, W.B. and Pincock, R.E. (1973). Compounds containing inverted carbon atoms. Synthesis and reactions of some 5-substituted 1,3-dehydroadamantanes. *J. Am. Chem. Soc.* 95: 2040–2041. (b) Pincock, R.E., Schmidt, J., Scott, W.B., and Torupka, E.J. (1972). Synthesis and reactions of strained hydrocarbons possessing inverted carbon atoms. Tetracyclo[3.3.1.$1^{3.7}.0^{1.3}$]decanes. *Can. J. Chem.* 50: 3958–3964.

64 (a) Fokin, A.A., Gunchenko, P.A., Yaroshinskii, A.I. et al. (1995). Oxidation of tetracyclo[4.3.1.$1^{4,8}.0^{1,4}$]undecane (1,4-cyclohomoadamantane). *Zh. Org. Khim.* 31: 796–797. (b) Fokin, A.A., Gunchenko, P.A., Yaroshinsky, A.I. et al. (1995). Oxidative addition to 3,6-dehydrohomoadamantane. *Tetrahedron Lett.* 36: 4479–4482. (c) Fokin, A.A., Gunchenko, P.A., Kulik, N.I. et al. (1996). NO_2^+-containing reagents in the electrophilic and oxidative addition to propellanic C-C bond. *Tetrahedron* 52: 5857–5866.

65 Mello, R., Cassidei, L., Fiorentino, M. et al. (1990). Oxidations by methyl(trifluoromethyl)dioxirane. 3. Selective polyoxyfunctionalization of adamantane. *Tetrahedron Lett.* 31: 3067–3070.

66 Tabushi, I., Aoyama, Y., Takahash, N. et al. (1973). Dichloromethyldiamantanes, homodiamantanones, and homodiamantane. *Tetrahedron Lett.* 107–110.

67 (a) Jones, S.R. and Mellor, J.M. (1976). Bridgehead functionalisation of saturated hydrocarbons with lead(IV) salts. *J. Chem. Soc. Perkin Trans.* 1: 2576–2581. (b) Jones, S.R. and Mellor, J.M. (1977). Mechanism of oxidation of saturated hydrocarbons by cobalt(III), manganese(III), and lead(IV) trifluoroacetates. *J. Chem. Soc. Perkin Trans.* 2: 511–517.

68 Neumann, R., Khenkin, A.M., and Dahan, M. (1995). Hydroxylation of alkanes with molecular oxygen catalyzed by a new ruthenium-substituted polyoxometalate, $[WZnRu_2^{III}(OH)(H_2O)(ZnW_9O_{34})_2]^{11-}$. *Angew. Chem., Int. Ed.* 34: 1587–1589.

69 Olah, G.A., Farooq, O., Wang, Q., and Wu, A.H. (1990). Synthetic methods and reactions. 147. $AlCl_3$-catalyzed dichlorophosphorylation of saturated hydrocarbons with PCl_3 in methylene chloride solution. *J. Org. Chem.* 55: 1224–1227.

70 Karlen, S.D., Ortiz, R., Chapman, O.L., and Garcia-Garibay, M.A. (2005). Effects of rotational symmetry order on the solid state dynamics of phenylene and diamantane rotators. *J. Am. Chem. Soc.* 127: 6554–6555.

71 Khusnutdinov, R.I., Shchadneva, N.A., Khisamova, L.F. et al. (2011). Amidation of adamantane and diamantane with acetonitrile and bromotrichloromethane in the presence of $Mo(CO)_6$ in aqueous medium. *Russ. J. Org. Chem.* 47: 1898–1900.

72 Khusnutdinov, R.I., Oshnyakova, T.M., Khalilov, L.M. et al. (2016). Selective hydroxylation of diamantane with 2,3,4,5,6-pentafluoroperbenzoic acid in the presence of molibdenum complexes. *Russ. J. Org. Chem.* 52: 1121–1125.

73 (a) Fokin, A.A., Gunchenko, P.A., Novikovsky, A.A. et al. (2009). Photoacetylation of diamondoids: selectivities and mechanism. *Eur. J. Org. Chem.* 5153–5161. (b) Tabushi, I., Kojo, S., Schleyer, P.v.R., and Gund, T.M. (1974). Selective functionalization of unactivated methine positions. 4-Acetyldiamantane. *J. Chem. Soc. Chem. Commun.* 591–591.

74 Chang, L., Wang, S., An, Q. et al. (2023). Resurgence and advancement of photochemical hydrogen atom transfer processes in selective alkane functionalizations. *Chem. Sci.* 14: 6841–6859.

75 Burtoloso, A.C.B., Dias, R.M.P., and Leonarczyk, I.A. (2013). Sulfoxonium and sulfonium ylides as diazocarbonyl equivalents in metal-catalyzed insertion reactions. *Eur. J. Org. Chem.* 2013: 5005–5016.

76 (a) Yurchenko, A.G., Kiriy, A.V., Likhotvorik, I.P. et al. (1992). New reactions of dimethylsulfoxonium methylide and dimsylsodium. *Ukr. Khim. Zh.* 58: 1106–1116. (b) Butova, E.D., Fokin, A.A., and Schreiner, P.R. (2007). Beyond the Corey reaction: one-step diolefination of cyclic ketones. *J. Org. Chem.* 72: 5689–5696.

77 Barabash, A.V., Didukh, N.A., Kibal'nyi, N.A. et al. (2014). Unusual aerobic oxidation of sterically hindered 1-diamantyl methyl ketone. *Russ. J. Org. Chem.* 50: 1690–1691.

78 Barabash, A.V., Butova, E.D., Kanyuk, I.M. et al. (2014). Beyond the Corey reaction II: dimethylenation of sterically congested ketones. *J. Org. Chem.* 79: 10669–10673.

79 Fokin, A.A., Pashenko, A.E., Bakhonsky, V.V. et al. (2017). Chiral building blocks based on 1,2-disubstituted diamantanes. *Synthesis* 49: 2003–2008.

80 Zhang, W.C. and Li, C.J. (2000). A direct retro-Barbier fragmentation. *J. Org. Chem.* 65: 5831–5833.

81 Bakhonsky, V.V., Pashenko, A.A., Becker, J. et al. (2020). Synthesis and antiproliferative activity of hindered, chiral 1,2-diaminodiamantane platinum(II) complexes. *Dalton Trans.* 49: 14009–14016.

82 Becker, J., Zhyhadlo, Y.Y., Butova, E.D. et al. (2018). Aerobic aliphatic hydroxylation reactions by copper complexes: a simple clip-and-cleave concept. *Chem. Eur. J.* 24: 15543–15549.

83 Schleyer, P.v.R., Lam, L.K.M., Raber, D.J. et al. (1970). Stereochemical inhibition of intramolecular 1,2-shifts. The intermolecular nature of hydride shifts in the adamantane series. *J. Am. Chem. Soc.* 92: 5246–5247.

84 (a) Geluk, H.W. and Keizer, V.G. (1973). Adamantanone. *Org. Synth.* 53: 8. (b) Geluk, H.W. and Keizer, V.G. (1988). Adamantanone. *Org. Synth.* 50-9: 48–50.

85 (a) Janku, J., Burkhard, J., and Vodička, L. (1981). Darstellung einiger 3-substituierter Diamantanderivate. *Z. Chem.* 21: 67–68. (b) Fokin, A.A., Zhuk, T.S., Pashenko, A.E. et al. (2009). Oxygen-doped nanodiamonds: synthesis and functionalizations. *Org. Lett.* 11: 3068–3071.

86 (a) Novoselov, E.F., Isaev, S.D., Yurchenko, A.G. et al. (1981). The reaction of ketones with hydroxylamine-O-sulfonic acid. *Zh. Org. Khim.* 17: 2558–2564. (b) Tinant, B., Declercq, J.P., Vanmeerssche, M. et al. (1986). Synthesis and identification of vicinal azaoxo-2(3)-homodiamantanes. *Bull. Soc. Chim. Belg.* 95: 619–630. (c) Olah, G.A., Wu, A.H., and Farooq, O. (1988). Preparation of 2-aryladamantanes and 3-aryldiamantanes by improved ionic hydrogenation of the corresponding tertiary alcohols with sodium borohydride-triflic acid or formic acid-triflic acid. *J. Org. Chem.* 53: 5143–5145. (d) Olah, G.A., Wu, A.H., and Farooq, O. (1989). Synthetic methods and reactions. 140. One-pot preparation of crowded olefins from hindered ketones with alkyllithiums and thionyl chloride. *J. Org. Chem.* 54: 1375–1378. (e) Yanku, I., Golovanyuk, A.N., Tsybulskii, A.V. et al. (1990). Synthesis of 10- and 11-oxa-2(3)-homodiamantane. *Zh. Org. Khim.* 26: 1366–1368. (f) Hosseini, A. and Schreiner, P.R. (2020). Direct exploitation of the ethynyl moiety in calcium carbide through sealed ball milling. *Eur. J. Org. Chem.* 2020: 4339–4346. (g) Khusnutdinov, R.I., Shchadneva, N.A., Mayakova, Y.Y., and Aminov, R.I. (2021). Condensation of diamantan-3-one with malononitrile and methyl and ethyl cyanoacetates in the presence of binder-free FeHY and NiHy zeolites. *Russ. J. Org. Chem.* 57: 950–953.

87 Vodička, L., Isaev, S.D., Burkhard, J., and Janku, J. (1984). Synthesis and reactions of hydroxydiamantanones. *Collect. Czechoslov. Chem. Commun.* 49: 1900–1906.

88 Zefirov, N.S., Averina, N.V., and Fomicheva, O.A. (1994). Bromination and oxidation of 2-oxaadamantane. *Khim. Geterotsikl. Soedin.* 608–612.

89 Alder, R.W., Carta, F., Reed, C.A. et al. (2010). Searching for intermediates in Prins cyclisations: the 2-oxa-5-adamantyl carbocation. *Org. Biomol. Chem.* 8: 1551–1559.

90 Gunchenko, P.A., Li, J., Liu, B.F. et al. (2018). Aerobic oxidations with N-hydroxyphthalimide in trifluoroacetic acid. *Mol. Catal.* 447: 72–79.

91 Recupero, F. and Punta, C. (2007). Free radical functionalization of organic compounds catalyzed by N-hydroxyphthalimide. *Chem. Rev.* 107: 3800–3842.

92 Stetter, H., Schwarz, M., and Hirschhorn, A. (1959). Über Verbindungen mit Urotropin-Struktur, XII. Monofunktionelle Adamantan-Derivate. *Chem. Ber.* 92: 1629–1635.

93 Raber, D.J., Fort, R.C., Wiskott, E. et al. (1971). Rearrangements in the adamantane series. Hydride shifts of the 2-(1-adamantyl)-2-propyl cation. *Tetrahedron* 27: 3–18.

94 Vodička, L., Janku, J., and Burkhard, J. (1983). Synthesis of diamantanedicarboxylic acids with the carboxy groups bonded at tertiary carbon atoms. *Collect. Czechoslov. Chem. Commun.* 48: 1162–1172.

95 Fokina, N.A., Tkachenko, B.A., Dahl, J.E.P. et al. (2012). Synthesis of diamondoid carboxylic acids. *Synthesis* 44: 259–264.

96 Bott, K. and Hellmann, H. (1966). Syntheses of carboxylic acids from 1,1-dichloroethylene. *Angew. Chem. Int. Ed.* 5: 870–874.

97 (a) Davis, R. and Untch, K.G. (1981). Direct one-step conversion of alcohols into nitriles. *J. Org. Chem.* 46: 2985–2987. (b) Barth, B.E.K., Tkachenko, B.A., Eussner, J.P. et al. (2014). Diamondoid hydrazones and hydrazides: sterically demanding ligands for Sn/S cluster design. *Organometallics* 33: 1678–1688.

98 Mlinarić-Majerski, K., Margeta, R., and Veljković, J. (2005). A facile and efficient one-pot synthesis of nitriles from carboxylic acids. *Synlett* 2089–2091.

99 Olah, G.A., Farooq, O., and Prakash, G.K.S. (1985). Synthetic methods and reactions. 116. Lewis acid-catalyzed preparation of bridge-head adamantanoid nitriles from their corresponding halides and trimethylsilyl cyanide. *Synthesis* 1140–1142.

100 Berndt, J.P., Erb, F.R., Ochmann, L. et al. (2019). Selective phthalimido-N-oxyl (PINO)-catalyzed C-H cyanation of adamantane derivatives. *Synlett* 30: 493–498.

101 Grillaud, M. and Bianco, A. (2015). Multifunctional adamantane derivatives as new scaffolds for the multipresentation of bioactive peptides. *J. Pept. Sci.* 21: 330–345.

102 Nagasawa, H.T., Elberling, J.A., and Shirota, F.N. (1973). 2-Aminoadamantane-2-carboxylic acid, a rigid, achiral, tricyclic alpha-amino acid with transport inhibitory properties. *J. Med. Chem.* 16: 823–826.

103 Wanka, L., Cabrele, C., Vanejews, M., and Schreiner, P.R. (2007). γ-Aminoadamantanecarboxylic acids through direct C-H bond amidations. *Eur. J. Org. Chem.* 1474–1490.

104 (a) Horvat, Š., Mlinarić-Majerski, K., Glavaš-Obrovac, L. et al. (2006). Tumor-cell-targeted methionine-enkephalin analogues containing unnatural amino acids: design, synthesis, and in vitro antitumor activity. *J. Med. Chem.* 49: 3136–3142. (b) Müller, J., Kirschner, R.A., Berndt, J.P. et al. (2019). Diamondoid amino acid-based peptide kinase A inhibitor analogues. *ChemMedChem* 14: 663–672.

105 (a) Müller, C.E., Wanka, L., Jewell, K., and Schreiner, P.R. (2008). Enantioselective kinetic resolution of trans-cycloalkane-1,2-diols. *Angew. Chem. Int. Ed.* 47: 6180–6183. (b) Hrdina, R., Müller, C.E., and Schreiner, P.R. (2010). Kinetic resolution of trans-cycloalkane-1,2-diols via Steglich esterification. *Chem. Commun.* 46: 2689–2690.

106 Stetter, H., Mayer, J., Schwarz, M., and Wulff, K. (1960). Über Verbindungen mit Urotropin-Struktur, XVI. Beiträge zur Chemie der adamantyl-(1)-derivate. *Chem. Ber. Recl.* 93: 226–230.

107 (a) Blaney, F., Johnston, D.E., McKervey, M.A. et al. (1974). Hydroxylation of diamantan-1-ol and diamantan-4-ol with fungus *Rhizopus nigricans*. *J. Chem. Soc., Chem. Commun.* 297–298. (b) Chern, Y.T. and Wang, J.J. (1995). Synthesis of 1,6-diaminodiamantane. *Tetrahedron Lett.* 36: 5805–5806.

108 Fokin, A.A., Merz, A., Fokina, N.A. et al. (2009). Synthetic routes to aminotriamantanes, topological analogues of the neuroprotector memantine. *Synthesis* 909–912.

109 Jirgensons, A., Kauss, V., Kalvinsh, I., and Gold, M.R. A practical synthesis of tert-alkylamines via the Ritter reaction with chloroacetonitrile. *Synthesis* 2000: 1709–1712.

110 (a) Prakash, G.K.S., Stephenson, M.A., Shih, J.G., and Olah, G.A. (1986). Preparation of secondary and tertiary cyclic and polycyclic hydrocarbon azides. *J. Org. Chem.* 51: 3215–3217. (b) Davis, M.C. and Nissan, D.A. (2006). Preparation of diamines of adamantane and diamantane from the diazides. *Synth. Commun.* 36: 2113–2119.

111 Codony, S., Valverde, E., Leiva, R. et al. (2019). Exploring the size of the lipophilic unit of the soluble epoxide hydrolase inhibitors. *Bioorg. Med. Chem.* 27: 115078.

112 Cahill, P.A. (1990). Unusual degree of selectivity in diamantane derivatizations. *Tetrahedron Lett.* 31: 5417–5420.

113 (a) Archibald, T.G. and Baum, K. (1988). Synthesis of polynitroadamantanes. Oxidations of oximinoadamantanes. *J. Org. Chem.* 53: 4645–4649. (b) Glaser, R., Steinberg, A., Šekutor, M. et al. (2011). Stereochemistry of 2,6-diaminoadamantane salts: transannular interactions. *Eur. J. Org. Chem.* 2011: 3500–3506.

114 Ochmann, L., Kessler, M.L., and Schreiner, P.R. (2022). Alkylphosphinites as synthons for stabilized carbocations. *Org. Lett.* 24: 1460–1464.

115 Schwertfeger, H., Wuertele, C., and Schreiner, P.R. (2010). Synthesis of diamondoid nitro compounds from amines with *m*-chloroperbenzoic acid. *Synlett* 493–495.

116 Davis, M.C., Dahl, J.E.P., and Carlson, R.M.K. (2008). Preparation of diisocyanates of adamantane and diamantane. *Synth. Commun.* 38: 1153–1158.

117 Hrdina, R., Metz, F.M., Larrosta, M. et al. (2015). Intramolecular C-H amination reaction provides direct access to 1,2-disubstituted diamondoids. *Eur. J. Org. Chem.* 6231–6236.

118 Larrosa, M., Zonker, B., Volkmann, J. et al. (2018). Directed C-H bond oxidation of bridged cycloalkanes catalyzed by palladium(II) acetate. *Chem. Eur. J.* 24: 6269–6276.

119 Zonker, B., Becker, J., and Hrdina, R. (2021). Synthesis of noradamantane derivatives by ring-contraction of the adamantane framework. *Org. Biomol. Chem.* 19: 4027–4031.

120 Bonsir, M., Kennedy, A.R., and Geerts, Y. (2022). Synthesis and structural properties of adamantane-substituted amines and amides containing an additional adamantane, azaadamantane or diamantane moiety. *ChemistryOpen* e202200031.

121 Meng, G.R., Zhang, J., and Szostak, M. (2021). Acyclic twisted amides. *Chem. Rev.* 121: 12746–12783.

122 (a) Kirby, A.J., Komarov, I.V., Wothers, P.D., and Feeder, N. (1998). The most twisted amide: structure and reactions. *Angew. Chem. Int. Ed.* 37: 785–786. (b) Komarov, I.V. (2001). Organic molecules with abnormal geometric parameters. *Usp. Khim.* 70: 1123–1151.

123 Winkler, F.K. and Dunitz, J.D. (1971). Non-planar amide group. *J. Mol. Biol.* 59: 169–182.

124 (a) Clay, J.P. (1951). A new method for the preparation of alkane phosphonyl dichlorides. *J. Org. Chem.* 16: 892–894. (b) Kinnear, A.M. and Perren, E.A.

(1952). Formation of organo-phosphorus compounds by the reaction of alkyl chlorides with phosphorus trichloride in the presence of aluminium chloride. *J. Chem. Soc.* 3437–3445.

125 Schwertfeger, H., Machuy, M.M., Würtele, C. et al. (2010). Diamondoid phosphines – selective phosphorylation of nanodiamonds. *Adv. Synth. Catal.* 352: 609–615.

126 Freedman, L.D. and Doak, G.O. (1957). The preparation and properties of phosphonic acids. *Chem. Rev.* 57: 479–523.

127 (a) Yurchenko, R.I., Peresypkina, L.P., and Fokina, N.A. (1992). 2-Hydroxyadamantane reaction with phosphoric acid chloride and phosphonic acid chloride in sulfuric acid. *Russ. J. Gen. Chem.* 62: 1912–1912. (b) Yurchenko, R.I. and Peresypkina, L.P. (1994). Phosphorilation of adamantane by chlorophosphites in sulfuric acid. *Russ. J. Gen. Chem.* 64: 1564–1564. (c) Erokhina, E.V., Shokova, E.A., Luzikov, Y.N., and Kovalev, V.V. (1995). Dichlorophosphrylation of adamantanols and 1-adamantylcarbinols in trifluoroacetic acid. *Synthesis* 851–854. (d) Shokova, E.A., Erokhina, E.V., and Kovalev, V.V. (1996). Phosphorylation of adamantane and its derivatives in trifluoroacetic acid. *Russ. J. Gen. Chem.* 66: 1666–1677.

128 (a) Fokin, A.A., Yurchenko, R.I., Tkachenko, B.A. et al. (2014). Selective preparation of diamondoid phosphonates. *J. Org. Chem.* 79: 5369–5373. (b) Moncea, O., Gunawan, M.A., Poinsot, D. et al. (2016). Defying stereotypes with nanodiamonds: stable primary diamondoid phosphines. *J. Org. Chem.* 81: 8759–8769.

129 Gunawan, M.A., Poinsot, D., Domenichini, B. et al. (2015). The functionalization of nanodiamonds (diamondoids) as a key parameter of their easily controlled self-assembly in micro- and nanocrystals from the vapor phase. *Nanoscale* 7: 1956–1962.

130 Prabagar, J., Cowley, A.R., and Brown, J.M. (2011). Electrophilic routes to tertiary adamantyl and diamantyl phosphonium salts. *Synlett* 2351–2354.

131 (a) Chen, L.Y., Ren, P., and Carrow, B.P. (2016). Tri(1-adamantyl)phosphine: expanding the boundary of electron-releasing character available to organophosphorus compounds. *J. Am. Chem. Soc.* 138: 6392–6395. (b) Carrow, B.P. and Chen, L. (2017). Tri(1-adamantyl) phosphine: exceptional catalytic effects enabled by the synergy of chemical stability, donicity, and polarizability. *Synlett* 28: 280–288.

132 Tkachenko, B.A., Fokina, N.A., Chernish, L.V. et al. (2006). Functionalized nanodiamonds part 3: thiolation of tertiary bridgehead alcohols. *Org. Lett.* 8: 1767–1770.

133 (a) Bersuker, I.B. (2013). Pseudo-Jahn-Teller effect – a two-state paradigm in formation, deformation, and transformation of molecular systems and solids. *Chem. Rev.* 113: 1351–1390. (b) Bersuker, I.B. (2001). Modern aspects of the Jahn-Teller effect theory and applications to molecular problems. *Chem. Rev.* 101: 1067–1114.

134 (a) Knight, L.B., King, G.M., Petty, J.T. et al. (1995). Electron spin resonance studies of the methane radical cations ($^{12,13}CH_4^+$, $^{12,13}CDH_3^+$, $^{12}CD_2H_2^+$, $^{12}CD_3H^+$, $^{12}CD_4^+$) in solid neon matrices between 2.5 and 11 K: analysis of

tunneling. *J. Chem. Phys.* 103: 3377–3385. (b) Wörner, H.J. and Merkt, F. (2009). Jahn-Teller effects in molecular cations studied by photoelectron spectroscopy and group theory. *Angew. Chem. Int. Ed.* 48: 6404–6424.

135 (a) Mondal, T. (2016). Origin of distinct structural symmetry of the neopentane cation in the ground electronic state compared to the methane cation. *Phys. Chem. Chem. Phys.* 18: 10459–10472. (b) Jacovella, U., Wörner, H.J., and Merkt, F. (2018). Jahn-Teller effect and large-amplitude motion in CH_4^+ studied by high-resolution photoelectron spectroscopy of CH_4. *J. Mol. Spectrosc.* 343: 62–75.

136 (a) Wörner, H.J., Qian, X., and Merkt, F. (2007). Jahn-Teller effect in tetrahedral symmetry: large-amplitude tunneling motion and rovibronic structure of CH_4^+ and CD_4^+. *J. Chem. Phys.* 126: 144305. (b) Wörner, H.J. and Merkt, F. (2007). Jahn-Teller effect in CH_3D^+ and CD_3H^+: conformational isomerism, tunneling-rotation structure, and the location of conical intersections. *J. Chem. Phys.* 126: 154304.

137 (a) Fokin, A.A., Schreiner, P.R., Gunchenko, P.A. et al. (2000). Oxidative single-electron transfer activation of σ-bonds in aliphatic halogenation reactions. *J. Am. Chem. Soc.* 122: 7317–7326. (b) Novikovskii, A.A., Gunchenko, P.A., Prikhodchenko, P.G. et al. (2011). Comparative theoretical and experimental analysis of hydrocarbon sigma-radical cations. *Russ. J. Org. Chem.* 47: 1293–1299. (c) Guerrero, A., Herrero, R., Quintanilla, E. et al. (2010). Single-electron self-exchange between cage hydrocarbons and their radical cations in the gas phase. *ChemPhysChem* 11: 713–721. (d) Crestoni, M.E. and Fornarini, S. (2012). Jahn-Teller distortion of hydrocarbon cations probed by infrared photodissociation spectroscopy. *Angew. Chem. Int. Ed.* 51: 7373–7375. (e) Richter, R., Wolter, D., Zimmermann, T. et al. (2014). Size and shape dependent photoluminescence and excited state decay rates of diamondoids. *Phys. Chem. Chem. Phys.* 16: 3070–3076.

138 Mella, M., Freccero, M., Soldi, T. et al. (1996). Oxidative functionalization of adamantane and some of its derivatives in solution. *J. Org. Chem.* 61: 1413–1422.

139 Gunchenko, P.A., Novikovskii, A.A., Byk, M.V., and Fokin, A.A. (2014). Structure and transformations of diamantane radical cation: theory and experiment. *Russ. J. Org. Chem.* 50: 1749–1754.

140 Pirali, O., Galue, H.A., Dahl, J.E. et al. (2010). Infrared spectra and structures of diamantyl and triamantyl carbocations. *Int. J. Mass Spectrom.* 297: 55–62.

141 Shubina, T.E. and Fokin, A.A. (2011). Hydrocarbon sigma-radical cations. *WIREs Comput. Mol. Sci.* 1: 661–679.

142 Bouwman, J., Horst, S., and Oomens, J. (2018). Spectroscopic characterization of the product ions formed by electron ionization of adamantane. *ChemPhysChem* 19: 3211–3218.

143 Boyer, A., Herve, M., Scognamiglio, A. et al. (2021). Time-resolved relaxation and cage opening in diamondoids following XUV ultrafast ionization. *Phys. Chem. Chem. Phys.* 23: 27477–27483.

144 Candian, A., Bouwman, J., Hemberger, P. et al. (2018). Dissociative ionisation of adamantane: a combined theoretical and experimental study. *Phys. Chem. Chem. Phys.* 20: 5399–5406.

145 Borovkov, V.I. and Molin, Y.N. (2004). Paramagnetic relaxation of adamantane radical cation in solution. *Chem. Phys. Lett.* 398: 422–426.

146 (a) de Petris, G., Rosi, M., Ursini, O., and Troiani, A. (2013). The oxidative mechanism in electrophilic C-H activation: the case of CH_2F_2 and CH_2Cl_2. *Chem. Asian J.* 8: 588–595. (b) Wang, Z.C., Wu, X.N., Zhao, Y.X. et al. (2010). Room-temperature methane activation by a bimetallic oxide cluster $AlVO_4^+$. *Chem. Phys. Lett.* 489: 25–29. (c) de Petris, G., Cartoni, A., Troiani, A. et al. (2010). Double C-H activation of ethane by metal-free $SO_2^{·+}$ radical cations. *Chem. Eur. J.* 16: 6234–6242. (d) de Petris, G., Troiani, A., Rosi, M. et al. (2009). Methane activation by metal-free radical cations: experimental insight into the reaction intermediate. *Chem. Eur. J.* 15: 4248–4252. (e) Shchapin, I.Y., Vasil'eva, V.V., Nekhaev, A.I., and Bagrii, E.I. (2006). On a possible radical-cation mechanism of the biomimetic oxidation of the saturated hydrocarbon 1,3-dimethyladamantane in a Gif-type system containing a Fe_2^+ salt, picolinic acid, and pyridine. *Kinet. Catal.* 47: 624–637. (f) Kirillova, M.V., Kuznetsov, M.L., Reis, P.M. et al. (2007). Direct and remarkably efficient conversion of methane into acetic acid catalyzed by amavadine and related vanadium complexes. A synthetic and a theoretical DFT mechanistic study. *J. Am. Chem. Soc.* 129: 10531–10545.

147 (a) Worley, S.D., Mateescu, G.D., McFarland, C.W. et al. (1973). Photoelectron spectra and MINDO-SCF-MO [modified intermediate neglect of differential overlap-self-consistent field-molecular orbital] calculations for adamantane and some of its derivatives. *J. Am. Chem. Soc.* 95: 7580–7586. (b) Schmidt, W. (1973). Photoelectron spectra of σ-bonded molecules, part 4. Photoelectron spectra of diamondoid molecules: adamantane, silamantane and urotropine. *Tetrahedron* 29: 2129–2134.

148 Tian, S.X., Kishimoto, N., and Ohno, K. (2002). Two-dimensional penning ionization electron spectroscopy of adamantanes and cyclohexanes: electronic structure of adamantane, 1-chloroadamantane, cyclohexane, and chlorocyclohexane and interaction potential $He^*(2^3S)$. *J. Phys. Chem. A* 106: 6541–6553.

149 (a) Crandall, P.B., Müller, D., Leroux, J. et al. (2020). Optical spectrum of the adamantane radical cation. *Astrophys. J. Lett.* 900: L20. (b) Xiong, T. and Saalfrank, P. (2019). Vibrationally broadened optical spectra of selected radicals and cations derived from adamantane: a time-dependent correlation function approach. *J. Phys. Chem. A* 123: 8871–8880.

150 Patzer, A., Schütz, M., Möller, T., and Dopfer, O. (2012). Infrared spectrum and structure of the adamantane cation: direct evidence for Jahn-Teller distortion. *Angew. Chem. Int. Ed.* 51: 4925–4929.

151 George, M.A.R., Förstel, M., and Dopfer, O. (2020). Infrared spectrum of the adamantane$^+$-water cation: hydration-induced C-H bond activation and free internal water rotation. *Angew. Chem. Int. Ed.* 59: 12098–12104.

152 Kranabetter, L., Martini, P., Gitzl, N. et al. (2018). Uptake and accommodation of water clusters by adamantane clusters in helium droplets: interplay between magic number clusters. *Phys. Chem. Chem. Phys.* 20: 21573–21579.

153 Goulart, M., Kuhn, M., Kranabetter, L. et al. (2017). Magic numbers for packing adamantane in helium droplets: cluster cations, dications, and trications. *J. Phys. Chem. C* 121: 10767–10772.

154 Duensing, F., Gruber, E., Martini, P. et al. (2021). Complexes with atomic gold ions: efficient bis-ligand formation. *Molecules* 26: 3484.

155 Crandall, P.B., Radloff, R., Förstel, M., and Dopfer, O. (2022). Optical spectrum of the diamantane radical cation. *Astrophys. J.* 104: 940.

156 Gali, A., Demjan, T., Voros, M. et al. (2016). Electron-vibration coupling induced renormalization in the photoemission spectrum of diamondoids. *Nat. Commun.* 7: 11327.

157 (a) Albini, A., Mella, M., and Freccero, M. (1994). A new method in radical chemistry: generation of radicals by photo-induced electron transfer and fragmentation of the radical cation. *Tetrahedron* 50: 575–607. (b) Protti, S., Fagnoni, M., Monti, S. et al. (2012). Activation of aliphatic C-H bonds by tetracyanobenzene photosensitization. A time-resolved and steady-state investigation. *RSC Adv.* 2: 1897–1904.

158 (a) Edwards, G.J., Jones, S.R., and Mellor, J.M. (1977). Anodic oxidation of substituted adamantanes. *J. Chem. Soc. Perkin Trans.* 2: 505–510. (b) Koch, V.R. and Miller, L.L. (1973). Anodic chemistry of adamantyl compounds. Scissible carbon, halogen, hydrogen, and oxygen substituents. *J. Am. Chem. Soc.* 95: 8631–8637.

159 Vincent, F., Tardivel, R., Mison, P., and Schleyer, P.v.R. (1977). Monoacetamidation de quelques hydrocarbures polycycliques par voie electrochimique. *Tetrahedron* 33: 325–330.

160 Shubina, T.E., Gunchenko, P.A., Yurchenko, A.G. et al. (2002). Structure and transformations of 1-alkyladamantane radical cations. *Theor. Exp. Chem.* 38: 8–13.

161 (a) George, M.A.R. and Dopfer, O. (2022). Opening of the diamondoid cage upon ionization probed by infrared spectra of the amantadine cation solvated by Ar, N_2, and H_2O. *Chem. Eur. J.* e202200577. (b) George, M.A.R. and Dopfer, O. (2022). Infrared spectrum of the amantadine cation: opening of the diamondoid cage upon ionization. *J. Phys. Chem. Lett.* 13: 449–454.

162 Ganguly, S., Gisselbrecht, M., Eng-Johnsson, P. et al. (2022). Coincidence study of core-ionized adamantane: site-sensitivity within a carbon cage? *Phys. Chem. Chem. Phys.* 24: 28994–29003.

163 Fokin, A.A., Butova, E.D., Chernish, L.V. et al. (2007). Simple preparation of diamondoid 1,3-dienes via oxetane ring opening. *Org. Lett.* 9: 2541–2544.

164 Butova, E.D., Barabash, A.V., Petrova, A.A. et al. (2010). Stereospecific consecutive epoxide ring expansion with dimethylsulfoxonium methylide. *J. Org. Chem.* 75: 6229–6235.

165 Fokin, A.A., Butova, E.D., Barabash, A.V. et al. (2013). Preparative synthesis of vinyl diamondoids. *Synth. Commun.* 43: 1772–1777.

166 Tae, E.L., Zhu, Z.D., and Platz, M.S. (2001). A matrix isolation, laser flash photolysis, and computational study of adamantene. *J. Phys. Chem. A* 105: 3803–3807.

167 Stepanov, F.N. and Sukhoverkhov, V.D. (1967). New fragmentation in the adamantane series. *Angew. Chem.* 6: 864.

168 Gund, T.M. and Schleyer, P.v.R. (1973). Fragmentation and ring-closure in diamantane system. *Tetrahedron Lett.* 1959–1962.

169 Valentin, L., Henss, A., Tkachenko, B.A. et al. (2015). Transition metal complexes with cage-opened diamondoid tetracyclo[7.3.1.^{14,12}.0^{2,7}] tetradeca-6.11-dien. *J. Coord. Chem.* 68: 3295–3301.

170 (a) Osawa, E., Majerski, Z., and Schleyer, P.v.R. (1971). Preparation of bridge-head alkyl derivatives by Grignard coupling. *J. Org. Chem.* 36: 205–207.
(b) Molle, G., Dubois, J.E., and Bauer, P. (1987). Contribution à l'étude des réactions d'alkylation et de polyalkylation de l'adamantane et de ses homologues. *Can. J. Chem.* 65: 2428–2433. (c) Yurchenko, A.G., Fedorenko, T.V., and Rodionov, V.N. (1985). Synthesis of 1-adamantylmagnesium bromide. *Zh. Org. Khim.* 21: 1673–1677.

171 Molle, G., Bauer, P., and Dubois, J.E. (1982). Formation of cage-structure organomagnesium compounds. Influence of the degree of adsorption of the transient species at the metal surface. *J. Org. Chem.* 47: 4120–4128.

172 (a) Molle, G., Dubois, J.E., and Bauer, P. (1978). Cage structure orgasometallic compounds: 1-Diamantyl, 1-twistyl, 1-triptycyl and 2-adamantyl lithium compounds. Synthesis and reactivity. *Tetrahedron Lett.* 3177–3180. (b) Molle, G., Bauer, P., and Dubois, J.E. (1983). High-yield direct synthesis of a new class of tertiary organolithium derivatives of polycyclic hydrocarbons. *J. Org. Chem.* 48: 2975–2981.

173 Samann, C., Dhayalan, V., Schreiner, P.R., and Knochel, P. (2014). Synthesis of substituted adamantylzinc reagents using a Mg-insertion in the presence of $ZnCl_2$ and further functionalizations. *Org. Lett.* 16: 2418–2421.

5
Diamondoid Self-Assembly

5.1 Adamantane-Containing SAMs on Surfaces

The surface stabilities and morphologies of self-assembled monolayers (SAMs) are governed primarily by London dispersion (LD) interactions (the attractive part of the van der Waals potential) [1] between the molecules within the monolayer and by the surface affinity of the adsorbate. For these reasons, long-chain *n*-alkyl derivatives are among the most suitable, and SAMs comprised of *n*-alkyl thiols on gold surfaces have been abundantly studied since their discovery in 1983 [2]. The Au(111) surface has many advantages because it is clean, inert, atomically flat, exhibits a hexagonal arrangement, and can be readily reconstructed. Because of their rigidity and high symmetry diamondoids [3] and their ethers [4] form clusters with preferred arrangements and clustering numbers, and the attention toward diamondoid SAMs [5] on Au(111) was stimulated by the fact that LD interactions between the diamondoid molecules are much stronger than between the alkyl chains of similar length [6]. Though LD interactions between diamondoids are generally lower than those of long-chain alkanes, the picture changes if the conformational landscape is taken into account (Figure 5.1a). Bringing diamondoid molecules into pairs requires no conformational changes, in contrast to linear alkanes, where a large decrease in conformational entropy decreases the favorable interaction free energy (Figure 5.1b).

MD simulations reveal that the entropy changes for diamondoid association exceed the differences in enthalpic stabilization for bringing together two linear alkane molecules (Figure 5.1c and d). Hence, when conformational factors are considered, the rigid cage molecules more effectively self-assemble because they "pay lower entropic penalties when assembling into contact pairs" [7]. That is in accord with recent studies on the self-association of hexane and **AD** at various temperatures and ionic strengths [8] and with the general view that entropy promotes local organization in rigid systems [9]. Additionally, the formation of diamondoid-containing monolayers occurs orders of magnitude faster than for most other adsorbates [10], and the resulting SAMs are more homogeneous and have fewer vacancies [11]. The adamantane moiety is almost spherical and much bulkier than alkyl chains, and this makes diamondoids good alternatives to the commonly employed molecules such as alkane thiols [12], whose SAM properties are not always reproducible for device applications because of tilting and structural

The Chemistry of Diamondoids: Building Blocks for Ligands, Catalysts, Materials, and Pharmaceuticals, First Edition. Andrey A. Fokin, Marina Šekutor, and Peter R. Schreiner.
© 2024 WILEY-VCH GmbH. Published 2024 by WILEY-VCH GmbH.

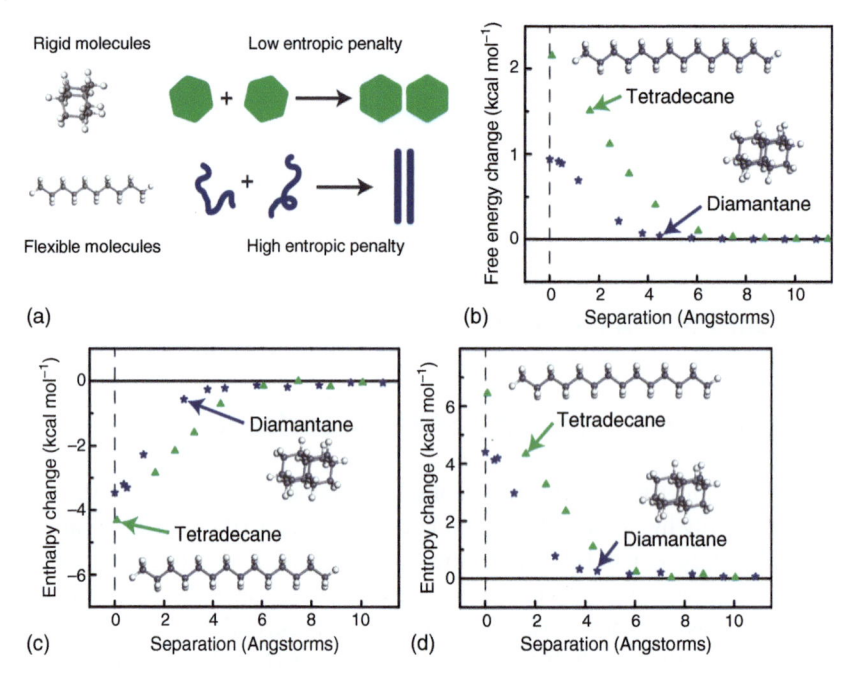

Figure 5.1 (a) The association of **AD** molecules into pairs requires little entropic penalty in contrast to linear alkanes, where a large decrease in conformational entropy occurs; (b) free energy change for the formation of a pair of tetradecane vs. a pair of **DIA** molecules and dissected enthalpy (c) and entropy (d) changes. Zero distance corresponds to the molecular contact in the crystalline state. Source: Reproduced from Ref. [7] with permission from the American Chemical Society, 2019.

irregularities. For bulky adamantane-containing molecules, little surface tilting is expected. A big advantage with respect to SAM homogeneity is that 1-substituted adamantane derivatives and Au(111) surfaces both display the same threefold symmetry, which leads to an increase in domain size [13]. Note that methyl thiol does not form a stable SAM, and 1-adamantanethiolate (**1SHAD**) is the only spherical molecule that does [10]. The first diamondoid SAM was constructed on an Au(111) surface, utilizing AdCH$_2$SSCH$_2$Ad as the precursor [6]. The scanning tunneling microscopy (STM) images display hexagonal 2D structures with a lattice constant of 0.65 ± 0.02 nm, which is close to the value in an adamantane crystal [14]. The surface structure is not homogeneous, and rotational isomerism was observed where at least three differently oriented hexagonal domains (A, B, and C) were identified (Figure 5.2, left). Remarkably, the distances between the adamantane fragments in the SAM ("2D crystal") are close to the nearest neighboring distance of the adamantane (**AD**) 3D-crystal, i.e., display the same lattice constants (ca. 0.665 nm) [16]. The vacancy islands, which are in the depth of gold surface appearing in the STM images as darker regions (Figure 5.2c) with a uniform distribution on the gold surface, are characteristic of diamondoid SAMs ("leopard-like pit structure"). Those result from the removal of selected gold atoms from the first Au layer [15]. Still, **1SHAD** SAMs contain much fewer defects than those observed for alkane thiolates.

5.1 Adamantane-Containing SAMs on Surfaces | **139**

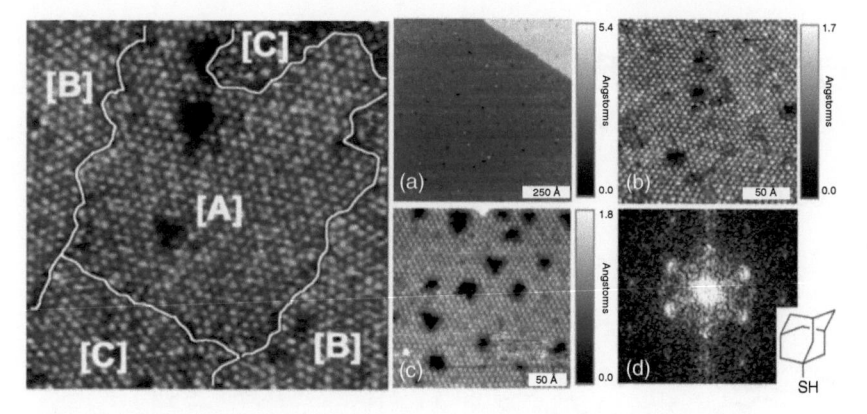

Figure 5.2 *Left*: Three groups of domains (A–C) of SAMs with different molecular alignment angles formed from AdCH$_2$SSCH$_2$Ad on an Au(111) surface with different angles of molecular alignments. White lines illustrate the domain boundaries. Source: Reproduced from Ref. [6] with permission from the American Chemical Society, 2002. *Right*: STM image of a hexagonally close-packed lattice of 1-adamantanethiolate (**1SHAD**) SAM on Au(111) at different resolutions and Fourier transform showing first-order and a few second-order reciprocal lattice points. Source: Reproduced from Ref. [15] with permission from the American Chemical Society, 2005.

Due to the lower surface affinity (the sulfur 2p-binding is ca. 0.16 eV less than that for dodecane thiol [17]), **1SHAD** may be easily replaced by alkyl thiols [18], which is a useful feature for the construction of temporary layers in soft lithography and microdisplacement nanoscale printing (Figure 5.3) [19].

The morphologies of 1-adamantanethiolate SAMs on an Au(111) surface are identical when the deposition is performed in a solution or under ultra-high vacuum (UHV) conditions, but the domain sizes increase when the deposition is performed at higher temperatures. Freshly prepared SAMs annealed at RT, 343, 353, and 373 K display domain sizes of 4–10, 8–20, 15–35, and 30–60 nm, respectively. For samples annealed at 373 K, the size of the ordered domains is limited by the dimension of the available Au(111) terraces, i.e., the domain boundaries vanish at higher temperatures [13]. X-ray photoelectron spectroscopy (XPS) studies show the presence of a densely packed, well-defined monolayer with the molecular axes orientated perpendicularly to the gold surface. Near edge X-ray absorption fine structure (NEXAFS) spectroscopy studies show that the sulfur–carbon bond tilts with a polar angle of about 30° [17]. The different conductivities of **1SHAD** and alkane thiols enable the deposition of metallic films, where the adamantane-containing SAM may be selectively desorbed and then replaced by metal films. Alternatively, metal films can be deposited directly onto the 1-adamantanethiolate regions, because the monolayers allow electron transfer from the supporting electrolyte to the electrode surface, in contrast to alkyl thiolates, which block this process [20]. The charge transfer through the 1-adamantanethiolate SAM also strongly depends on the type of anion present [21]. The technology that utilizes **1SHAD** for displacement printing was developed at the microscale [22]. Scanning electron micrographs (Figure 5.4) demonstrate that absolute coverage increases with contact time when **1SHAD** is replaced by 1-dodecanethiol (**C12SH**) used as a molecular ink deposited with a stamp.

140 | *5 Diamondoid Self-Assembly*

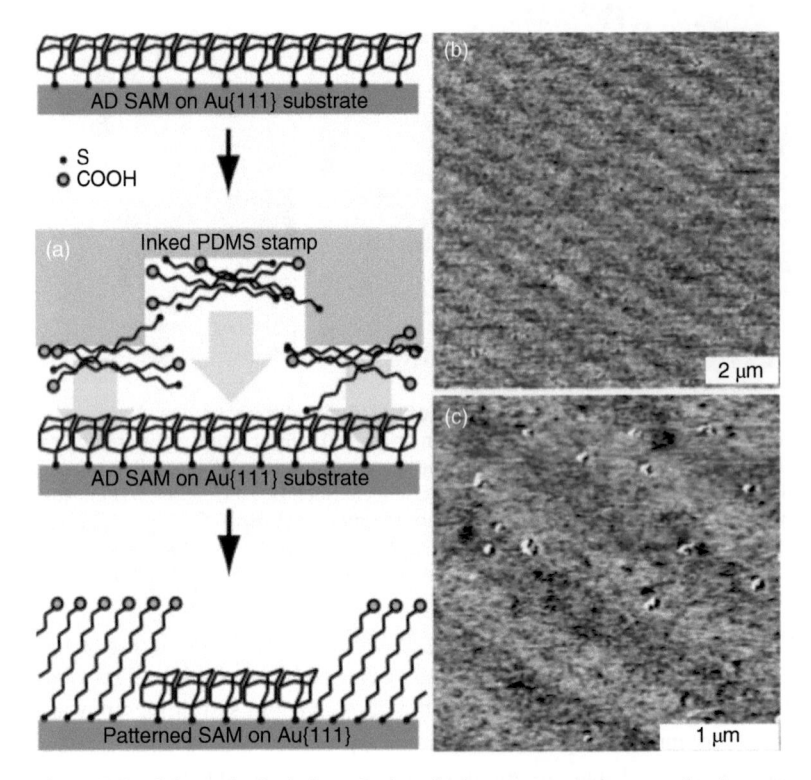

Figure 5.3 Schematic depiction of microdisplacement printing on a 1-adamantanethiolate self-assembled monolayer with an 11-mercaptoundecanoic acid-inked stamp (a). Lateral force microscopy images of patterned Au(111) made by microdisplacement printing using a 1200 lines/mm poly(dimethylsiloxane) stamp inked with a 25 mM 11-mercaptoundecanoic acid solution (b and c). Source: Reproduced from Ref. [19] with permission from the American Chemical Society, 2008.

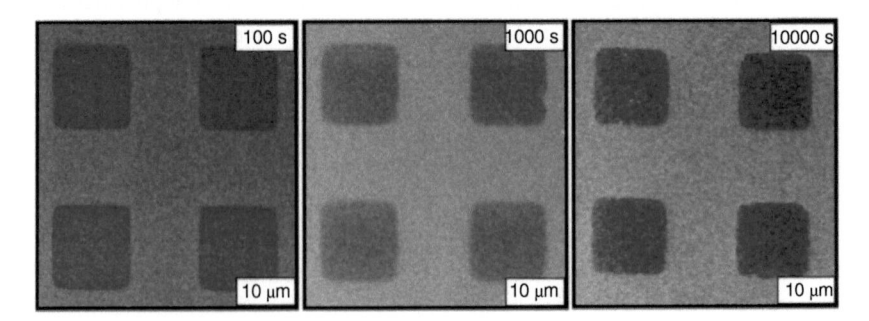

Figure 5.4 Scanning electron micrographs of SAMs produced by microdisplacement printing of **1SHAD** (lighter regions) replaced by 1-dodecanethiol deposited with a stamp (darker regions). Source: Reproduced from Ref. [22] with permission from the American Chemical Society, 2013.

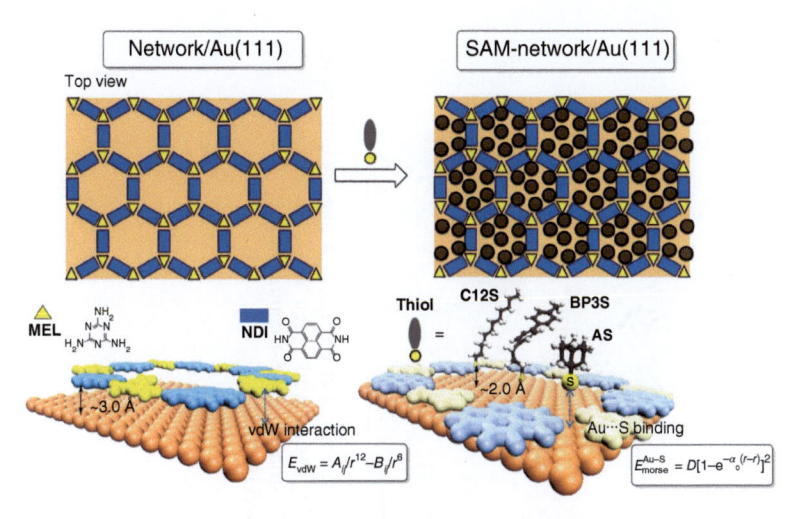

Figure 5.5 *Left*: The hydrogen-bonded hexagonal honeycomb network of naphthalene tetracarboxylic diimide (**NDI**) and melamine (**MEL**) lying flat on an Au(111) template. *Right*: 1-adamantanethiolate (**ASH**), ω-(4′-methylbiphenyl-4-yl)propane thiol (**BP3SH**), and 1-dodecane thiol (**C12SH**), deposited on the Au(111) surface within the surface-supported porous network. Reproduced from Ref. [23] with permission from the American Chemical Society, 2012.

The potential of weakly coordinating **1SHAD** in the construction of SAMs with controlled topology was demonstrated utilizing a porous supramolecular platform constructed from 1,3,5-triazine-2,4,6-triamine (melamine, **MEL**) and perylene-3,4,9,10-tetracarboxylic diimide (**NDI**) lying flat on the gold surface (Figure 5.5) [23]. In contrast to **1SHAD**, the deposition of guest thiols ω-(4′-methylbiphenyl-4-yl)propane thiol (**BP3SH**) and **C12SH** into the pore template is affected by the chain flexibility and results in two typical (standing-up and lying-down) SAM orientations. Such a hexagonal honeycomb platform is perfectly stable when **1SHAD** is absorbed in the pores, while in the presence of **C12SH** the platform is slowly displaced [24]. Such hybrid systems can provide control on the length scale with high precision that is difficult to achieve otherwise. Co-deposition of 1,7-diadamantane thioperylene-3,4,9,10-tetracarboxylic diimide (Ad-S)(2)-PTCDI with **MEL** yields a honeycomb network that gives a well-ordered monolayer adsorbed on Au(111) [25].

Alternatively, the triazine/perylene porous network may be covered by ω-(4′-methylbiphenyl-4-yl)ethane thiol (**BP**, Figure 5.6), and then modified by Cu deposited via an electrochemical underpotential deposition. The replacement of the triazine/perylene network by **1SHAD** then results in a binary **1SHAD/BP** SAM with a well-defined structure. The **1SHAD** can then easily be removed, thereby giving a gold surface covered with **BP**/Cu hexagonal nanoislands [26].

Certain hopes for SAM construction were associated with secondary adamantane thiol **2SHAD** [27]. Cyclic voltammetry measurements suggest that the intermolecular interactions in **2SHAD** SAMs are weaker than those in **C12SH** SAMs but stronger than for **1SHAD**. This may result in more close packing of **2SHAD** on gold surfaces

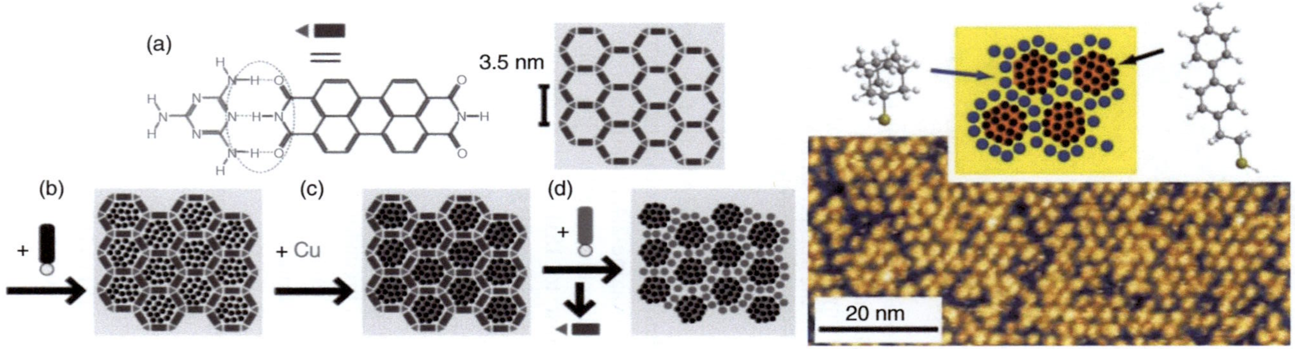

Figure 5.6 Construction of binary ω-(4′-methylbiphenyl-4-yl)ethane thiol/**1SHAD** SAM. (a) The hydrogen bonding motif of 1,3,5-triazine-2,4,6-triamine and perylene-3,4,9,10-tetracarboxylic diimide forming the hexagonal porous network on Au(111); (b) filling the network pores with (ω-(4′-methylbiphenyl-4-yl)ethanethiol; (c) underpotential deposition of copper; (d) substitution of the 3,5-triazine-2,4,6-triamine and perylene-3,4,9,10-tetracarboxylic diimide backbone by **1SHAD**. Source: Reproduced from Ref. [26], Wiley-VCH, 2010.

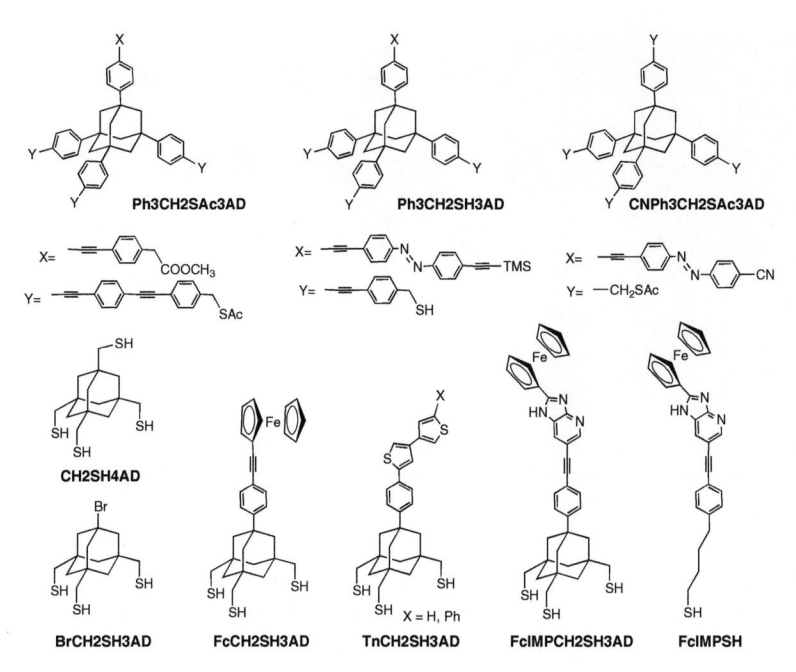

Figure 5.7 Adamantane-based tripodal building blocks utilized for self-assembly on gold surfaces.

as the average number of molecules per unit area is slightly higher than for **1SHAD**. As a result, SAMs formed with **2SHAD** are more resistant against displacement with **C12SH**.

Note that despite substantial efforts in thiol/Au material construction, severe limitations in the preparation of devices still exist. The primary problem arises from the relatively low Au—S bond dissociation energy (ca. 50 kcal mol^{-1}) that reduces the practical potential of monothiols. With increasing coordination sites (e.g., number of thiol groups in the substrate), the binding energy toward the Au surface increases dramatically, and tri-thiols thus far display the slowest desorption rate from Au surfaces [28]. Hence, substrates with multiple attachment points are much more practical, and rigid thiol substrates form very stable adsorbates, making molecular tripods the most promising [29]. One of the first rigid molecular platforms with multiple attachment points was based on the adamantane framework, which is an ideal tripodal C_3-symmetric building block. The Au-binding properties of the adamantane-based molecular tripod **CH2SH4AD** (Figure 5.7) were recognized in 2002 [30]. Sulfur-containing tripods with three legs such as **Ph3CH2SAc3AD** [31], **Ph3CH2SH3AD** [32], **CNPh3CH2SAc3AD** [33], and **FcIMPCH2SH3AD** [34] were used for the modification of Au surfaces, and the first nanoscale SAM formed from 1-bromo-3,5,7-tris(mercaptomethyl)adamantane (**BrCH2SH3AD**) on Au(111) was visualized in 2006 [35].

IR and STM studies suggest a three-point adsorption of **BrCH2SH3AD** with the formation of close-packed hexagonal lattices with the nearest spacing of 8.65 Å (Figure 5.8). In this case, the surface density of the adsorbed sulfur atom is close to

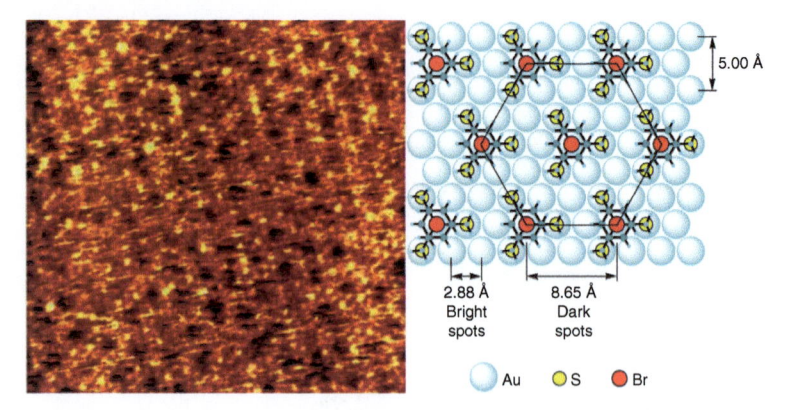

Figure 5.8 *Left*: STM image of a SAM of **BrCH2SH3AD** on an Au(111) surface. *Right*: Top view of a possible arrangement of **BrCH2SH3AD** in the SAM. The unit cell is indicated by a hexagon. Source: Reproduced from Ref. [35] with permission from the American Chemical Society, 2006.

that of *n*-alkane thiol SAMs. The stability of such SAMs approaches that of alkane thiols, as determined from electrochemical reductive desorption experiments [35].

A more detailed follow-up study of the **BrCH2SH3AD** adamantane tripod deposited on Au(111) under ultrahigh vacuum conditions [36] confirmed the formation of highly ordered SAMs with a three-point contact with the substrate (Figure 5.9, left). The self-assembly of three achiral **BrCH2SH3AD** units creates chiral trimers (Figure 5.9, middle) that then form chiral hexagonal substructures (Figure 5.9, right). The surface-induced SAM chirality was confirmed by STM, where mirror image configurations were observed. The chirality is due to the methylene groups of the anchoring legs forming both clockwise and anticlockwise complexes. In contrast to monothiols, where SAM structures are determined primarily by the noncovalent interactions between the hydrocarbon moieties, the contribution of the S•••S attractions between the CH_2S legs is substantial in the case of tris-thiolated tripods. It was suggested that such interactions are not able to resist high bias voltages like adamantane tripod SAMs, which are stable at 1 V but decompose at higher bias voltages [37].

SAMs with redox properties have the potential for use in molecular photovoltaic and light-emitting devices. Densely packed SAMs derived from adamantane-containing tripods helped overcome one of the main drawbacks in this field because they helped shield the electroactive group from significant steric and/or electrostatic interactions that would alter its properties. For instance, adamantane-based tripods with electron-rich aromatic moieties such as in **FcCH2SH3AD** [29b, 38, 39] (Figure 5.7) or bisthiophene **TnCH2SH3AD** [40] (Figure 5.7) have their redox-active groups spatially isolated. SAMs formed by **FcCH2SH3AD** on Au(111) are structurally very close to those formed by **BrCH2SH3AD** [41] and display "ideal redox behavior" [39]. The surface electrochemistry of dithiophene **TnCH2SH3AD** (Figure 5.7) showed an irreversible cathodic wave in cyclic voltammetry due to reductive desorption [40]. While ferrocene-terminated alkane thiols have near-ideal behavior only at low concentrations of ferrocene that avoid ferrocene–ferrocene

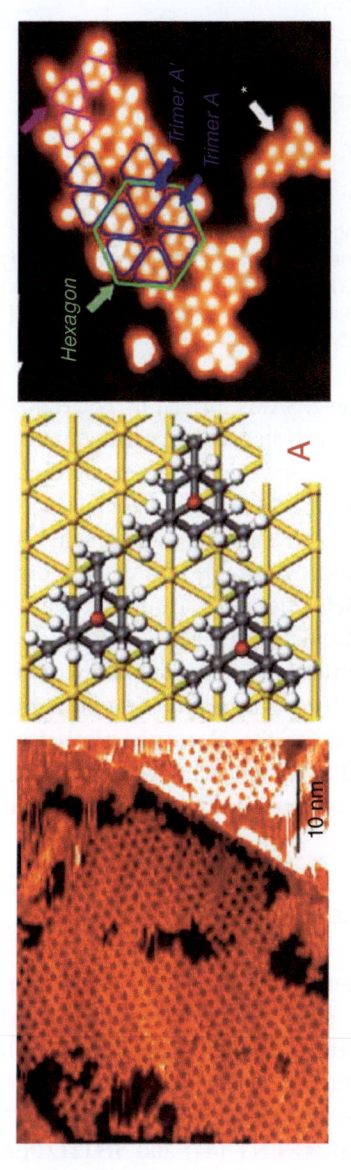

Figure 5.9 *Left*: STM images of **BrCH2SH3AD** on Au(111). *Middle*: Trimers of **BrCH2SH3AD** that form a chiral substructure determined by the favorable CH$_2$ group packing. *Right*: The STM image of an island with a trimer of **BrCH2SH3AD**, where hexagon units are indicated. Source: Reproduced from Ref. [36] with permission from the American Chemical Society, 2007.

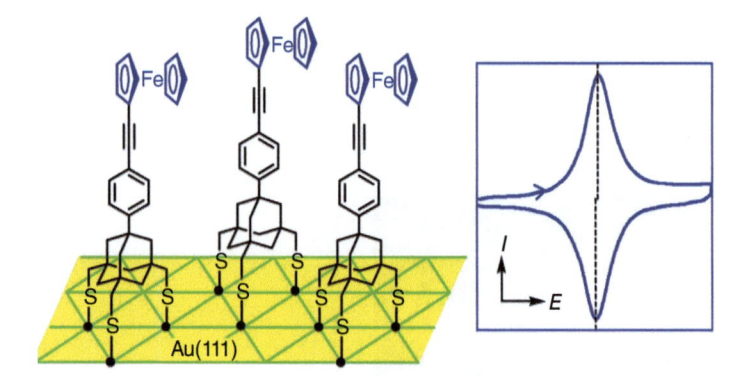

Figure 5.10 *Left*: A possible arrangement of the molecules in the SAM of **FcCH2SH3AD** where the ferrocenyl groups are oriented toward the same direction; *Right*: Cyclic voltammograms recorded with **FcCH2SH3AD** surface-modified gold working electrode display almost ideal Nernstian behavior. Source: Reproduced from Ref. [39] with permission from the American Chemical Society, 2013.

interactions within the monolayer, SAMs formed by **FcCH2SH3AD** (Figure 5.10, left) display almost perfect Nernst-shaped cyclic voltammograms (Figure 5.10, right) and sustain 1000 cycles, thereby displaying extraordinary robustness [39]. Due to multiple bindings, immersing SAM in butane thiol solution does not change the electrochemical properties. The adamantane fragments in SAMs formed by **FcCH2SH3AD** isolate the ferrocenyl groups at a defined distance of 16 Å from the gold surface, which is a highly desirable feature for the construction of rigid junctions between a device molecule and a metal electrode.

The SAM from the ferrocene-linked imidazopyridine trithiol tripod **FcIMPCH2SH3AD** (Figure 5.7) on Au has recently been shown [34] to be an efficient ion-sensitive electrode for the electrochemical detection of Pb^{2+}. In contrast, the monothiol **FcIMPSH** (Figure 5.7) with the same molecular height showed an unsatisfactory ionic response due to poor binding to the Au surface.

5.2 Higher Diamondoids for SAM Formation

Thiols with diamondoids larger than adamantane attracted much attention as potential building blocks for diamond-like 2D electronics. 1-Diamantyl thiol (**1SHDIA**) and its 4-isomer (**4SHDIA**) were first prepared already in 1990 [42], and the higher diamondoid thiols (Figure 5.11), namely, triamantane thiols (**9SHTRIA** and **3SHTRIA**), [121]tetramantane thiols (**6SH121TET** and **2SH121TET**) [43], and pentamantane thiols (**SH1234PENT** [44] and **SH1212PENT** [45]), are now available synthetically from their corresponding diamondoid bromides.

As mentioned above, **1SHAD** imposes steric restrictions due to the bulky nature of the adamantane cage, which induces suboptimal S—Au bonding. The medial positions of higher diamondoids are more sterically congested and decrease the stability of SAMs even further [17]. At the same time, LD interactions grow with the

5.2 Higher Diamondoids for SAM Formation | 147

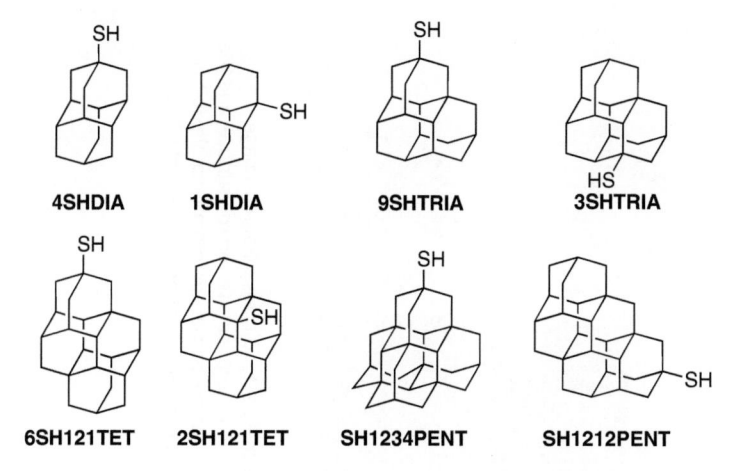

Figure 5.11 Diamondoid thiols available synthetically. Source: From Refs. [43–45].

size of the molecule and play an important stabilizing role in SAM formation. Such dichotomy prompted studies of SAMs built from higher diamondoid thiols. It is not surprising that **4SHDIA** forms (Figure 5.12a and b) hexagonally packed SAMs on Au(111) structurally similar to those observed for **1SHAD**, as both molecules possess a C_3-symmetry axis and have identical horizontal cross sections [46]. Conversely, the on-surface behavior of sterically congested **1SHDIA** is clearly different where four kinds of molecular packings are observed in the STM image (Figure 5.12c). The metastable SAMs coexist because less symmetric **1SHDIA** may aggregate effectively in various orientations that determine, in combination with relatively low surface binding energy and the less hindered rotation around the C—S bond, the dynamic behavior. The observed rotationally metastable states coexist at ambient temperatures as the rotation does not change the 2D cell dimensionality, i.e., partial conformational flexibility interchanges the morphology but conserves the SAM.

The behavior of **4SHDIA** and 4,9-diamantyl dithiol (**49SHDIA**) on metal (Au, Ag, Cu) (111) surfaces was studied with low-temperature STM [47]. **4SHDIA** was detected both as individual single molecules (Figure 5.13a) and as self-assembled islands (Figure 5.13b) on the Au(111) surface. Individual molecules appear as discs, indicating an intact rotating molecule (Figure 5.13a). **4SHDIA** SAMs show triangles that arise from the three hydrogen atoms highlighted in green in Figure 5.13b-iii and are identical to those previously described [46] (Figure 5.12b). Adsorbed **4SHDIA** desorbs from the surface under heating above 160 °C, leaving many individual S atoms on the Au(111), thereby indicating C—S bond dissociation during thermal layer decomposition. While the topological triangular shape of SAMs formed on Au(111) is similar to those found for Ag(111), on Cu(111) surfaces **4SHDIA** lies almost down so that the Cu atoms linked to molecules stick together to form "Y" shaped trimers instead of highly ordered SAMs. Dithiol **49SHDIA** displays clearly different behavior as it lies down on the Au(111) surface, and after thermal annealing gives higher-order linear disulfide-bridged head-to-tail nanodiamond chains (Figure 5.13, bottom) that display some flexibility upon STM manipulation.

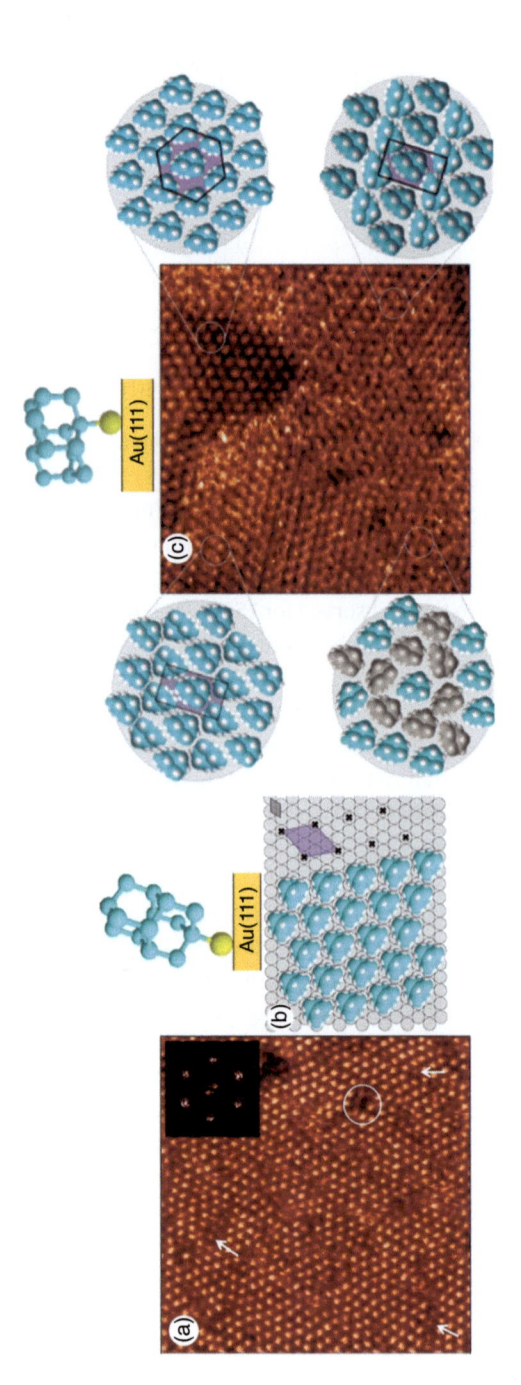

Figure 5.12 (a) Constant-current STM image of a 4-diamantyl thiol (**4SHDIA**) monolayer on Au(111), where the inset indicates a Fourier-image characteristic for hexagonal packing. (b) Schematic model of **4SHDIA** packing on Au(111), where crosses denote adsorption sites on the surface, and the gray circles represent surface Au atoms; (c) STM image of a 1-diamantyl thiol (**1SHDIA**) monolayer on Au(111), where partial conformational flexibility interchanges the morphology but conserves the SAM. Source: Reproduced from Ref. [46] with permission from the American Chemical Society, 2019.

Figure 5.13 *Top*: STM images of single molecules (A) and self-assembled islands (B) of 4-diamantyl thiol (**4SHDIA**) on Au(111) surface and their line profiles (A-ii and B-ii) over **4SHDIA** molecule with suggested configurations (A-iii and B-iii). *Bottom*: STM images of 4,9-diamantyl dithiol (**49SHDIA**) polymers on Au(111) surface after annealing at 199.6 °C (i, overview image, 43 × 43 nm and, ii, zoomed-in image 5 × 5 nm). Source: Reproduced from Ref. [47] with permission from the American Chemical Society, 2021.

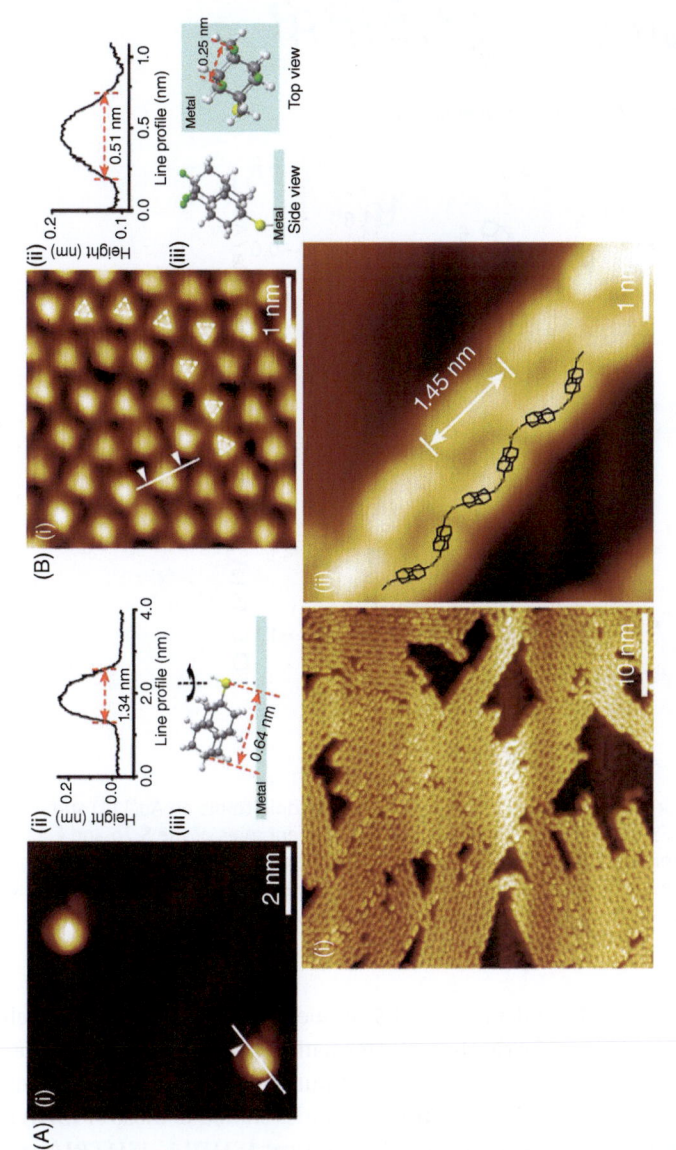

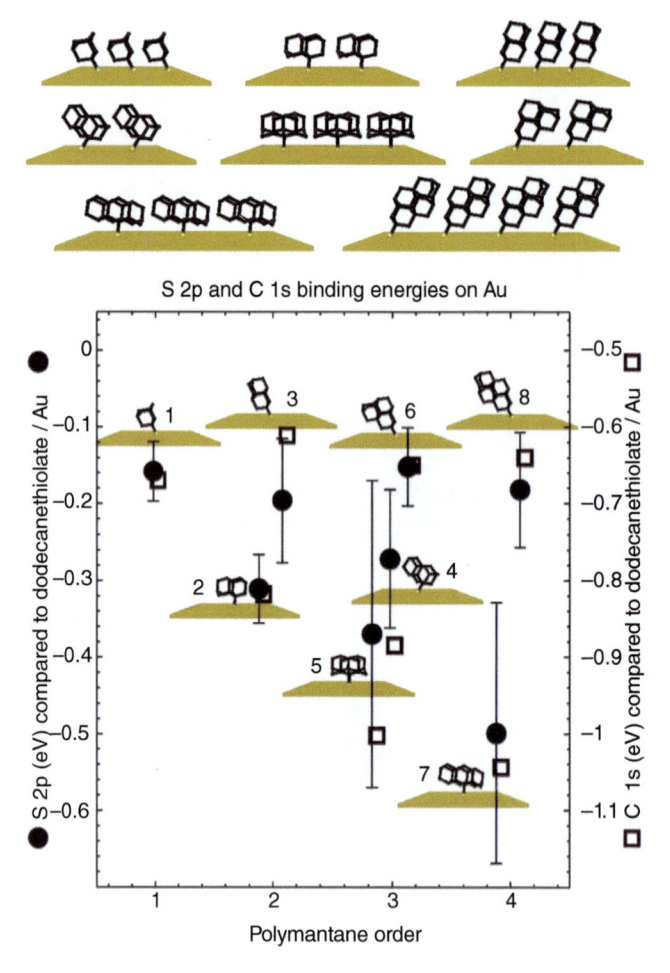

Figure 5.14 *Top*: Orientation of apical and medial diamondoid thiols on Au(111) surfaces based on the NEXAFS data. *Bottom*: The core-level binding energies of the S 2p and C 1s photoelectrons of thiolated diamondoids relative to **C12SH** on Au(111). The S 2p binding energies are depicted with filled circles, using the left axis, while the C 1s binding energies are depicted with open boxes on the right. Source: Reproduced from Ref. [17] with permission from the American Chemical Society, 2008.

Comparative angular-dependent NEXAFS studies of diamondoid thiol self-association on Au(111) also clearly show the formation of SAMs with the thiolate binding energies strongly depending on the substitution pattern (Figure 5.14, top) [17]. All apical diamondoid thiols have surface affinities similar to those of **1SHAD**. As expected, more sterically hindered medial derivatives **1SHDIA**, **3SHTRIA**, and **2SH121TET** are bound more weakly (Figure 5.14, bottom). For instance, the most sterically hindered **2SH121TET** has a ca. 0.6 eV lower binding energy than **1SHAD** (note that **1SHAD** is 0.16 eV less strongly bound compared to **C12SH**). The correlation between the binding energies and the S—C polar tilt angles is instructive. While for apical **6SH121TET** the S—C polar angle with respect to the surface is close to that of **1SHAD** (30°), more hindered **2SH121TET** displays a tilt angle of only 10°. The optimal combination of high surface affinity and increased dispersion binding makes apical thiol **6SH121TET** the most promising studied building block for nanoelectronic diamondoid applications, and this finding initiated a number of electronic and on-surface studies of this molecule (see below).

Remarkably, the SAM topologies formed by **6SH121TET** on silver and gold display almost identical NEXAFS pictures, demonstrating the dominant role of intermolecular interactions between the hydrocarbon moieties within the SAMs [45]. The self-association of diamondoids, including pentamantanes **SH1212PENT** and **SH1234PENT,** was also tested on silver surfaces, for which XPS data confirmed the formation of SAMs that mimic the structure of a hydrogen-terminated diamond surface. This allows the construction of new diamondoid electronic devices, especially electron emitters. The electron-emitting properties of natural diamond due to its negative electron affinity (NEA) were discovered in 1979 [48] and are associated with the hydrogen termination of the surface (Figure 5.15, left) [49]. For diamondoids, the NEA was theoretically predicted in 2005, asserting that surface coatings with diamondoids, viewed as H-terminated nanodiamond particles, could be a simple and economic method for producing electron emitters [51]. The first experimental proof that this indeed is true was provided in 2007 with a **6SH121TET** SAM on a gold surface [50]. The electronic excitation of gold causes strong electron emission with up to 68% of all electrons being emitted within a single energy peak with a full-width half-max (FWHM) around 0.3 eV (Figure 5.15, right). This is a strong indication of the NEA of the material and the first experimental proof that a metal surface covered by diamondoids indeed mimics this property of hydrogen-terminated natural diamond. Recently, it was demonstrated that the deposition of **6SH121TET** reduces the work function of Au(111) from ~5.3 to 1.60 ± 0.3 eV, corresponding to an increase in the emission intensity by a factor of over 13 000 [52]. This effect

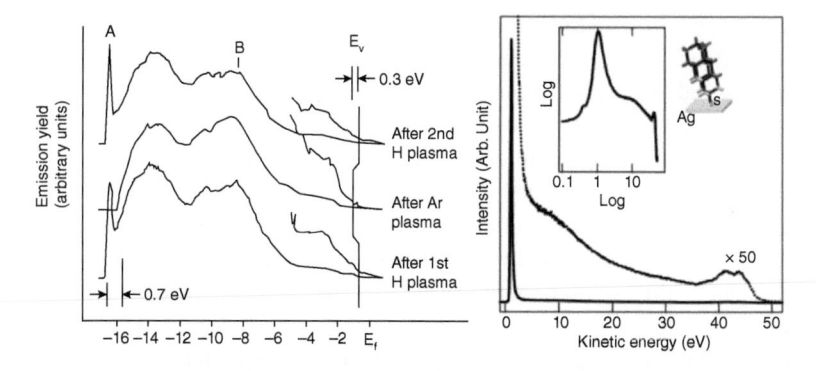

Figure 5.15 *Left*: Negative electron affinity (NEA) of a hydrogen-terminated diamond surface shown experimentally for the natural IIb-type diamond when treated with hydrogen plasma. After the removal of hydrogen with argon plasma the emission peak disappears and reappears after repeated hydrogen plasma reduction. Source: Reproduced from Ref. [49] with permission from AIP Publishing, 1993. *Right*: Monochromatic electron emission peak from a gold surface modified with [121]tetramantane-6-thiol (**6SH121TET**). Source: Reproduced from Ref. [50] with permission from AAAS, 2007.

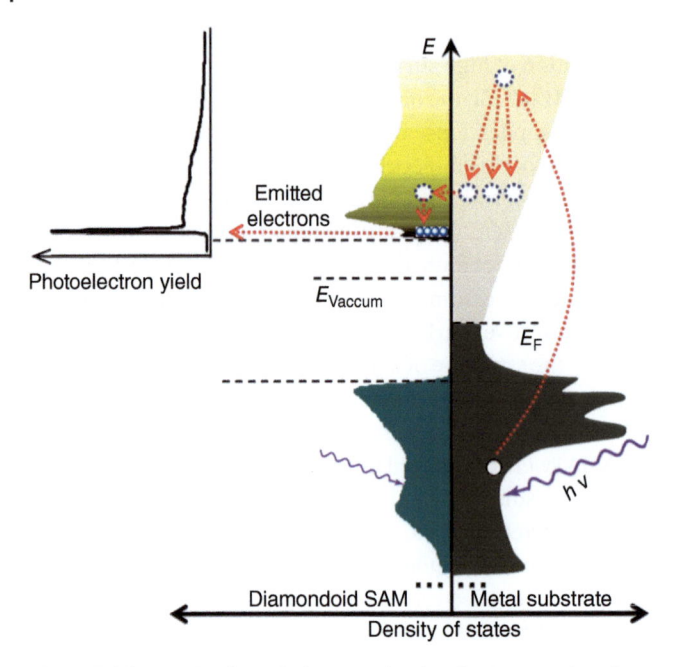

Figure 5.16 Mechanism of electron ejection from a metal surface modified by diamondoids. Excitation of the surface causes electron transfer to the diffuse diamondoid LUMO, which lies above the vacuum level. This leads to efficient electron emission from the surface. Source: Reproduced from Ref. [50] with permission from AAAS, 2007.

seems to be a general property of diamondoid cages. Photoelectron spectra of adamantane films on Si(111) show a peak at low-kinetic energy that also can be attributed to NEA [53]. Deposition of **1SHAD** also changes the emission properties of CdSe nanocrystals. In a comparative study of photoluminescence properties of nanocrystals capped with **C12SH**, thiophenol, and **1SHAD** even a partial ligand substitution increases the overall quantum yields due to delocalizing the excitonic hole, leading to a red-shift of both core and surface emissions [54].

The energy diagram (Figure 5.16) illustrates how X-ray absorption leads to monochromatic emission from diamondoid SAMs on metal substrates. The metal is irradiated and some of the excited electrons are transferred to the unoccupied high-energy electronic levels of the diamondoid. Relaxation eventually leads to the occupation of electrons in the highly delocalized and diffuse [51, 55] diamondoid LUMO that is above the vacuum energy level [56]. The electrons are then emitted freely with high quantum yields from this discrete energy level (Figure 5.16, top left).

The sharp emission peak and the virtual absence of a secondary electron tail outside the monochromatic peak could be attributed to efficient phonon scattering within the diamondoid and an unusually short electron mean free path [57]. These effects allow electrons to accumulate quickly in the diamondoid LUMO. The fact that the photon energy changes the intensity rather than the position of the NEA peak is strong evidence for electron emission from the diamondoid LUMO, and only little electron scattering from the uncovered metal surface [57, 58]. However,

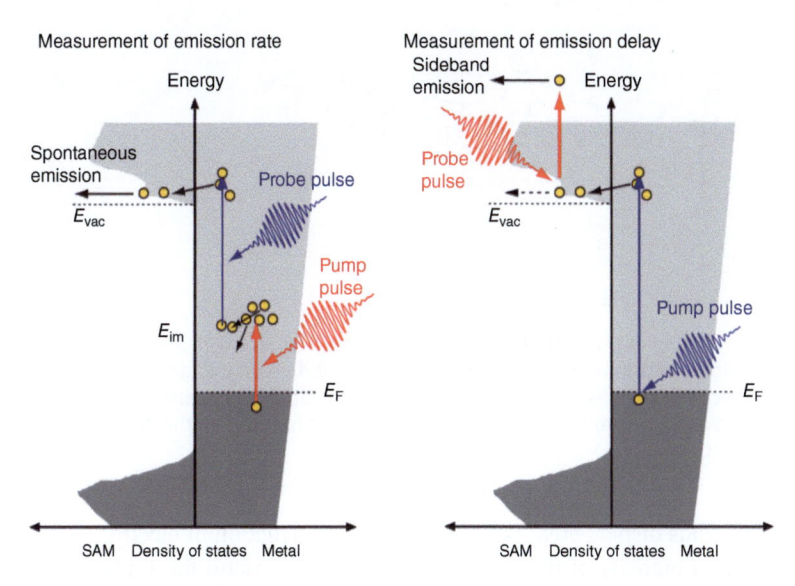

Figure 5.17 Two-photon photoemission experiments from negative-electron affinity layers with two different pulse sequences. *Left*: Generation of a hot electron gas in the metal substrate followed by a second pulse exciting the electrons to levels above the diamondoid LUMO, leading to monochromatic emission. *Right*: If the electron is located on the LUMO when the probe pulse is fired, the electron absorbs a photon and appears in a sideband at higher kinetic energy shifted by the probe photon energy from the main peak. Source: Reproduced from Ref. [62] with permission from AIP Publishing, 2017.

a detailed analysis of the electronic properties of on-surface **6SH121TET** [59] and parent **121TET** [60] shows that in the ground state, the LUMO is located below the vacuum level, and such species should not display NEA in a vertical electron transfer processes. The relevant orbital could be filled only in an adiabatic process, e.g., through the formation of geometrically relaxed diamondoid radical anions. Only in such species can the highest singly occupied molecular orbital (SOMO) be shifted above the vacuum level [59]. This agrees well with the orbital picture in which the LUMO is located on the diamondoid surface outside the carbon cage and mostly describes C—H antibonding [51]. Note that while the molecule increases in size during this excitation process, bond stretching occurs mostly in the C—C bonds [61]. The substantial lifetime of the electrons in the surface-attached diamondoid SOMO was confirmed by two-photon photoemission (TPPE) experiments (Figure 5.17). TPPE for **6SH121TET** on a silver surface confirms that the electrons are mostly excited in the metallic substrate and transferred to the diamondoid, thus leading to an anionic state [58]. Spontaneous emission from the SOMO of thus formed diamondoid radical anion leads to an intense monochromatic electron spectrum. As the first pulse leads directly to the metal LUMO occupation, the second (probe) pulse can then be used to clock the emission (Figure 5.17, left). The second experiment (Figure 5.17, right) in fact leads to the excitation and ionization of the geometrically relaxed anionic diamondoid structure, whose lifetime is long enough to be detectable. In latter case, the electron is in the LUMO when the probe pulse is fired.

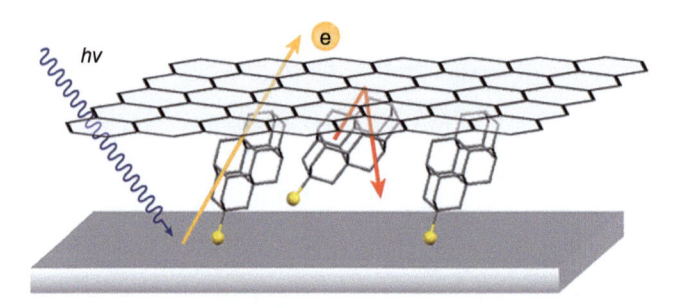

Figure 5.18 The coating of the **6SH121TET** SAM on the gold surface with a graphene monolayer with high electron transparency serves as a robust diamondoid desorption barrier. Such protection provides a fourfold enhancement of stability compared to the bare diamondoid SAM. Source: Reproduced from Ref. [64] with permission from the American Chemical Society, 2018.

Thus, diamondoids display great potential for the construction of electron emitters. The monochromaticity of the NEA peak is especially useful for applications ranging from aberration-free electron imaging to low-emittance photoinjectors. However, the utilization of diamondoids for practical devices is limited because of the low surface affinity of tertiary diamondoid thiols (see above). This decreases the lifetimes of devices and narrows the working temperatures (the monolayers are stable only up to 100 °C) [63]. The stability of cathodes, however, is a general problem in electron emitter technology and there are many work-arounds. One consists of covering the surface with cesium bromide. While this increases the lifetime of the **6SH121TET** monolayer on gold (at least fivefold) [63], the observed energy spread is too large (FWHM = 1.06 eV) and is higher than the measured energy spread of CsBr films deposited directly onto inorganic materials; it is also much higher than FWHM = 0.38 eV measured for **6SH121TET** on gold without CsBr. Obviously, the CsBr monolayer causes strong electron scattering and widens the emission peak of a **6SH121TET** SAM on gold.

An alternative solution is graphene coating of the **6SH121TET** SAM on gold (Figure 5.18), providing much higher thermal stability (significant decomposition starts only at 280 °C) and even increasing the monochromaticity of the photoelectrons up to FWHM = 0.12 eV [64]. This type of protection, however, leads to a different SAM degradation process through photoionization of on-surface diamondoid molecules and subsequent reaction of thus formed radical cations with the graphene coating. The structures and transformations of ionized diamondoids have been well studied and described in previous chapters in detail [65].

Alternatively, very stable (up to 350 °C) SAMs were obtained through covalently bound diamondoid phosphonates [66]. Although the secondary electron emission tail is too high for the use of such materials for high-resolution devices, alternative applications with these metal-oxide coatings are straightforward [67]. In addition, theory predicts that *N*-heterocyclic carbenes (NHCs) containing diamondoids are an alternative to thiol-based SAMs that will be much more stable under harsh conditions [68].

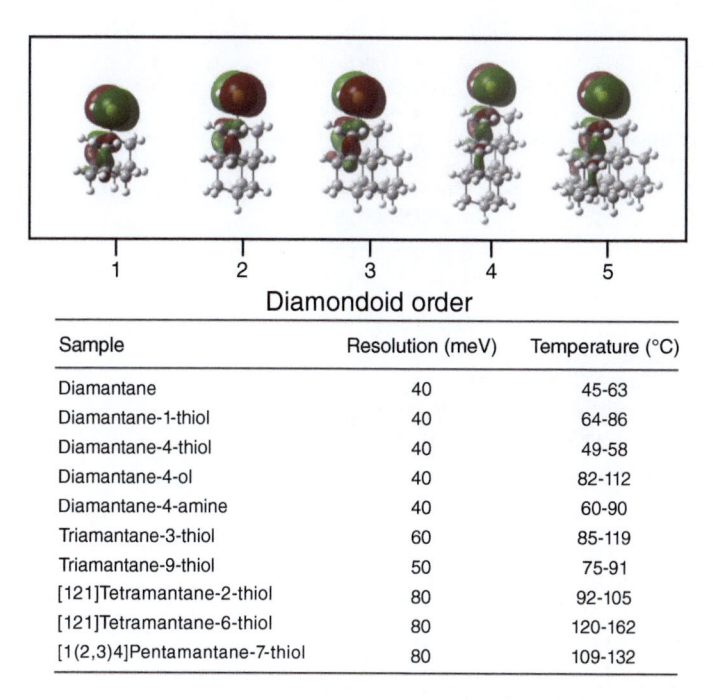

Sample	Resolution (meV)	Temperature (°C)
Diamantane	40	45-63
Diamantane-1-thiol	40	64-86
Diamantane-4-thiol	40	49-58
Diamantane-4-ol	40	82-112
Diamantane-4-amine	40	60-90
Triamantane-3-thiol	60	85-119
Triamantane-9-thiol	50	75-91
[121]Tetramantane-2-thiol	80	92-105
[121]Tetramantane-6-thiol	80	120-162
[1(2,3)4]Pentamantane-7-thiol	80	109-132

Figure 5.19 *Top*: The HOMOs of apical diamondoid thiols are localized mostly on the thiol group. *Bottom*: Comparison of the ionization potentials of diamondoids and their thiol derivatives. Source: Reproduced from Ref. [70] with permission from AIP Publishing, 2013.

The high potential of diamondoid thiols for nanoelectronic applications prompted a number of studies on the electronic properties of the isolated species. The first two excited states of diamondoid thiols correspond to transitions from the sulfur lone pair $n-\sigma^*_{SH}$ and $n-\sigma^*_{SC}$ for the transition to the second excited state. Gas-phase deep UV absorption spectra of diamondoid thiols are almost independent of diamondoid size and only slightly dependent on the position of the substituent in the cage: experimentally and theoretically, a redshift of 0.4–0.5 eV for the medial 2-isomers with respect to apical-isomers was observed [69]. The electronic properties of diamondoid thiols are dominated by the thiol group where the HOMO is located (Figure 5.19, top); the unoccupied states remain nearly unchanged upon thiolation [71]. The situation may change with increasing diamondoid size when the HOMO is delocalized over the cage instead of being localized on the functional group. The diamondoid ionization potentials and those of their corresponding thiols are size dependent. For instance, the IP of **9SHTRIA** is already close to that of unsubstituted triamantane (Figure 5.19, bottom) [70]. This was partially confirmed by a more recent comparative study on the electronic properties of pristine and substituted diamondoids [72].

A Cu(111) surface was identified as the metal template for on-surface reactions of diamantane 4,9-dicarboxylic acid (**49COOHDIA**) [73]. Namely, deposition under high vacuum conditions followed by thermal annealing resulted in a cascade of self-assembled structures (Figure 5.20a), depending mainly on the applied

156 | *5 Diamondoid Self-Assembly*

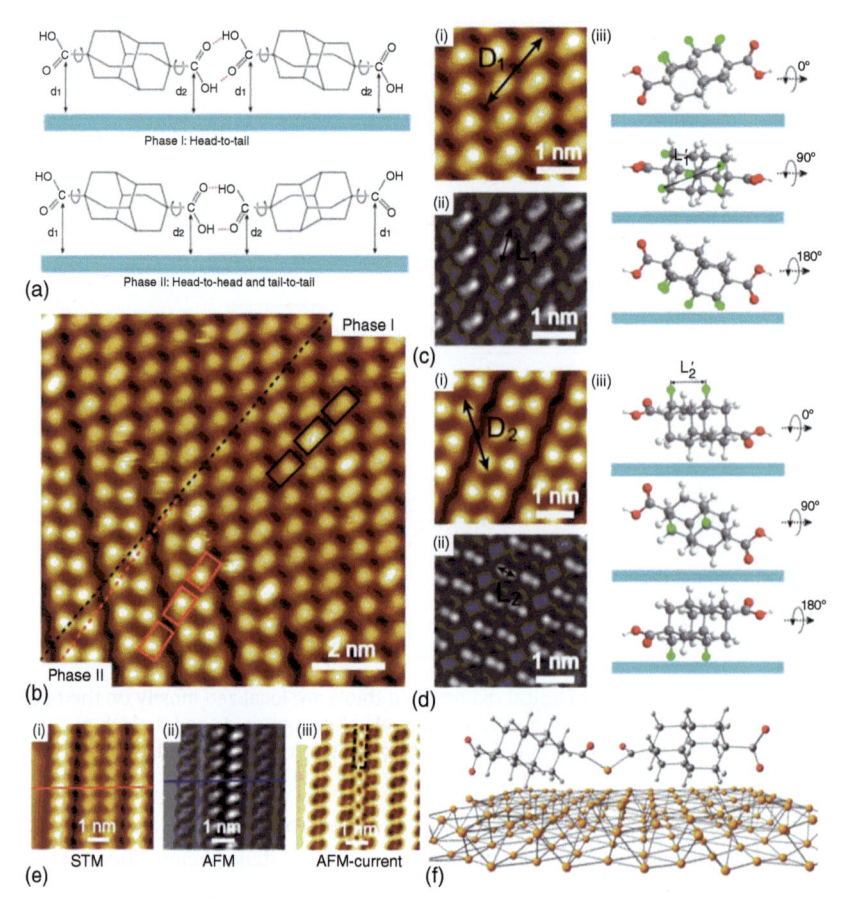

Figure 5.20 (a) Head-to-tail (phase I) and head-to-head/tail-to-tail (phase II) hydrogen-bonding modes of self-assembled structure of 4,9-diamantane dicarboxylic acid (**49COOHDIA**) on a Cu(111) surface; (b) STM and AFM constant height image showing both hydrogen-bonding phases I and II; (c) and (d) STM and AFM constant height images as well as their spatial configurations with top hydrogen atoms marked in green; (e) STM images (i) and AFM images (ii), and simultaneous AFM current images (iii) of double **49COOHDIA** chains consisting of suspended single Cu atoms formed after annealing; (f) computed orientation of two dehydroxylated **49COOHDIA** molecules forming double chains with a copper atom in the middle. Source: Reproduced from Ref. [74] with permission from the American Chemical Society, 2019.

temperature [74]. First, the hydrogen bonding motif emerged, characterized by uniform self-assembly of **49COOHDIA** molecules on the copper surface via intermolecular hydrogen bonding between the carboxylic acid groups for which two different hydrogen-bonding orientation modes (Figure 5.20a) were detected based on STM and atomic force microscopy (AFM) images (Figure 5.20b–d). Heating causes stepwise dehydrogenation and deoxygenation of **49COOHDIA** on the metal surface and at 193 °C higher-order diamondoid double chains form with a single suspended copper atom anchored by two diamantane frameworks and located above the metal surface with a periodic distance of 0.67 ± 0.01 nm

(Figure 5.20e and f). The manufacturing of such one-atom-wide copper chains may have implications for the construction of quantum-scale electronic circuits with controlled band structure and charge/spin transport.

5.3 Pristine Diamondoids on Surfaces

Unfunctionalized diamondoids are able to form well-ordered self-assembled structures on noble metal surfaces at low temperatures. Due to different contributions to the LUMO, the CH and CH$_2$ sites were clearly differentiated by STM topography [60]. Infrared scanning tunneling microscopy (IR-STM) in combination with DFT computations allowed the determination of the preferential orientation of **AD** on an Au(111) surface [75] and to distinguish between **AD•••AD** and **AD•••gold** interactions based on the changes in frequencies and IR intensities. The first single-molecule study was performed for isomeric tetramantanes on Au(111) using STM and spectroscopy and allowed to differentiate between monolayer structures formed by **121TET** and individual enantiomers of chiral **123TET** (Figure 5.21). Both hydrocarbons form hexagonal SAMs, but with distinctly different IR spectral signatures [78]. High-resolution STM and atomic-resolution AFM of **121TET** on Cu(111) and Au(111) surfaces display remarkable agreement between the experimentally determined SAM structure and computations conducted without taking the metal surface into account [76].

This clearly demonstrates that SAM formation is dominated by intermolecular dispersion forces rather than substrate-surface interactions. Unprecedented submolecular resolution achieved for SAMs on Cu(111) provided access to the absolute configuration of **123TET** by direct visual inspection (Figure 5.21d–i) [77].

5.4 Other Applications of Diamondoid SAM Materials

Diamondoid thiols were tested as noble metal catalyst modifiers as coating of a heterogeneous catalyst that may increase its selectivity and efficiency. **1SHAD** is thereby superior to alkane thiols due to better surface control. Indeed, the hydrogenation of ethylene is 17 times faster on Pd/Al$_2$O$_3$ coated with **1SHAD** than with 1-octadecanethiol attached to the metal surface [79]. The selectivity of Pd/Al$_2$O$_3$ catalysts in competition between hydrogenation and isomerization reactions utilizing 1-hexene shows that modification with **1SHAD** does not alter the properties of the catalyst [80]. In contrast, coating increases the rate of the hydrodeoxygenation of benzylic alcohol on Pd/Al$_2$O$_3$ [81]. Modifying this catalyst with thiols substantially increases the selectivity in the aerobic oxidation of alcohols to aldehydes, presumably due to the reduction of catalyst poisoning due to the formation of carbonaceous materials. Coating with **1SHAD** increases the selectivity relative to octadecane thiol [82].

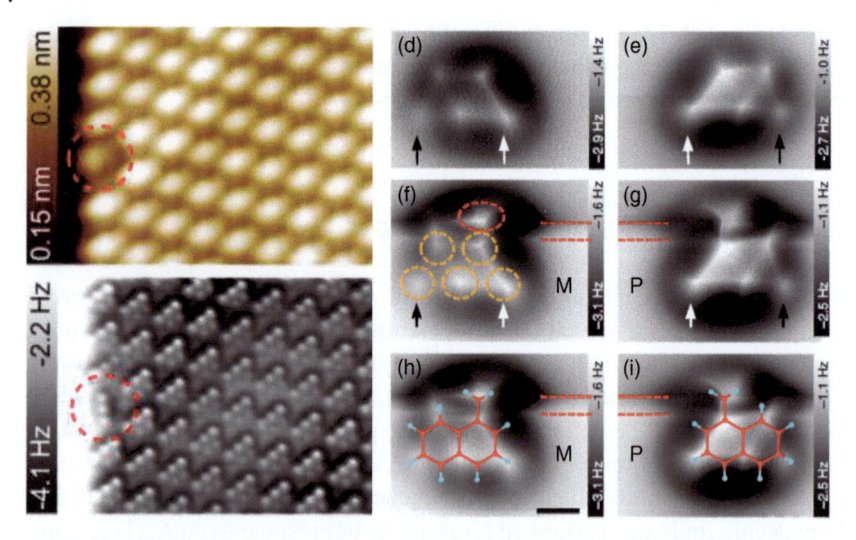

Figure 5.21 *Left up*: High-resolution STM topography image of self-assembled **121TET** on an Au(111) surface. *Left bottom*: High-resolution AFM frequency shift image of self-assembled **121TET** on an Au(111) surface display submolecular resolution. Source: Reproduced from Ref. [76] with permission from the American Chemical Society, 2017. *Right up:* Constant-height AFM scans of two enantiomers of **123TET** self-assembled on an Au(111) surface. White and black arrows indicate brighter (higher) and darker (lower) sides of the molecules, respectively. *Right, middle*: Both images reveal a bright halo, which is either located on the right or the left side of the Olympic ring pattern (dashed orange circles). *Right, bottom*: Corresponding AFM scans with overlaid molecular structures of (*M*)-**123TET** and (*P*)-**123TET**. For clarity, only the plane containing the two specific hydrogens is plotted. Source: Reproduced from Ref. [77] with permission from Springer Nature, 2018.

An adamantane trithiol anchor improves SAM-based electrochemical biosensor stability and enables efficient electron transfer for DNA signaling [12, 83]. However, no substantial advantages over flexible trithiol anchors were found.

The ability of diamondoid SAMs to reduce the electron spread of the surface was used to increase the spatial resolution of X-ray photoemission electron microscopy (X-PEEM) to 10 nm [84]. Coating with diamondoid thiols reduces chromatic aberration without expensive and complex corrective optics (Figure 5.22). Additional benefits arise from the fact that high-contrast images could thus be obtained at low photon flux and low acceleration voltages. This is especially important for studies of materials that are sensitive to X-ray irradiation, such as polymers or biomolecules that may degrade under standard conditions.

The combination of diamondoids with fullerene C_{60} enabled the construction of the first all-hydrocarbon p–σ–n type monomolecular rectifier [85]. Due to NEA, the diamondoid part plays the role of the electron n-donor while the C_{60} moiety with its positive electron affinity acts as the p-component and the hexene bridge operates as a σ-insulator of the junction (Figure 5.23). Such a diamantane-fullerene hybrid **DIAC60** was prepared through [4+2]-addition of 4-butadienyl diamantane (**DIA13DIEN**) [87] to C_{60} [86].

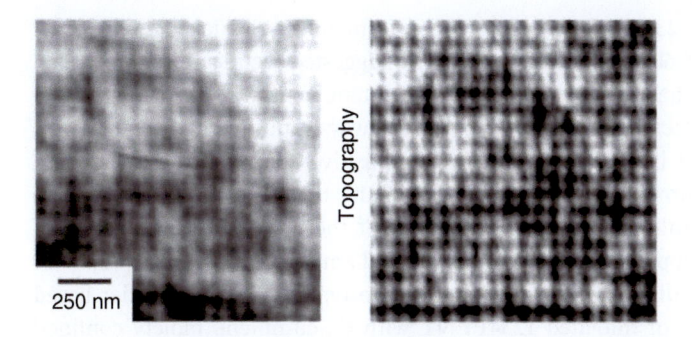

5.4 Other Applications of Diamondoid SAM Materials | 159

Figure 5.22 X-ray photoemission electron microscopy (X-PEEM) topography of Co/Pd bit patterned media with a period of 60 nm/(35 nm bits/25 nm spacing). The left image was obtained without and the right with diamondoid coating. Source: Reproduced from Ref. [84] with permission from AIP Publishing, 2012.

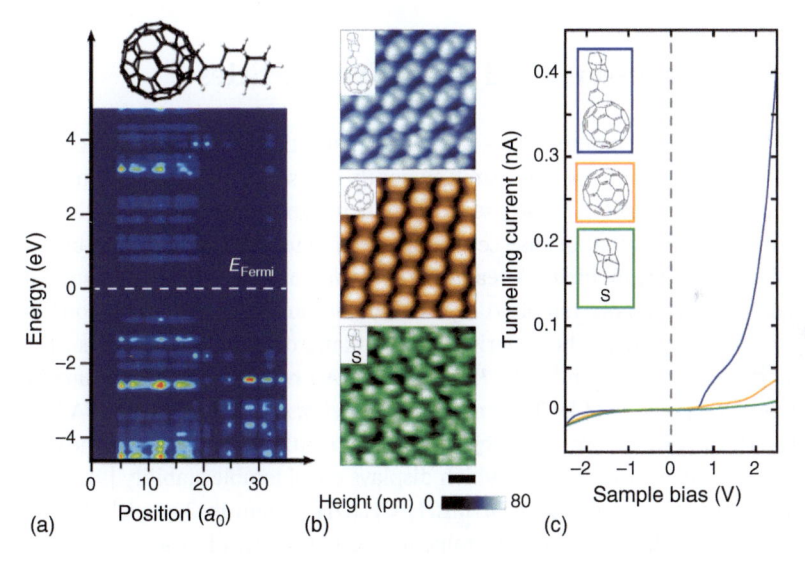

Figure 5.23 (a) DFT color map of the projected density of states of the diamantane-fullerene hybrid **DIAC60** integrated over the x, y plane as a function of the z-coordinate (molecular axis in Bohr units (a_0)); (b) 5 × 5 nm² STM images of **DIAC60** (top), C_{60} (middle), and **4SHDIA** (bottom) on Au(111) surface; (c) current–voltage characteristics measured for the adlayers illustrated in (b). Source: Reproduced from Ref. [86] with permission from Springer Nature, 2014.

Adduct **DIAC60** forms a well-ordered SAM on the gold surface with the fullerene contacting the gold; the diamantane moiety is clearly visible in the STM (Figure 5.23b). The I/V characteristics of such a SAM display pronounced current rectification in contrast to the separate constituents C_{60} and **4SHDIA** (Figure 5.23c). Surprisingly, the current through SAM of **DIAC60** has the opposite bias than predicted by theory. While **DIAC60** comprises a suitable orbital distribution where the HOMO and LUMO of the fullerene are located close to the Fermi level of the metal, they lie within the band gap of the diamondoid, i.e., they are not shifted relative to the donor (Figure 5.23a). As the latter is a requirement for classic single-molecule rectifiers [85], **DIAC60** behaves differently. The suggested rectification mechanism is based on the fact that hydrogen atoms of diamantane moiety carry partial positive charge and the molecule is repelled and rotates away from the positively charged STM tip. This effect increases with a growing positive charge of the tip causing reduction of the current due to an increase of the tip–**DIAC60** distance. Some other adamantane-fullerene hybrids were used for the preparation of a family of superstructures with potential in photovoltaics and anti-wetting coatings [88].

Another possible diamondoid-containing p–σ–n-type molecular rectifier based on the combination of thiolated **1234PENT** with C_n-cumulene moiety confined between two gold electrodes was studied theoretically [89]. The rectification is determined by the occupancy of the carbon 2p states in the cumulene chain for which computations predict large rectification for odd n-systems. As expected, almost no rectification is observed for the dithiolate derivative of **1234PENT** ($n = 0$).

5.5 Adamantane-stabilized Metal Nanoparticles

Ligand-stabilized metal nanoparticles have great potential in medicine, catalysis, and nanoelectronics, where most research currently focuses on gold-thiol chemistry [90]. Adamantane thiol **1SHAD** was used for producing gold nanoparticles with controlled shapes and sizes that display several interesting features due to their bulkiness, rigidity, and hydrophobicity. Already, the first comparative study demonstrated that **1SHAD**-stabilized gold nanoparticles are smaller and more homogeneous in size (Figure 5.24, top) than those derived from alkane or cycloalkane thiols [91]. Most importantly, bulky ligands like **1SHAD** yield Au nanoparticles with narrow and discrete sizes. From MALDI-TOF mass spectrometry, $Au_{30}(SAd)_{18}$, $Au_{39}(SAd)_{23}$, but not $Au_{25}(SAd)_{18}$ were observed, suggesting that the bulky ligands control the particle sizes. The cluster $Au_{30}(SAd)_{18}$, which displays considerable stability [92], was isolated from $HAuCl_4 \cdot 3H_2O$/**1SHAD**/$NaBH_4$/CH_3OH system in 20% yield (based on Au atoms). It is slightly soluble in toluene and chloroform but almost insoluble in tetrahydrofuran, acetone, acetonitrile, and dichloromethane. From the single crystal X-ray (CXR) structure geometry (Figure 5.24, bottom), the $Au_{30}(SAd)_{18}$ is best viewed as a quasi-D_{3d} symmetric Au_{18} inner core protected by six dimeric $Au_2(SAd)_3$ staples of C_3-symmetry.

A body-centered cubic structured gold nanocluster $Au_{38}S_2(AdS)_{20}$ with the sulfide S—S bridge was also characterized structurally [93] and contrasts with the face-centered cubic structure, which gold clusters usually adopt. Such an unusual cluster topology with the —S—S— bridge undergoes the S—C bond cleavage upon heating. The S—S bridge, however, survives the oxygen atmosphere, revealing $Au_{38}S_2(SAd)_{20}$ as a useful catalyst for aerobic oxidations as it may be recovered and reused (Figure 5.25) [95]. The stability substantially increases through complexation with cyclodextrins, where $Au_{38}S_2(SAd)_{20}$ can accumulate two β-cyclodextrin molecules. It was experimentally shown that while a pure nanocluster is not stable in the presence of TBHP, its cyclodextrin conjugate is stable [96]. Electrochemical

5.5 Adamantane-stabilized Metal Nanoparticles | 161

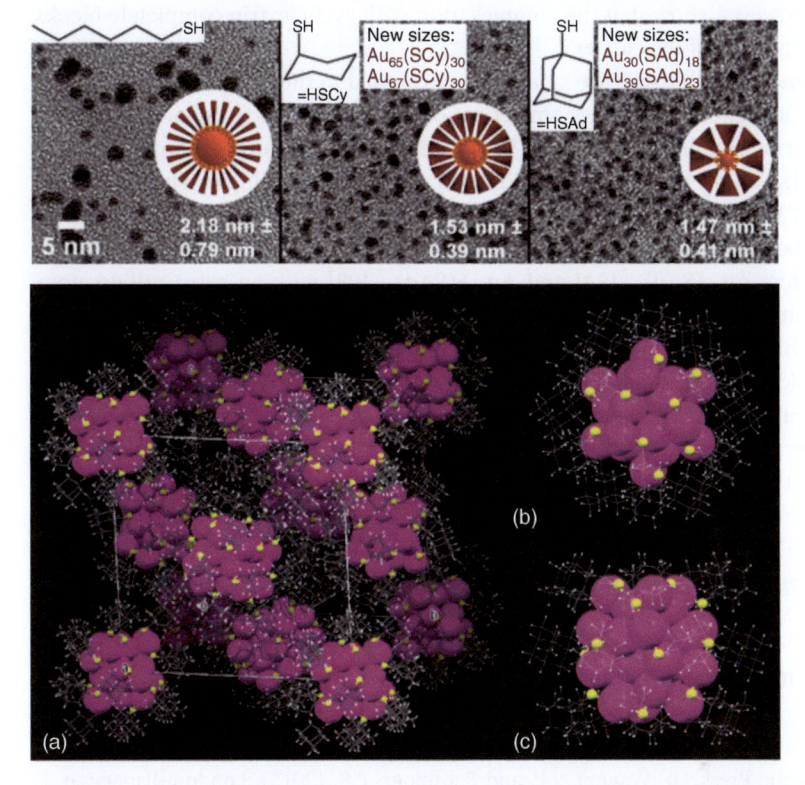

Figure 5.24 *Top*: Thiol-stabilized gold nanoparticles give smaller sizes and size distributions with **1SHAD** as the ligand. Source: Reproduced from Ref. [91] with permission from the American Chemical Society, 2012. *Bottom*: X-ray structural analysis of the $Au_{30}(SAd)_{18}$ nanocluster: (a) unit arrangement; (b) top view; (c) side view. Labels: magenta = Au, yellow = S, gray = C, white = H. The adamantane fragments are in wireframe mode. Source: Reproduced from Ref. [92], Wiley-VCH, 2016.

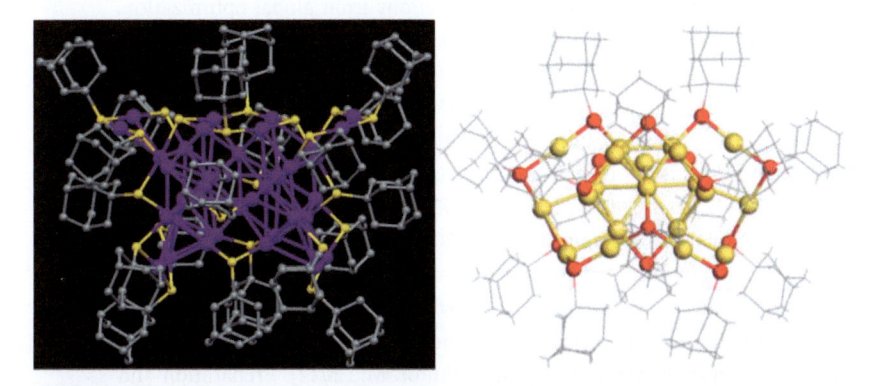

Figure 5.25 *Left*: X-ray crystal structures of gold nanocluster $Au_{38}S_2(AdS)_{20}$ (hydrogen atoms omitted for clarity). Source: Adapted from Ref. [93], Wiley-VCH, 2016. *Right*: Highly symmetric $Au_{21}(SAd)_{15}$ nanocluster. Source: Reproduced from Ref. [94] with permission from the American Chemical Society, 2008.

studies, however, show that the complexation with cyclodextrin completely blocks the redox activity of the cluster.

The trends in stabilizing the small gold cluster led to the successful preparation, isolation, and CXR structure characterization of the $Au_{24}(SAd)_{16}$ nanocluster [97]. The superior stability of $Au_{24}(SAd)_{16}$ was explained by the steric hindrance in $Au_{23}(SAd)_{16}$ on the one hand, and the increase in metal bonding in $Au_{25}(SAd)_{16}$ on the other. Yet, both clusters were observed in the ESI mass spectra of the reaction mixture before the crystallization of the Au_{24} cluster. Two different but related nanoclusters $Au_{21}(SAd)_{15}$ [94] and $Au_{21}S(SAd)_{15}$ [98] with an extra sulfido-atom were recently characterized by X-ray diffraction. While the gold core of $Au_{21}(SAd)_{15}$ displays high symmetry (Figure 5.25), the presence of additional sulfido-atom in $Au_{21}S(SAd)_{15}$ breaks the symmetry [98b]. Photoluminescence experiments demonstrate its anisotropic features in excellent agreement with theoretical and crystallographic analysis [98a].

Only little is known about the stabilization of silver nanoparticles with diamondoid thiolate ligands [90a]. The first cubic bimetallic Pt—Ag nanocluster $PtAg_{28}(1AdS)_{18}(PPh_3)_4$, which exhibits largely enhanced photoluminescence was prepared via a re-aggregation of $PtAg_{24}(SPhMe_2)_{18}$ in the presence of **1SHAD** and PPh_3 [99].

References

1 (a) Israelachvili, J. (2011). *Intermolecular and Surface Forces*, 704. Elsevier Inc., Academic Press. (b) Wagner, J.P. and Schreiner, P.R. (2015). London dispersion in molecular chemistry-reconsidering steric effects. *Angew. Chem. Int. Ed.* 54: 12274–12296.

2 Nuzzo, R.G. and Allara, D.L. (1983). Adsorption of bifunctional organic disulfides on gold surfaces. *J. Am. Chem. Soc.* 105: 4481–4483.

3 (a) Hernandez-Rojas, J. and Calvo, F. (2019). The structure of adamantane clusters: atomistic vs. coarse-grained predictions from global optimization. *Front. Chem.* 7: 573. (b) Goulart, M., Kuhn, M., Kranabetter, L. et al. (2017). Magic numbers for packing adamantane in helium droplets: cluster cations, dications, and trications. *J. Phys. Chem. C* 121: 10767–10772. (c) Kranabetter, L., Martini, P., Gitzl, N. et al. (2018). Uptake and accommodation of water clusters by adamantane clusters in helium droplets: interplay between magic number clusters. *Phys. Chem. Chem. Phys.* 20: 21573–21579.

4 (a) Alić, J., Stolar, T., Štefanic, Z. et al. (2023). Sustainable synthesis of diamondoid ethers by high-temperature ball milling. *ACS Sustain. Chem. Eng.* 11: 617–624. (b) Alić, J., Messner, R., Alešković, M. et al. (2023). Diamondoid ether clusters in helium nanodroplets. *Phys. Chem. Chem. Phys.* 25: 11951–11958. (c) Alić, J., Biljan, I., Štefanić, Z., and Šekutor, M. (2022). Preparation and characterization of non-aromatic ether self-assemblies on a HOPG surface. *Nanotechnology* 33: 355603.

5 (a) Hohman, J.N., Claridge, S.A., Kim, M., and Weiss, P.S. (2010). Cage molecules for self-assembly. *Mater. Sci. Eng. Rep.* 70: 188–208. (b) Clay, W.A.,

Dahl, J.E.P., Carlson, R.M.K. et al. (2015). Physical properties of materials derived from diamondoid molecules. *Rep. Prog. Phys.* 78: 016501. (c) Drexler, C.I., Causey, C.P., and Mullen, T.J. (2015). 1-Adamantanethiol as a versatile nanografting tool. *Scanning* 37: 6–16.

6 Fujii, S., Akiba, U., and Fujihira, M. (2002). Geometry for self-assembling of spherical hydrocarbon cages with methane thiolates on Au(111). *J. Am. Chem. Soc.* 124: 13629–13635.

7 King, E.M., Gebbie, M.A., and Melosh, N.A. (2019). Impact of rigidity on molecular self-assembly. *Langmuir* 35: 16062–16069.

8 Bogunia, M., Liwo, A., Czaplewski, C. et al. (2022). Influence of temperature and salt concentration on the hydrophobic interactions of adamantane and hexane. *J. Phys. Chem. B* 126: 634–642.

9 Klishin, A.A. and van Anders, G. (2020). When does entropy promote local organization? *Soft Matter* 16: 6523–6531.

10 Korolkov, V.V., Allen, S., Roberts, C.J., and Tendler, S.J.B. (2010). Subsecond self-assembled monolayer formation. *J. Phys. Chem. C* 114: 19373–19377.

11 Jobbins, M.M., Raigoza, A.F., and Kandel, S.A. (2011). Adatoms at the sulfur-gold interface in 1-adamantanethiolate mono layers, studied using reaction with hydrogen atoms and scanning tunneling microscopy. *J. Phys. Chem. C* 115: 25437–25441.

12 Claridge, S.A., Liao, W.S., Thomas, J.C. et al. (2013). From the bottom up: dimensional control and characterization in molecular monolayers. *Chem. Soc. Rev.* 42: 2725–2745.

13 Azzam, W., Bashir, A., and Shekhah, O. (2011). Thermal study and structural characterization of self-assembled monolayers generated from diadamantane disulfide on Au(111). *Appl. Surf. Sci.* 257: 3739–3747.

14 (a) Reynolds, P.A. (1978). An X-ray diffuse-scattering study of the ordered, cubic, F{4}3m phase of adamantane (tricyclo[3.3.1.1^{3,7}]decane). *Acta Crystallogr. A* 34: 242–249. (b) Amoureux, J.P., Bee, M., and Damien, J.C. (1980). Structure of adamantane, $C_{10}H_{16}$, in the disordered phase. *Acta Crystallogr. B* 36: 2633–2636.

15 Dameron, A.A., Charles, L.F., and Weiss, P.S. (2005). Structures and displacement of 1-adamantanethiol self-assembled monolayers on Au{111}. *J. Am. Chem. Soc.* 127: 8697–8704.

16 Ulman, A. (1996). Formation and structure of self-assembled monolayers. *Chem. Rev.* 96: 1533–1554.

17 Willey, T.M., Fabbri, J.D., Lee, J.R.I. et al. (2008). Near-edge X-ray absorption fine structure spectroscopy of diamondoid thiol monolayers on gold. *J. Am. Chem. Soc.* 130: 10536–10544.

18 (a) Mullen, T.J., Dameron, A.A., and Weiss, P.S. (2006). Directed assembly and separation of self-assembled monolayers via electrochemical processing. *J. Phys. Chem. B* 110: 14410–14417. (b) Dameron, A.A., Mullen, T.J., Hengstebeck, R.W. et al. (2007). Origins of displacement in 1-adamantanethiolate self-assembled monolayers. *J. Phys. Chem. C* 111: 6747–6752. (c) Mullen, T.J., Dameron, A.A., Saavedra, H.M. et al. (2007). Dynamics of solution displacement in 1-adamantanethiolate self-assembled monolayers. *J. Phys. Chem. C*

111: 6740–6746. (d) Saavedra, H.M., Barbu, C.M., Dameron, A.A. et al. (2007). 1-adamantanethiolate monolayer displacement kinetics follow a universal form. *J. Am. Chem. Soc.* 129: 10741–10746.

19 (a) Dameron, A.A., Hampton, J.R., Smith, R.K. et al. (2005). Microdisplacement printing. *Nano Lett.* 5: 1834–1837. (b) Srinivasan, C., Mullen, T.J., Hohman, J.N. et al. (2007). Scanning electron microscopy of nanoscale chemical patterns. *ACS Nano* 1: 191–201.

20 Mullen, T.J., Zhang, P.P., Srinivasan, C. et al. (2008). Combining electrochemical desorption and metal deposition on patterned self-assembled monolayers. *J. Electroanal. Chem.* 621: 229–237.

21 Park, J.H., Hwang, S., and Kwak, J. (2010). Nanosieving of anions and cavity-size-dependent association of cyclodextrins on a 1-adamantanethiol self-assembled mono layer. *ACS Nano* 4: 3949–3958.

22 Schwartz, J.J., Hohman, J.N., Morin, E.I., and Weiss, P.S. (2013). Molecular flux dependence of chemical patterning by microcontact printing. *ACS Appl. Mater. Interfaces* 5: 10310–10316.

23 Wen, J. and Ma, J. (2012). Modulating morphology of thiol-based monolayers in honeycomb hydrogen-bonded nanoporous templates on the Au(111) surface: simulations with the modified force field. *J. Phys. Chem. C* 116: 8523–8534.

24 Madueno, R., Raisanen, M.T., Silien, C., and Buck, M. (2008). Functionalizing hydrogen-bonded surface networks with self-assembled monolayers. *Nature* 454: 618–621.

25 Raisanen, M.T., Slater, A.G., Champness, N.R., and Buck, M. (2012). Effects of pore modification on the templating of guest molecules in a 2D honeycomb network. *Chem. Sci.* 3: 84–92.

26 Silien, C., Raisanen, M.T., and Buck, M. (2010). A supramolecular network as sacrificial mask for the generation of a nanopatterned binary self-assembled monolayer. *Small* 6: 391–394.

27 Kim, M., Hohman, J.N., Morin, E.I. et al. (2009). Self-assembled monolayers of 2-adamantanethiol on Au{111}: control of structure and displacement. *J. Phys. Chem. A* 113: 3895–3903.

28 Srisombat, L.O., Zhang, S.S., and Lee, T.R. (2010). Thermal stability of mono-, bis-, and tris-chelating alkanethiol films assembled on gold nanoparticles and evaporated "flat" gold. *Langmuir* 26: 41–46.

29 (a) Valasek, M., Lindner, M., and Mayor, M. (2016). Rigid multipodal platforms for metal surfaces. *Beilstein J. Nanotechnol.* 7: 374–405. (b) Kitagawa, T., Kawano, T., Hase, T. et al. (2018). Electron-transfer properties of phenyleneethynylene linkers bound to gold via a self-assembled monolayer of molecular tripod. *Molecules* 23: 2893.

30 Kittredge, K.R., Minton, M.A., Fox, M.A., and Whitesell, J.K. (2002). Alpha-helical polypeptide films grown from sulfide or thiol linkers on gold surfaces. *Helv. Chim. Acta* 85: 788–798.

31 (a) Li, Q., Rukavishnikov, A.V., Petukhov, P.A. et al. (2003). Nanoscale tripodal 1,3,5,7-tetrasubstituted adamantanes for AFM applications. *J. Org. Chem.* 68: 4862–4869. (b) Li, Q., Jin, C.S., Petukhov, P.A. et al. (2004). Synthesis of

well-defined tower-shaped 1,3,5-trisubstituted adamantanes incorporating a macrocyclic trilactam ring system. *J. Org. Chem.* 69: 1010–1019.

32 (a) Takamatsu, D., Yamakoshi, Y., and Fukui, K. (2006). Photoswitching behavior of a novel single molecular tip for noncontact atomic force microscopy designed for chemical identification. *J. Phys. Chem. B* 110: 1968–1970. (b) Takamatsu, D., Fukui, K., Aroua, S., and Yamakoshi, Y. (2010). Photoswitching tripodal single molecular tip for noncontact AFM measurements: synthesis, immobilization, and reversible configurational change on gold surface. *Org. Biomol. Chem.* 8: 3655–3664.

33 Wagner, S., Leyssner, F., Kordel, C. et al. (2009). Reversible photoisomerization of an azobenzene-functionalized self-assembled monolayer probed by sum-frequency generation vibrational spectroscopy. *Phys. Chem. Chem. Phys.* 11: 6242–6248.

34 Kitagawa, T., Hanai, N., Kawano, T. et al. (2023). Metal-ion sensor composed of self-assembled monolayer of amine ligand formed by the use of molecular tripod. *J. Phys. Org. Chem.* 36: e4493.

35 Kitagawa, T., Idomoto, Y., Matsubara, H. et al. (2006). Rigid molecular tripod with an adamantane framework and thiol legs. Synthesis and observation of an ordered monolayer on Au(111). *J. Org. Chem.* 71: 1362–1369.

36 Katano, S., Kim, Y., Matsubara, H. et al. (2007). Hierarchical chiral framework based on a rigid adamantane tripod on Au(111). *J. Am. Chem. Soc.* 129: 2511–2515.

37 Katano, S., Kim, Y., Kitagawa, T., and Kawai, M. (2008). Self-assembly and scanning tunneling microscopy tip-induced motion of ferrocene adamantane trithiolate adsorbed on Au(111). *Jpn. J. Appl. Phys.* 47: 6156–6159.

38 Weidner, T., Zharnikov, M., Hossbach, J. et al. (2010). Adamantane-based tripodal thioether ligands functionalized with a redox-active ferrocenyl moiety for self-assembled monolayers. *J. Phys. Chem. C* 114: 14975–14982.

39 Kitagawa, T., Matsubara, H., Komatsu, K. et al. (2013). Ideal redox behavior of the high-density self-assembled monolayer of a molecular tripod on a Au(111) surface with a terminal ferrocene group. *Langmuir* 29: 4275–4282.

40 Kitagawa, T., Matsubara, H., Okazaki, T., and Komatsu, K. (2014). Electrochemistry of the self-assembled monolayers of dyads consisting of tripod-shaped trithiol and bithiophene on gold. *Molecules* 19: 15298–15313.

41 Katano, S., Kim, Y., Kitagawa, T., and Kawai, M. (2013). Tailoring electronic states of a single molecule using adamantane-based molecular tripods. *Phys. Chem. Chem. Phys.* 15: 14229–14233.

42 Cahill, P.A. (1990). Unusual degree of selectivity in diamantane derivatizations. *Tetrahedron Lett.* 31: 5417–5420.

43 Tkachenko, B.A., Fokina, N.A., Chernish, L.V. et al. (2006). Functionalized nanodiamonds part 3: thiolation of tertiary bridgehead alcohols. *Org. Lett.* 8: 1767–1770.

44 Fokin, A.A., Schreiner, P.R., Fokina, N.A. et al. (2006). Reactivity of [1(2,3)4] pentamantane (T_d-pentamantane): a nanoscale model of diamond. *J. Org. Chem.* 71: 8532–8540.

45 Willey, T.M., Lee, J.R.I., Fabbri, J.D. et al. (2009). Determining orientational structure of diamondoid thiols attached to silver using near-edge X-ray absorption fine structure spectroscopy. *J. Electron Spectrosc. Relat. Phenom.* 172: 69–77.

46 Lopatina, Y.Y., Vorobyova, V.I., Fokin, A.A. et al. (2019). Structures and dynamics in thiolated diamantane derivative monolayers. *J. Phys. Chem. C* 123: 27477–27482.

47 Feng, K., Solel, E., Schreiner, P.R. et al. (2021). Diamantanethiols on metal surfaces: spatial configurations, bond dissociations, and polymerization. *J. Phys. Chem. Lett.* 12: 3468–3475.

48 Cohen, M.L. (1980). Quantum photoyield of diamond (111) – a stable negative-affinity emitter – comment. *Phys. Rev. B* 22: 1095–1095.

49 (a) van der Weide, J. and Nemanich, R.J. (1993). Argon and hydrogen plasma interactions on diamond (111) surfaces – electronic states and structure. *Appl. Phys. Lett.* 62: 1878–1880. (b) Xu, N.S. and Huq, S.E. (2005). Novel cold cathode materials and applications. *Mater. Sci. Eng. R Rep.* 48: 47–189. (c) Bandis, C. and Pate, B.B. (1995). Photoelectric emission from negative-electron-affinity diamond (111) surfaces: exciton breakup versus conduction-band emission. *Phys. Rev. B* 52: 12056–12071. (d) Rameau, J.D., Smedley, J., Muller, E.M. et al. (2011). Properties of hydrogen terminated diamond as a photocathode. *Phys. Rev. Lett.* 106: 137602. (e) van der Weide, J., Zhang, Z., Baumann, P.K. et al. (1994). Negative-electron-affinity effects on the diamond (100) surface. *Phys. Rev. B* 50: 5803–5806.

50 Yang, W.L., Fabbri, J.D., Willey, T.M. et al. (2007). Monochromatic electron photoemission from diamondoid monolayers. *Science* 316: 1460–1462.

51 Drummond, N.D., Williamson, A.J., Needs, R.J., and Galli, G. (2005). Electron emission from diamondoids: a diffusion quantum Monte Carlo study. *Phys. Rev. Lett.* 95: 096801.

52 Narasimha, K.T., Ge, C., Fabbri, J.D. et al. (2016). Ultralow effective work function surfaces using diamondoid monolayers. *Nat. Nanotechnol.* 11: 267–273.

53 Meevasana, W., Supruangnet, R., Nakajima, H. et al. (2009). Electron affinity study of adamantane on Si(111). *Appl. Surf. Sci.* 256: 934–936.

54 Jethi, L., Mack, T.G., Krause, M.M. et al. (2016). The effect of exciton-delocalizing thiols on intrinsic dual emitting semiconductor nanocrystals. *Chem. Phys. Chem* 17: 665–669.

55 Sasagawa, T. and Shen, Z.X. (2008). A route to tunable direct band-gap diamond devices: electronic structures of nanodiamond crystals. *J. Appl. Phys.* 104: 073704.

56 Drummond, N.D. (2007). Diamondoids display their potential. *Nat. Nanotechnol.* 2: 462–463.

57 Clay, W.A., Liu, Z., Yang, W.L. et al. (2009). Origin of the monochromatic photoemission peak in diamondoid monolayers. *Nano Lett.* 9: 57–61.

58 Roth, S., Leuenberger, D., Osterwalder, J. et al. (2010). Negative-electron-affinity diamondoid monolayers as high-brilliance source for ultrashort electron pulses. *Chem. Phys. Lett.* 495: 102–108.

59 Zhang, W.H., Gao, B., Yang, J.L. et al. (2009). Electronic structure of [121]tetramantane-6-thiol on gold and silver surfaces. *J. Chem. Phys.* 130: 054705.

60 Wang, Y., Kioupakis, E., Lu, X. et al. (2008). Spatially resolved electronic and vibronic properties of single diamondoid molecules. *Nat. Mater.* 7: 38–42.
61 Marsusi, F., Sabbaghzadeh, J., and Drummond, N.D. (2011). Comparison of quantum Monte Carlo with time-dependent and static density-functional theory calculations of diamondoid excitation energies and Stokes shifts. *Phys. Rev. B* 84: 245315.
62 Gallmann, L., Jordan, I., Worner, H.J. et al. (2017). Photoemission and photoionization time delays and rates. *Struct. Dyn.* 4: 061502.
63 Clay, W.A., Maldonado, J.R., Pianetta, P. et al. (2012). Photocathode device using diamondoid and cesium bromide films. *Appl. Phys. Lett.* 101: 241605.
64 Yan, H., Narasimha, K.T., Denlinger, J. et al. (2018). Monochromatic photocathodes from graphene-stabilized diamondoids. *Nano Lett.* 18: 1099–1103.
65 (a) Fokin, A.A. and Schreiner, P.R. (2002). Selective alkane transformations via radicals and radical cations: insights into the activation step from experiment and theory. *Chem. Rev.* 102: 1551–1593. (b) Shubina, T.E. and Fokin, A.A. (2011). Hydrocarbon σ-radical cations. *WIREs Comput. Mol. Sci.* 1: 661–679. (c) Patzer, A., Schütz, M., Möller, T., and Dopfer, O. (2012). Infrared spectrum and structure of the adamantane cation: direct evidence for Jahn–Teller distortion. *Angew. Chem. Int. Ed.* 51: 4925–4929. (d) Xiong, T., Wlodarczyk, R., Gallandi, L. et al. (2018). Vibrationally resolved photoelectron spectra of lower diamondoids: a time-dependent approach. *J. Chem. Phys.* 148: 044310.
66 Li, F.H., Fabbri, J.D., Yurchenko, R.I. et al. (2013). Covalent attachment of diamondoid phosphonic acid dichlorides to tungsten oxide surfaces. *Langmuir* 29: 9790–9797.
67 Pujari, S.P., Scheres, L., Marcelis, A.T.M., and Zuilhof, H. (2014). Covalent surface modification of oxide surfaces. *Angew. Chem. Int. Ed.* 53: 6322–6356.
68 Adhikari, B., Meng, S., and Fyta, M. (2016). Carbene-mediated self-assembly of diamondoids on metal surfaces. *Nanoscale* 8: 8966–8975.
69 Landt, L., Bostedt, C., Wolter, D. et al. (2010). Experimental and theoretical study of the absorption properties of thiolated diamondoids. *J. Chem. Phys.* 132: 144305.
70 Rander, T., Staiger, M., Richter, R. et al. (2013). Electronic structure tuning of diamondoids through functionalization. *J. Chem. Phys.* 138: 024310.
71 Landt, L., Staiger, M., Wolter, D. et al. (2010). The influence of a single thiol group on the electronic and optical properties of the smallest diamondoid adamantane. *J. Chem. Phys.* 132: 024710.
72 Sarap, C.S., Adhikari, B., Meng, S. et al. (2018). Optical properties of single- and double-functionalized small diamondoids. *J. Phys. Chem. A* 122: 3583–3593.
73 Blaney, F., Johnston, D.E., McKervey, M.A., and Rooney, J.J. (1975). Diamondoid rearrangements in chlorosulphonic acid. A highly regioselective route to apically disubstituted diamantanes. *Tetrahedron Lett.* 99–100.
74 Gao, H.Y., Šekutor, M., Liu, L.C. et al. (2019). Diamantane suspended single copper atoms. *J. Am. Chem. Soc.* 141: 315–322.

75 Sakai, Y., Nguyen, G.D., Capaz, R.B. et al. (2013). Intermolecular interactions and substrate effects for an adamantane monolayer on a Au(111) surface. *Phys. Rev. B* 88: 235407.

76 Ebeling, D., Šekutor, M., Stiefermann, M. et al. (2017). London dispersion directs on-surface self-assembly of [121]tetramantane molecules. *ACS Nano* 11: 9459–9466.

77 Ebeling, D., Šekutor, M., Stiefermann, M. et al. (2018). Assigning the absolute configuration of single aliphatic molecules by visual inspection. *Nat. Commun.* 9: 2420.

78 Pechenezhskiy, I.V., Hong, X., Nguyen, G.D. et al. (2013). Infrared spectroscopy of molecular submonolayers on surfaces by infrared scanning tunneling microscopy: Tetramantane on Au(111). *Phys. Rev. Lett.* 111: 126101.

79 Schoenbaum, C.A., Schwartz, D.K., and Medlin, J.W. (2013). Controlling surface crowding on a Pd catalyst with thiolate self-assembled monolayers. *J. Catal.* 303: 92–99.

80 Pang, S.H., Lien, C.H., and Medlin, J.W. (2016). Control of surface alkyl catalysis with thiolate monolayers. *Catal. Sci. Technol.* 6: 2413–2418.

81 Lien, C.H. and Medlin, J.W. (2014). Promotion of activity and selectivity by alkanethiol monolayers for Pd-catalyzed benzyl alcohol hydrodeoxygenation. *J. Phys. Chem. C* 118: 23783–23789.

82 Hao, P.X., Pylypenko, S., Schwartz, D.K., and Medlin, J.W. (2016). Application of thiolate self-assembled monolayers in selective alcohol oxidation for suppression of Pd catalyst deactivation. *J. Catal.* 344: 722–728.

83 (a) Phares, N., White, R.J., and Plaxco, K.W. (2009). Improving the stability and sensing of electrochemical biosensors by employing trithiol-anchoring groups in a six-carbon self-assembled monolayer. *Anal. Chem.* 81: 1095–1100. (b) Maier, F.C., Sarap, C.S., Dou, M.F. et al. (2019). Diamondoid-functionalized nanogaps: from small molecules to electronic biosensing. *Eur. Phys. J. Spec. Top.* 227: 1681–1692.

84 Ishiwata, H., Acremann, Y., Scholl, A. et al. (2012). Diamondoid coating enables disruptive approach for chemical and magnetic imaging with 10 nm spatial resolution. *Appl. Phys. Lett.* 101: 163101.

85 Aviram, A. and Ratner, M.A. (1974). Molecular rectifiers. *Chem. Phys. Lett.* 29: 277–283.

86 Randel, J.C., Niestemski, F.C., Botello-Mendez, A.R. et al. (2014). Unconventional molecule-resolved current rectification in diamondoid-fullerene hybrids. *Nat. Commun.* 5: 4877.

87 Fokin, A.A., Butova, E.D., Chernish, L.V. et al. (2007). Simple preparation of diamondoid 1,3-dienes via oxetane ring opening. *Org. Lett.* 9: 2541–2544.

88 Zhou, S.J., Wang, L., Chen, M.J. et al. (2017). Superstructures with diverse morphologies and highly ordered fullerene C-60 arrays from 1:1 and 2:1 adamantane-C-60 hybrid molecules. *Nanoscale* 9: 16375–16385.

89 Tawfik, S.A., Cui, X.Y., Ringer, S.P., and Stampfl, C. (2016). Enhanced oscillatory rectification and negative differential resistance in pentamantane diamondoid-cumulene systems. *Nanoscale* 8: 3461–3466.

90 (a) Yu, H.Z., Rao, B., Jiang, W. et al. (2019). The photoluminescent metal nanoclusters with atomic precision. *Coord. Chem. Rev.* 378: 595–617. (b) Aikens, C.M. (2018). Electronic and geometric structure, optical properties, and excited state behavior in atomically precise thiolate-stabilized noble metal nanoclusters. *Acc. Chem. Res.* 51: 3065–3073. (c) Jin, R.C., Zeng, C.J., Zhou, M., and Chen, Y.X. (2016). Atomically precise colloidal metal nanoclusters and nanoparticles: fundamentals and opportunities. *Chem. Rev.* 116: 10346–10413.

91 Krommenhoek, P.J., Wang, J.W., Hentz, N. et al. (2012). Bulky adamantanethiolate and cyclohexanethiolate ligands favorsmaller gold nanoparticles with altered discrete sizes. *ACS Nano* 6: 4903–4911.

92 Higaki, T., Liu, C., Zeng, C.J. et al. (2016). Controlling the atomic structure of Au_{30} nanoclusters by a ligand-based strategy. *Angew. Chem. Int. Ed.* 55: 6694–6697.

93 Liu, C., Li, T., Li, G. et al. (2015). Observation of body-centered cubic gold nanocluster. *Angew. Chem. Int. Ed.* 54: 9826–9829.

94 Chen, S., Xiong, L., Wang, S.X. et al. (2016). Total structure determination of $Au_{21}(S\text{-}Adm)_{15}$ and geometrical/electronic structure evolution of thiolated gold nanoclusters. *J. Am. Chem. Soc.* 138: 10754–10757.

95 Li, Z.M., Liu, C., Abroshan, H. et al. (2017). $Au_{38}S_2(SAdm)_{20}$ photocatalyst for one-step selective aerobic oxidations. *ACS Catal.* 7: 3368–3374.

96 Yan, C.Y., Liu, C., Abroshan, H. et al. (2016). Surface modification of adamantane-terminated gold nanoclusters using cyclodextrins. *Phys. Chem. Chem. Phys.* 18: 23358–23364.

97 Crasto, D., Barcaro, G., Stener, M. et al. (2014). $Au_{24}(SAdm)_{16}$ nanomolecules: X-ray crystal structure, theoretical analysis, adaptability of adamantane ligands to form $Au_{23}(SAdm)_{16}$ and $Au_{25}(SAdm)_{16}$, and its relation to $Au_{25}(SR)_{18}$. *J. Am. Chem. Soc.* 136: 14933–14940.

98 (a) Jones, T.C., Sementa, L., Stener, M. et al. (2017). $Au_{21}S(SAdm)_{15}$: crystal structure, mass spectrometry, optical spectroscopy, and first-principles theoretical analysis. *J. Phys. Chem. C* 121: 10865–10869. (b) Fortunelli, A., Sementa, L., Thanthirige, V.D. et al. (2017). $Au_{21}S(SAdm)_{15}$: an anisotropic gold nanomolecule. Optical and photoluminescence spectroscopy and first-principles theoretical analysis. *J. Phys. Chem. Lett.* 8: 457–462.

99 Kang, X., Zhou, M., Wang, S.X. et al. (2017). The tetrahedral structure and luminescence properties of bi-metallic $Pt_1Ag_{28}(SR)_{18}(PPh_3)_4$ nanocluster. *Chem. Sci.* 8: 2581–2587.

6

Growing Diamond Structures from Diamondoids Via Seeding

Currently, diamondoids appear to be the most promising building blocks for nanodiamond synthesis [1]. The formation of diamond from **AD** and many other organic molecules under high temperature and high pressure (HT–HP) conditions was disclosed in 1965 [2] and has since been studied intensely [3]. Alternatively, diamond growth from the gas phase at low pressures through plasma-induced chemical vapor deposition (CVD) techniques had already been developed in the middle of the 1950s [4]. The mechanism for diamond formation is complex and still debated [5], also with respect to utilizing diamondoids as models [6]. The idea that higher diamondoids and some other cage hydrocarbons may serve as nucleation centers for diamond growth on arbitrary surfaces was put forward in 1983 (Figure 6.1) by Matsumoto and Matsui [7] based on the behavior of diamond particles formed in CVD experiments. Additional support for this idea comes from the formation of higher diamondoids *en route* on-surface diamond growth [8] involving diamondoidyl radicals [9]. Diamondoid-seeded diamond growth may result in the formation of diamond films, nanodiamond particles, or higher diamondoids [10]. All these processes are mechanistically related and are discussed below.

6.1 Diamondoid-Promoted Growth of Diamond Under HT-HP or CVD Conditions

The epitaxial growth of polycrystalline diamond on various surfaces attracted much attention due to its high potential in several fields of technology, especially in electronics. Although over 10,000 original and 200 review articles have been published since 1990 in this area, the subject is still intensely explored [11]. The major challenge is the growth of orientationally well-defined, highly dense, and low-stress diamond films; this depends mostly on the nucleation mechanism. Numerous surface pre-treatments (from simple mechanic scratching to pre-deposition of various inorganic and organic materials) result in low-quality diamond films as they are neither homogeneous nor structurally well-defined to satisfy the requirements for nanoelectronic applications. The HT–HP conditions for diamond growth from graphite require pressures of 15 GPa (150 kbars = 148,038.5 atm) at temperatures

The Chemistry of Diamondoids: Building Blocks for Ligands, Catalysts, Materials, and Pharmaceuticals, First Edition. Andrey A. Fokin, Marina Šekutor, and Peter R. Schreiner.
© 2024 WILEY-VCH GmbH. Published 2024 by WILEY-VCH GmbH.

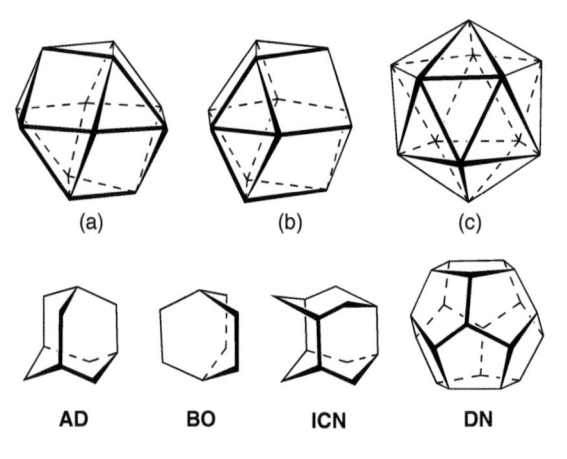

Figure 6.1 The correspondence first suggested by Matsumoto and Matsui between (a) cubo-octahedron (a single crystal), (b) twinned cubo-octahedron, (c) icosahedron diamond particles and cage compounds adamantane (**AD**), bicyclo[2.2.2]octane (**BO**), iceane (**ICN**), and dodecahedrane (**DN**).

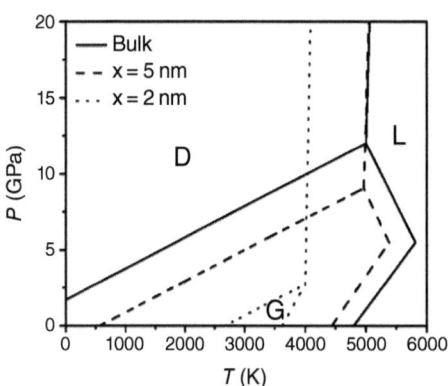

Figure 6.2 T–P phase diagrams for graphite (G) transition to diamond (D) and where both phases melt to form the liquid phase (L). The position of the G/D/L triple point is size-dependent and shifts toward lower temperatures and pressures with decreasing the crystal size. The solid, dash, and dot lines denote the bulk diamond and nanodiamond particles of 5 and 2 nm size. Source: Reproduced from Ref. [12] with permission from the American Chemical Society, 2008.

in the range of 2000–3000 K. The graphite/diamond/liquid triple transition point is size-dependent and shifts towards lower temperatures and pressures with decreasing the crystal size (Figure 6.2). The presence of organic material, e.g., diamondoids or various catalysts lowers the temperature threshold for diamond formation. The first attempt to use diamondoids as a source for HT–HP diamond growth showed that at least at 15 GPa and 2000 K there is no advantage in using **AD** over many other organic materials such as paraffin wax or polyethylene [2]. All these starting materials form soft white powders where, based on X-ray diffraction data, a diamond crystalline phase is present. Such harsh reaction conditions (**AD** decomposes above 500 °C) [7] triggered the question of whether diamond forms prior to the completion of graphitization, i.e., directly from the organic precursors or through the crystallization of graphite in the fluid medium. Note that diamondoid crystals are characterized by low compressibility and structurally remain intact at least up to 20 GPa pressure without heating [13].

In 1992, a detailed comparative study of the behavior of three potential diamond precursors, namely camphene, **AD**, and fluorene under slightly milder conditions (from 8 GPa and 1500 K), was published [14]. The obtained samples were investigated by X-ray and electron diffraction, Raman scattering, transmission electron microscopy (TEM), and scanning electron microscopy (SEM). The fact that the formation of diamond from **AD** took place at 8 GPa and 1600 K, but graphite was observed already at 1100 K at the same pressure, strongly suggests complete graphitization of the organic material prior to diamond formation. Additionally, the presence of transition metals may catalyze diamond formation. Recent studies [15] excluded the participation of metals and suggested that at 8 GPa and 1520 K, "the process of graphite and nanodiamond formation occurs simultaneously under the decomposition of adamantane" [15a], contradicting the finding that a gradual increase in temperature shows that already at 800 °C only graphite is present in the reaction mixture [3a]. Notably, the above experimental conditions are still much milder than those required for quantitative conversion of graphite to diamond (12.5 GPa and 3000 °C) [16]. In the presence of GeI_4 in tetracosane media, nanodiamond formation with **AD** as a seed was observed at relatively low pressures and temperatures (3.5 GPa and 500 °C) [17]. The control experiments without metal catalysts only yielded amorphous carbon material, thus showing that at moderate temperatures **AD** may indeed be used as a source for catalytic nanodiamond growth. Diamondoids higher than **AD** were tested as a carbon source for diamond growth under HT-HP conditions without a catalyst [18], and the lowest temperature that yielded diamond was 900 K at ~20 GPa, whereas heating up to 2000 K is needed to obtain diamond at lower pressures (12 GPa).

The mechanism of diamondoid-to-diamond conversion was studied based on model experiments where graphite, initially formed from diamondoids at lower temperatures, was subjected to the reaction conditions under which diamond forms from diamondoids. The absence of diamond suggests that the diamondoid-to-diamond transformation is a direct process and may omit graphite formation. This is also supported by the fact that reactions proceeded differently for different diamondoids, of which **TRIA** displays the best results and forms diamond under slightly milder reaction conditions. The latter is possibly due to the presence of quaternary carbon atoms that help the formation of a cubic diamond lattice in the initial stages. Though it was stated that experiments conducted with octadecane instead of diamondoids [18] "confirmed that higher temperatures were necessary for diamond formation", no experimental data were presented at comparable temperatures and pressures.

Extraordinary high volatility and only moderate thermal stability hamper the utilization of **AD** as seed material for conventional CVD diamond growth, where temperatures up to 1000 °C are usually required [19]. Attempts to apply lower temperatures (530 °C) using microwave plasma (MP) CVD techniques gave the expected high-quality diamond film on silicon substrates; however, **AD** still first converts either to graphite or amorphous carbon and only then grows into diamond films [20]. This was confirmed independently by showing that surface pretreatment with **AD** or graphite gives similar results [21]. The absorption of **AD** on a Pt-covered Si/SO_2 surface prior to MW-CVD deposition at 700 °C allowed to follow the **AD** decomposition dynamics and to study the surface morphology by the combined SEM, XPS, and TEM. In contrast to the above, **AD** transforms into nanodiamond as well as to some unidentified carbon particles at the early stage of deposition before the formation of a graphite phase [22]. Preparation of a silicon pre-surface by dipping it into a solution of **AD** in high-boiling organic solvents (ethylene glycol or diethylene glycol) was also tested [23]. Presumably due to reducing the evaporation of **AD**, the quality of diamond films is much better with diethylene glycol than with those obtained with ethylene glycol. This was estimated from the relative intensities of the fully symmetric diamond Raman stretching mode at 1332 cm^{-1} (not present in diamondoids [24]) vs. the absorption around 1510 cm^{-1} attributed to residual graphitic material [23]. Pre-dissolving of **AD** in high-boiling point solvents appeared to be useful for diamond growth on surfaces different from silicon. Good-quality and highly crystalline diamond films were obtained from **AD**-coated sapphire substrate at high temperatures (700 °C) through MW-CVD deposition in the presence of H_2/CH_4 [25]. SEM, XRD, and Raman spectroscopy showed that dipping into ethylene glycol assists diamond nucleation; however, the question of whether **AD** directly transforms to diamond or first decomposes into carbon-containing species remains open.

Further studies aimed at diamond growth on freshly reduced silicon surfaces at temperatures just below the **AD** decomposition limit. Such Si surfaces were covered with **AD** via sublimation (Figure 6.3), and subsequent diamond growth at 475 °C in a CH_4/H_2 plasma gave films with good crystallinity as judged from SEM and Raman spectroscopic measurements. XPS analysis confirmed that diamond nucleation and growth occur on silicon carbide SiC surfaces rather than on clean Si surfaces [26].

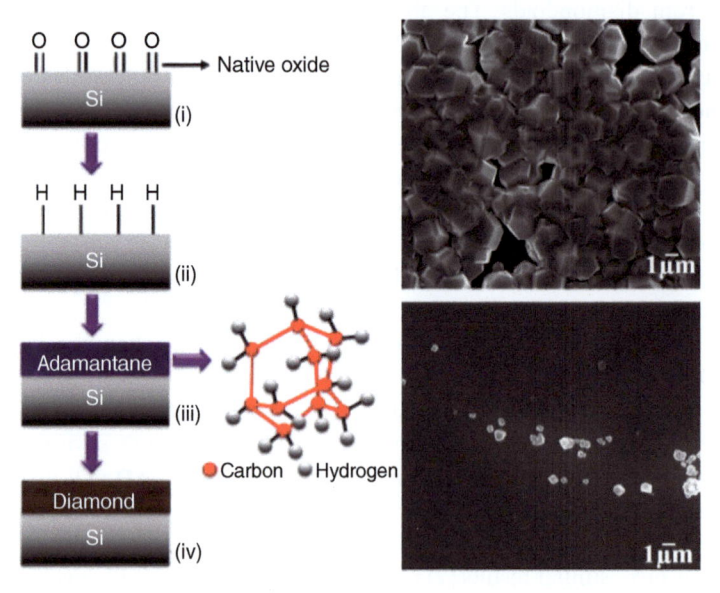

Figure 6.3 *Left*: Schematic diagram showing diamond synthesis in four steps; (i) silicon substrate with native oxide layer; (ii) HF/NH$_4$F solution treatment to remove native oxide from the Si surface; (iii) adamantane deposition on silicon surface by sublimation; and (iv) diamond growth by microwave plasma CVD; *Right*: SEM image of diamond obtained with (top) and without (bottom) **AD** pre-deposition. Source: Reproduced from Ref. [26] with permission from Elsevier, 2010.

6.1 Diamondoid-Promoted Growth of Diamond Under HT-HP or CVD Conditions | 175

Utilization of low-pressure MPs allows even further reduction of the temperature (to ca. 150 °C), completely protecting **AD** from decomposition [27]. The diamond films thus obtained display superior quality (Figure 6.4) over deposits using graphite-cluster diamond (GCD) as a seeding agent. This confirmed for the first time the possibility of true **AD**-enhanced diamond nucleation where the diamondoid seed acts as a direct diamond precursor.

Nanodiamond particles form from **AD** as a precursor and seed in barrier discharge microplasmas under high pressure in supercritical xenon [28]. Transmission electron microscopy analysis of the lattice of the synthesized particles revealed crystal structures similar to those observed in nanodiamonds (Figure 6.5).

Alternatively, the **AD** cage may be chemically bonded to the silicon surface that prevents its migration or evaporation: 2,2-divinyladamantane [29] was attached to a pristine silicon surface via a photochemical hydrosilylation [30]. The process was initially carried out at 250 °C in the presence of CH_4 to prevent etching of the seeds and

Figure 6.4 *Top*: Optical micrographs of diamond films deposited on borosilicate glass through low-temperature (150 °C) microwave plasma CVD; (a) a film grown on the substrate using graphite-cluster diamond (GCD) seed. (b) The same but with **AD** seeding. *Bottom*: Sharp characteristic 1333 cm^{-1} peak in the UV Raman scattering spectrum of a nanocrystalline diamond film grown at 150 °C. Source: Reproduced from Ref. [27] with permission from the American Chemical Society, 2010.

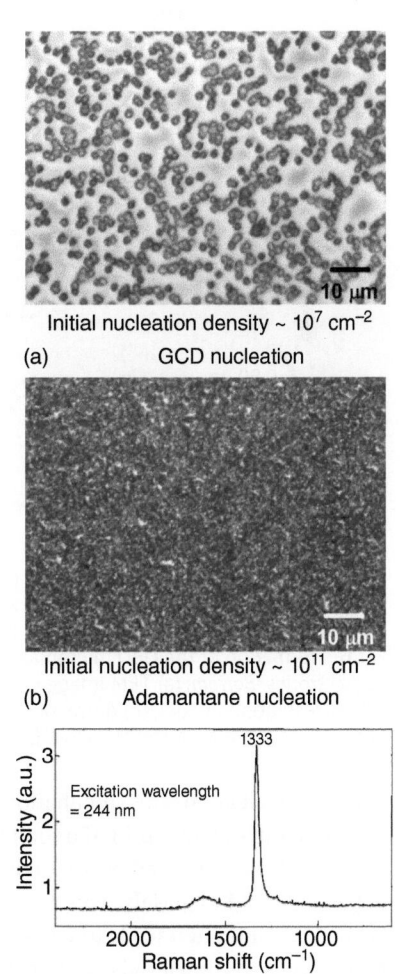

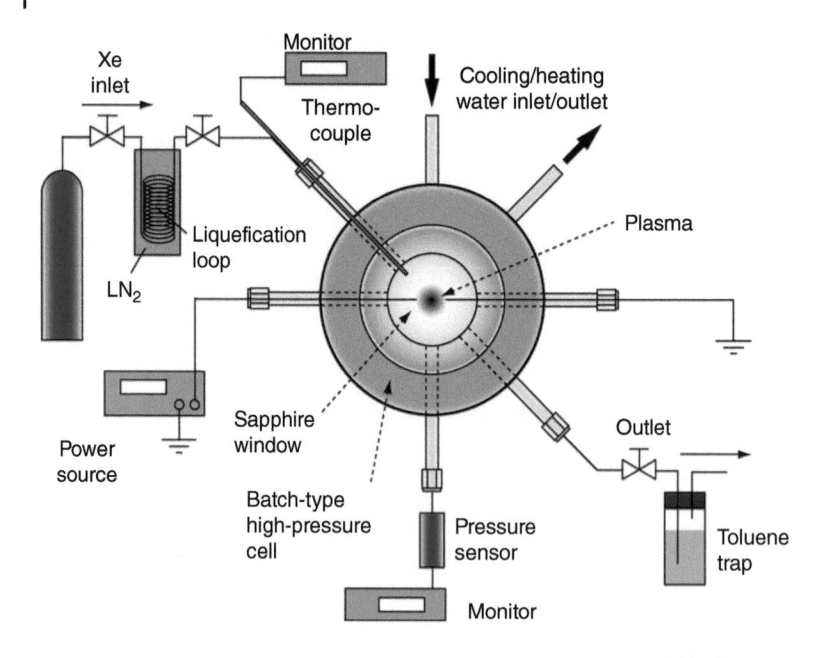

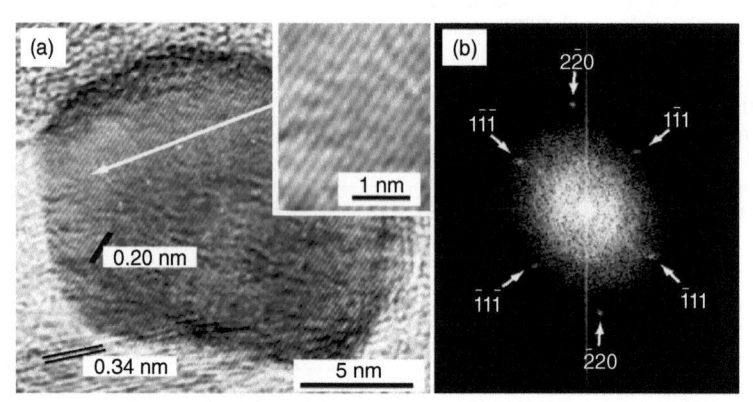

Figure 6.5 *Top*: Experimental setup of a high-pressure cell used for generating sheet-like dielectric barrier discharge (DBD) microplasma in supercritical xenon. To achieve supercritical conditions, Xe was first condensed in a cold liquefaction loop and then introduced into the high-pressure cell until reaching conditions close to the critical point. For the formation of the DBD microplasma, a tungsten-mesh and Ag paste were employed as electrodes. *Bottom*: (a) TEM image of a synthesized nanodiamond particle (inset showing a high-resolution image); (b) diffraction pattern obtained by Fourier transformation of the image. Source: Reproduced from Ref. [28] with permission from Elsevier, 2010.

to create nucleation sites (incubation) followed by diamond growth at 700–900 °C. This gave high-quality thin diamond films (Figure 6.6).

AD itself may be used as a carbon source for CVD growth that often results in doped diamond materials. This usually requires very high deposition temperatures (>1000 °C) and often leads to less ordered films [31]. Utilization of pure **AD** gives highly defective materials that may even display ferromagnetic properties due

Grafting   Incubation   Growth

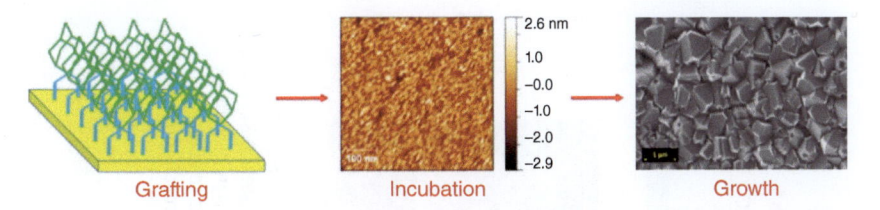

Figure 6.6 Formation of high-quality diamond thin films from photochemical grafting of a silicon surface with 2,2-divinyladamantane, followed by low-temperature incubation and high-temperature diamond growth. Scanning electron micrographs (right) display the formation of high-quality diamond film. Source: Reproduced from Ref. [30] with permission from the American Chemical Society, 2016.

to the presence of residual radical centers (sometimes called "dangling bonds", vide supra) [32]. Other successful attempts to grow defect diamond utilizing **AD** as a carbon source included the preparation of nitrogen- [33] and iodine-doped [34] diamond films. Comparative studies show that CVD films obtained from low-volatile 1,3-dibromoadamantane display superior stabilities and permittivities vs. films obtained from pure **AD**, which give a material with a high concentration of sp^2-defects [35].

6.2 Higher Diamondoids for Diamond Nucleation

Due to their lower volatilities, higher diamondoids and their derivatives are viewed as more attractive than **AD** for surface nucleation seeding, with the potential of incorporation of color centers into diamond material with a variety of possible applications [36]. Diamondoid phosphonates attract special attention due to the fact that some have extremely high melting points and remarkable thermal stabilities [37]. The largest synthesized diamondoid phosphonate 7-dichlorophosphoryl-[1(2,3)4]pentamantane (**POCLPENT**, Figure 6.7a) was used as a seed for nanodiamond crystal and diamond film CVD growth on silicon carbide. This resulted in high-quality fluorescent diamond nanoparticles and diamond films with bonding between the heteroepitaxial diamond layer and the substrate [38]. **POCLPENT** was first covalently attached to a pre-oxidized silicon carbide surface through P—O—Si bonding (Figure 6.7b). The nucleation and growth steps were performed at 450 and 830 °C, respectively, and gave in the presence of CH_4/H_2 high-quality diamond crystals based on SEM data (Figure 6.7c, d). Hybrid diamond-SiC microdome structures (diamond microcrystals) with diameters ~2 μm were first grown on 150 nm thick 3C-SiC epitaxial film on Si, followed by removal of the sacrificial Si layer with XeF_2 (Figure 6.7, right, 1–3). The images of the thus obtained diamond-3C-SiC microdomes were obtained from SEM (Figure 6.7, right, d, e). Thus constructed semiconductor-based quantum photonic structures contain negatively charged silicon vacancy color centers and are viewed as diamond qubit states with the potential for constructing solid-state quantum memory devices.

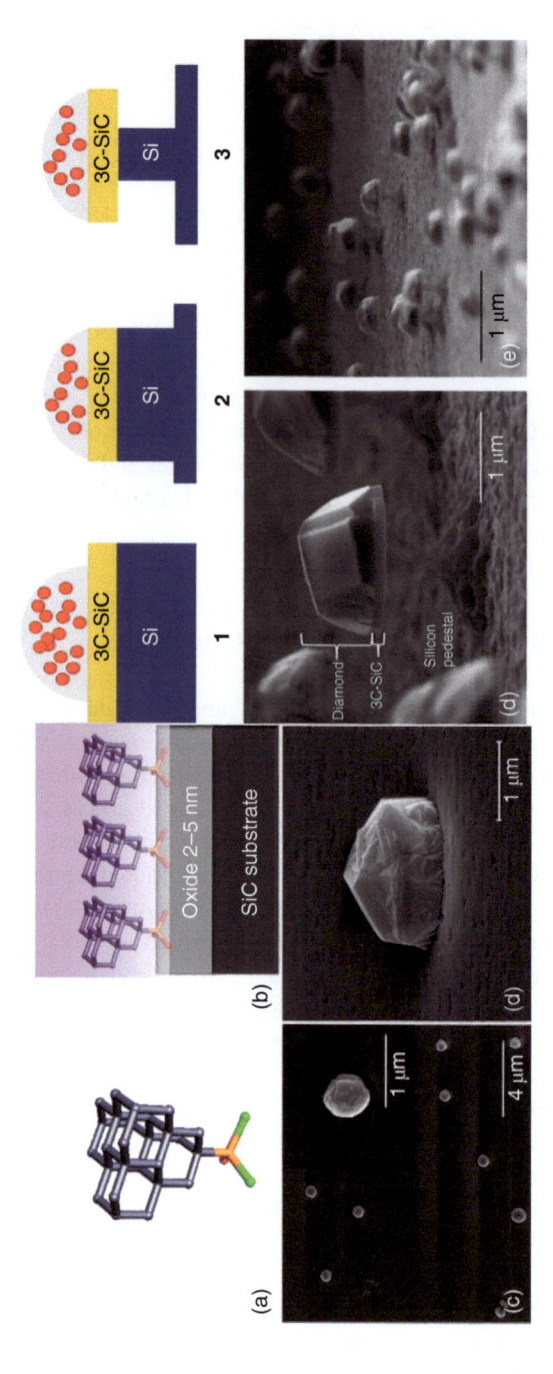

Figure 6.7 *Left:* (a) Molecular structure of 7-dichlorophosphoryl-[1(2,3)4]pentamantane (**1(2,3)4PENTPOCl2**) formed on top of a SiC substrate; (b) **1(2,3)4PENTPOCl2** on an oxide layer formed on top of a SiC substrate; (c) scanning electron micrograph (SEM) of 500 nm diameter nanodiamonds grown on a 4H-SiC substrate; (d) SEM of a micrometer-size diamond on a 3C-SiC substrate. *Right:* 1–3. Process flow for hybrid diamond–SiC microdome structure fabrication through hard mask pattern transfer. The red dots represent the silicon vacancy centers in the diamond nanocrystals; (d) SEM image of a typical microdome structure. High-quality diamond as well as the heteroepitaxial interface are visible; (e) SEM image of an ensemble of microdome structures. Source: Reproduced from Ref. [38] with permission from the American Chemical Society, 2016.

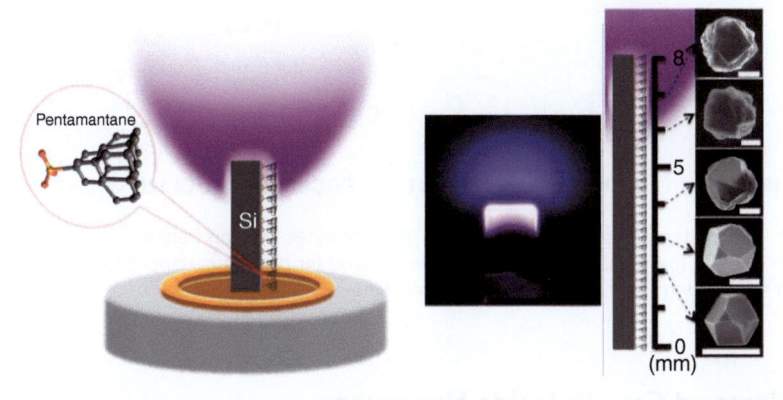

Figure 6.8 *Left*: Schematic illustration and a photo of the vertical substrate microwave plasma CVD-diamond growth seeded with [1(2,3)4]pentamantane chemically bonded to an oxidized silicon wafer surface via a phosphonyl dichloride function. The hydrogen plasma is concentrated on the top edge. *Right*: Diamond morphologies as they change along the Si-wafer substrate, where the highest-quality nanodiamonds form 2–3 mm from the bottom of the wafer. Source: Reproduced from Ref. [6] with permission from the American Chemical Society, 2017.

Further attempts with higher diamondoids as seeds concentrated on improving the diamond deposition conditions (Figure 6.8) [6e]. For instance, [1(2,3)4]pentamantane chlorophosphonate (**POCLPENT**) was attached to an oxidized silicon surface as above, but the silicon wafer substrate with the attached diamondoid was orientated perpendicularly on a molybdenum substrate holder, thereby playing the role of a plasma antenna. The MP growth in a CH_4/H_2 atmosphere was performed at relatively low temperatures of 350 °C, i.e., below the diamondoid decomposition temperature. The vertical location of the substrate (Figure 6.8, left) enabled the longitudinal growth that displayed a systematic trend of crystal morphologies and diamond particle densities. The highest-quality diamond nanoparticles formed 2–3 mm from the bottom of the wafer (Figure 6.8, right) and showed single crystal faceting and a very sharp characteristic 1332 cm^{-1} peak in the Raman spectrum. The largest but least pure (sp^2-contaminated) diamonds and diamond-like nanoparticles were found to grow at the top of the wafer. All this adds to the understanding of the fundamental aspects of CVD-diamond growth in controlled hydrogen plasmas.

Various diamondoid phosphonates were recently [6d] used as seeds on silicon surfaces, revealing key mechanistic details of diamond CVD growth. Even under quite harsh plasma CH_4/H_2 CVD growing conditions at 750 °C, the nucleation efficiency appears to depend on the diamondoid size (Figure 6.9). This provides important information about the critical nucleus size, i.e., when particles start to grow deterministically. Two of the pentamantane isomers employed exceed the critical size, suggesting that the critical nucleus size value is between 22 and 26 carbon atoms, or about 0.8 nm particle diameter. The nucleus size appears to be the dominant factor for nucleation, while symmetry plays a secondary role as the nucleation behavior of highly symmetric **POCLPENT** and less symmetric [12(1)3]pentamantane (**POCL12(1)3PENT**) chlorophosphonates (Figure 6.9) is comparable.

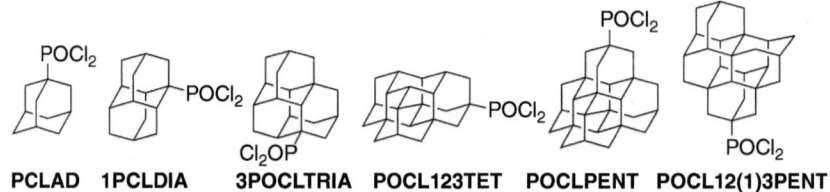

PCLAD 1PCLDIA 3POCLTRIA POCL123TET POCLPENT POCL12(1)3PENT

Figure 6.9 Molecular structures of diamondoid phosphonates ranging from 10 (adamantane) to 26 (pentamantane) carbon atoms utilized for plasma-enhanced CVD deposition seeded with diamondoid phosphonates. Source: From Ref. [6].

6.3 Diamond Growth Inside Nanotubes

The interest in 1D-diamond nanoscale materials (nanowires/rods/threads) arises from their unique mechanical and electronic properties with expectedly valuable practical potential [39]. In particular, the stiffness of diamond nanowires is expected to be higher than that of carbon nanotubes (CNT) [40], and the electronic band gap is smaller than that of diamond [41]. The existing methods for the preparation of diamond nanowires [39b, 42] give materials that are significantly thicker than the range of interest (<10 nm diameter) [43], while even slightly narrower tubes are difficult to make. An example is the direct hydrogenation of CNTs in hydrogen plasmas that provides diamond nanorods with an outsized diameter of 4–8 nm [44]. CNTs offer an *inner* space for diamond growth as the currently only viable approach to ultrathin nanowires of 1–3 nm diameter. This size defines the limit of the thermodynamic stability of diamond nanowires [45] and occupies a region where electronic band gap tunability is expected [41]. Furthermore, the formation of diamond from carbon materials confined in a CNT is thermodynamically more favorable than the growth on flat surfaces [46]. The possibility of diamond growth from methane inside a CNT was confirmed experimentally in 2009 with the example of diamond nanorods as thin as 2.1 nm [47].

Diamondoids are viable seeding materials for diamond growth as **AD** can be encapsulated inside CNTs, and the density functional theory (DFT) computations predicted that diamondoids enter CNTs spontaneously [48]. The experimental evidence that **AD** efficiently inserts into singe-wall CNTs with diameters of 1.3–1.6 nm was obtained in 2011 [49] through high-resolution TEM-imaging (Figure 6.10a) as well as the experimental ^{13}C NMR shifts and IR- and Raman spectra of the encapsulated material [49]. Notably, unsuccessful attempts to encapsulate **AD** in smaller nanotubes with a diameter of less than 1.2 nm are in agreement with DFT computations [51] and MD simulations [52]. The first attempt to chemically transform **AD** inside a CNT was made in 2012 [50]. The sharp peak in the Raman spectrum at 1857 cm^{-1} observed after thermolysis of **AD** encapsulated in a CNT was attributed to carbon chains (Figure 6.10b) interacting with a nanotube wall. Thus, **AD** decomposes inside the CNT, but there were no signs of diamond formation. The presence of hydrogen suppresses the decomposition, and **AD** inside the CNT

6.3 Diamond Growth Inside Nanotubes | 181

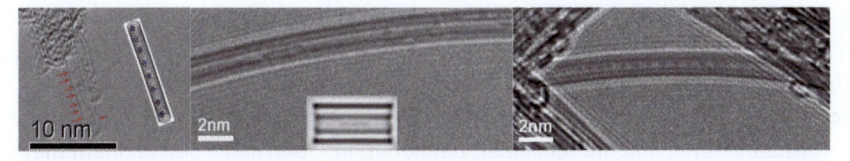

Figure 6.10 HR-TEM images and computer simulations of: (*Left*) Single-wall carbon nanotube filled with **AD** molecules (inset: molecular mechanics simulations of eight **AD** molecules inside a (10,10) single-wall CNT with a diameter of 1.38 nm). Source: Reproduced from Ref. [49] with permission from Elsevier, 2011. (*middle*) Adamantane@CNT sample annealed at 10^{-7} Torr vacuum, showing carbon chains inside the nanotube. (*right*) Adamantane@CNT sample annealed under a hydrogen atmosphere containing intact guest molecules. Source: Reproduced from Ref. [50] with permission of the American Chemical Society, 2012.

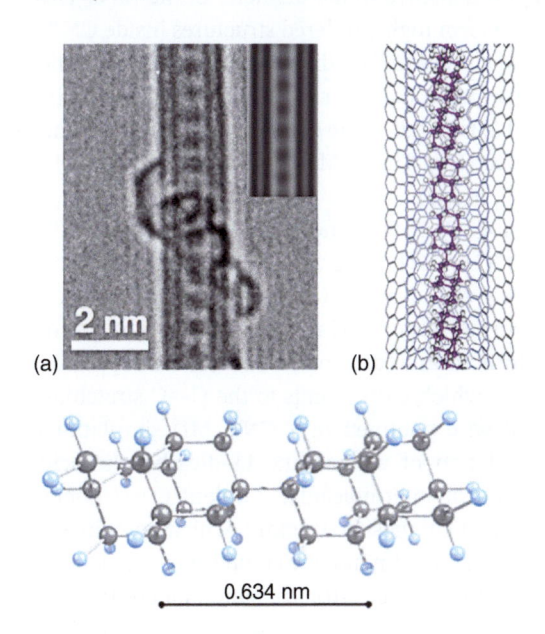

Figure 6.11 *Top*: (a) High-resolution transmission electron microscopy image of linear-chain nanodiamond polymer (well-aligned dots) inside double-wall CNTs obtained from the vaporized 4,9-dibromodiamantane at high vacuum and temperature; inset: simulated TEM image of a diamantane polymer confined in a nanotube with a center-to-center distance of 6.35 ± 0.2 Å. (b) Corresponding DFT structural model of diamantane polymer linked via apical positions. *Bottom*: Covalently bound diamantane dimer optimized at the B3LYP-D3(BJ)/cc-pVTZ level of theory reveals a center-to-center distance of 6.34 Å that agrees well with the experiment. Source: Reproduced from Ref. [55], Wiley-VCH, 2015.

remains intact (Figure 6.10c). DFT computations suggest that the self-condensation of **AD** inside a CNT requires much higher pressures [53].

This situation changed radically when substituted diamondoid derivatives were encapsulated inside CNTs. Filling of a single-walled CNT with 1-adamantyl methanol gives the desired encapsulation, whereas 1-bromoadamantane dissociates as TEM images confirm the presence of individual bromine atoms inside CNTs [54].

Such dissociation during the NT filling at relatively low temperatures (~160 °C) provides the possibility to couple confined carbon-centered radical intermediates. In the case of dibromo derivatives this may cause polymerization inside the CNT, which was successfully demonstrated with 4,9-dihalodiamantanes (Fig 6.11) [55]. The formation of a one-dimensional nanodiamond polymer (polydiamantane) was achieved under moderate heating of 4,9-dibromodiamantane at 175 °C inside a multi-walled CNT. The corresponding dichloro derivative reacts at higher temperatures, in agreement with the higher BDE of the C—Cl bond.

Higher diamondoids may form well-ordered structures inside CNTs [56]; particularly densely packed and ordered structures were predicted by MD simulations for substituted diamondoids such as 4,9-dihydroxydiamantane, where the degree of encapsulation and ordering is highly sensitive to the diameter of the tube. The ability of diamantane 4,9-derivatives to form highly ordered structures inside CNTs was experimentally confirmed by TEM studies [57] on diamantane-4,9-dicarboxylic acid (Figure 6.12). While a CNT with an inner diameter of <0.8 nm remains empty, increasing the diameter to 1.0 nm and 1.3 nm led to the encapsulation of single and double arrays of guest molecules, respectively. The TEM images are consistent with their simulated structures (Figure 6.12, insets).

Remarkably, when a CNT with $D_{inner} \approx 1.3$ nm with encapsulated diamantane-4,9-dicarboxylic acid was annealed at 600 °C for 12 h under a flow of hydrogen, carbon nanowires with 0.78 nm diameter were found inside (Figure 6.13) [57]. No such structures were observed after the thermolysis of CNTs with an inner diameter of 1 nm with single guest arrays. The formation of diamond nanowires is supported by a new Raman feature at 1245 cm^{-1}, which corresponds to the C—C stretching vibrational mode of a nanodiamond wire confined in a CNT. MD simulations support the notion of the decarboxylation of diamantane-4,9-dicarboxylic acid inside the CNT prior to the formation of intermolecular covalent C—C bonds (Figure 6.13, bottom). Thus, the obtained composite structure, which combines a nanotube shell with an insulator core (diamond nanowire) represents a new class of robust carbon material with potential in nanoelectronics, nano-optoelectronics, and nano-electromechanics [58].

Synthesis of higher diamondoids from **AD** as a precursor and seed is possible under atmospheric pressure under conditions typical for conventional CVD diamond growth in argon–methane–hydrogen plasmas. GC–MS analyses indicate the formation of **DIA** (a standard sample was used as a reference) together with a mixture of isomeric methyl adamantanes [59]. This shows that **DIA** can, in principle, be obtained from **AD** in atmospheric pressure plasmas; however, the quantities of material thus obtained were on the order of a few nanograms, which is one to two orders of magnitude less than the amounts obtained in supercritical liquids [60] (vide infra). Such low yields are due to negligible reactant concentrations in the gas phase under atmospheric pressure in plasma microreactors (Figure 6.14).

The yields of **DIA** increase substantially with increasing H$_2$ pressure, whereas CH$_4$ has no influence on the yields [61]. Monitoring of the relative abundance of **DIA** and 2-methyl adamantane in the course of the reaction indicates that the latter cannot be a precursor for **DIA** formation as these two components display opposite

6.3 Diamond Growth Inside Nanotubes | 183

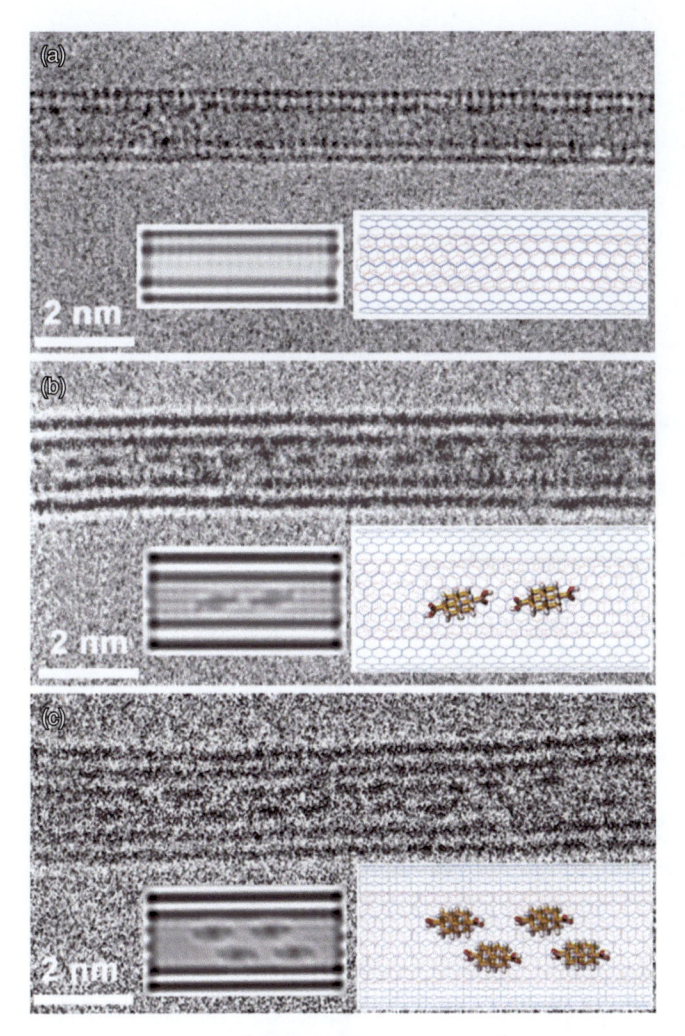

Figure 6.12 Experimental and simulated high-resolution transmission electron microscopy images of (a) empty carbon nanotube ($D_{inner} < 0.8$ nm), (b) linear arrays of diamantane-4,9-dicarboxylic acid inside carbon nanotube with $D_{inner} \approx 1$ nm, and (c) multiple arrays of diamantane-4,9-dicarboxylic acid inside carbon nanotube with $D_{inner} \approx 1.3$ nm. Source: Reproduced from Ref. [57], Wiley-VCH, 2013.

abundances depending on the H_2/CH_4 ratio (Figure 6.15). The fact that the increase in the amount of CH_4 did not influence the **DIA** yield suggests that CH_3-radical is not essential and, rather, other carbon-containing particles C_x ($x = 1-3$) are involved in the homologenation process.

Under conventional CVD conditions, no diamondoids higher than **DIA** were detected. In contrast, under high electron energies achieved through nanosecond pulsed discharge in a helium atmosphere with **AD** as the precursor, a mixture of sp^2- and sp^3-carbon nanoparticles forms based on Raman spectral data [62]. Under such conditions, most of the **AD** precursor decomposed into various types

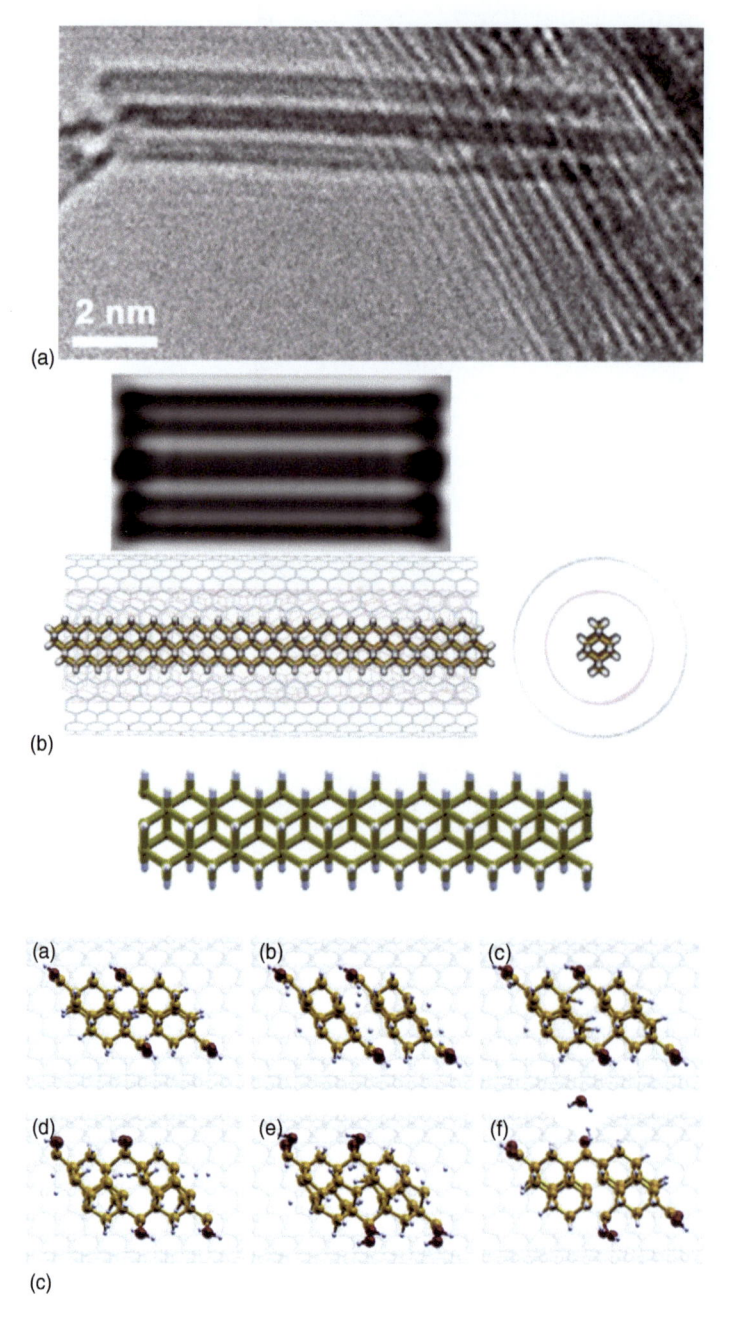

184 | *6 Growing Diamond Structures from Diamondoids Via Seeding*

Figure 6.13 *Top*: (a) High-resolution transmission electron microscopy image of a diamond-based nanowire inside a multiwalled CNT; (b) simulated image of a diamond nanowire inside a CNT along with the corresponding structural model; (c) structural model of a diamond nanowire. *Bottom*: Snapshots of molecular dynamics simulations representing key steps leading to the fusion of diamantane-4,9-dicarboxylic acid molecules inside of an (8,8) CNT. Source: Reproduced from Ref. [57], Wiley-VCH, 2013.

6.3 Diamond Growth Inside Nanotubes

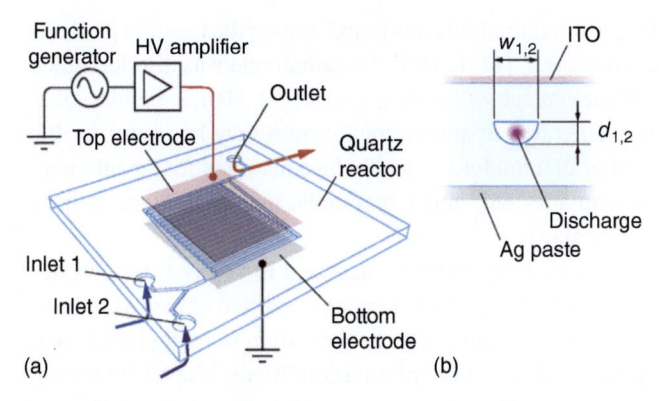

Figure 6.14 Experimental setup and cross-section of plasma microreactor. (a) A three-dimensional view of a quartz microreactor containing 20 microchannels used for the different types of source gases. **AD** is introduced into the microreactor by mixing it with the source gases and by flowing it through the reactor. (b) Cross-section of a single microreactor channel with indium-tin oxide (ITO) and silver (Ag) electric contacts on the top and bottom, respectively. Source: Reproduced from Ref. [61] with permission from Elsevier, 2015.

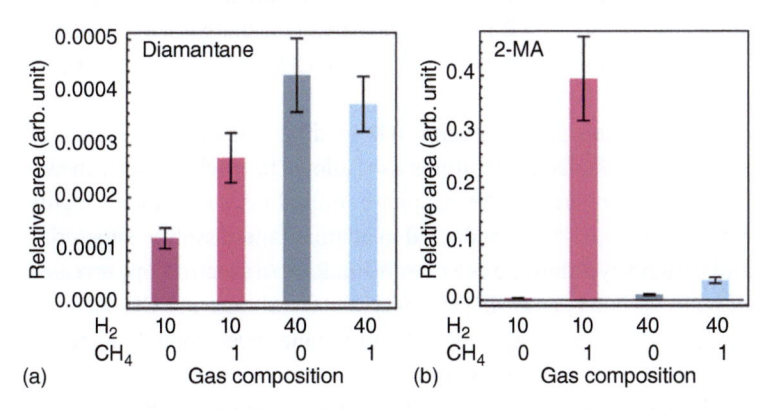

Figure 6.15 Selected ion monitoring peak intensities of diamantane (a) and 2-methyl adamantane (**2 MA**, b) for the four different CH_4/H_2 plasma gas ratios, with adamantane used as a nucleus for diamondoid growth. Source: Reproduced from Ref. [61] with permission from Elsevier, 2015.

of carbonaceous materials. Other examples include the transformation of **AD** to **DIA** in supercritical xenon [63] and, using pulsed laser ablation plasmas, in supercritical CO_2 [64]. In the latter case, however, relatively high reaction temperatures (ca. 500 °C) are required. Quantitative analysis of the reaction products in supercritical xenon plasmas provided a maximum quantity of synthesized **DIA** of about 15 μg from the best experiment [60]. In a typical setup, **AD** dissolved in a high-pressure cell and then flowed through a capillary reactor driven by xenon gas. The electrode consists of a silica capillary coated with ITO and a tungsten wire inserted into the capillary. It should be noted that in all cases the products from plasmas, generated in supercritical fluids, consisted of complex hydrocarbon mixtures where individual components were identified only through their characteristic mass spectra. Along with **DIA**, a large set of higher diamondoids was found among the reaction products based on GC–MS data. Amazingly, [1231241(2)3]decamantane was also identified based on the presence of the molecular ion $m/z = 456$ [64]. Hence, besides **DIA**, low-temperature plasmas generated in supercritical xenon may be useful for the preparation of a set of higher diamondoids, namely hexamantane [65], pentamantanes, [1231241(2)3]decamantane [66], and remarkably, undecamantane [60], at room temperature or slightly above.

The formation of diamondoids from **AD** was suggested to proceed through two steps, namely the abstraction of hydrogen atoms and the addition of methyl radicals to the intermediate hydrocarbon radicals [60]. The formation of methyl radicals is attributed to the decomposition of **AD** under plasma conditions. The role of methyl radicals as key players was suggested based on the formation of methylated diamondoids. Note that the suggested mechanism contradicts the experimental observation [61] that 2-methyl adamantane is not an intermediate in the homologenation process and, rather, C_x-particles are the key players (vide supra). In full agreement, supercritical plasma optical emission spectroscopy shows the presence of dicarbon C_2 [28] together with the CH-radical [59], which also may take part in diamond growth. The optical emission spectrum of **AD** in dielectric barrier discharge (DBD) plasmas indicates characteristic bands of C_2 ($A^3\Pi_g \to X^3\Pi_u$) and CH ($A^2\Delta \to X^2\Pi$) at 516 and 431 nm, respectively [59].

The formation of diamond particles using **AD** seeding was most successful at moderate temperatures in supercritical fluids as it allows to reach higher concentrations of reactants as compared to the gas phase. Pulsed laser ablation of highly oriented pyrolytic graphite [67] was conducted in adamantane-dissolved supercritical CO_2 with and without cyclohexane as a co-solvent. Raman spectroscopy revealed the presence of hydrocarbons possessing sp^3-hybridized carbons, and GC–MS confirmed the formation of diamondoids up to decamantane. Even though all of the above methods are still far from being synthetically useful, they add to our understanding of the role of diamondoids in diamond formation under a variety of reaction conditions.

References

1 Mayerhoefer, E. and Krueger, A. (2022). Surface control of nanodiamond: from homogeneous termination to complex functional architectures for biomedical applications. *Acc. Chem. Res.* 55: 3594–3604.

2 Wentorf, R.H. (1965). The behavior of some carbonaceous materials at very high pressures and high temperatures. *J. Phys. Chem.* 69: 3063–3069.

3 (a) Ekimov, E.A., Kondrina, K.M., Mordvinova, N.E. et al. (2019). High-pressure, high-temperature synthesis of nanodiamond from adamantane. *Inorg. Mater.* 55: 437–442. (b) Handschuh-Wang, S., Wang, T., and Tang, Y.B. (2021). Ultrathin diamond nanofilms-development, challenges, and applications. *Small* 17: 2007529.

4 Angus, J.C. and Hayman, C.C. (1988). Low-pressure, metastable growth of diamond and "diamondlike" phases. *Science* 241: 913–921 and references therein.
5 Kudryavtsev, O.S., Ekimov, E.A., Romshin, A.M. et al. (2018). Structure and luminescence properties of nanonodiamonds produced from adamantane. *Phys. Status Solidi A* 215: 1800252.
6 (a) Plaisted, T.A. and Sinnott, S.B. (2001). Hydrocarbon thin films produced from adamantane-diamond surface deposition: molecular dynamics simulations. *J. Vac. Sci. Technol. A* 19: 262–266. (b) Merkle, R.C. and Freitas, R.A. (2003). Theoretical analysis of a carbon-carbon dimer placement tool for diamond mechanosynthesis. *J. Nanosci. Nanotechnol.* 3: 319–324. (c) Dolmatov, V.Y., Myllymaki, V., and Vehanen, A. (2013). A possible mechanism of nanodiamond formation during detonation synthesis. *J. Superhard Mater.* 35: 143–150. (d) Gebbie, M.A., Ishiwata, H., McQuade, P.J. et al. (2018). Experimental measurement of the diamond nucleation landscape reveals classical and nonclassical features. *Proc. Natl. Acad. Sci. U. S. A.* 115: 8284–8289. (e) Tzeng, Y.K., Zhang, J., Lu, H.Y. et al. (2017). Vertical-substrate MPCVD epitaxial nanodiamond growth. *Nano Lett.* 17: 1489–1495. (f) Dahl, J.E.P., Moldowan, J.M., Wei, Z. et al. (2010). Synthesis of higher diamondoids and implications for their formation in petroleum. *Angew. Chem. Int. Ed.* 49: 9881–9885.
7 Matsumoto, S. and Matsui, Y. (1983). Electron microscopic observation of diamond particles grown from the vapour phase. *J. Mater. Sci.* 18: 1785–1793.
8 Piekarczyk, W. (1999). Crystal growth of CVD diamond and some of its peculiarities. *Cryst. Res. Technol.* 34: 553–563.
9 Feldman, K.S., Campbell, R.F., West, T.R. et al. (1999). Modeling chemical vapor deposition (CVD) diamond film growth with diamantane-derived radicals in solution: permissive evidence in support of the Garrison-Brenner mechanism for incorporation of carbon into the dimer sites of the {100} diamond surface. *J. Organomet. Chem.* 64: 7612–7617.
10 Mandal, S. (2021). Nucleation of diamond films on heterogeneous substrates: a review. *RSC Adv.* 11: 10159–10182.
11 Stauss, S. and Terashima, K. (2016). *Diamondoids: Synthesis, Properties, and Applications*, 242. Jenny Stanford Publishing.
12 Yang, C.C. and Li, S. (2008). Size-dependent temperature-pressure phase diagram of carbon. *J. Phys. Chem. C* 112: 1423–1426.
13 Yang, F., Lin, Y., Baldini, M. et al. (2016). Effects of molecular geometry on the properties of compressed diamondoid crystals. *J. Phys. Chem. Lett.* 7: 4641–4647.
14 Onodera, A., Suito, K., and Morigami, Y. (1992). High-pressure synthesis of diamond from organic compounds. *Proc. Jpn. Acad., Ser. B, Phys. Biol. Sci.* 68: 167–171.
15 (a) Ekimov, E.A., Kudryavtsev, O.S., Mordvinova, N.E. et al. (2018). High-pressure synthesis of nanodiamonds from adamantane: myth or reality? *ChemNanoMat* 4: 269–273. (b) Kudryavtsev, O.S., Bagramov, R.H., Pasternak, D.G. et al. (2023). Raman fingerprints of ultrasmall nanodiamonds produced from adamantane. *Diam. Relat. Mater.* 133: 109770.

16 Bundy, F.P. (1963). Direct conversion of graphite to diamond in static pressure apparatus. *J. Chem. Phys.* 38: 631–643.

17 Liang, J.X., Ender, C.P., Zapata, T. et al. (2020). Germanium iodide mediated synthesis of nanodiamonds from adamantane "seeds" under moderate high-pressure high-temperature conditions. *Diam. Relat. Mater.* 108: 108000.

18 Park, S., Abate, I.I., Liu, J. et al. (2020). Facile diamond synthesis from lower diamondoids. *Sci. Adv.* 6: eaay9405.

19 Williams, O.A. (2011). Nanocrystalline diamond. *Diam. Relat. Mater.* 20: 621–640.

20 (a) Tiwari, R.N. and Chang, L. (2010). Chemical precursor for the synthesis of diamond films at low temperature. *Appl. Phys. Express* 3: 045501. (b) Tiwari, R.N. and Chang, L. (2010). Growth, microstructure, and field-emission properties of synthesized diamond film on adamantane-coated silicon substrate by microwave plasma chemical vapor deposition. *J. Appl. Phys.* 107: 103305.

21 Atakan, B., Lummer, K., and Kohse-Höinghaus, K. (1999). Diamond deposition in acetylene-oxygen flames: nucleation and early growth on molybdenum substrates for different pretreatment procedures. *Phys. Chem. Chem. Phys.* 1: 3151–3156.

22 Tiwari, R.N., Tiwari, J.N., Chang, L., and Yoshimura, M. (2011). Enhanced nucleation and growth of diamond film on Si by CVD using a chemical precursor. *J. Phys. Chem. C* 115: 16063–16073.

23 Chen, Y.C. and Chang, L. (2013). Chemical vapor deposition of diamond on silicon substrates coated with adamantane in glycol chemical solutions. *RSC Adv.* 3: 1514–1518.

24 May, P.W., Ashworth, S.H., Pickard, C.D.O. et al. (1998). Interactive Raman spectra of adamantane, diamantane and diamond, and their relevance to diamond film deposition. *PhysChemComm* 1: 1–10.

25 Chen, Y.C. and Chang, L. (2014). Chemical vapor deposition of diamond on an adamantane-coated sapphire substrate. *RSC Adv.* 4: 18945–18950.

26 Tiwari, R.N., Tiwari, J.N., and Chang, L. (2010). The synthesis of diamond films on adamantane-coated Si substrate at low temperature. *Chem. Eng. Sci.* 158: 641–645.

27 Tsugawa, K., Ishihara, M., Kim, J. et al. (2010). Nucleation enhancement of nanocrystalline diamond growth at low substrate temperatures by adamantane seeding. *J. Phys. Chem. C* 114: 3822–3824.

28 Kikuchi, H., Stauss, S., Nakahara, S. et al. (2010). Development of sheet-like dielectric barrier discharge microplasma generated in supercritical fluids and its application to the synthesis of carbon nanomaterials. *J. Supercrit. Fluids* 55: 325–332.

29 Giraud, L., Huber, V., and Jenny, T. (1998). 2,2-Divinyladamantane: a new substrate for the modification of silicon surfaces. *Tetrahedron* 54: 11899–11906.

30 (a) Giraud, A., Jenny, T., Leroy, E. et al. (2001). Chemical nucleation for CVD diamond growth. *J. Am. Chem. Soc.* 123: 2271–2274. (b) Mandal, S., Thomas, E.L.H., Jenny, T.A., and Williams, O.K. (2016). Chemical nucleation of diamond films. *ACS Appl. Mater. Interfaces* 8: 26220–26225.

31 Sangphet, S., Siriroj, S., Sriplai, N. et al. (2018). Enhanced ferromagnetism in mechanically exfoliated CVD-carbon films prepared by using adamantane as precursor. *Appl. Phys. Lett.* 112: 242406.
32 Jia, H., Shinar, J., Lang, D.P., and Pruski, M. (1993). Nature of the native-defect ESR and hydrogen-dangling-bond centers in thin diamond films. *Phys. Rev. B* 48: 17595–17598.
33 Zeze, D.A., North, D.R., Brown, N.M.D., and Anderson, C.A. (2000). Comparison of C_xN_y: H films obtained by deposition using magnetron sputtering or an inductively coupled plasma. *Surf. Interface Anal.* 29: 369–376.
34 Umeno, A., Noda, M., Uchida, H., and Takeuchi, H. (2008). Deposition of DLC film from adamantane by using pulsed discharge plasma CVD. *Diam. Relat. Mater.* 17: 684–687.
35 Shirafuji, T., Nishimura, Y., Tachibana, K., and Ishii, H. (2009). Plasma-enhanced chemical vapor deposition of carbon films using dibromoadamantane. *Thin Solid Films* 518: 993–1000.
36 Alkahtani, M.H., Alghannam, F., Jiang, L. et al. (2018). Fluorescent nanodiamonds: past, present, and future. *Nanophotonics* 7: 1423–1453.
37 (a) Fokin, A.A., Yurchenko, R.I., Tkachenko, B.A. et al. (2014). Selective preparation of diamondoid phosphonates. *J. Organomet. Chem.* 79: 5369–5373. (b) Li, F.H., Fabbri, J.D., Yurchenko, R.I. et al. (2013). Covalent attachment of diamondoid phosphonic acid dichlorides to tungsten oxide surfaces. *Langmuir* 29: 9790–9797.
38 Zhang, J., Ishiwata, H., Babinec, T.M. et al. (2016). Hybrid group IV nanophotonic structures incorporating diamond silicon-vacancy color centers. *Nano Lett.* 16: 212–217.
39 (a) Fitzgibbons, T.C., Guthrie, M., Xu, E.S. et al. (2015). Benzene-derived carbon nanothreads. *Nat. Mater.* 14: 43–47. (b) Yu, Y.; Wu, L. Z.; Zhi, J. F., Diamond nanowires: fabrication, structure, properties and applications. *Novel Aspects of Diamond: From Growth to Applications*, Yang, N., 2015; 121, 123–164; (c) Szunerits, S., Coffinier, Y., and Boukherroub, R. (2015). Diamond nanowires: a novel platform for electrochemistry and matrix-free mass spectrometry. *Sensors* 15: 12573–12593. (d) Marutheeswaran, S. and Jemmis, E.D. (2018). Adamantane-derived carbon nanothreads: high structural stability and mechanical strength. *J. Phys. Chem. C* 122: 7945–7950.
40 Shenderova, O., Brenner, D., and Ruoff, R.S. (2003). Would diamond nanorods be stronger than fullerene nanotubes? *Nano Lett.* 3: 805–809.
41 Barnard, A.S., Russo, S.P., and Snook, I.K. (2003). Electronic band gaps of diamond nanowires. *Phys. Rev. B* 68: 235407.
42 (a) Yu, Y., Wu, L.Z., and Zhi, J.F. (2014). Diamond nanowires: fabrication, structure, properties, and applications. *Angew. Chem. Int. Ed.* 53: 14326–14351. (b) Hsu, C.H. and Xu, J. (2012). Diamond nanowire – a challenge from extremes. *Nanoscale* 4: 5293–5299.
43 (a) Vlasov, I.L., Lebedev, O.I., Ralchenko, V.G. et al. (2007). Hybrid diamond-graphite nanowires produced by microwave plasma chemical vapor deposition. *Adv. Mater.* 19: 4058–4062. (b) Hsu, C.H., Cloutier, S.G., Palefsky, S.,

and Xu, J. (2010). Synthesis of diamond nanowires using atmospheric-pressure chemical vapor deposition. *Nano Lett.* 10: 3272–3276.

44 Sun, L.T., Gong, J.L., Zhu, D.Z. et al. (2004). Diamond nanorods from carbon nanotubes. *Adv. Mater.* 16: 1849–1853.

45 Barnard, A.S. and Snook, I.K. (2004). Phase stability of nanocarbon in one dimension: nanotubes versus diamond nanowires. *J. Chem. Phys.* 120: 3817–3821.

46 (a) Liu, Q.X., Wang, C.X., Li, S.W. et al. (2004). Nucleation stability of diamond nanowires inside carbon nanotubes: a thermodynamic approach. *Carbon* 42: 629–633. (b) Liu, Q.X., Wang, C.X., Yang, Y.H., and Yang, G.W. (2004). One-dimensional nanostructures grown inside carbon nanotubes upon vapor deposition: a growth kinetic approach. *Appl. Phys. Lett.* 84: 4568–4570.

47 Shang, N.G., Papakonstantinou, P., Wang, P. et al. (2009). Self-assembled growth, microstructure, and field-emission high-performance of ultrathin diamond nanorods. *ACS Nano* 3: 1032–1038.

48 McIntosh, G.C., Yoon, M., Berber, S., and Tomanek, D. (2004). Diamond fragments as building blocks of functional nanostructures. *Phys. Rev. B* 70: 045401.

49 Yao, M.G., Stenmark, P., Abou-Hamad, E. et al. (2011). Confined adamantane molecules assembled to one dimension in carbon nanotubes. *Carbon* 49: 1159–1166.

50 Zhang, J.Y., Feng, Y.Q., Ishiwata, H. et al. (2012). Synthesis and transformation of linear adamantane assemblies inside carbon nanotubes. *ACS Nano* 6: 8674–8683.

51 Yao, Z., Liu, C.J., Lv, H., and Liu, B.B. (2016). The stable orientations analysis of linearly arrayed $C_{10}H_{16}$ molecules in single-walled carbon nanotube by using the multiple-molecule model. *Chin. J. Phys.* 54: 424–432.

52 Li, Y., Yao, Z., Liu, B.B. et al. (2019). Molecular dynamics simulation of the oscillatory behaviour and vibrational analysis of an adamantane molecule encapsulated in a single-walled carbon nanotube. *Philos. Mag.* 99: 401–418.

53 Yao, Z., Liu, C.J., and Lv, H. (2018). The structure and interaction mechanisms of $C_{10}H_{16}$@(13, 0)SWCNT under high pressure. *Int. J. Mod. Phys. B* 32: 1850054.

54 Tonkikh, A.A., Rybkovskiy, D.V., Orekhov, A.S. et al. (2016). Optical properties and charge transfer effects in single-walled carbon nanotubes filled with functionalized adamantane molecules. *Carbon* 109: 87–97.

55 Nakanishi, Y., Omachi, H., Fokina, N.A. et al. (2015). Template synthesis of linear-chain nanodiamonds inside carbon nanotubes from bridgehead-halogenated diamantane precursors. *Angew. Chem. Int. Ed.* 54: 10802–10806.

56 Legoas, S.B., dos Santos, R.P.B., Troche, K.S. et al. (2011). Ordered phases of encapsulated diamondoids into carbon nanotubes. *Nanotechnology* 22: 315708.

57 Zhang, J., Zhu, Z., Feng, Y. et al. (2013). Evidence of diamond nanowires formed inside carbon nanotubes from diamantane dicarboxylic acid. *Angew. Chem. Int. Ed.* 52: 3717–3721.

58 Wang, H.D., Li, B., and Yang, J.L. (2017). Electronic, optical, and mechanical properties of diamond nanowires encapsulated in carbon nanotubes: a first-principles view. *J. Phys. Chem. C* 121: 3661–3672.

59 Stauss, S., Ishii, C., Pai, D.Z. et al. (2014). Diamondoid synthesis in atmospheric pressure adamantane-argon-methane-hydrogen mixtures using a continuous flow plasma microreactor. *Plasma Sources Sci. Technol.* 23: 035016.

60 Oshima, F., Stauss, S., Inose, Y., and Terashima, K. (2014). Synthesis and investigation of reaction mechanisms of diamondoids produced using plasmas generated inside microcapillaries in supercritical xenon. *Jpn. J. Appl. Phys.* 53: 010214.

61 Ishii, C., Stauss, S., Kuribara, K. et al. (2015). Atmospheric pressure synthesis of diamondoids by plasmas generated inside a microfluidic reactor. *Diam. Relat. Mater.* 59: 40–46.

62 Stauss, S., Pai, D.Z., Shizuno, T., and Terashima, K. (2014). Nanosecond pulsed electric discharge synthesis of carbon nanomaterials in helium at atmospheric pressure from adamantane. *IEEE Trans. Plasma Sci.* 42: 1594–1601.

63 Nakahara, S., Stauss, S., Miyazoe, H. et al. (2010). Pulsed laser ablation synthesis of diamond molecules in supercritical fluids. *Appl. Phys. Express* 3: 096201.

64 Nakahara, S., Stauss, S., Kato, T. et al. (2011). Synthesis of higher diamondoids by pulsed laser ablation plasmas in supercritical CO_2. *J. Appl. Phys.* 109: 123304.

65 Stauss, S., Miyazoe, H., Shizuno, T. et al. (2010). Synthesis of the higher-order diamondoid hexamantane using low-temperature plasmas generated in supercritical xenon. *Jpn. J. Appl. Phys.* 49: 070213.

66 Shizuno, T., Miyazoe, H., Saito, K. et al. (2011). Synthesis of diamondoids by supercritical xenon discharge plasma. *Jpn. J. Appl. Phys.* 50: 030207.

67 Stauss, S., Nakahara, S., Kato, T. et al. (2013). Synthesis of higher diamondoids by pulsed laser ablation plasmas in supercritical fluids. In: *Graphene, Carbon Nanotubes, and Nanostructures: Techniques and Applications* (ed. J.E. Morris and K. Iniewski), 211–245.

7

Diamondoid Polymers

Incorporation of bulky, rigid, and robust adamantane units into polymers is quite attractive, as this often leads to the formation of materials with improved stability, higher glass-transition temperature (T_g), increased chain stiffness, decreased crystallinity, higher solubility in organic solvents, lower dielectric constants, and high mass densities [1]. As the large-scale production of adamantane-containing polymers is hampered by the relatively high costs of the monomers, research mostly concentrates on the preparation of materials, for which costs play a secondary role. The market for such polymers is rich and covers specialty polymers for coatings and biomedical applications, construction of dielectric, very robust and hydrolytically stable thermoresistant materials. Very recent works on adamantane-containing polymers mostly concentrate on the preparation of self-healing elastomers [2], dielectric polymers [1n], highly transparent [3] and porous materials [1d, 4], including MOFs [5], adamantyl-modified polyethylene [6], cross-linked [7], memory-effect [8], biodegradable [9], thermo- [10] and acid-responsive [11] polymers as well as various copolymers [12], hydrogels [13], macrocycles [14], dendrimers [15], and organic frameworks [16]. Recently, it was also shown that polyvinyl adamantane (**pVa**) may be useful as a template for nanodiamond synthesis [17]. Currently, a large variety of adamantane-containing polymers are prepared and tested, and the number of publications in this area has continuously and rapidly risen over the last decade (Figure 7.1).

Diamondoids higher than adamantane are quite attractive polymer building blocks as they allow to tailor porosity, mechanical, dielectric, and optical properties [18]. This research mostly concentrates on diamantane and diadamantyl derivatives, while higher diamondoids were tested only as modifiers of polymer composites [19]. Herein, we primarily concentrate on polymers containing diamantane and on comparative studies of adamantane and diamantane polymers in order to disclose the potential of diamondoid assemblies and higher diamondoids in polymer chemistry. The polymers obtained from monomers based on 1-adamantyl-1-adamantane, mono-, and difunctionalized diamantanes will be discussed separately, but we will not cover the vast number of patents.

The Chemistry of Diamondoids: Building Blocks for Ligands, Catalysts, Materials, and Pharmaceuticals,
First Edition. Andrey A. Fokin, Marina Šekutor, and Peter R. Schreiner.
© 2024 WILEY-VCH GmbH. Published 2024 by WILEY-VCH GmbH.

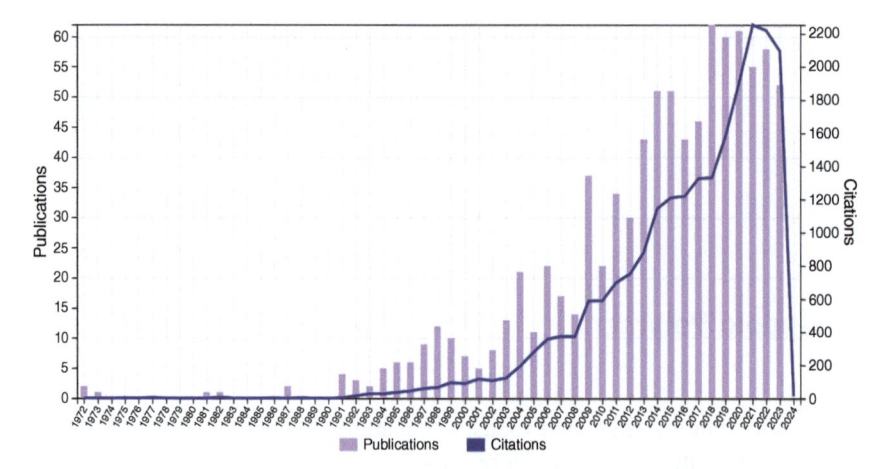

Figure 7.1 The publication activity (apart from patent sources) based on Web of Science (WoS) utilizing the "adamantane+polymer" keyword (as of December 31, 2023).

7.1 Polymers Based on 1-Adamantyl-1-adamantane (1ADAD)

The homopolymer based on 3,3′-diethynyl-1,1′-biadamantane (**ETHYNADAD**), which may be prepared from **BrADADBr** by an effective heavy-metal-free procedure (Scheme 7.1), exhibits slightly higher thermal stability than analogous polymers prepared from 1,3-diethynyladamantane. Neither polymer displays detectable signs of decomposition up to 400 °C. Since these materials are pure hydrocarbons, do not contain heteroatoms, and may be prepared through simple thermal polymerization, they are excellent high-temperature polymeric dielectrics [20].

Scheme 7.1 The preparation of the 3,3′-diethynyl-1,1′-biadamantane (**ETHYNADAD**) monomer from **BrADADBr** by an effective heavy-metal-free procedure.

Recently, the derivatives of **1ADAD** were utilized as monomers in the construction of microporous materials [4d]. The 3D-building blocks **6BrPhADAD** and **4BrPhAD** obtained via Sonogashira-Hagihara coupling (Figure 7.2) with 4,4′-diethynylbiphenyl gave the corresponding microporous frameworks **p6BrPhADAD** and **p4BrPhAD**.

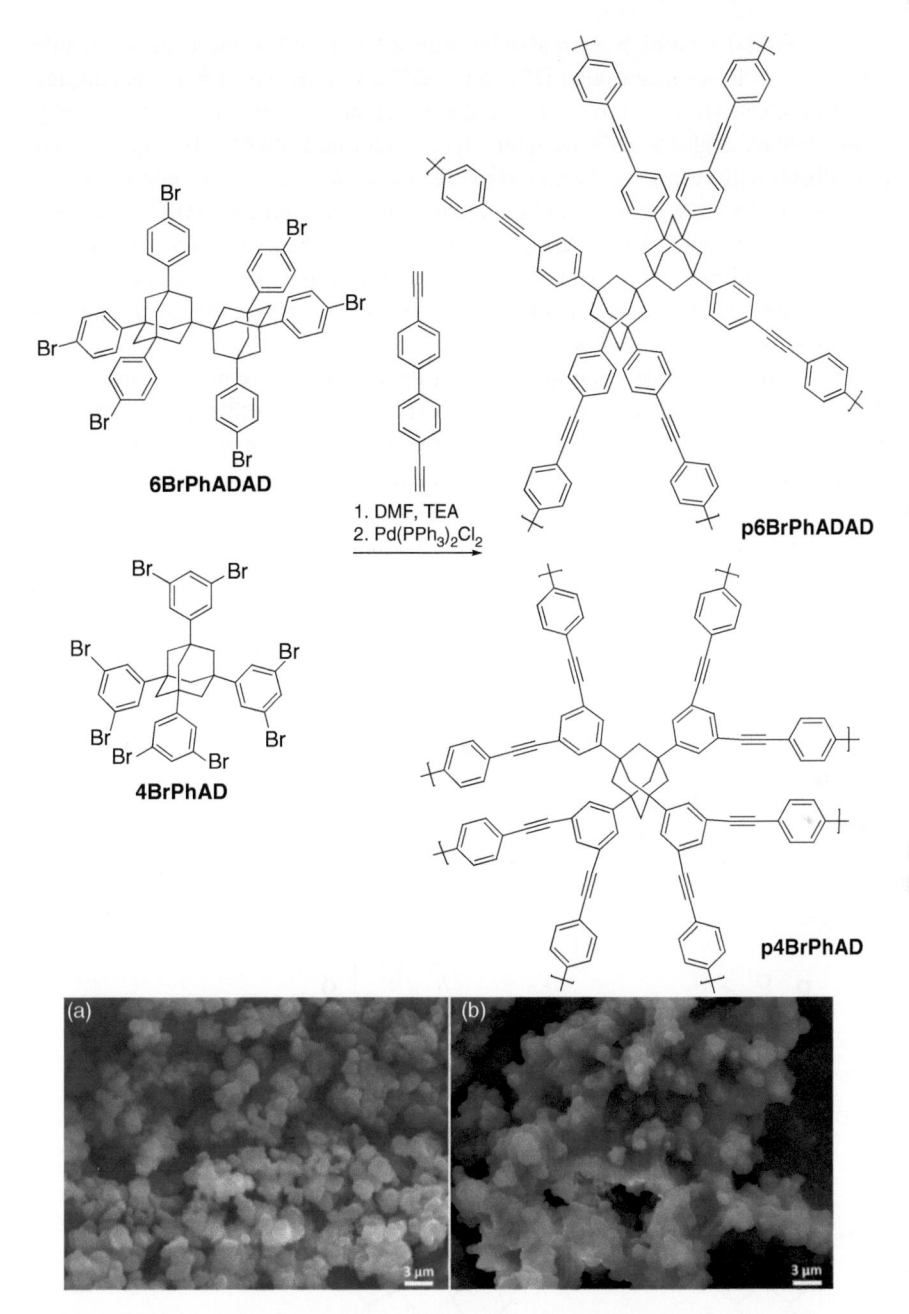

7.1 Polymers Based on 1-Adamantyl-1-adamantane (1ADAD) | **195**

Figure 7.2 Preparation (top) and SEM images (bottom) of microporous materials **p4BrPHAD** (a) and **p6BrPhADAD** (b). Source: Reproduced from Ref. [4] with permission from Elsevier, 2018.

Both networks display high thermal stability up to 400 °C and are inert toward organic solvents as well as concentrated HCl and NaOH solutions. These polymers display excellent selectivity for CO_2 over N_2 gas adsorptions and exhibit high hydrophobicity (water contact angles >165°). Diadamantyl polymer **p6BePhADAD** outperformed **p4BrPhAD** with respect to thermal stability (408 vs. 395 °C at 5% weight loss), surface area (642 vs. 536 $m^2\ g^{-1}$), and gas uptake. However, **p6BePhADAD** has a less regular structure than **p4BrPhAD**, which contains more distinct spherical fragments (Figure 7.2) and, as a result, displays higher gas adsorption selectivities. These microporous materials [4b] as well as their cross-polymers [21] have the potential to be efficient absorbent alternatives for gases or toxic vapors.

The diadamantyl moiety was also used for the preparation of novel coordination clusters for metal oxide-organic frameworks (MOOFs) [22], utilizing 1,2,4-triazole linkers (Figure 7.3) [23]. While 1,3-bis-(1,2,4-triazol-4-yl)adamantane (**TRIAZAD**) serves as a triangular building block, the diadamantyl derivative **TRIAZADAD** is seen as an elongated linear tecton. Fluoride anions cause the disintegration of metal oxide chains and promote the insertion of diamondoid-based bidentate linkers. Such building blocks could be useful in the construction of 2D and 3D fluoride MOOF networks.

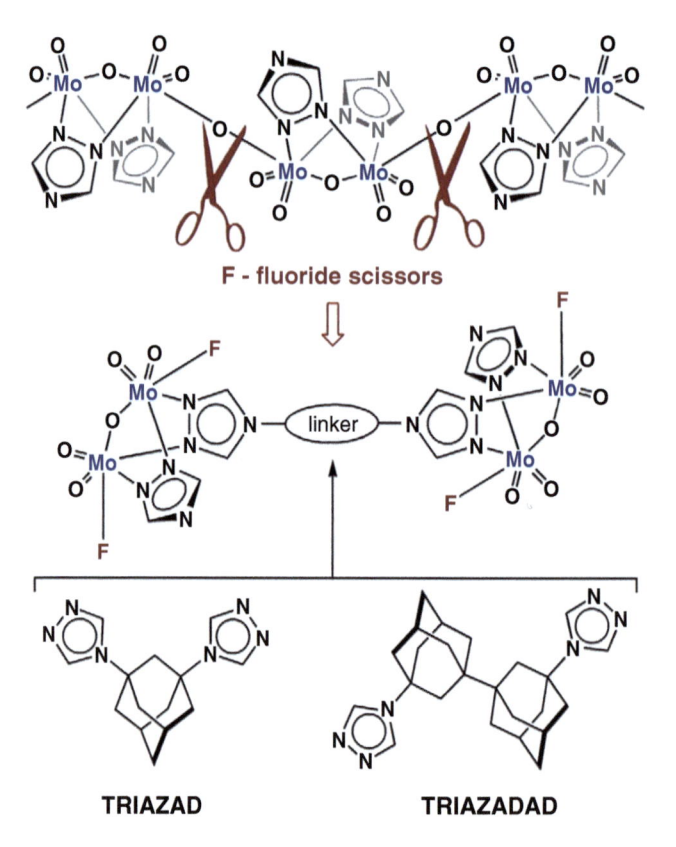

Figure 7.3 Adamantyl and diadamantyl triazoles as linker coordination clusters for the construction of polymeric metal oxide frameworks. Source: Adopted from Ref. [23] with permission from Elsevier, 2011.

7.2 Polymers Based on Monofunctionalized Diamantanes

Polymeric maleimides usually attract attention mostly due to high thermal stability, but only a few have been commercialized because of the toxicity of their monomers [24]. Currently, maleimides mostly target so-called "smart"[1] self-healing polymers. Polymerization of easily available *N*-(1)-diamantyl maleimide (**1DIAMALEA**, Scheme 7.2) gives polymaleimide **pDmal**, which exhibits a high glass transition temperature and excellent thermal stability [25]. However, the yields and molecular weights of the polymer are much lower than those of other maleimides due to steric hindrance in the propagation step [26]. Less hindered apical diamantane *N*-maleimide has not prepared and tested yet.

Apart from many applications in medicine [27], methacrylates display a light transmittance that is higher than that of many other polymers or glasses [28] and are therefore highly preferred for optical applications. The extraordinarily high refractive index of diamond ($n_d = 2.42$) in combination with its low light

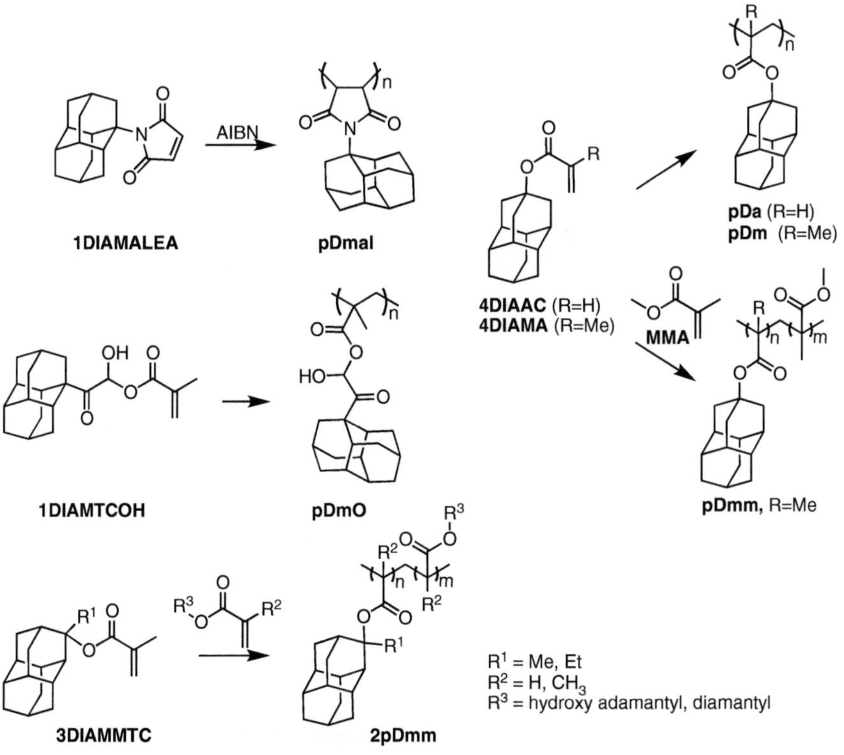

Scheme 7.2 Diamantane-based polymeric maleimides and acrylates.

---

1 It is rather disappointing to have to observe the demise of proper scientific language by the use of entirely inappropriate terms such as "smart" polymers. Of course, there is nothing "smart" in dead matter; it may have been designed and prepared in a "smart" way. But that would only be worth mentioning if "stupid" polymers or ways of preparing them also existed.

198 | *7 Diamondoid Polymers*

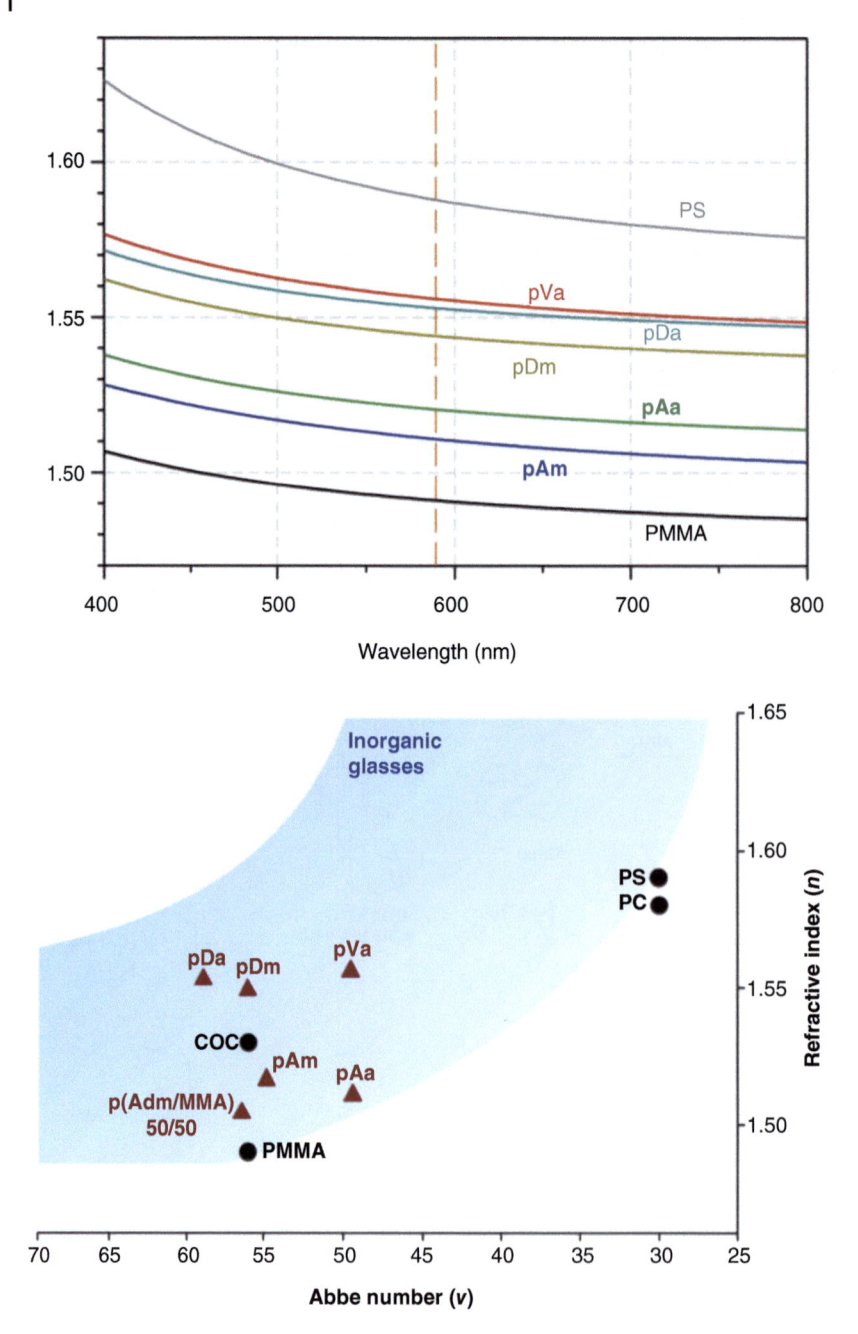

Figure 7.4 *Top*: Refractive index vs. wavelength for diamondoid-containing acrylate polymers. *Bottom*: Refractive index vs. Abbe number for diamondoid-containing acrylate polymers mapping on inorganic glass vs. some commodity plastics (**pAd** – polyadamantyl acrylate, **pAm** – polyadamantyl methacrylate; **PMMA**-poly[methylmethacrylate], **PC** - polycarbonate, and **PS** - polystyrene; for other abbreviations, see text). Source: Reproduced from Ref. [29], Wiley-VCH, 2013.

dispersion (high Abbe numbers) prompted numerous studies on the incorporation of diamondoids into polymethacrylates (PMCs). In particular, the inclusion of an adamantane moiety improves the thermal and mechanical properties, increases the transparency in the UV/VIS region, but decreases the dielectric constant of such polymers as compared to traditional PMC [12d, e].

The optical properties of polymers derived from 4-diamantyl acrylate (**4DIAAC**, Scheme 7.2) and 4-diamantyl methacrylate (**4DIAMA**) were compared with those of the corresponding adamantyl polymers as well as with other commodity plastics such as polystyrene (**PS**), polycarbonate (**PC**), and cyclic olefin copolymers (**COC**, Figure 7.4) [29]. While both diamantane and adamantane acrylates and methacrylates all show optical characteristics comparable to those of inorganic glasses (Figure 7.4, right), the diamantane-containing polymers combine high refractive indices and high Abbe numbers that are very rare in the polymer world. The applications of saturated polymers such as polyvinyl adamantane (**pVa**), whose optical properties approach acrylates, are limited due to their very low degree of polymerization [30]. All of this makes diamondoid acrylates valuable as specialty optical plastics, and even more superior properties are expected for materials obtained from higher diamondoids such as triamantane [29]. The problem, however, exists with the manufacturing of **pDm** as its T_g is above the polymer's decomposition point.

Copolymerization of **4DIAMA** with methacrylate gives polymers **pDmm** (Scheme 7.2) with substantially increased T_g values relative to **PMMA** [31]. Modifications with diamondoids provide a very simple method to improve the thermal properties of **PMMC**; however, the optical properties of such materials were not reported. Very high refractive indices (up to 1.56) are also characteristic of the copolymers **pDmO** derived from 1-diamantyl carboxylic acid derivative **1DIAMTCOH**. Physicochemical and optical studies display clear advantages of **pDmO** over isobornyl methacrylate homo-polymers [32]. Alternatively, 3-diamantyl acrylates **3DIAMMTC** (Scheme 7.2) were tested as polymer platforms for photoresist applications with substituted derivatives used in various copolymerizations with lactone monomers to achieve better solubility in solvents usually used in photoresist technologies. Such **2pDmm** polymers display photoresist properties as well as T_g and polydispersities similar to those found for the corresponding adamantane polymers [33].

7.3 Polymers Based on Difunctionalized Diamantanes

The first diamantane-based polymer was prepared in 1991 by thermal or catalytic polymerization of 4,9-diethynyldiamantane (**49DIAACET**, Figure 7.5, left) [34]. This polymer survives heating up to 500 °C, which is substantially higher (Figure 7.5, right) than that of the corresponding adamantane-based homopolymer derived from 1,3-diethynyladamantane (**13ADACET**). Less regular polymers

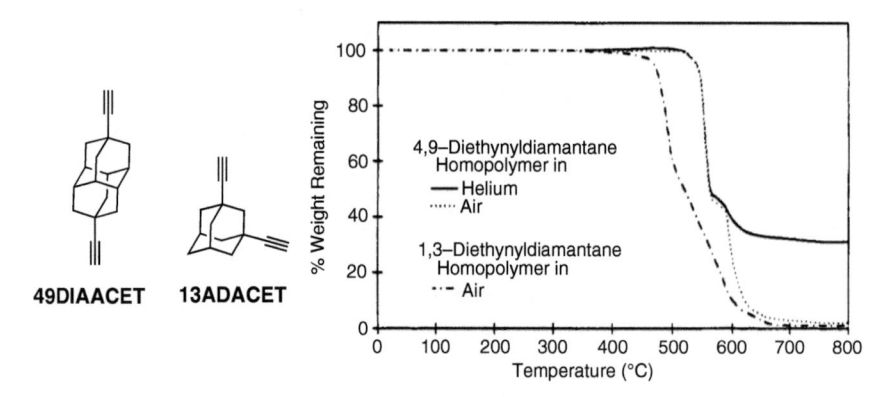

Figure 7.5 Structure of diamantane and adamantane diacetylenic monomers and comparison of thermal stability of their homopolymers from TGA. Source: Reproduced from Ref. [34] with permission from the American Chemical Society, 1991.

Scheme 7.3 Polyesters based on diamantane dichlorodicarboxyates **16COCLDIA** and **49COCLDIA**.

prepared from the polymerization of relatively inexpensive mixtures of 1,6-, 1,4-, and 4,9-diethynyldiamantanes give a material that is substantially less thermally stable.

Polyesters are another class of thermostable polymers that may utilize dicarboxylic diamantane derivatives as building blocks. For polyesters based on 1,6-diamantane dichlorodicarboxyate **16COCLDIA** (Scheme 7.3), the highest melting temperatures were observed for the **A** group of polymers **16De**, while the highest quality transparent films were constructed from the **B** group [35]. The latter polymers also display

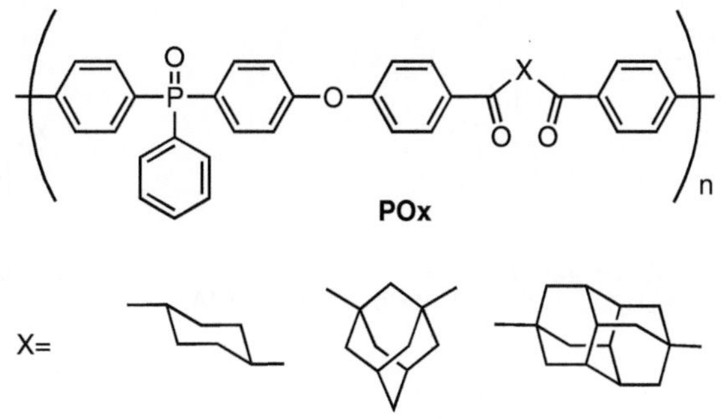

Figure 7.6 Structure, thermal glass transition (T_g), and decomposition (T_d) temperatures of triphenylphosphine oxide polymers incorporating cycloaliphatic/cage hydrocarbon moieties. Source: Adopted from Ref. [37], Wiley-VCH, 2004.

POx

X=			
T_g, °C	215	192	239
T_d, °C	445	490	515

good mechanical and dielectric properties, are soluble in *o*-chlorophenol, and can be cast into transparent, tough, and flexible films with tensile strength at break up to 80.5 MPa, an elongation break up to 15.1%, and an initial modulus up to 1.6 GPa. The properties of polyesters (**16De, C**) derived from 1,6-diamantane dichlorocarboxylates **16COClDIA** were further compared with those of **49De** obtained from 4,9-dichlorocarboxylates (**49COClDIA**) [36], but both display almost identical thermal and mechanical properties.

Incorporation of the diamantane moiety also substantially increases the T_g of polyesters and the thermal stability T_d of polyaryene ether ketone triphenylphosphine oxide polymers **POx** (Figure 7.6). From organic solvents, such polymers form tough, free-standing films that are transparent in the 300–800 nm range and are claimed to have potential applications for space thermal control coatings to give polymers stable up to 500 °C, especially if a diamantane moiety is incorporated [37].

The outstanding properties of polyamides, in particular, aramide (kevlar) initiated studies on the incorporation of diamondoids into polyamides mostly by replacing the aromatic fragment with topologically related diamantane derivatives. The simplest diamantane monomers are diacids (**COOHDIA**) and diamines (**NH2DIA**) as well as their derivatives modified by aromatic spacers (**PhOPhCOOHDIA, PhNH2DIA, and PhOPhNH2DIA**, Figure 7.7).

The first aromatic polyamide modified with a diamondoid was prepared in 1995 [38] through polycondensation of diamantane 1,6-dicarboxylic acid (**16COOHDIA**) with simple aromatic amides to form the polyamide **16Dcarbam** (Figure 7.8). Similar work was performed for diamantane 4,9-dicarboxylate to give polyamide **49Dcarbam** [38, 39]. While **16Dcarbam** polymers of group **A** display poor solubility and difficulties to cast, group **B** polymers form colorless films with tensile strengths up to 82.9 MPa, elongation to break values of up to 14%, initial moduli of up to 1.6 GPa, and T_ds up to ca. 400 °C [38]. Polyamides **B** show rather high G' values (about 10^8 Pa), even at temperatures higher than 350 °C. Polyamides based on bis-apical diamantanes (**49Dcarbam**) display superior thermomechanical properties. The polymers of group **B** conserve good mechanical properties even at temperatures of 450 °C, elongation to break values up to 38%, and initial modulus up to 1.9 GPa. Generally, polymers derived from 4,9-diamantane dicarboxylic acid

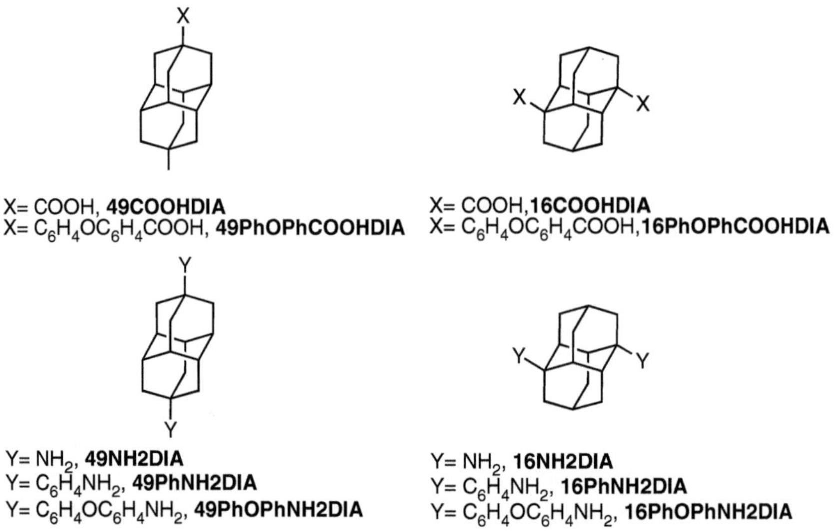

7 Diamondoid Polymers

X= COOH, **49COOHDIA**
X= C$_6$H$_4$OC$_6$H$_4$COOH, **49PhOPhCOOHDIA**

X= COOH, **16COOHDIA**
X= C$_6$H$_4$OC$_6$H$_4$COOH, **16PhOPhCOOHDIA**

Y= NH$_2$, **49NH2DIA**
Y= C$_6$H$_4$NH$_2$, **49PhNH2DIA**
Y= C$_6$H$_4$OC$_6$H$_4$NH$_2$, **49PhOPhNH2DIA**

Y= NH$_2$, **16NH2DIA**
Y= C$_6$H$_4$NH$_2$, **16PhNH2DIA**
Y= C$_6$H$_4$OC$_6$H$_4$NH$_2$, **16PhOPhNH2DIA**

Figure 7.7 Diamantane monomers such as diacids and diamines tested for the preparation of polyamides.

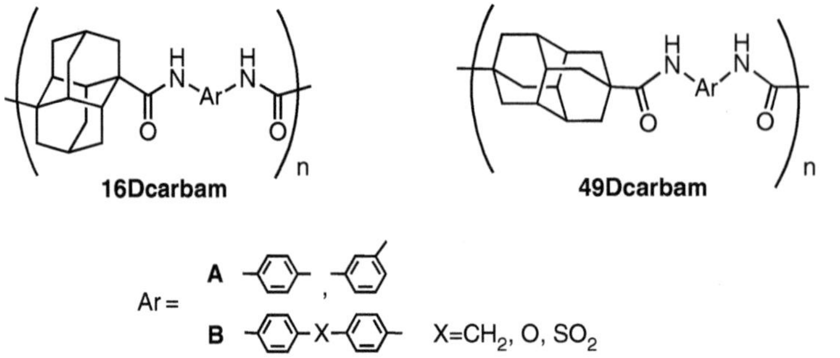

16Dcarbam

49Dcarbam

Ar = A: ⌬—, —⌬
B: ⌬—X—⌬   X=CH$_2$, O, SO$_2$

Figure 7.8 Diamantane-containing polyamides.

(**49COOHDIA**) have a markedly lower tendency to crystallize than those based on 1,6-diamantane dicarboxylates **16COOHDIA** [38].

Alternatively, **16Dcarbam** and **49Dcarbam** polyamides (Figure 7.8) were prepared from the corresponding acyl chlorides through various polymerization techniques; however, the mechanical and thermal properties of thus obtained polymers were poorer than those prepared from the polymerization of free dicarboxylic acids **16COOHDIA** and **49COOHDIA** with simple aromatic amides [40].

Diamantane amines in combination with aryl dicarboxylic acids also provided a number of diamondoid polyamides. Utilizing aminophenyl as well as aminophenoxy diamantane 1,6- and 4,9-derivatives the families of polyamides **DPham** [41] and **DPhOPham** were prepared (Figure 7.9) [42]. The **49DPham** polymer displays tensile strength up to 92.8 MPa, elongation to break modulus 26.7%, and an initial modulus of 2.1 Gpa, as well as rather high G' values (above 10^8 Pa), even at temperatures exceeding 400 °C [41]. The thermomechanical properties

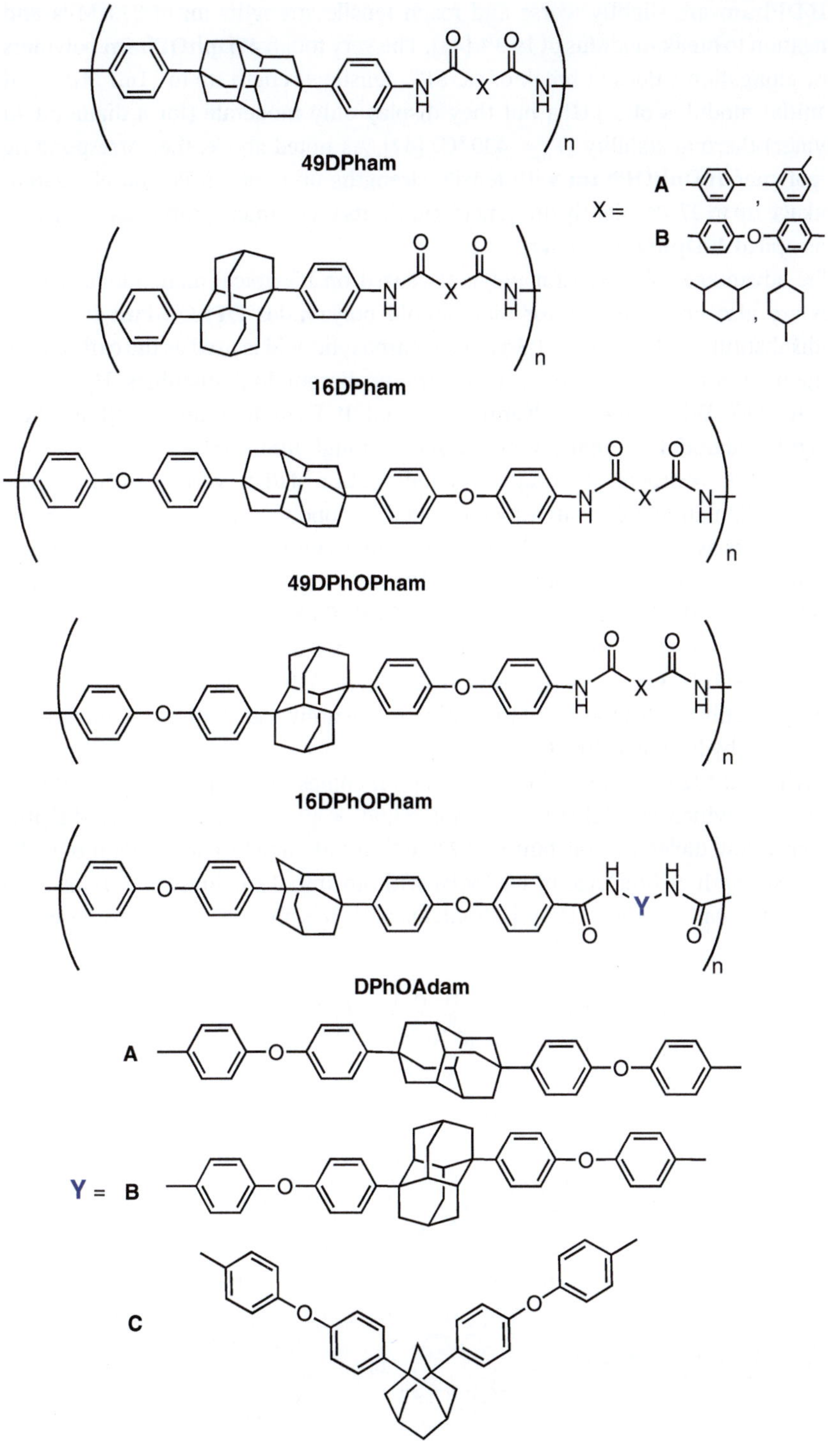

Figure 7.9 Series of hybrid diamantane/aryl polyamides.

of **16DPham** are slightly worse and reach tensile strengths up to 87.8 MPa and elongation to break modulus of 19.3% [43]. The very tough **49DphOPham** polymers show elongation values at break of ca. 61%, tensile strength up to 116.1 MPa, and an initial modulus of 2.1 GPa but they display only moderate (for a diamondoid polymer) thermal stability (T_d = 430 °C) [42]. As noted above, the corresponding 1,6-polymer **16DphOPham** with tensile strengths up to 90.2 MPa and elongation breakage up to 27.7% slightly underperforms in its mechanical properties compared to bis-apical **49DphOPham** [44].

The advantages observed for polymers based on bis-apical diamantane derivatives are also characteristic for diamondoid polyamides **DPhOAdam** in which 1,3-disubstituted adamantane 1,3-phenoxycarboxylic acid is used as the carboxylate component for polymerization with various diamondoid diamines H_2NYNH_2 (Figure 7.9). While linear polyamides **A** and **B** form high-quality films, their triangular adamantane analog **C** does not form tough films [45].

Polyimides are produced through the condensation of dianhydrides and diamines. Diamino diamantane derivatives were tested in copolymerizations with aromatic dianhydrides as alternatives to aliphatic polyimides with the hope of improved processing through increased solubility. 4,9-Diamino and aminophenyl diamantanes (**49NH2DIA, 49PhNH2DIA,** and **49PhOPHNH2DIA**) were copolymerized with a series of aromatic tetracarboxylic dianhydrides to give the corresponding polyimides **49Dpim, 49DPhpim,** and **49DPhoPhpim** (Figure 7.10). The processing involved the ring-opening polyaddition of the amine to the anhydride with subsequent thermal cyclodehydration at 300 °C [46].

Polyimides **49Dpim** are problematic when it comes to film processing, and only polymer **A**, which is soluble in *m*-cresol, could be processed. Polyimide **49Dpim** (**A**) has a low dielectric constant of 2.7 F m^{-1}, a low moisture absorption of 0.2%, tensile strength at breakage of 68.4 MPa with an initial modulus of 2.0 Gpa, and is stable to heat up to 430 °C [46]. Problems with the processing of films based on

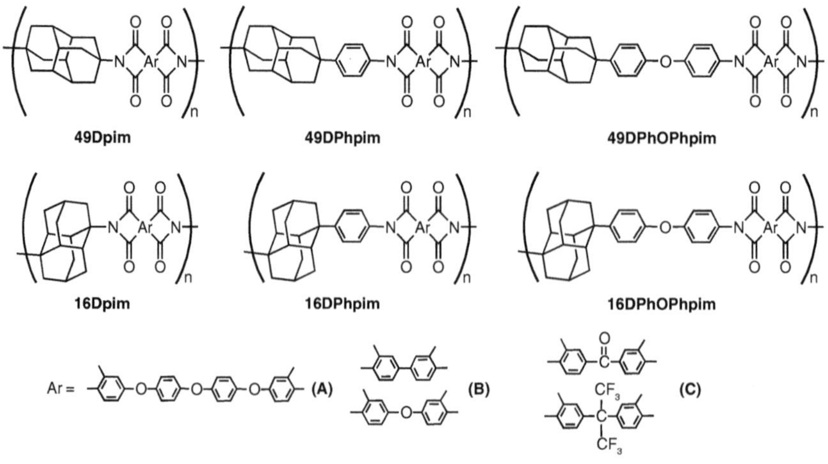

Figure 7.10 Diamantane polyimides.

49NH2DIA prompted further modifications of the amine component. Polyimides **49DPhpim** (Figure 7.10) based on bis-4,9-aminophenyl diamantane **49PhNH2DIA** (except polymer based on **A**) display many outstanding properties such as high tensile strengths up to 127 MPa, markedly low dielectric constants ranging from 2.53 to 2.72 and extremely high temperature stability (530 °C for polymer **B**). This class of polyimides was recommended as high-temperature materials for aviatic applications [47]. Very similar physical properties were observed for polyimides derived from well-soluble polymers **49DPhOPhpim** based on 4,9-bis-4-(4-aminophenoxy)phenyl diamantane **49PhOPhNH2DIA** [48].

Polyimides derived from 1,6-diaminodiamantane display higher solubilities than those obtained from bis-apical 4,9-diaminodiamantane. Polyimides **16Dpim** (Fig 7.10) combine high thermal stability with low coefficients of thermal expansion (CTE) and do not decompose up to 400 °C; polymer **B** is the most thermally stable. Higher solubility was measured for **16DPhpim** polymers, where fluorinated **C** and nonfluorinated **A** display superior production potential. For instance, **16DPhpim** (**C**) having a number-average molecular weight of 15 000 g mol^{-1} (though slightly lower than in commercial polyimides) is soluble in common organic solvents such as THF or chloroform. Polymer **16DPhpim** (**B**) displays the highest decomposition temperature of 526 °C [49]. The dielectric constants, water uptakes, and thermomechanical properties of **16DPhpim** are close to those observed for **49DPhpim** and 1,6-bis-4-(4-aminophenoxy)phenyl diamantane-based polymers **16DPhOPhpim** (Figure 7.10) [50].

Other diamantane-based polyimides were obtained from 4,9-bis[4(3,4-dicarboxyphenoxy)-phenyl]diamantane dianhydride (**49ANHDIA**) and aromatic and diamondoid diamines (Scheme 7.4), and their properties were compared to polyimide produced from 4,4′-oxydianiline (ODA) and pyromellitic dianhydride (PMDA), which is the most commonly used dielectric material in microelectronics [51].

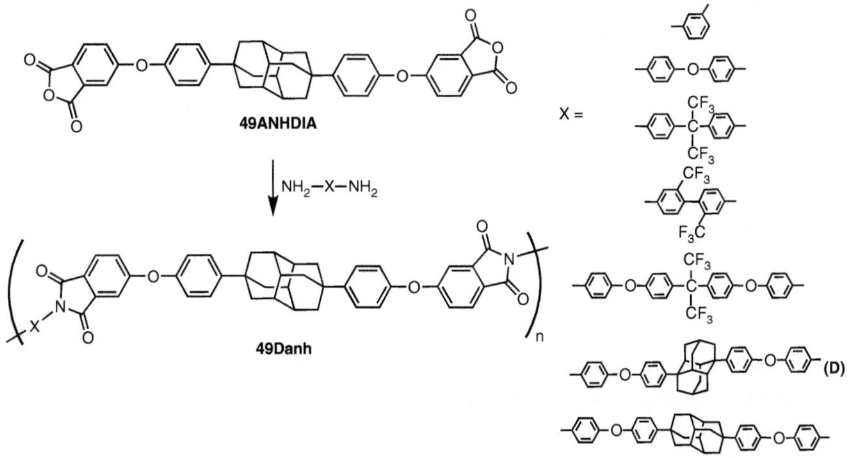

Scheme 7.4 Polyimides obtained from 4,9-bis[4(3,4-dicarboxyphenoxy)-phenyl] diamantane dianhydride (**49ANHDIA**) and aromatic and aromatic/diamondoid diamines.

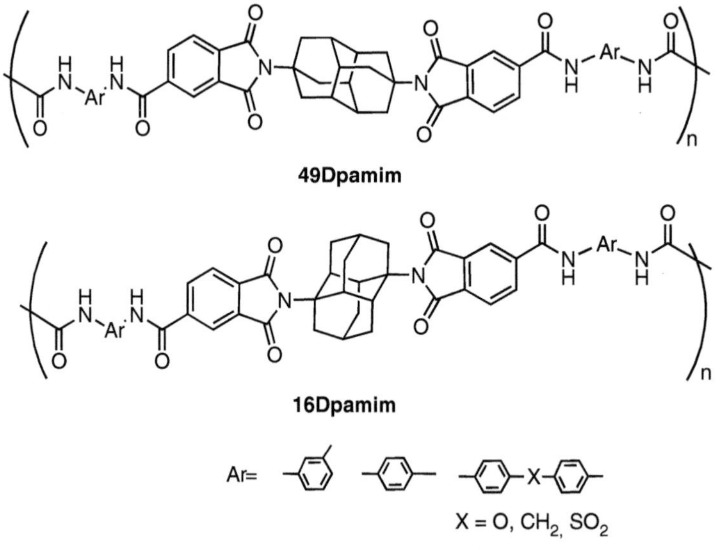

7 Diamondoid Polymers

Figure 7.11 Poly(amide-imide) diamantane copolymers.

Diamondoid polyimides all display much lower dielectric constants and hydroscopicities in combination with much higher hydrolytic temperature stabilities than some standard model polymers. The best property combination was observed for polymer **49Danh (D)**, which makes it promising for electronic applications as a thin film dielectric since it combines good mechanical properties with outstanding thermal stability, low dielectric constant, excellent hydrolytic resistance, and low moisture absorption.

Poly(amide-imide)s are another type of copolymer that combines high thermal stability with satisfactory processability. The syntheses and characterizations of polymers of this type containing 4,9-diamantane (**49Dpamim**) [52] and 1,6-diamantane (**16Dpamim**) [53] moieties in the main chain were described (Figure 7.11). Although such polymers display good solubility, may be cast into flexible films, and maintain good mechanical properties up to 350 °C, this polymer class does not reach the key characteristics of the best diamantane polyamides and polyimide described above.

Other polymers based on diamantane are limited to fluorescent materials. Polymers composed of oligomeric bis(2-ethylhexyl)-p-phenylenevinylene (BEH-PPV) monomers with rigid morphologically directing adamantane and diamantane groups (morphons) were prepared using Sonogashira cross-coupling polymerization. Single-molecule fluorescence spectroscopy, thin-film absorption, atomic force and fluorescence spectroscopies were used to study the morphological properties of the polychromophore polymers, clearly showing that the most ordered structures are those utilizing diamantane morphons [54]. The fluorescence excitation polarization experiments (Figure 7.12) indicate the alignment of the chromophores within a single polymer chain. Higher anisotropy (M, Figure 7.12) values correspond to higher degrees of chromophore alignment within the polymer, which is characteristic of an anisotropic structure.

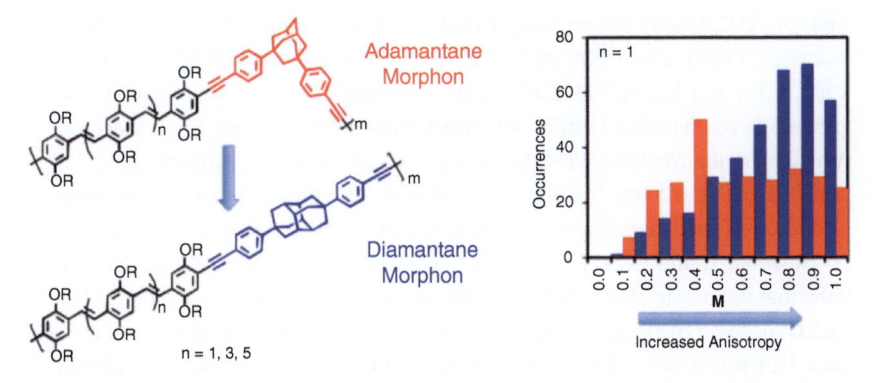

Figure 7.12 Histograms of fluorescence intensity modulation depth (M) for polymers containing bent adamantane (red) and linear diamantane (blue) morphons. Source: Reproduced from Ref. [54] with permission from the American Chemical Society, 2016.

Such control of topology and chromophore size offers powerful insights for further improvement of optoelectronic devices based on conjugated polymers using morphons to achieve controlled energy transfer.

We conclude that many classes of diamantane (cross)polymers have been prepared and tested to date. In many cases, such polymers display superior properties over conventional polymer building blocks, especially with respect to the optical properties (acrylates) and thermal stabilities (polyamides). In general, polymers based on apical diamantane derivatives display superior properties over those based on medial derivatives.

References

1 (a) Novikov, S.S., Khardin, A.P., Novakov, I.A., and Radchenko, S.S. (1974). Synthesis of adamantane-containing polyamides. *Vysokomol. Soed. B* 16: 155–156. (b) Khardin, A.P. and Radchenko, S.S. (1982). Adamantane polymer derivatives. *Usp. Khim.* 51: 480–506. (c) Novakov, I.A. and Orlinson, B.S. (2005). Polymers based on adamantane derivatives: synthesis, properties, and application. *Polym. Sci. C* 47: 50–73. (d) Nasrallah, H. and Hierso, J.C. (2019). Porous materials based on 3-dimensional T_d-directing functionalized adamantane scaffolds and applied as recyclable catalysts. *Chem. Mater.* 31: 619–642. (e) Grillaud, M. and Bianco, A. (2015). Multifunctional adamantane derivatives as new scaffolds for the multipresentation of bioactive peptides. *J. Pept. Sci.* 21: 330–345. (f) Muller, T. and Bräse, S. (2014). Tetrahedral organic molecules as components in supramolecular architectures and in covalent assemblies, networks and polymers. *RSC Adv.* 4: 6886–6907. (g) Kassab, R.M., Jackson, K.T., El-Kadri, O.M., and El-Kaderi, H.M. (2011). Nickel-catalyzed synthesis of nanoporous organic frameworks and their potential use in gas storage applications. *Res. Chem. Intermed.* 37: 747–757. (h) Nozaki, K. (2010). Material innovations for 193-nm resists. *J. Photopolym. Sci. Technol.* 23: 795–801. (i) Gupta, K.C., Sutar, A.K.,

and Lin, C.C. (2009). Polymer-supported Schiff base complexes in oxidation reactions. *Coord. Chem. Rev.* 253: 1926–1946. (j) Newkome, G.R., Kim, H.J., Choi, K.H., and Moorefield, C.N. (2004). Synthesis of neutral metallodendrimers possessing adamantane termini: supramolecular assembly with β-cyclodextrin. *Macromolecules* 37: 6268–6274. (k) Wang, S.J., Oldham, W.J., Hudack, R.A., and Bazan, G.C. (2000). Synthesis, morphology, and optical properties of tetrahedral oligo(phenylenevinylene) materials. *J. Am. Chem. Soc.* 122: 5695–5709. (l) Ishizone, T. and Goseki, R. (2018). Synthesis of polymers carrying adamantyl substituents in side chain. *Polym. J. (Tokyo, Jpn.)* 50: 805–819. (m) Jiang, W.Z. and Guo, J.W. (2020). pH-sensitive micelles of adamantane-based random copolymer. *Mater. Lett.* 260: 126889. (n) Ree, B.J., Kobayashi, S., Heo, K. et al. (2019). Nanoscale film morphology and property characteristics of dielectric polymers bearing monomeric and dimeric adamantane units. *Polymer* 169: 225–233.

2 Nomimura, S., Osaki, M., Park, J. et al. (2019). Self-healing alkyl acrylate-based supramolecular elastomers cross-linked via host-guest interactions. *Macromolecules* 52: 2659–2668.

3 Bai, J.W., Liu, Z.P., Zhang, J.H., and Zhong, F.C. (2019). Synthesis and characterization of novel high transparency polymer films bearing adamantanol groups. *Colloids Surf. A Physicochem. Eng. Asp.* 578: 123594.

4 (a) Lv, P.X., Dong, Z.X., Dai, X.M., and Qiu, X.P. (2019). High-T-g porous polyimide films with low dielectric constant derived from spiro-(adamantane-2,9'[2',7'-diamino]-fluorene). *J. Appl. Polym. Sci.* 136: 47313. (b) Jiang, W.Z., Yue, H.B., Shuttleworth, P.S. et al. (2019). Adamantane-based micro- and ultra-microporous frameworks for efficient small gas and toxic organic vapor adsorption. *Polymers* 11: 486. (c) Bhunia, A., Boldog, I., Moller, A., and Janiak, C. (2013). Highly stable nanoporous covalent triazine-based frameworks with an adamantane core for carbon dioxide sorption and separation. *J. Mater. Chem. A* 1: 14990–14999. (d) Li, X., Guo, J.W., Tong, R. et al. (2018). Microporous frameworks based on adamantane building blocks: synthesis, porosity, selective adsorption and functional application. *React. Funct. Polym.* 130: 126–132.

5 (a) Zhang, Y.N., Wang, Y., and Dang, B.J. (2017). A series of novel metal-organic frameworks assembled from adamantane dicarboxylates and flexible 1,2-bis(imidazole) ethane ligand: syntheses, structures, and properties. *Inorg. Nano-Met. Chem.* 47: 438–445. (b) Takahashi, K., Hoshino, N., Noro, S.I. et al. (2014). A crystal structures, dielectric, and CO_2-adsorption properties of one-dimensional cu(II)$_2$(adamantane-1-carboxylate)$_4$(pyrazine)$_\infty$ coordination polymers with polar ligands. *Sci. Adv. Mater.* 6: 1417–1424.

6 Friebel, J., Ender, C.P., Mezger, M. et al. (2019). Synthesis of precision poly(1,3-adamantylene alkylene)s via acyclic diene metathesis polycondensation. *Macromolecules* 52: 4483–4491.

7 Fu, S.Q., Zhu, J.P., and Chen, S.J. (2018). Tunable shape memory polyurethane networks cross-linked by 1,3,5,7-tetrahydroxyadamantane. *Macromol. Res.* 26: 1035–1041.

8 (a) Fu, S.Q., Zhu, J.P., Zou, F.X. et al. (2018). A novel adamantane-based polyurethane with shape memory effect. *Mater. Lett.* 229: 44–47. (b) Miyamae, K., Nakahata, M., Takashima, Y., and Harada, A. (2015). Self-healing, expansion-contraction, and shape-memory properties of a preorganized supramolecular hydrogel through host-guest interactions. *Angew. Chem. Int. Ed.* 54: 8984–8987.

9 Maity, S., Choudhary, P., Manjunath, M. et al. (2015). A biodegradable adamantane polymer with ketal linkages in its backbone for gene therapy. *Chem. Commun.* 51: 15956–15959.

10 Van Guyse, J.F.R., Bera, D., and Hoogenboom, R. (2021). Adamantane functionalized poly(2-oxazoline)s with broadly tunable LCST-behavior by molecular recognition. *Polymers* 13: 374.

11 Wen, W.Q., Guo, C., and Guo, J.W. (2021). Acid-responsive adamantane-cored amphiphilic block polymers as platforms for drug delivery. *Nanomaterials* 11: 188.

12 (a) Liu, B.W., Zhou, H., Zhou, S.T. et al. (2014). Synthesis and self-assembly of CO_2-temperature dual stimuli-responsive triblock copolymers. *Macromolecules* 47: 2938–2946. (b) Koike, K., Araki, T., and Koike, Y. (2015). A highly transparent and thermally stable copolymer of 1-adamantyl methacrylate and styrene. *Polym. Int.* 64: 188–195. (c) Matsumoto, A. and Sumihara, T. (2017). Thermal and mechanical properties of random copolymers of diisopropyl fumarate with 1-adamantyl and bornyl acrylates with high glass transition temperatures. *J. Polym. Sci. A* 55: 288–296. (d) Tsai, C.W., Wang, J.C., Li, F.N. et al. (2016). Synthesis of adamantane-containing methacrylate polymers: characterization of thermal, mechanical, dielectric and optical properties. *Mater. Express* 6: 220–228. (e) Tsai, C.W., Wu, K.H., Wang, J.C., and Shih, C.C. (2017). Synthesis, characterization, and properties of petroleum-based methacrylate polymers derived from tricyclodecane for microelectronics and optoelectronics applications. *J. Ind. Eng. Chem.* 53: 143–154. (f) Zhang, L., Qiu, G.R., Liu, F.F. et al. (2018). Controlled ROMP synthesis of side-chain ferrocene and adamantane-containing diblock copolymer for the construction of redox-responsive micellar carriers. *React. Funct. Polym.* 132: 60–73. (g) Noh, H., Myung, S., Kim, M.J., and Yang, S.K. (2019). Stimuli-responsive supramolecular assemblies via self-assembly of adamantane-containing block copolymers. *Polymer* 175: 65–70. (h) Miao, Z.C., Shi, J.T., Liu, T.J. et al. (2019). Adamantane-modified graphene oxide for cyanate ester resin composites with improved properties. *Appl. Sci.* 9: 881. (i) Ding, L., Li, J., Jiang, R.Y. et al. (2019). Noncovalently connected supramolecular metathesis graft copolymers: one pot synthesis and self-assembly. *Eur. Polym. J.* 112: 670–677. (j) Hill, C.J., McDonald, A.G., and Roll, M.F. (2021). Dienes and diamondoids: poly(2-1-adamantyl −1,3-butadiene) and random copolymers with isoprene via redox-emulsion polymerization and their hydrogenation. *J. Appl. Polym. Sci.* 138: e50711.

13 (a) Kakuta, T., Takashima, Y., Nakahata, M. et al. (2013). Preorganized hydrogel: self-healing properties of supramolecular hydrogels formed by polymerization of host-guest-monomers that contain cyclodextrins and hydrophobic guest groups.

Adv. Mater. 25: 2849–2853. (b) Kakuta, T., Takashima, Y., and Harada, A. (2013). Highly elastic supramolecular hydrogels using host-guest inclusion complexes with cyclodextrins. *Macromolecules* 46: 4575–4579.

14 (a) Vainer, A.Y., Dyumaev, K.M., Kovalenko, A.M. et al. (2017). 1,3,5,7-Tetrasubstituted adamantanes as frameworks in the design of assemblies of porphyrin macrocycles. *Dokl. Chem.* 474: 129–132. (b) Tominaga, M., Kunitomi, N., Ohara, K. et al. (2019). Hollow and solid spheres assembled from functionalized macrocycles containing adamantane. *J. Org. Chem.* 84: 5109–5117.

15 (a) Lamanna, G., Russier, J., Menard-Moyon, C., and Bianco, A. (2011). HYDRAmers: design, synthesis and characterization of different generation novel Hydra-like dendrons based on multifunctionalized adamantane. *Chem. Commun.* 47: 8955–8957. (b) Grillaud, M., de Garibay, A.P.R., and Bianco, A. (2016). Polycationic adamantane-based dendrons form nanorods in complex with plasmid DNA. *RSC Adv.* 6: 42933–42942.

16 Wang, C., Wang, Y., Ge, R.L. et al. (2018). A 3D covalent organic framework with exceptionally high iodine capture capability. *Chem. Eur. J.* 24: 585–589.

17 Spohn, M., Alkahtani, M.H.A., Leiter, R. et al. (2018). Poly(1-vinyladamantane) as a template for nanodiamond synthesis. *ACS Appl. Nano Mater.* 1: 6073–6080.

18 Yang, K. and Guo, J.W. (2019). Three-dimensional nanoporous organic frameworks based on rigid unites. *Mater. Lett.* 236: 155–158.

19 Ghosh, A., Sciamanna, S.F., Dahl, J.E. et al. (2007). Effect of nanoscale diamondoids on the thermomechanical and morphological behaviors of polypropylene and polycarbonate. *J. Polym. Sci. B* 45: 1077–1089.

20 Malik, A.A., Archibald, T.G., Baum, K., and Unroe, M.R. (1992). Thermally stable polymers based on acetylene-terminated adamantanes. *J. Polym. Sci., Part A-1: Polym. Chem.* 30: 1747–1754.

21 Wen, W.Q., Shuttleworth, P.S., Yue, H.B. et al. (2020). Exceptionally stable microporous organic frameworks with rigid building units for efficient small gas adsorption and separation. *ACS Appl. Mater. Interfaces* 12: 7548–7556.

22 Lysenko, A.B., Senchyk, G.A., Lincke, J. et al. (2010). Metal oxide-organic frameworks (MOOFs), a new series of coordination hybrids constructed from molybdenum(VI) oxide and bitopic 1,2,4-triazole linkers. *J. Chem. Soc. Dalton Trans.* 39: 4223–4231.

23 Senchyk, G.A., Lysenko, A.B., Krautscheid, H., and Domasevitch, K.V. (2011). "Fluoride molecular scissors": a rational construction of new Mo(VI) oxofluorido/1,2,4-triazole MOFs. *Inorg. Chem. Commun.* 14: 1365–1368.

24 Dolci, E., Froidevaux, V., Joly-Duhamel, C. et al. (2016). Maleimides as a building block for the synthesis of high performance polymers. *Polym. Rev.* 56: 512–556.

25 Wang, J.E., Chern, Y.T., and Chung, M.A. (1996). Synthesis and characterization of new poly(*N*-1-adamantylmaleimide) and poly(*N*-1-diamantylmaleimide). *J. Polym. Sci. A* 34: 3345–3354.

26 Barabash, A.V., Didukh, N.A., Kibal'nyi, N.A. et al. (2014). Unusual aerobic oxidation of sterically hindered 1-diamantyl methyl ketone. *Russ. J. Org. Chem.* 50: 1690–1691.

27 (a) Jaeblon, T. (2010). Polymethylmethacrylate: properties and contemporary uses in orthopaedics. *J. Am. Acad. Orthop. Surg.* 18: 297–305. (b) Goldman, A. and Wollina, U. (2019). Polymethylmethacrylate-induced nodules of the lips: clinical presentation and management by intralesional neodymium:YAG laser therapy. *Dermatol. Ther.* 32: e12755.

28 Ali, U., Abd Karim, K.J.B., and Buang, N.A. (2015). A review of the properties and applications of poly (methyl methacrylate) (PMMA). *Polym. Rev.* 55: 678–705.

29 Robello, D.R. (2013). Moderately high refractive index, low optical dispersion polymers with pendant diamondoids. *J. Appl. Polym. Sci.* 127: 96–103.

30 Žuanić, M., Majerski, Z., and Janović, Z. (1981). Poly(1-vinyladamantane). *J. Polym. Sci. C* 19: 387–389.

31 Sinkel, C., Agarwal, S., Fokina, N.A., and Schreiner, P.R. (2009). Synthesis, characterization, and property evaluations of copolymers of diamantyl methacrylate with methyl methacrylate. *J. Appl. Polym. Sci.* 114: 2109–2115.

32 Takano, T., Lin, Y.C., Shi, F.G. et al. (2010). Novel methacrylated diamondoid to produce high-refractive index polymer. *Opt. Mater.* 32: 648–651.

33 Padmanaban, M., Chakrapani, S., Lin, G. et al. (2007). Novel diamantane polymer platform for resist applications. *J. Photopolym. Sci. Technol.* 20: 719–728.

34 Malik, A.A., Archibald, T.G., Baum, K., and Unroe, M.R. (1991). New high-temperature polymers based on diamantane. *Macromolecules* 24: 5266–5268.

35 Chern, Y.T. (1995). Synthesis and properties of new polycyclic polyesters from 1,6-diamantanedicarboxylic acyl chloride and aromatic diols. *Macromolecules* 28: 5561–5566.

36 Chern, Y.T. and Huang, C.M. (1998). Synthesis and characterization of new polyesters derived from 1,6- or 4,9-diamantanedicarboxylic acyl chlorides with aryl ether diols. *Polymer* 39: 2325–2329.

37 Dang, T.D., Dalton, M.J., Venkatasubramanian, N. et al. (2004). Synthesis and characterization of polyaryeneetherketone triphenylphosphine oxides incorporating cycloaliphatic/cage hydrocarbon structural units. *J. Polym. Sci. A* 42: 6134–6142.

38 Chern, Y.T. and Wang, W.L. (1995). Synthesis and properties of new polyamides based on diamantane. *Macromolecules* 28: 5554–5560.

39 Chern, Y.T. and Wang, W.L. (1996). Synthesis and characterization of tough polycyclic polyamides containing 4,9-diamantyl moieties in the main chain. *J. Polym. Sci. A* 34: 1501–1509.

40 Chern, Y.T., Fang, J.S., and Kao, S.C. (1995). Preparation of polyamides derived from 1,6-diamantane dicarboxylic chloride by solution polycondensation and interfacial polycondensation. *J. Polym. Sci. A* 33: 2833–2840.

41 Chern, Y.T. (1998). Synthesis of polyamides derived from 4,9-bis(4-aminophenyl)diamantane. *Polymer* 39: 4123–4127.

42 Chern, Y.T. and Wang, W.L. (1998). Synthesis and characterization of tough polyamides derived from 4,9-bis 4-(4-aminophenoxy)phenyl diamantane. *Polymer* 39: 5501–5506.

43 Chern, Y.T. and Wang, W.L. (1998). High alpha transitions of new polyamides based on diamantane. *J. Polym. Sci. A* 36: 1257–1263.

44 Chern, Y.T. and Wang, W.L. (1998). Synthesis and characterization of new polyamides derived from 1,6-bis 4-(4-aminophenoxy)phenyl diamantane. *J. Polym. Sci. A* 36: 2185–2192.

45 Chern, Y.T., Shiue, H.C., and Kao, S.C. (1998). Synthesis and characterization of new polyamides containing adamantyl and diamantyl moieties in the main chain. *J. Polym. Sci. A* 36: 785–792.

46 Chern, Y.T. and Huang, C.M. (1998). Synthesis and characterization of new polyimides derived from 4,9-diaminodiamantane. *Polymer* 39: 6643–6648.

47 Chern, Y.T. and Shiue, H.C. (1998). High subglass transition temperatures and low dielectric constants of polyimides derived from 4,9-bis(4-aminophenyl) diamantane. *Chem. Mater.* 10: 210–216.

48 Chern, Y.T. and Shiue, H.C. (1997). Low dielectric constants of soluble polyimides derived from the novel 4,9-bis 4-(4-aminophenoxy)phenyl diamantane. *Macromolecules* 30: 5766–5772.

49 Chern, Y.T. (1998). High subglass transitions appearing in the rigid polyimides derived from the novel 1,6-bis(4-aminophenyl)diamantane. *Macromolecules* 31: 1898–1905.

50 Chern, Y.T. (1998). Low dielectric constant polyimides derived from novel 1,6-bis 4-(4-aminophenoxy)phenyl diamantane. *Macromolecules* 31: 5837–5844.

51 Chern, Y.T. and Wang, J.J. (2009). Hydrolytic stability and high T_g of polyimides derived from the novel 4,9-bis[4-(3,4-dicarboxyphenoxy)phenyl]-diamantane dianhydride. *J. Polym. Sci. A* 47: 1673–1684.

52 Chern, Y.T., Huang, C.M., and Huang, S.C. (1998). Synthesis and characterization of new poly(amide-imide)s containing 4,9-diamantane moieties in the main chain. *Polymer* 39: 2929–2934.

53 Chern, Y.T. (1996). Preparation and properties of new polyimides derived from 1,6-diaminodiamantane. *J. Polym. Sci. A* 34: 125–131.

54 Zhu, X.J., Shao, B.Y., Vanden Bout, D.A., and Plunkett, K.N. (2016). Directing the conformation of oligo(phenylenevinylene) polychromophores with rigid, nonconjugatable morphons. *Macromolecules* 49: 3838–3844.

8

Diamondoids in Catalysis

Catalysis is another field that has benefited from the incorporation of diamondoids. A recent increase in literature examples of diamondoid-containing catalysts demonstrates the general potential of cage compounds in ligand design with applicability for a wide range of organic reactions. Certainly, the most prevalent diamondoid used in catalysis so far has been adamantane, and its lipophilicity and bulkiness have been a key feature in many diverse ligands, as described in the excellent review by Agnew-Francis and Williams [1]. The advantage of adamantane and other diamondoids lies in their rigidity and richness with C—H bonds, resulting in significant polarizability of the cage moieties, leading to significant attractive London dispersion (LD) effects [2]. The electron delocalization in the intermolecular region enables the extension of electron density beyond the formally assigned molecular boundaries, where CH•••HC and CH•••C interactions are stronger than usually assumed [3]. Their high polarizabilities make diamondoids excellent dispersion energy donors (DEDs) [4]. Along with high electron-donor capacity, which is an important feature of promising ligands for use in catalysis, all of these effects increase the electron density on the metal center in, e.g., an organometallic catalyst. Consequently, a metal center is more prone to oxidative addition, which makes the catalyst more effective and more reactive. Thus, the introduction of diamondoid ligands is a straightforward method for tuning catalyst activity. Additional benefits are varying cages, topologies, attachment points, and substitution patterns.

Broadly documented catalytic systems incorporating diamondoid cages are organometallic catalysts with diamondoid phosphines as ligands [5]. Diamondoid phosphine ligands form unsaturated phosphine complexes in terms of coordination, accelerating the reaction rate while at the same time facilitating oxidative addition, which enables the use of milder reaction conditions and less active and often much cheaper metals.

Before diving deeper into examples of such catalysts, we first need to mention the use of the Tolman cone angle [6], a practical tool for assessing the viability of phosphines in catalytic applications [7]. The Tolman angle (θ) is defined as the apex angle of the smallest possible cone originating at the metal center and with its edges lying on the van der Waals radius spheres of the outermost ligand atoms (Figure 8.1). Tolman's approach was originally developed for monodentate phosphine ligands

The Chemistry of Diamondoids: Building Blocks for Ligands, Catalysts, Materials, and Pharmaceuticals, First Edition. Andrey A. Fokin, Marina Šekutor, and Peter R. Schreiner.
© 2024 WILEY-VCH GmbH. Published 2024 by WILEY-VCH GmbH.

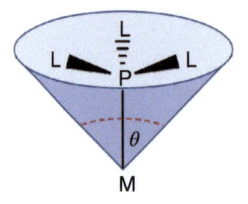

Figure 8.1 Tolman cone angle of a monodentate phosphine ligand.

bound to a nickel center but has since been used for palladium and other metals as coordination centers of the catalyst.

Since simple calculations of Tolman angles in a catalyst are a somewhat rough method with limited accuracy, even when using X-ray structural data [8], in recent years improvements to the approach have been made, most notably by the Allen group [9]. This modification of the procedure includes taking the structure of the whole complex (either from X-ray analysis or computations) and determining the exact cone angle of any ligand therein without any approximations. The ligand is placed into the most accurate possible cone that encompasses all of its atoms, and the method is in principle applicable for any type of ligand bound to any metal center. A rule of thumb for Tolman angle calculation states that the larger the angle value is, the better the ligand is for catalysis. For alkyl phosphines, the trend is apparent: more branched and bulkier hydrocarbons incorporated into the phosphines that serve as ligands form larger cone angles and are typically better scaffolds in ligand design (Table 8.1).

The first preparation of **AD2PH** marked the start of the consideration of diamondoid phosphines as ligands [10], since their condensation with alkyl halides

Table 8.1 Exact cone angles for selected organometallic complexes with monodentate phosphine ligands obtained by the Allen method.

Ligand	Exact cone angle (Pd)/°	Exact cone angle (Ni)/°	Exact cone angle (Pt)/°
PMe$_3$	120.4	125.8	127.7
PMe$_2$(*i*-Pr)	147.6	154.0	150.9
PMe$_2$(*t*-Bu)	146.0	152.6	150.2
PMe(*i*-Pr)$_2$	159.2	166.3	157.4
PMe(*t*-Bu)$_2$	174.4	182.2	179.0
P(*i*-Pr)$_3$	169.0	176.6	173.2
P(*t*-Bu)$_3$	187.6	196.3	191.8
PAD$_3$ [a]	188.0	–	–
PAD$_2$(4-DIA) [a]	188.0	–	–
PAD(4-DIA)$_2$ [a]	187.9	–	–
P(4-DIA)$_3$ [a]	187.8	–	–
PPh$_3$	170.0	177.6	174.6

a) Unpublished results.
Source: Data taken from Ref. [10].

AD2PH   **4DIA2PH**   **9TRIA2PH**

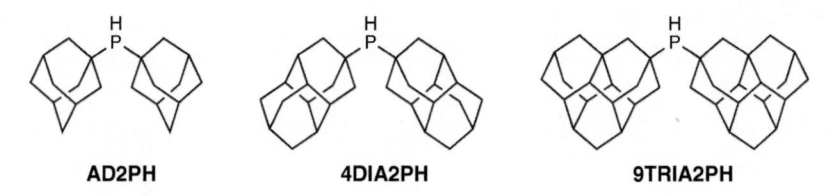

Figure 8.2 Secondary diamondoid phosphines used for the preparation of bulky trialkylphosphine ligands. Source: Adapted from Ref. [11].

is facile and therefore provides an ideal synthetic starting point for the preparation of bulky trialkylphosphines. Later, larger secondary diamondoid phosphines were also prepared for the same purpose (Figure 8.2) [11].

As predicted by the cone angle considerations, phosphines bearing bulky alkyl cage moieties proved to be excellent ligands, especially for C—C and C—N bond formation reactions. For example, the Beller group described a Suzuki C—C coupling reaction of poorly reactive aryl chlorides with arylboronic acids catalyzed by palladium with **nBuAD2P** as a ligand (Scheme 8.1) [5a, 12]. The ligand performed better in this reaction than any other catalyst system known at the time and was superior in terms of activity and productivity, managing to catalyze reactions with high yields (above 80%), even with deactivated aryls, and achieving high catalyst turnover numbers (up to 20 000). The same group expanded the application of **nBuAD2P** to the Heck reaction, again with poorly reactive aryl chlorides, and successfully obtained styrenes as products of coupling with terminal alkenes (Scheme 8.1) [13]. The use of **nBuAD2P** as a ligand was so successful that it became widely commercially available (cataCXium®), and some of its derivatives also enjoyed widespread application (**nBuAD2PHI** marketed as cataCXium HI and **BnAD2P** marketed as cataCXium ABn).

Further usefulness of **nBuAD2P** as a ligand for metal-catalyzed reactions is demonstrated in the preparation of oxindoles (Scheme 8.2) [14]. Namely, *ortho*-methylated carbamoyl chlorides can be readily converted to oxindoles in good yields, and in the case of *ortho*-carbocycles like cyclopropanes, the obtained products are synthetically very useful spirooxindoles. Note that here the selective C (sp^3)–H activation occurred in the presence of a competitive C (sp^2)–H bond, which makes this synthetic approach highly appealing, especially when applied for the preparation of structurally more complex molecules.

Other applications of the cataCXium class of ligands include Buchwald–Hartwig amination reactions [15]. Since these reactions often benefit from the presence of a bulky ligand, the adamantane cages significantly assist in the oxidative addition steps and also promote the amine substrate binding and reductive elimination of the product (Scheme 8.3). The fact that the reaction of unactivated aryl chlorides with sterically hindered amines is more effective [15a] with bulky adamantane ligands may be associated with additional LD interactions between the reactants [2]. In addition to using **nBuAD2PHI**, the applicability of various alkylated **AlkAD2PHI** ligands (alkyl = methyl, *i*-butyl, *i*-propyl, allyl, 2-methoxyethyl, and benzyl) was also tested, and *n*-Bu and allyl substituted phosphines were found to be the best-performing derivatives for the studied reaction.

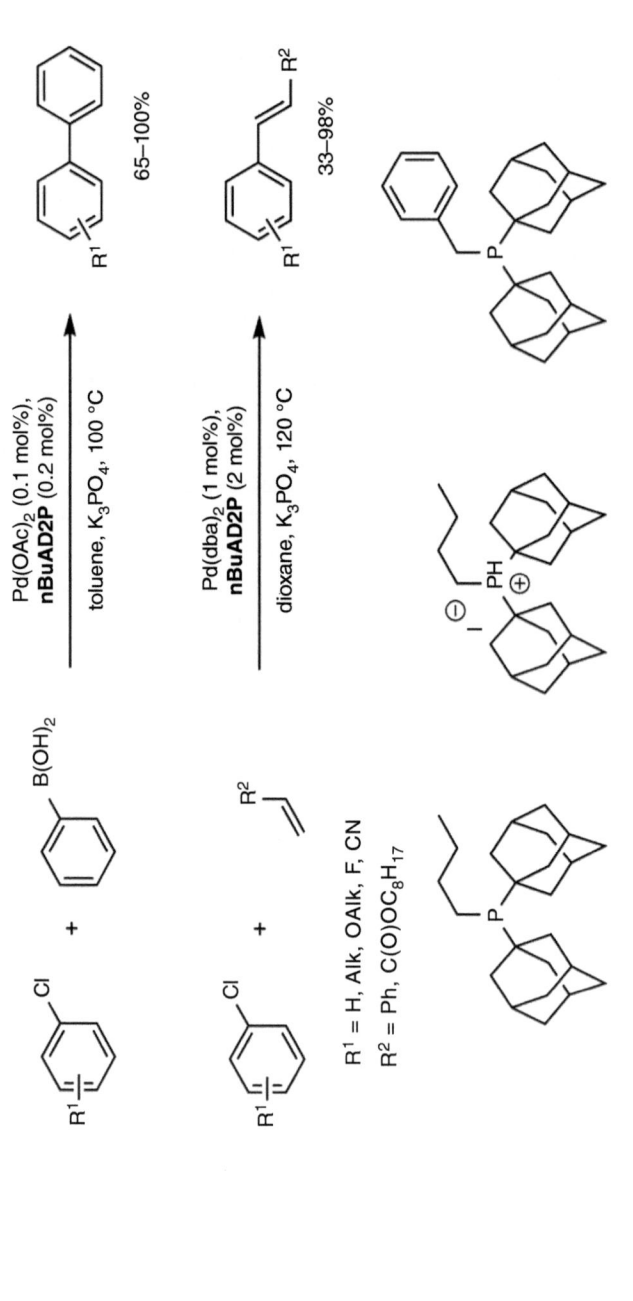

Scheme 8.1 Pd-catalyzed Suzuki coupling (top) and Heck reaction (middle) with a **nBuAD2P** ligand and similar cataCXium® ligands (bottom); dba = dibenzylideneacetone. Source: Adapted from Refs. [5a, 12, 13].

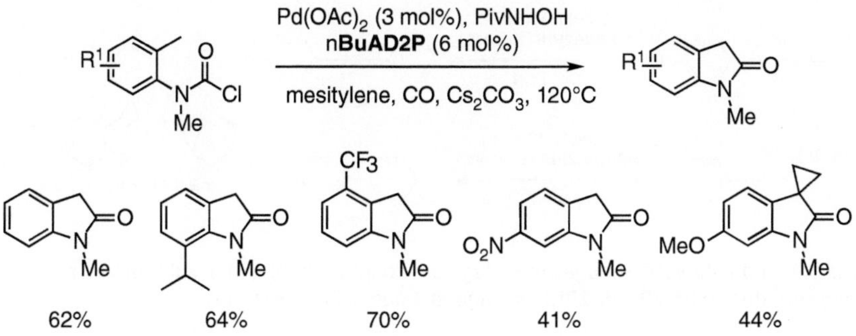

Scheme 8.2 Application of **nBuAD2P** in the synthesis of oxindoles. Source: From Ref. [14b].

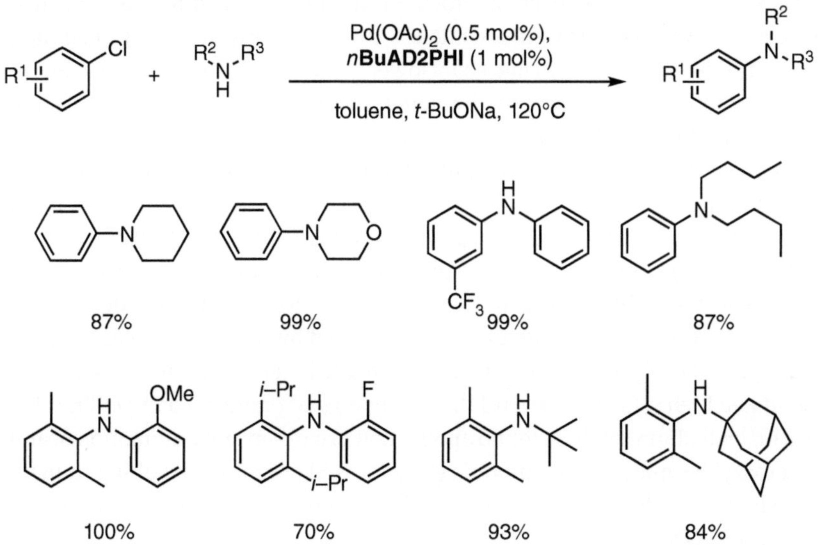

Scheme 8.3 Pd-catalyzed Buchwald-Hartwig amination with a **nBuAD2PHI** ligand. Source: From Ref. [15a].

The diamantane containing **nBuDIA2PHI** ligand, a larger analogue of cataCXium HI, was tested in Sonogashira–Hagihara and Suzuki–Miyaura coupling reactions (Scheme 8.4) [11]. In the performed Sonogashira coupling between phenylacetylene and 4-bromofluorobenzene the reaction yielded 70% of the product, which was a marked improvement when compared to the analogous reaction using PPh$_3$ as a co-catalyst that had a yield of only 38%. In the case of the Suzuki coupling between 4-bromotoluene and phenylboronic acid, the reaction proceeded quantitatively, while without **nBuDIA2PHI** only partial conversion was achieved in addition to the formation of biphenyl as a side product. These two examples demonstrate the superiority of phosphines consisting of larger diamondoid cages as co-ligands in the catalysis of today common C—C bond formation reactions.

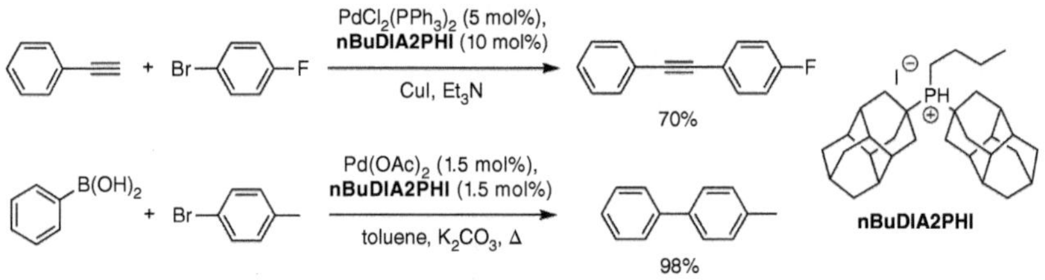

Scheme 8.4 Pd-catalyzed Sonogashira–Hagihara coupling (top) and Suzuki–Miyaura (bottom) couplings with **nBuDIA2PHI** as a ligand. Source: From Ref. [11].

Aside from the cataCXium class of ligands, their previously mentioned synthetic precursor **AD2PH** also has potential for application in Pd-catalyzed cross-coupling reactions such as the Heck reaction. However, it showed mixed results: activated aryl chlorides gave excellent yields, while electron-rich derivatives and heteroaryl chlorides afforded products with markedly lower yields [16]. Nevertheless, since secondary phosphines are less sterically hindered than their tertiary counterparts, the relatively high efficiency of **AD2PH** is a somewhat surprising feat and can probably be ascribed to the beneficial and still significant steric bulk of the two present adamantane cages.

The next class of diamondoid phosphine ligands includes the DalPhos derivatives (Scheme 8.5). Here again two adamantane cages play a central part in the structural features of the ligands, and they were also first tested in the demanding Buchwald-Hartwig-type reactions extensively studied by the Stradiotto group [18]. In general, DalPhos ligands in these reactions enabled high yields, often above 90%, and facilitated high selectivity toward the monoarylated product. Later DalPhos ligands within palladium catalysts found application in carbonylative amination of aryl and heteroaryl bromides with ammonia (Scheme 8.5) or amines, resulting in moderate product yields for primary amines and somewhat lower yields for secondary amines with **PyrDalPhos** [17].

Second-generation phosphine biphenyl ligands with two adamantane moieties **BIPHAD2P** also found application in the preparation of aryl ethers from aryl halides and phenols and similar alkyl aryl ethers from aryl halides and alkyl alcohols (Scheme 8.6) [19]. A wide variety of ethers was thus prepared in very high yields with a catalyst system that tolerated many functional groups, including amines.

Application of bulky phosphine ligands is not limited to palladium coordination chemistry, and recent examples of the corresponding gold complexes testify to that. DalPhos derivatives were used for the preparation of the catalytically active gold(I) compounds (Scheme 8.7). The first presented Au complex (**MorDalPhosAuCl**) promoted amide addition to triple bonds followed by spontaneous cyclization, affording oxazoles in a [3 + 2] annulation reaction [20]. Tricoordinated gold complexes are rare in homogeneous gold catalysis, and the reactivity in this case strongly depends on the engagement of the non-phosphorus heteroatom in the catalytic process. The second, even bulkier-depicted Au complex converted alkynes into enol ethers, imines, and ketones.

MorDalPhos
1%

MeDalPhos
10%

PyrDalPhos
49%

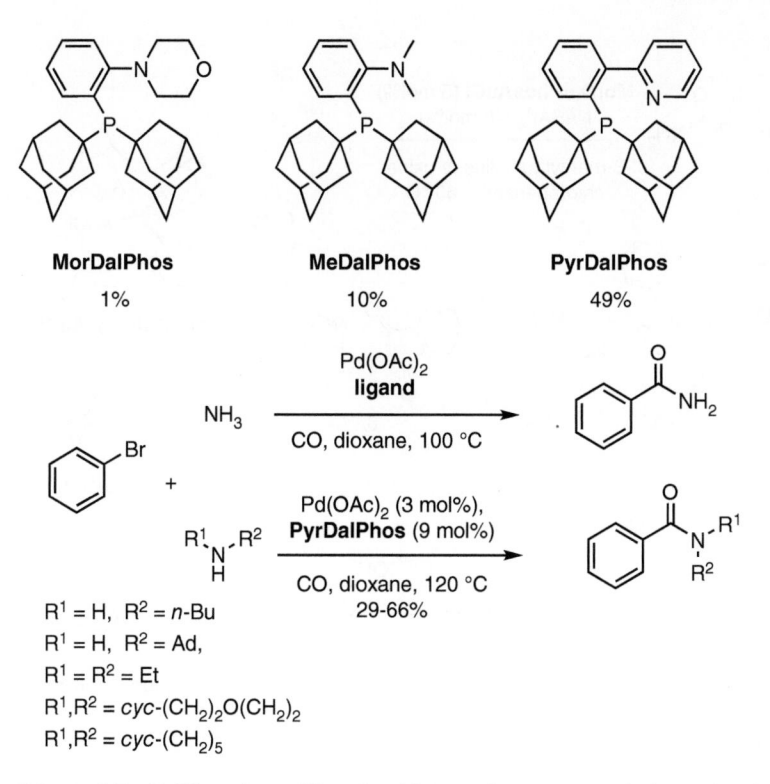

Reaction 1: PhBr + NH$_3$ → benzamide (PhC(O)NH$_2$)
Conditions: Pd(OAc)$_2$, ligand, CO, dioxane, 100 °C

Reaction 2: PhBr + R^{1}R^{2}NH → PhC(O)NR^{1}R^2
Conditions: Pd(OAc)$_2$ (3 mol%), PyrDalPhos (9 mol%), CO, dioxane, 120 °C, 29–66%

R^1 = H, R^2 = n-Bu
R^1 = H, R^2 = Ad,
R^1 = R^2 = Et
R^1,R^2 = cyc-(CH$_2$)$_2$O(CH$_2$)$_2$
R^1,R^2 = cyc-(CH$_2$)$_5$

Scheme 8.5 DalPhos class of ligands with two adamantane moieties and their application in a carbonylative amination. Source: From Ref. [17].

Reaction: R^1-ArX + R^2-ArOH → R^1-Ar-O-Ar-R^2
Conditions: Pd(OAc)$_2$ (2 mol%), BIPHAD2P (3 mol%), toluene, K$_3$PO$_4$, 100 °C

73%   87%   **BIPHAD2P**

Scheme 8.6 Pd-catalyzed synthesis of aryl ethers with biphenyl diamondoid phosphine ligand **BIPHAD2P**. Source: From Ref. [19a].

Secondary phosphine oxides [22] are also useful as ligands and recently adamantane and diamantane derivatives were used as air-stable pre-catalysts for the C—H arylation of oxazoles [23] and oxazolines [23, 24]. Comparable yields were observed for respective 1-adamantyl and 4-diamantyl complexes. These oxides readily self-assemble in the presence of palladium and form complexes, which are in essence bidentate ligands stabilized with hydrogen bonds (Scheme 8.8).

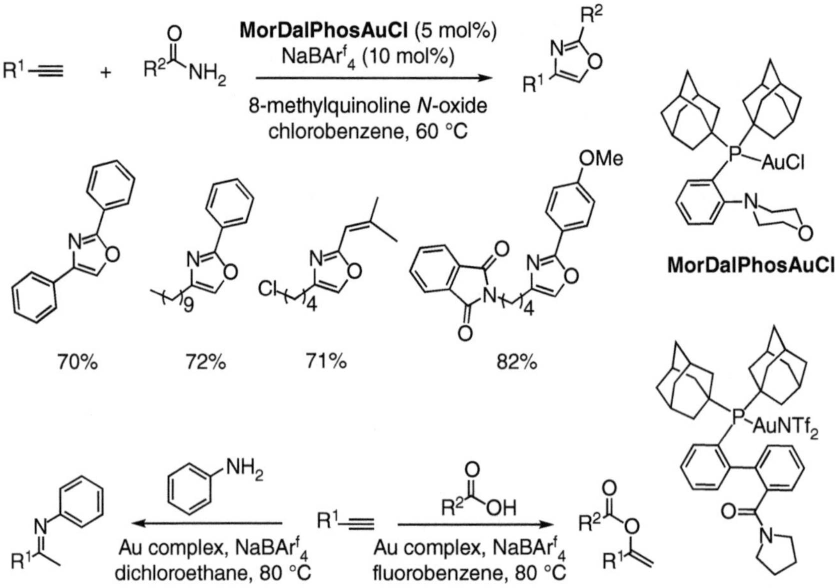

Scheme 8.7 Synthesis of aryl ethers (top)[20] and imines and enol ethers (bottom) using diamondoid phosphine Au complexes. Source: From Refs. [20, 21].

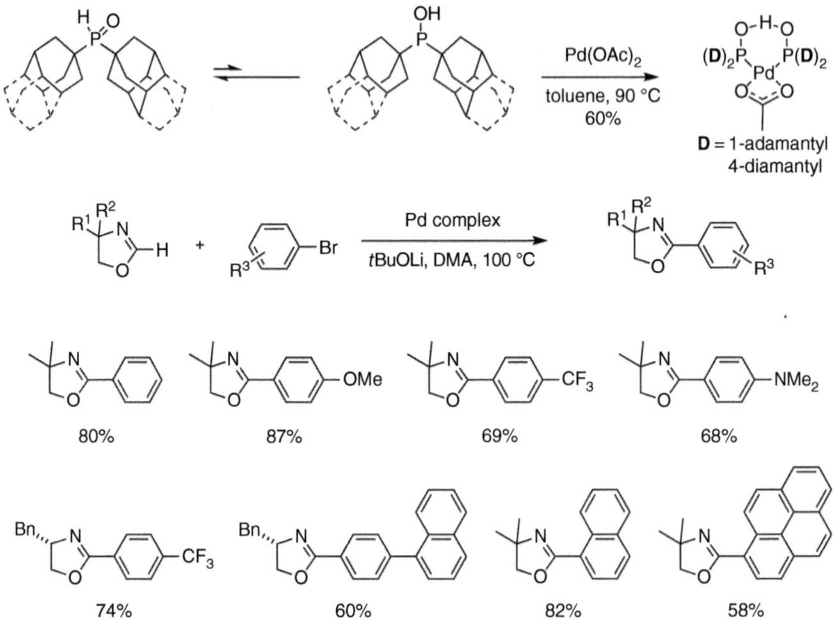

Scheme 8.8 Complex formation by self-assembly of secondary 1-adamantyl and 4-diamantyl phosphine oxides with palladium (top) and its use for C–H arylations of oxazolines (bottom). Source: From Refs. [23, 24].

In case of bulky **DIA2POH**, C—H arylation of electron-rich oxazolines with aryl halides proceeds without the need for a directing group in the aryl precursors. Moreover, the scope of this catalytic Pd complex is truly versatile since the reaction proceeded with aryl bromides containing substituents in the *ortho*-, *meta*-, and *para*-positions. On the other side of the reactant spectrum, monosubstituted oxazolines as well as those having *gem*-disubstitution also proved to be viable substrates for this reaction. Remarkably, aryl chlorides were also applicable reactants for this arylation as well as a number of useful *N*- and *S*-heteroaryl halides, including pyridine, indole, quinoline, and thiophene derivatives.

A sterically demanding triadamantyl phosphine (**AD3P**) has also recently been prepared and applied in a Suzuki–Miyaura reaction (Scheme 8.9) [5g]. Pd-catalyzed cross-coupling reactions with this ligand involving aryl chlorides and boronic acids proceeded with good yields and high turnover values and demonstrated the utility of such highly crowded catalytic scaffolds. The use of **AD3P** was compared favorably not only against P(*t*Bu)$_3$, PAd$_2$(*n*Bu), or PCy$_3$ but also to state-of-the-art pre-catalysts such as XPhos-Pd G3 or PEPPSIIPr. The structural analysis displays that the planarization of phosphorus does not play a role, but rather that the Taft polarizability parameter (σ_α) of Ad (− 0.95) compared to *t*Bu (− 0.75) determines the decisive electron donation from phosphorus to metal.

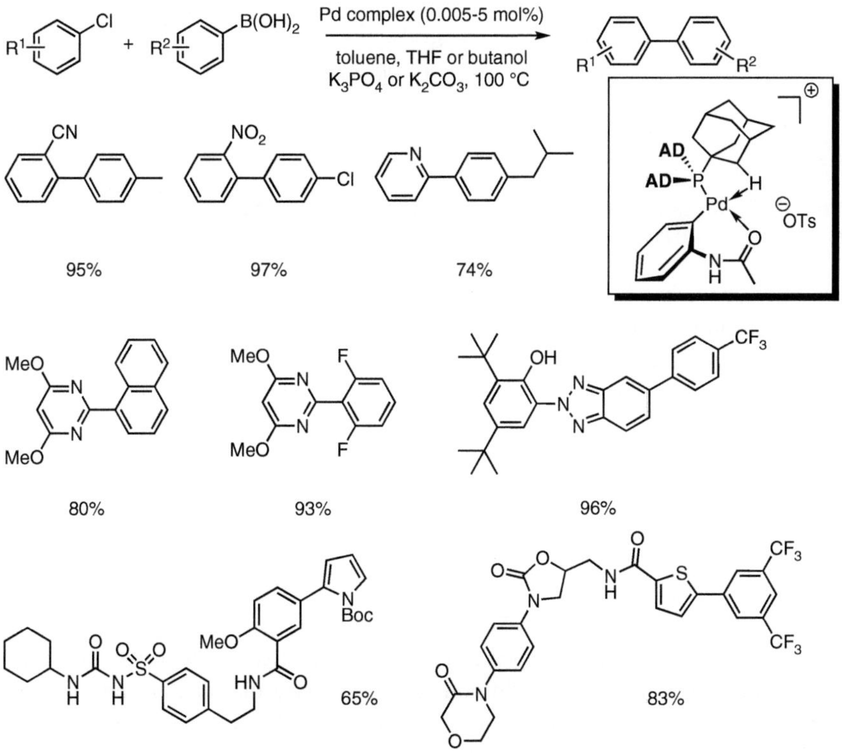

Scheme 8.9 Suzuki–Miyaura coupling of chloroarenes with a palladium triadamantyl phosphine complex. Source: From Ref. [5g].

The role of chiral phosphine ligands in catalysis should also not be neglected, and their role in asymmetric, metal-catalyzed processes is often important. A wise selection of the appropriate ligand determines the isomer ratio of the products and thereby governs the usefulness of a reaction for the preparation of chiral compounds of high optical purity. Diamondoid phosphines found applicability in this niche as well, as exemplified by the depicted asymmetric hydrogenation reaction that readily afforded protected amino acids (Scheme 8.10) [25]. Chelation of the adamantyl version of 1,2-*bis*(alkylmethylphosphino)ethane to a rhodium center produced an asymmetric five-membered ring having C_2-symmetry and consequently enabled asymmetric hydrogenation of α-(acylamino)acrylic derivatives. Note that the directing power of the bulky adamantane cages proved to be substantial and afforded *ee* values of up to 99.9%.

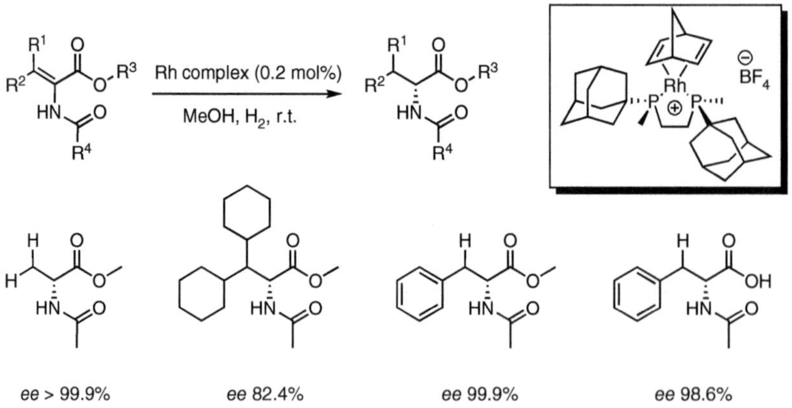

Scheme 8.10 Rh-catalyzed asymmetric hydrogenation for enantioselective preparation of amino acids. Source: From Ref. [25].

Recently, there has been a renewed interest in primary phosphines as ligands for coordination with the catalyst metal centers. Despite primary phosphines being notorious for their poor stability, air sensitivity, and highly flammable (even explosive!) nature, primary phosphines of larger diamondoids proved to be exceptionally manageable and useful for catalytic applications [26]. For example, the introduction of the phosphine group on the diamantane cage renders the resulting ligand surprisingly air-stable compared to typical phosphines (Figure 8.3). It appears that the increase in cage size and molecular weight decreases the pyrophoricity and lowers air sensitivity of such compounds since primary diamantane phosphines demonstrate higher air stability than the analogous **1ADPH2**. It follows that the increase in bulkiness around the phosphine functional group induces overall resistance toward air oxidation, a property that can be exploited in the design of future primary phosphine ligands. In addition, when primary diamondoid phosphines are introduced to the cage scaffold already containing a heteroatom, an increase in stability is also observed.

Since primary diamantane phosphines with electron-withdrawing functional groups at remote apical cage positions proved to be sufficiently stable for catalytic

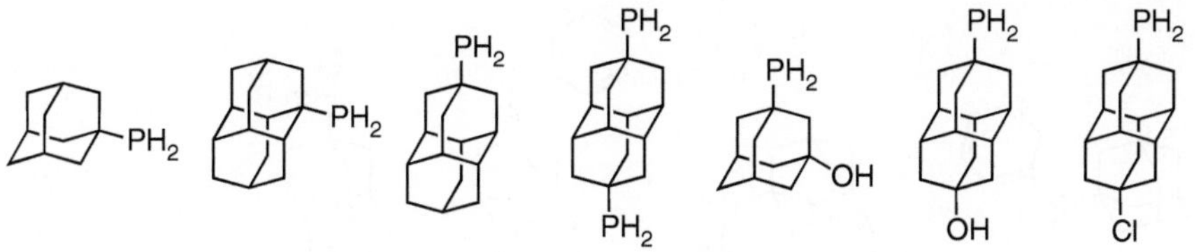

Figure 8.3 Relatively stable primary diamondoid phosphines. Source: From Ref. [26].

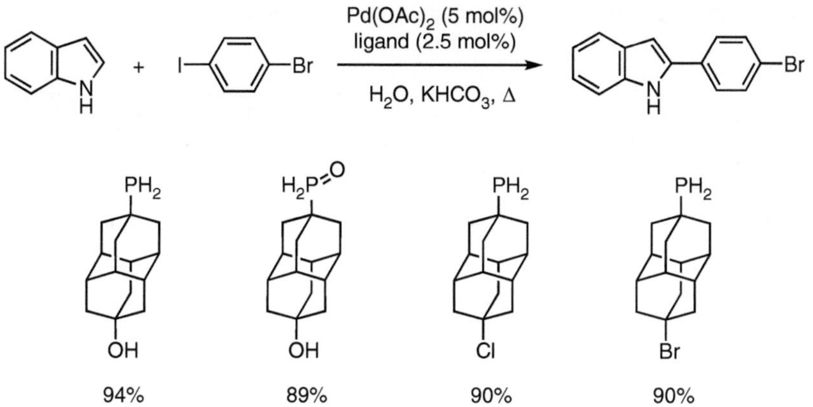

Scheme 8.11 C2-arylation of an unprotected indole using **DIA** phosphines. Source: From Ref. [27].

applications, they were tested in the C2—H arylation of unprotected indoles (Scheme 8.11) [27]. In this aerobic, "on-water", palladium-catalyzed reaction, all studied primary phosphines afforded excellent yields with good selectivity. Note that the diamantane phosphine oxide also proved to be a good ligand, and it is therefore highly likely that this species is actually an intermediate in the primary phosphine catalytic cycle since the reaction of the corresponding primary phosphine performs poorly under strictly anaerobic conditions. Overall, the presence of trivalent phosphorous appears to be essential for an efficient "on water" indole arylation. Differently substituted aryl iodides were also tested as substrates, and it was found that electron-donating and electron-withdrawing groups were tolerated at the *para*- and *ortho*-positions on iodoarenes, making this catalytic system very versatile and promising for further applications.

Diamondoid phosphines incorporating one cage scaffold and one or more aromatic substituents were also recently prepared (Figure 8.4) [26]. While still not tested for their catalytic potential, they are an indication that the development of new synthetic strategies toward affording a rich selection of diamondoid phosphines is on the rise, promising an exciting future for catalytic screening of new bulky derivatives.

The utility of diamondoids in catalysis is not limited to phosphorous compounds. Another important class of useful diamondoid ligands are carbenes, and the corresponding *N*-heterocyclic carbenes (NHC) deserve a special mention since they are highly sterically demanding and electron-rich compounds. The first such isolated

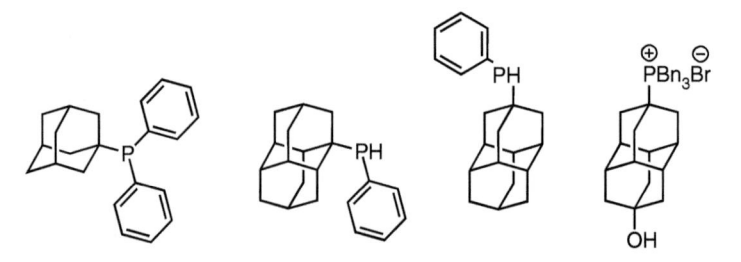

Figure 8.4 Examples of substituted diamondoid phosphine derivatives.

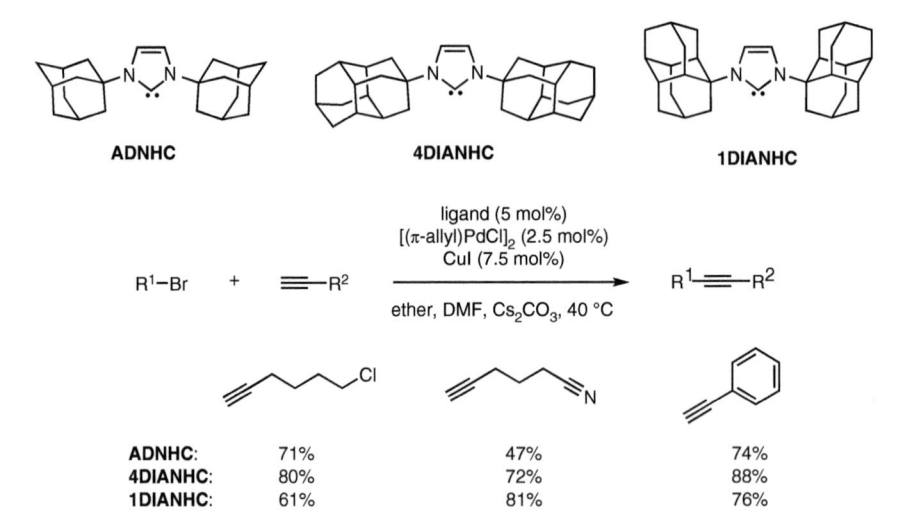

Scheme 8.12 Sonogashira coupling using diamondoid NHC ligands. Source: From Ref. [29].

carbene (the "Arduengo-Carbene", **ADNHC**) was obtained by deprotonation of the 1,3-diadamantyl disubstituted imidazolium salt [28]. Almost two decades later, two diamantane NHC analogues were also successfully prepared (derivatives with apical and medial cage substitutions) [29]. These bench-stable NHC ligands were used for the Pd-catalyzed Sonogashira coupling reaction between unactivated alkyl bromides and alkynes (Scheme 8.12). The screening demonstrated that bulkier diamantane NHCs performed better than the parent adamantane compounds, confirming the proposed importance of steric bulk for such catalytic reactions, even though it was not mentioned that much of the bulk effect is not "steric hindrance" but rather due to LD attractions.

The **ADNHC** ligand also finds use for alkynylation of C(sp^3)–H bonds in aliphatic amides with protected acetylene [30] and for the preparation of benzo-fused cyclobutanes from aldehydes (Scheme 8.13) [31]. In the latter reaction, the **ADNHC** ligand was demonstrated to be critical for achieving high selectivity (99 : 1), i.e., for affording a four-membered ring and not a non-fused product like some other tested NHC ligands. Thus, this procedure is characterized by its broad scope and exceptional site

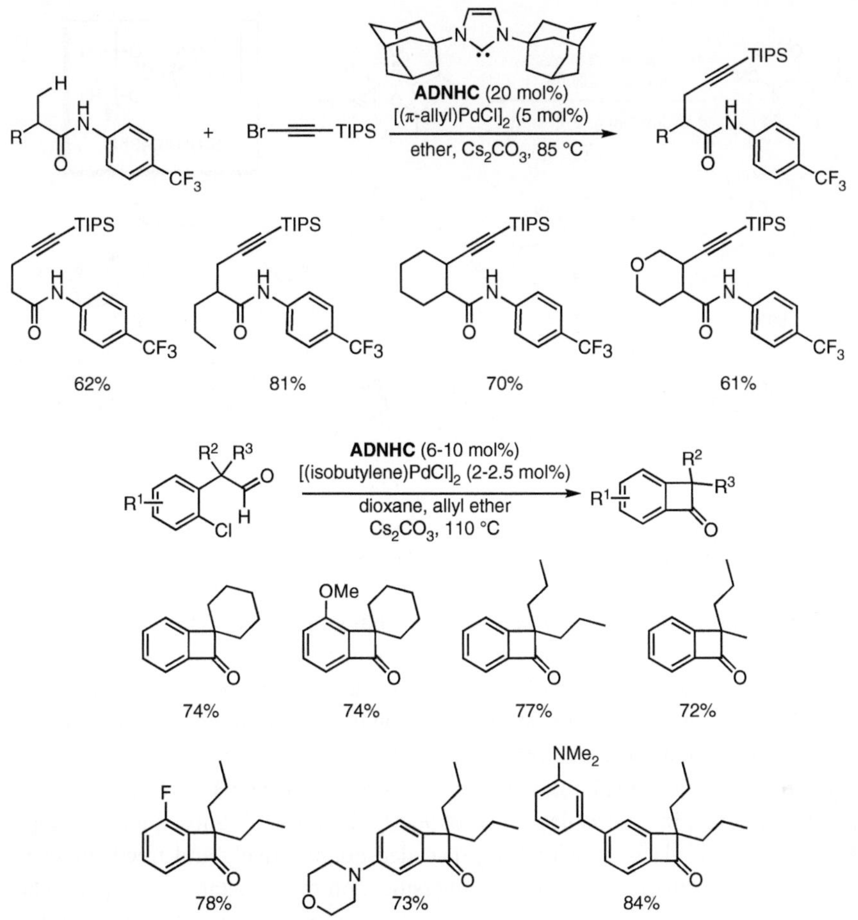

Scheme 8.13 Alkynylation of C(sp³)–H bonds (top) and Pd-catalyzed cyclobutane synthesis (bottom) using the **ADNHC** ligand. Source: From Refs. [30, 31].

selectivity, mainly because the ligand cage backbone dictates the selectivity pattern of the reaction.

Diamondoid carboxylates are another type of compound used for complex preparation with various metals that can in turn be very useful for catalytic applications. For example, the dirhodium tetracarboxylate adamantane complex readily promotes cycopropanation and cyclopropenation reactions (Scheme 8.14) [32]. The authors evaluated the reactivity of a wide range of terminal alkene substrates, and this catalytic reaction proceeded in general with excellent yields and high stereoselectivity. As for the 1,1-disubstituted alkenes, while being compatible with the used reaction conditions, the observed diastereoselectivity was unfortunately somewhat poor.

A recent example of a dirhodium tetracarboxylate adamantane complex with a free amino group available for post-functionalization showcases the targeted design of a site-selective nitrenoid insertion catalyst containing a diamondoid ligand [33]. Namely, the most widely used methods for the preparation of such Rh(II)

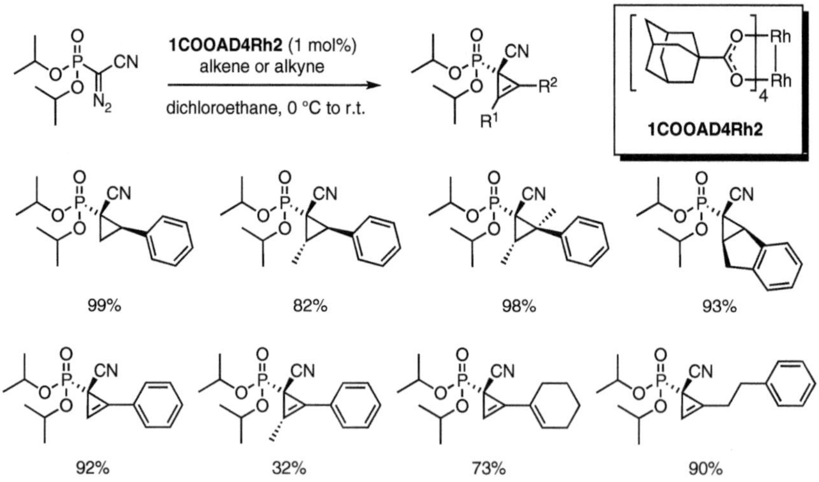

Scheme 8.14 Cycopropanation and cyclopropenation with a dirhodium tetracarboxylate adamantyl complex **1COOAD4Rh2**. Source: From Ref. [32b].

complexes start with the metal acetate by exchanging the acetates with carboxylate ligands. Here, however, the exchanged ligands were not the final structural scaffolds of the catalyst but rather underwent additional post-functionalization, increasing the diversity scope of the functional groups attached to the ligand core. The active complex was prepared through condensation of carboxybenzyl-(Cbz)-protected 1-aminoadamantane 3-carboxylic acid **3COOH1NH2AD** with the parent $Rh_2(OAc)_4$ in an excellent yield of 96% (Scheme 8.15). Subsequent cleavage of the Cbz-group via hydrogenation proceeded quantitatively and freed the cage amino group for further buildup and diversification of the catalytic complex. The presence of the bulky adamantane cage in this complex is crucial for its stability since a sterically crowded ligand shields the rhodium center from internal amine coordination and thereby suppresses rapid complex decomposition.

Control of site-selectivity is a big challenge in Rh(II)-mediated aziridinations, as insertion pathways in principle favor sites that stabilize the positive charge. However, sterically demanding catalysts can alter the reaction selectivity by pushing for insertions at sterically more accessible sites. With this goal in mind, a urea-functionalized adamantyl dirhodium tetracarboxylate **3COOH1CARBCF3ADRh2** catalyst was tested in a remote site-selective aziridination reaction of farnesol carbamate (Scheme 8.16). Note that hydrogen bonding through the 3,5-bis(trifluoromethyl)phenyl moiety [34] present on the other side of the urea linker crafted well-defined spatial relationships and so greatly directed the catalyst's site preference by not depending solely on the ligand bulk as a governing factor. Farnesol carbamate, with its many double bonds and allylic sites available for C—H insertions and nitrenoid additions, thereby acting as a hydrogen-bond acceptor, is a perfect model system for assessing the site-selectivity of the prepared catalyst, and the obtained result did not disappoint. The double bond nitrenoid addition with efficient H-bond donor 2,2,2-trichloroethyl sulfamate proceeded with

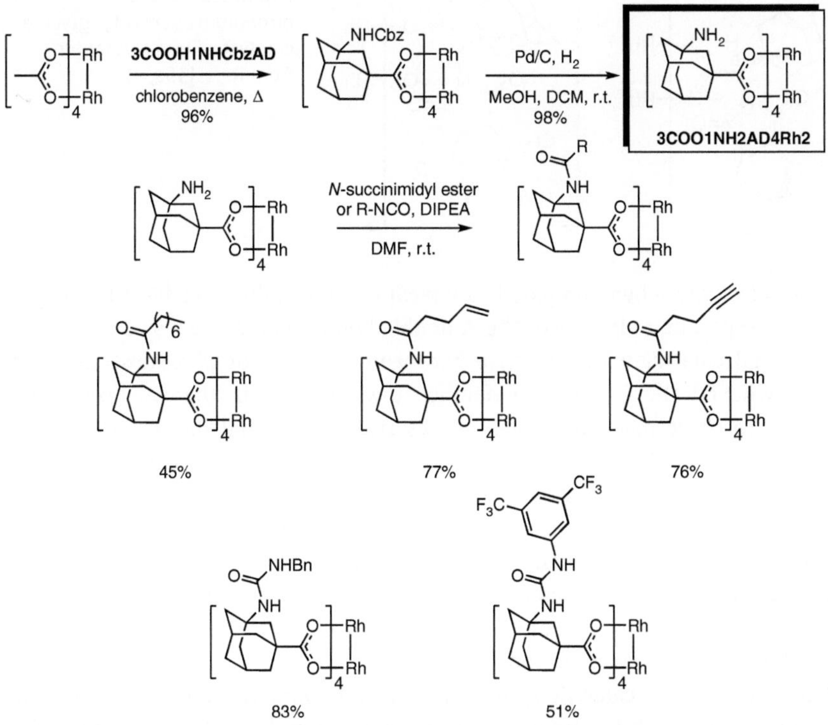

Scheme 8.15 Preparation of aminoadamantane dirhodium tetracarboxylate complex 3COO1NH2AD4Rh2 and its post-functionalization. Source: From Ref. [33].

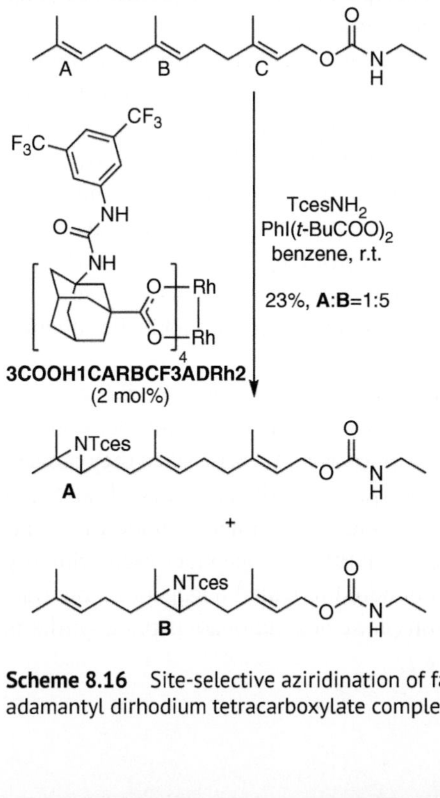

Scheme 8.16 Site-selective aziridination of farnesol carbamate using urea-functionalized adamantyl dirhodium tetracarboxylate complex. Source: From Ref. [33].

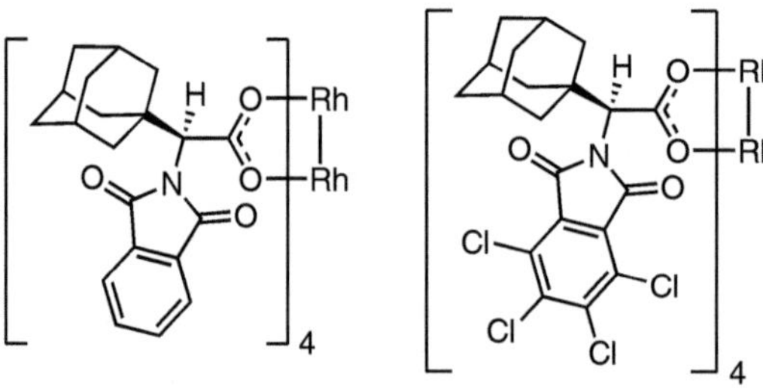

Figure 8.5 Versatile dirhodium adamantyl glycine catalysts. Source: Adapted from Refs. [35a, e].

reverse selectivity when compared to typical commercially available Rh catalysts (e.g., Rh$_2$[esp]$_2$) that often favor the A double bond. That is, the tested adamantyl bifunctional catalyst with an active hydrogen-bonding complex-substrate interaction switched the insertion preference toward bond B, thereby overcoming farnesol's intrinsic reactivity. A telling observation was that the reaction with a similar sterically bulky catalyst that could not engage in remote hydrogen bonding resulted in a product ratio in favor of the intrinsically preferred A bond insertion product. Aziridination of double bond C was not observed for any catalyst tested in this study. This example demonstrates the ability of functionalized diamondoid dirhodium tetracarboxylate complexes to achieve unique selectivities in aziridinations of polyenes by utilizing a combination of steric bulk and noncovalent interactions for site-selectivity control.

Similar dirhodium catalytic systems but with a coordinated adamantyl glycine substructure (Figure 8.5) were effectively used for a number of C—C bond-forming reactions that proceeded through the insertion of the dirhodium carbenoid into various activated intra- and intermolecular C—H bonds [35]. The yields, scope, and asymmetric induction of these reactions were typically excellent, regardless whether the catalyst was the parent or the tetrachloro-derivative.

The expansion of catalytic systems toward larger, robust nanoparticles containing metal cores adorned with diamondoid carboxylates or amines was recently reported [36]. The ruthenium nanoparticles in question were used for selective hydrogenation of alkynes to alkenes on a phenylacetylene substrate (Figure 8.6). The function of cage moieties in such catalytic systems is to enable structural stabilization and size control of discrete ruthenium atom conglomerates. The hydrogenations of phenylacetylene proceeded to afford styrene with ethylbenzene and ethylcyclohexane as additional reduction products when higher conversions were reached. Consequently, the prepared ruthenium nanoparticles can be used for facile and mild hydrogenation of diverse unsaturated substrates.

Up to now, we only considered organometallic catalysts; however, diamondoids found applications in the field of organocatalysis as well. The recent focus on catalytic systems not containing heavy metals comes as no surprise since they provide a clean and environmentally friendly alternative to traditional catalysts [37]. Especially practical examples of organocatalysts containing an adamantane cage scaffold in the backbone of their structure are nonnatural oligopeptides used primarily for acyl transfer reactions [38]. These diamondoid-modified oligopeptide catalysts are readily available through either solution-phase or solid-phase peptide synthesis

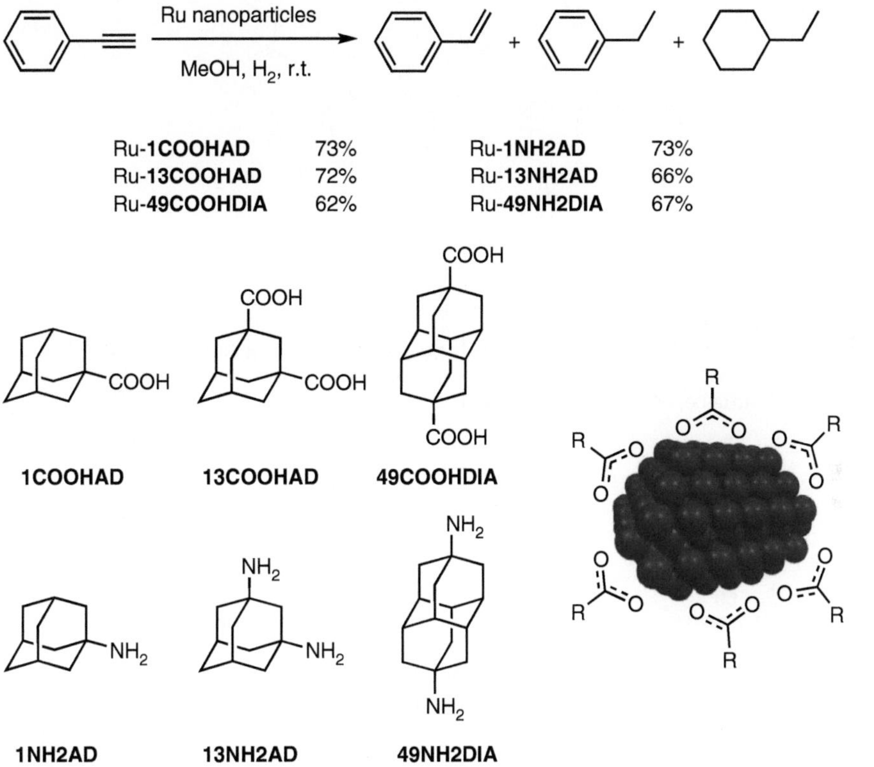

Figure 8.6 Hydrogenation of phenylacetylene catalyzed by diamondoid/Ru nanoparticles. Source: From Ref. [36].

procedures, and their huge advantage is their biodegradability, which makes them environmentally friendly.

Since oligopeptides with nucleophilic catalytic moieties in their structures proved to be efficient for kinetic resolutions of racemic substrates and in desymmetrizations of *meso*-compounds through acyl group transfer, the inclusion of an additional rigid adamantane scaffold into such organocatalysts was a promising venue of investigation since it would provide a far more lipophilic and structurally less flexible peptide structure. Indeed, such a tetrapeptide (Boc-L-Pmh-^AGly-L-Cha-L-Phe-OMe) that contained π-methyl histidine (Pmh) as a catalytic moiety and a **3COOH1NH2AD** subunit built directly into the peptide chain backbone proved to be an exceptionally useful organocatalyst (Figure 8.7) [40].

It should be mentioned that this tetrapeptide does not form a stable secondary structure stabilized by intramolecular hydrogen bonding like many other commonly designed oligopeptide catalysts, which in turn has an effect on the stereochemistry of the conducted reactions [41]. In other words, the stereochemistry of the resulting product is governed by the nature of the intermediate species, a charged acylium ion complex with the peptide catalyst, meaning that the catalyst-substrate interaction directly gives rise to reaction selectivity. The described adamantane

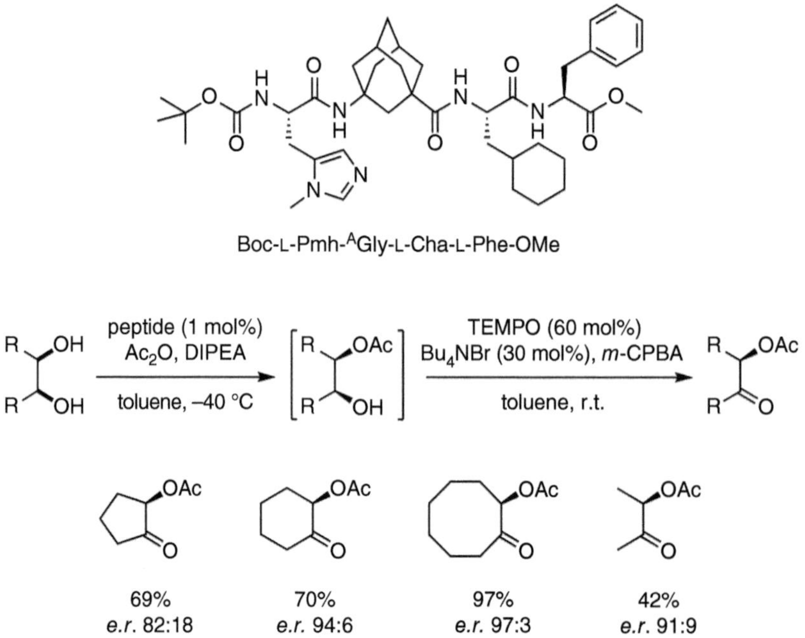

Boc-L-Pmh-^AGly-L-Cha-L-Phe-OMe

Figure 8.7 Adamantane-containing tetrapeptide-catalyzed desymmetrization of *meso*-alkane-1,2-diols and subsequent TEMPO oxidation of the mono-acylated intermediate. Source: From Ref. [39].

tetrapeptide catalyst was first applied for enantioselective kinetic resolution of *trans*-cycloalkane-1,2-diols [40a] and later used in a perfected desymmetrization of *meso*-alkane-1,2-diols through acylation and oxidation via TEMPO that led to useful α-acetoxy ketone building blocks in moderate to excellent yields and high enantiomeric ratios [39].

Inspired by such a resounding success of this one-pot catalytic desymmetrization procedure, the authors went one step further and designed a new generation of an oligopeptide catalyst: one that would incorporate the TEMPO moiety directly in its structure. By preparing this multicatalyst oligopeptide (Boc-L-Pmh-^AGly-L-Cha-L-Phe-NH-TEMPO) that was capable of performing the entire reaction sequence independently, a nontrivial multicatalyst system bearing two orthogonal catalytic moieties was prepared (Figure 8.8) [42]. What is even more remarkable is that this efficient multicatalyst showed increased oxidation power when compared to TEMPO itself. The same group later developed other similar oligopeptides while searching for efficient organocatalysts for epoxidation of alkenes and preparation of enantiomerically enriched *trans*-1,2-alkanediols and monoacetylated diols [43].

Systematic screening of various similar peptides incorporating only slight structural variations revealed other promising candidates belonging to this series of organocatalysts [44] as well as demonstrating their use in other types of catalyzed reactions, e.g., oxidative esterification [45] and enantioselective Dakin–West

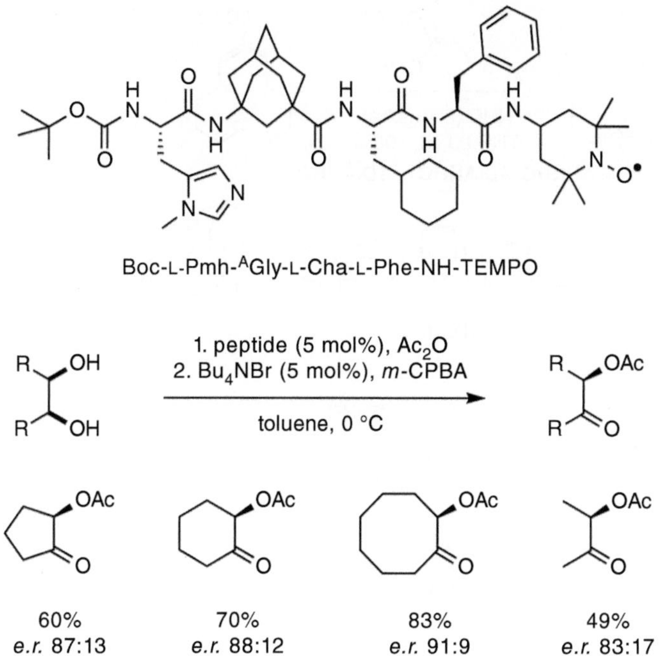

Boc-L-Pmh-^AGly-L-Cha-L-Phe-NH-TEMPO

Figure 8.8 Oligopeptide-based multicatalyst system with TEMPO moiety capable of independent one-pot desymmetrization and oxidation of *meso*-alkane-1,2-diols. Source: From Ref. [42].

reaction [46]. Despite many literature examples, the field of oligopeptide-based multicatalyst systems incorporating a rigid diamondoid backbone still hides many venues waiting to be explored.

Diamondoid NHCs mentioned previously were also applied as organocatalysts in the silylation of cyclododecanone with trimethylsilylketene acetal, affording the corresponding silyl enol ether (Figure 8.9) [29]. All three NHCs showed good reactivity and facilitated full product conversion while needing quite low catalyst loading. Further applicability of the **ADNHC** in organocatalysis was found in transesterification reactions [47], trifluoromethylations of aldehydes and ketones [48], in conversions of ketones into TMS or TBS silyl enol ethers [49], click reactions affording 1,2,3-triazoles [50], and many more.

While not exhaustive, this chapter presents the current trends of using diamondoid derivatives in catalysis. The key features crucial for their application as ligands and catalysts are their bulkiness and polarizability, and the presented examples demonstrate how steric factors (repulsive *and* attractive!) often contribute to the stabilization of ligands, e.g., diamondoid phosphines, and can enable high enantioselectivity of the catalyzed reaction, e.g., hydrogenations. However, a word of caution is needed: in some reactions, the mentioned beneficial effect of bulkiness can actually transform into steric crowding. If the space around the coordination center becomes too tight due to the branching or size of the ligands, the substrate approach to the

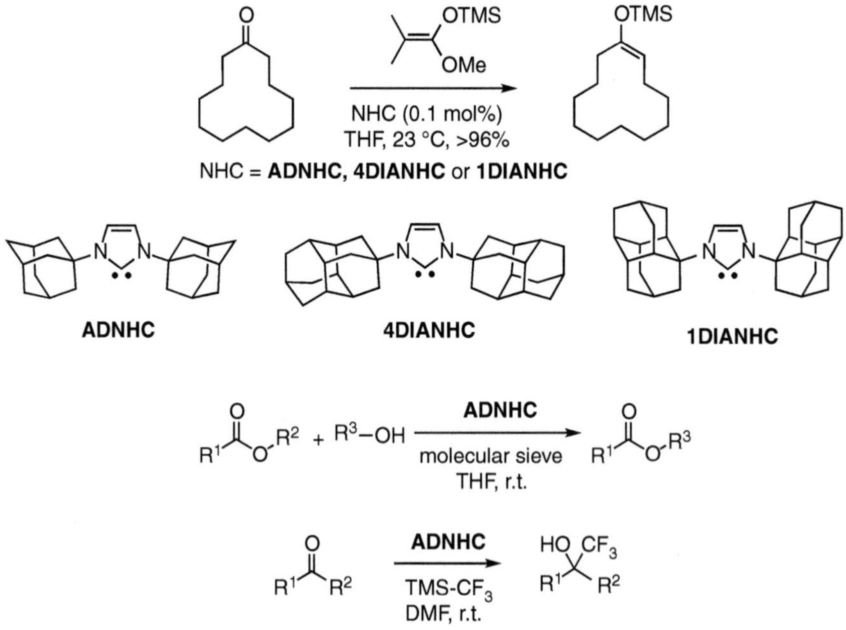

Figure 8.9 Silyl enol ether formation catalyzed by *N*-heterocyclic diamondoid carbenes (top) and examples of other organocatalytic reactions. Source: From Refs. [29, 47c, 48].

catalyst becomes impeded, and the reaction may become sluggish or even stop completely. Some benefits arise from the introduction of ligands with partial flexibility ("flexible steric bulk" concept) [51]. Even though diamondoids represent rigid structures, the combination of differently substituted cages or cages of various topologies in one ligand may lead to better fitting of the constituents. Such a concept of "controlled steric bulk" in combination with the high polarizability of the cages should be kept in mind in diamondoid catalyst design, as it may improve the reactivity and selectivity of many catalytic processes. It is our opinion that diamondoid ligands are still insufficiently explored, especially the higher diamondoid variants of the ligands already proven to be successful, and we are convinced that many new exciting catalysts and their applications will be discovered in the future.

References

1 Agnew-Francis, K.A. and Williams, C.M. (2016). Catalysts containing the adamantane scaffold. *Adv. Synth. Catal.* 358: 675–700.
2 Wagner, J.P. and Schreiner, P.R. (2015). London dispersion in molecular chemistry-reconsidering steric effects. *Angew. Chem. Int. Ed.* 54: 12274–12296.
3 (a) Echeverria, J., Aullon, G., Danovich, D. et al. (2011). Dihydrogen contacts in alkanes are subtle but not faint. *Nat. Chem.* 3: 323–330. (b) Fokin, A.A., Gerbig, D., and Schreiner, P.R. (2011). σ/σ- and π/π-interactions are equally important: multilayered graphanes. *J. Am. Chem. Soc.* 133: 20036–20039. (c) Feng, A.N.,

Zhou, Y., Al-Shebami, M.A.Y. et al. (2022). σ-σ Stacked supramolecular junctions. *Nat. Chem.* 14: 1158–1164.

4 Grimme, S., Huenerbein, R., and Ehrlich, S. (2011). On the importance of the dispersion energy for the thermodynamic stability of molecules. *ChemPhysChem* 12: 1258–1261.

5 (a) Zapf, A., Ehrentraut, A., and Beller, M. (2000). A new highly efficient catalyst system for the coupling of nonactivated and deactivated aryl chlorides with arylboronic acids. *Angew. Chem. Int. Ed.* 39: 4153–4155. (b) Stambuli, J.P., Stauffer, S.R., Shaughnessy, K.H., and Hartwig, J.F. (2001). Screening of homogeneous catalysts by fluorescence resonance energy transfer. Identification of catalysts for room-temperature Heck reactions. *J. Am. Chem. Soc.* 123: 2677–2678. (c) Sather, A.C., Lee, H.G., De la Rosa, V.Y. et al. (2015). A fluorinated ligand enables room-temperature and regioselective Pd-catalyzed fluorination of aryl triflates and bromides. *J. Am. Chem. Soc.* 137: 13433–13438. (d) Guerriero, A. and Gonsalvi, L. (2021). From traditional PTA to novel CAP: a comparison between two adamantane cage-type aminophosphines. *Inorg. Chim. Acta* 518: 120251. (e) Yang, Y.C., Lin, Y.C., and Wu, Y.K. (2019). Palladium-catalyzed cascade arylation of vinylogous esters enabled by tris(1-adamantyl)phosphine. *Org. Lett.* 21: 9286–9290. (f) Carrow, B.P. and Chen, L. (2017). Tri(1-adamantyl) phosphine: exceptional catalytic effects enabled by the synergy of chemical stability, donicity, and polarizability. *Synlett* 28: 280–288. (g) Chen, L.Y., Ren, P., and Carrow, B.P. (2016). Tri(1-adamantyl)phosphine: expanding the boundary of electron-releasing character available to organophosphorus compounds. *J. Am. Chem. Soc.* 138: 6392–6395.

6 Tolman, C.A. (1977). Steric effects of phosphorus ligands in organometallic chemistry and homogeneous catalysis. *Chem. Rev.* 77: 313–348.

7 (a) Bunten, K.A., Chen, L.Z., Fernandez, A.L., and Poe, A.J. (2002). Cone angles: Tolman's and Plato's. *Coord. Chem. Rev.* 233: 41–51. (b) Brown, T.L. and Lee, K.J. (1993). Ligand steric properties. *Coord. Chem. Rev.* 128: 89–116. (c) Brown, T.L. (1992). A molecular mechanics model of ligand effects. 3. A new measure of ligand steric effects. *Inorg. Chem.* 31: 1286–1294.

8 Müller, T.E. and Mingos, D.M.P. (1995). Determination of the Tolman cone angle from crystallographic parameters and a statistical analysis using the crystallographic Data Base. *Transit. Met. Chem.* 20: 533–539.

9 Bilbrey, J.A., Kazez, A.H., Locklin, J., and Allen, W.D. (2013). Exact ligand cone angles. *J. Comput. Chem.* 34: 1189–1197.

10 (a) Goerlich, J.R. and Schmutzler, R. (1993). Di-1-adamantylphosphine: a sterically highly hindered secondary phosphine – synthesis and reactions. *Phosphorus Sulfur Silicon Relat. Elem.* 81: 141–148. (b) Goerlich, J. R.; Schmutzler, R. Organophosphorus compounds with tertiary alkyl substituents. A convenient method for the preparation of di-1-adamantylphosphine and di-1-adamantylchlorophosphine. *Phosphorus Sulfur Silicon Relat. Elem.* 1995, 102, 211–215.

11 Schwertfeger, H., Machuy, M.M., Würtele, C. et al. (2010). Diamondoid phosphines – selective phosphorylation of nanodiamonds. *Adv. Synth. Catal.* 352: 609–615.

12 Tewari, A., Hein, M., Zapf, A., and Beller, M. (2004). General synthesis and catalytic applications of di(1-adamantyl)alkylphosphines and their phosphonium salts. *Synthesis* 935–941.

13 Ehrentraut, A., Zapf, A., and Beller, M. (2000). A new efficient palladium catalyst for Heck reactions of deactivated aryl chlorides. *Synlett* 1589–1592.

14 (a) Tsukano, C., Okuno, M., and Takemoto, Y. (2013). Synthesis of spirooxindoles from carbamoyl chlorides via cyclopropyl methine C(sp^3)-H activation using palladium catalyst. *Chem. Lett.* 42: 753–755. (b) Tsukano, C., Okuno, M., and Takemoto, Y. (2012). Palladium-catalyzed amidation by chemoselective C(sp^3)-H activation: concise route to oxindoles using a carbamoyl chloride precursor. *Angew. Chem. Int. Ed.* 51: 2763–2766.

15 (a) Ehrentraut, A., Zapf, A., and Beller, M. (2002). A new improved catalyst for the palladium-catalyzed amination of aryl chlorides. *J. Mol. Cat. A* 182: 515–523. (b) Tewari, A., Hein, M., Zapf, A., and Beller, M. (2005). Efficient palladium catalysts for the amination of aryl chlorides: a comparative study on the use of phosphium salts as precursors to bulky, electron-rich phosphines. *Tetrahedron* 61: 9705–9709.

16 Schnyder, A., Aemmer, T., Indolese, A.E. et al. (2002). First application of secondary phosphines, as supporting ligands for the palladium-catalyzed Heck reaction: efficient activation of aryl chlorides. *Adv. Synth. Catal.* 344: 495–498.

17 Alsabeh, P.G., Stradiotto, M., Neumann, H., and Beller, M. (2012). Aminocarbonylation of (hetero)aryl bromides with ammonia and amines using a palladium/DalPhos catalyst system. *Adv. Synth. Catal.* 354: 3065–3070.

18 (a) Lundgren, R.J., Sappong-Kumankumah, A., and Stradiotto, M. (2010). A highly versatile catalyst system for the cross-coupling of aryl chlorides and amines. *Chem. Eur. J.* 16: 1983–1991. (b) Lundgren, R.J. and Stradiotto, M. (2010). Palladium-catalyzed cross-coupling of aryl chlorides and tosylates with hydrazine. *Angew. Chem. Int. Ed.* 49: 8686–8690. (c) Lundgren, R. J.; Peters, B. D.; Alsabeh, P. G.; Stradiotto, M. A P,N-ligand for palladium-catalyzed ammonia arylation: Coupling of deactivated aryl chlorides, chemoselective arylations, and room temperature reactions. *Angew. Chem. Int. Ed.* 2010, 49, 4071–4074; (d) Alsabeh, P.G., Lundgren, R.J., McDonald, R. et al. (2013). An examination of the palladium/Mor-DalPhos catalyst system in the context of selective ammonia monoarylation at room temperature. *Chem. Eur. J.* 19: 2131–2141.

19 (a) Aranyos, A., Old, D.W., Kiyomori, A. et al. (1999). Novel electron-rich bulky phosphine ligands facilitate the palladium-catalyzed preparation of diaryl ethers. *J. Am. Chem. Soc.* 121: 4369–4378. (b) Gowrisankar, S., Sergeev, A.G., Anbarasan, P. et al. (2010). A general and efficient catalyst for palladium-catalyzed C-O coupling reactions of aryl halides with primary alcohols. *J. Am. Chem. Soc.* 132: 11592–11598.

20 Luo, Y.D., Ji, K.G., Li, Y.X., and Zhang, L.M. (2012). Tempering the reactivities of postulated alpha-oxo gold carbenes using bidentate ligands: implication of tricoordinated gold intermediates and the development of an expedient bimolecular assembly of 2,4-disubstituted oxazoles. *J. Am. Chem. Soc.* 134: 17412–17415.

21 Wang, Y.Z., Wang, Z.X., Li, Y.X. et al. (2014). A general ligand design for gold catalysis allowing ligand-directed anti-nucleophilic attack of alkynes. *Nat. Commun.* 5: 3470.

22 Ackermann, L. (2006). Air- and moisture-stable secondary phosphine oxides as preligands in catalysis. *Synthesis* 1557–1571.

23 Ackermann, L., Barfusser, S., Kornhaass, C., and Kapdi, A.R. (2011). C-H bond arylations and benzylations on oxazol(in)es with a palladium catalyst of a secondary phosphine oxide. *Org. Lett.* 13: 3082–3085.

24 Ghorai, D., Muller, V., Keil, H. et al. (2017). Secondary phosphine oxide preligands for palladium-catalyzed C-H (hetero)arylations: efficient access to pybox ligands. *Adv. Synth. Catal.* 359: 3137–3141.

25 Imamoto, T., Watanabe, J., Wada, Y. et al. (1998). P-chiral bis(trialkylphosphine) ligands and their use in highly enantioselective hydrogenation reactions. *J. Am. Chem. Soc.* 120: 1635–1636.

26 Moncea, O., Gunawan, M.A., Poinsot, D. et al. (2016). Defying stereotypes with nanodiamonds: stable primary diamondoid phosphines. *J. Org. Chem.* 81: 8759–8769.

27 Moncea, O., Poinsot, D., Fokin, A.A. et al. (2018). Palladium-catalyzed C2-H arylation of unprotected (N-H)-indoles "on water" using primary diamantyl phosphine oxides as a class of primary phosphine oxide ligands. *ChemCatChem* 10: 2915–2922.

28 Arduengo, A.J., Harlow, R.L., and Kline, M. (1991). A stable crystalline carbene. *J. Am. Chem. Soc.* 113: 361–363.

29 Richter, H., Schwertfeger, H., Schreiner, P.R. et al. (2009). Thieme chemistry journal awardees – where are they now? Synthesis of diamantane-derived N-heterocyclic carbenes and applications in catalysis. *Synlett* 193–197.

30 He, J., Wasa, M., Chan, K.S.L., and Yu, J.Q. (2013). Palladium(0)-catalyzed alkynylation of C(sp^3)-H bonds. *J. Am. Chem. Soc.* 135: 3387–3390.

31 Flores-Gaspar, A., Gutierrez-Bonet, A., and Martin, R. (2012). N-heterocyclic carbene dichotomy in Pd-catalyzed acylation of aryl chlorides via C−H bond functionalization. *Org. Lett.* 14: 5234–5237.

32 (a) Cotton, F.A. and Thompson, J.L. (1984). Preparation and structural characterization of three tetracarboxylato dirhodium (Rh−Rh) compounds with bulky ligands. *Inorg. Chim. Acta* 81: 193–203. (b) Lindsay, V.N.G., Fiset, D., Gritsch, P.J. et al. (2013). Stereoselective Rh$_2$(S-IBAZ)$_4$-catalyzed cyclopropanation of alkenes, alkynes, and allenes: asymmetric synthesis of diacceptor cyclopropylphosphonates and alkylidenecyclopropanes. *J. Am. Chem. Soc.* 135: 1463–1470.

33 Berndt, J.P., Radchenko, Y., Becker, J. et al. (2019). Site-selective nitrenoid insertions utilizing postfunctionalized bifunctional rhodium(II) catalysts. *Chem. Sci.* 10: 3324–3329.

34 Lippert, K.M., Hof, K., Gerbig, D. et al. (2012). Hydrogen-bonding thiourea organocatalysts: the privileged 3,5-bis(trifluoromethyl)phenyl group. *Eur. J. Org. Chem.* 2012: 5919–5927.

35 (a) Reddy, R.P., Lee, G.H., and Davies, H.M.L. (2006). Dirhodium tetracarboxylate derived from adamantylglycine as a chiral catalyst for carbenoid reactions. *Org. Lett.* 8: 3437–3440. (b) Davies, H.M.L. and Beckwith, R.E.J. (2003). Catalytic enantioselective C-H activation by means of metal-carbenoid-induced C-H insertion. *Chem. Rev.* 103: 2861–2903. (c) Davies, H.M.L. and Denton, J.R. (2009). Application of donor/acceptor-carbenoids to the synthesis of natural products. *Chem. Soc. Rev.* 38: 3061–3071. (d) Reddy, R.P. and Davies, H.M.L. (2006). Dirhodium tetracarboxylates derived from adamantylglycine as chiral catalysts for enantioselective C-H aminations. *Org. Lett.* 8: 5013–5016. (e) Davies, H.M.L. and Manning, J.R. (2008). Catalytic C-H functionalization by metal carbenoid and nitrenoid insertion. *Nature* 451: 417–424.

36 Min, Y.Y., Nasrallah, H., Poinsot, D. et al. (2020). 3D ruthenium nanoparticle covalent assemblies from polymantane ligands for confined catalysis. *Chem. Mater.* 32: 2365–2378.

37 Wende, R.C. and Schreiner, P.R. (2012). Evolution of asymmetric organocatalysis: multi- and retrocatalysis. *Green Chem.* 14: 1821–1849.

38 Müller, C.E. and Schreiner, P.R. (2011). Organocatalytic enantioselective acyl transfer onto racemic as well as *meso* alcohols, amines, and thiols. *Angew. Chem. Int. Ed.* 50: 6012–6042.

39 Müller, C.E., Zell, D., and Schreiner, P.R. (2009). One-pot desymmetrization of *meso*-1,2-hydrocarbon diols through acylation and oxidation. *Chem. Eur. J.* 15: 9647–9650.

40 (a) Müller, C.E., Wanka, L., Jewell, K., and Schreiner, P.R. (2008). Enantioselective kinetic resolution of trans-cycloalkane-1,2-diols. *Angew. Chem. Int. Ed.* 47: 6180–6183. (b) Hrdina, R., Müller, C.E., and Schreiner, P.R. (2010). Kinetic resolution of *trans*-cycloalkane-1,2-diols via Steglich esterification. *Chem. Commun.* 46: 2689–2690.

41 (a) Prochazkova, E., Kolmer, A., Ilgen, J. et al. (2016). Uncovering key structural features of an enantioselective peptide-catalyzed acylation utilizing advanced NMR techniques. *Angew. Chem. Int. Ed.* 55: 15754–15759. (b) Nowag, J., Brauser, M., Steuernagel, L. et al. (2024). Quantifying intermolecular interactions in asymmetric peptide organocatalysis as a key toward understanding selectivity. *J. Am. Chem. Soc.* 146: 170–180.

42 Müller, C.E., Hrdina, R., Wende, R.C., and Schreiner, P.R. (2011). A multicatalyst system for the one-pot desymmetrization/oxidation of *meso*-1,2-alkane diols. *Chem. Eur. J.* 17: 6309–6314.

43 (a) Hrdina, R., Müller, C.E., Wende, R.C. et al. (2012). Enantiomerically enriched *trans*-diols from alkenes in one pot: a multicatalyst approach. *Chem. Commun.* 48: 2498–2500. (b) Alachraf, M.W., Wende, R.C., Schuler, S.M.M. et al. (2015). Functionality, effectiveness, and mechanistic evaluation of a multicatalyst-promoted reaction sequence by electrospray ionization mass spectrometry. *Chem. Eur. J.* 21: 16203–16208.

44 Müller, C.E., Zell, D., Hrdina, R. et al. (2013). Lipophilic oligopeptides for chemo- and enantioselective acyl transfer reactions onto alcohols. *J. Org. Chem.* 78: 8465–8484.
45 (a) Hofmann, C., Schuler, S.M.M., Wende, R.C., and Schreiner, P.R. (2014). En route to multicatalysis: kinetic resolution of *trans*-cycloalkane-1,2-diols via oxidative esterification. *Chem. Commun.* 50: 1221–1223. (b) Hofmann, C., Schümann, J.M., and Schreiner, P.R. (2015). Alcohol cross-coupling for the kinetic resolution of diols via oxidative esterification. *J. Org. Chem.* 80: 1972–1978.
46 Wende, R.C., Seitz, A., Niedek, D. et al. (2016). The enantioselective Dakin-West reaction. *Angew. Chem. Int. Ed.* 55: 2719–2723.
47 (a) Grasa, G.A., Singh, R., Scott, N.M. et al. (2004). Reactivity of a *N*-heterocyclic carbene, 1,3-di-(1-adamantyl) imidazol 2-ylidene, with a pseudo-acid: structural characterization of Claisen condensation adduct. *Chem. Commun.* 2890-2891. (b) Grasa, G.A., Guveli, T., Singh, R., and Nolan, S.P. (2003). Efficient transesterification/acylation reactions mediated by *N*-heterocyclic carbene catalysts. *J. Org. Chem.* 68: 2812–2819. (c) Grasa, G.A., Kissling, R.M., and Nolan, S.P. (2002). *N*-heterocyclic carbenes as versatile nucleophilic catalysts for transesterification/acylation reactions. *Org. Lett.* 4: 3583–3586.
48 Song, J.J., Tan, Z.L., Reeves, J.T. et al. (2005). *N*-heterocyclic carbene catalyzed trifluoromethylation of carbonyl compounds. *Org. Lett.* 7: 2193–2196.
49 Song, J.H., Tan, Z.L., Reeves, J.T. et al. (2008). N-heterocyclic carbene-catalyzed silyl end ether formation. *Org. Lett.* 10: 877–880.
50 Collinson, J.M., Wilton-Ely, J., and Diez-Gonzalez, S. (2013). Reusable and highly active supported copper(I)-NHC catalysts for click chemistry. *Chem. Commun.* 49: 11358–11360.
51 Altenhoff, G., Goddard, R., Lehmann, C.W., and Glorius, F. (2003). An *N*-heterocyclic carbene ligand with flexible steric bulk allows Suzuki cross-coupling of sterically hindered aryl chlorides at room temperature. *Angew. Chem. Int. Ed.* 42: 3690–3693.

9

Medicinal Compounds

When considering bioactive compounds in the context of diamondoid chemistry, the first fact that comes to mind is the ubiquity of the adamantane structure in various pharmacophore candidates and drugs [1] (Figure 9.1). The higher diamondoids are by far not as prevalent. The attractiveness of the adamantane cage for medicinal chemists has been established over the years owing to the straightforward accessibility of adamantane derivatives and their moderately low price, facile introduction into structures of vastly different parent lead compounds, attractive chemical properties, lipophilicity of the cage subunit that enables easier membrane crossing, known metabolic profiles, etc. [1b].

When a lipophilic boost to improve the action of the active substance is needed, the obvious choice is to reach for the adamantane scaffold and incorporate it into the next generation of derivatives. However, the greatest strength when using the adamantane cage in such a way can also be a source of its weakness; the adamantane subunit simply becomes an add-on group that is typically attached to the terminal position of the drug candidate. Although at times very efficient, such black box treatment of the adamantane substituent can often lead to a narrow scan of the structural landscape and may result in a failure to consider alternative options in the design of new drug candidates. Nevertheless, the body of literature describing different bioapplications of adamantane-containing derivatives is truly impressive but out of the scope of this work. Interested readers are pointed toward an exhaustive review concerning this topic [1b]. In continuation of this review, only selected examples of bioactive diamondoid compounds bearing an adamantane cage will be addressed in this Chapter.

Unnatural amino acids are useful building blocks in the design of peptide-based bioactive compounds. The introduction of such amino acids can cause changes in the peptide's secondary structure and influence the mode of action of a potential drug. Diamondoid amino acids were therefore prepared with the goal to improve the pharmacokinetic profile of the resulting pharmacophores [3]. Various adamantane-containing amino acids were prepared, including derivatives substituted at the admantyl bridge and bridgehead position as well as amino acids with alkyl spacers (Figure 9.2) [4]. The bis-apical diamantane amino acid was also described [5].

The Chemistry of Diamondoids: Building Blocks for Ligands, Catalysts, Materials, and Pharmaceuticals,
First Edition. Andrey A. Fokin, Marina Šekutor, and Peter R. Schreiner.
© 2024 WILEY-VCH GmbH. Published 2024 by WILEY-VCH GmbH.

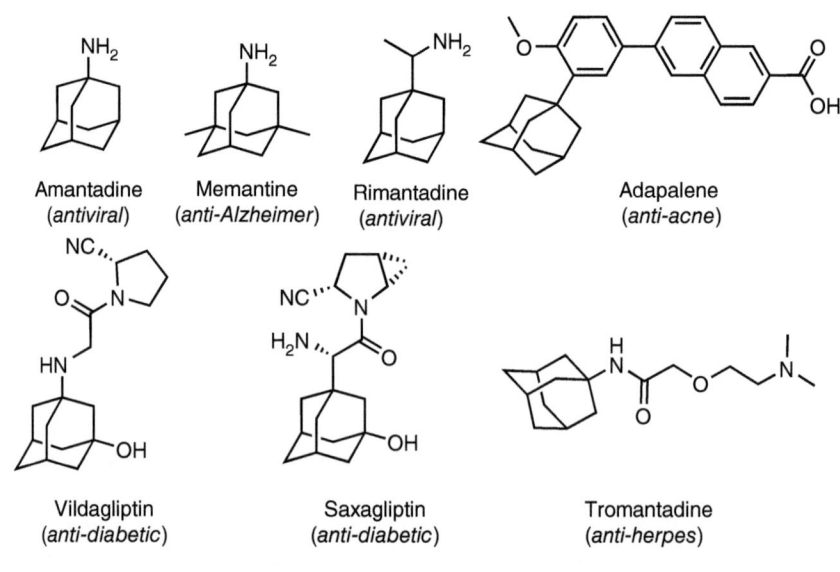

Figure 9.1 Adamantane derivatives with clinical use. Source: From Ref. [2].

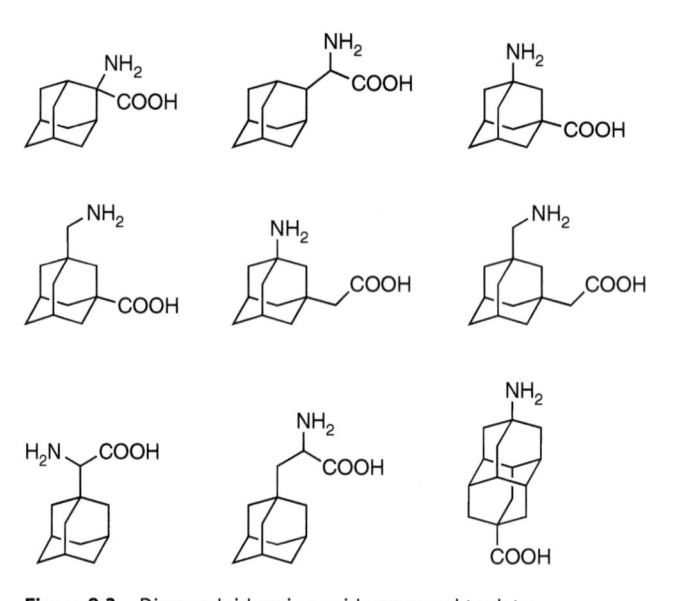

Figure 9.2 Diamondoid amino acids prepared to date.

Incorporation of non-proteinogenic adamantane amino acids into di- to hexapeptides and into pentapeptide analogues of enkephalin (Tyr-Gly-Gly-Phe-Met, or Leu instead of Met) afforded derivatives that were active against several malignant tumor cell lines [4a]. Additionally, the metabolic stability of these adamantane-containing peptides increased when compared to the parent compounds [4b]. Gramicidin S (cyclo-[D-Phe-Pro-Val-Orn-Leu]$_2$) analogues incorporating adamantane amino acids demonstrated performance that was also superior to the parent molecule

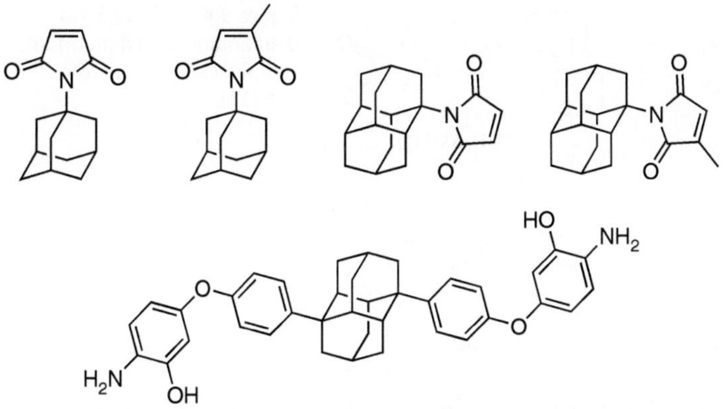

Figure 9.3 Diamondoid maleimide and citraconimide derivatives (top) and highly active DPD (bottom) that possess antiproliferative activity. Source: From Refs. [9, 10a].

when screened for antibacterial activity [6]. Similarly, the design of a neurogenic peptide incorporating an adamantane amino acid moiety afforded an active compound capable of enhancing cognition and improving learning capabilities and memory in mice [7]. Furthermore, when introducing adamantane amino acids into native peptide kinase inhibitors that target the enzyme protein kinase A, some of the obtained polypeptides showed better binding affinity toward the enzyme than the native substrate [8]. This study also demonstrated that computationally aided design facilitates easier selection of a suitable position for the unnatural amino acid in the peptide sequence and therefore helps with the screening for promising inhibition candidates.

The use of diamondoid derivatives for antiproliferative purposes is common, and examples of bioactive molecules with the diamantane moiety that act against cancer cells include maleimide and citraconimide derivatives [9] (Figure 9.3) and 1,6-bis[4-(4-amino-3-hydroxyphenoxy)phenyl] diamantane (DPD) and related derivatives [10].

Recently [11], chiral ligand 1,2-diaminodiamantane in both of its enantiomeric forms was used for the preparation of new platinum (II) complexes with chloride (**DIAPTCl2**) and oxalate (**DIAPTOX**) ligands (Figure 9.4), and the efficacy was compared to cisplatin Pt(NH$_2$)$_2$Cl$_2$.

The binding of diamondoid-based platinum complexes to nucleotides was tested for both enantiomers with guanosine monophosphate (GMP) and deoxyguanosine monophosphate. The interaction occurs at a similar or faster rate for both isomers compared to cisplatin, despite greatly increased steric demand. The platinum complexes were tested against the human ovarian cancer cell line A2780 and its cisplatin-resistant variant A2780cis, where (*R*,*R*)-**DIAPTCL2** shows superior cytotoxicity compared to cisplatin for both cancer cell lines [11]. These findings highlight the potential of 1,2-diaminodiamantane as a viable pharmacophore.

Diamondoids were also applied as probes for cytochrome P450 enzymes that are involved in the metabolism of xenobiotics such as drugs, pollutants, carcinogens, and superfluous endogenous compounds. It is known that diamantane binds

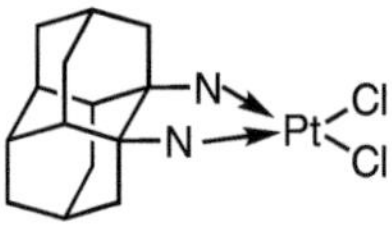

Figure 9.4 1,2-Diamino diamantane-based platinum complexes. Source: From Ref. [11].

(S,S)-**DIAPTCL2**

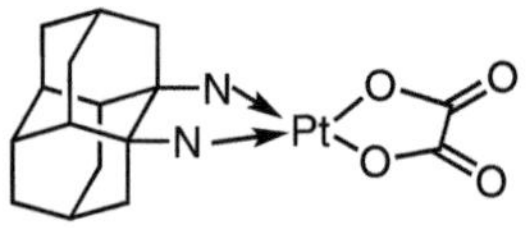

(S,S)-**DIAPTOX**

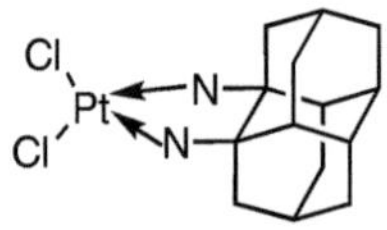

(R,R)-**DIAPTCL2**

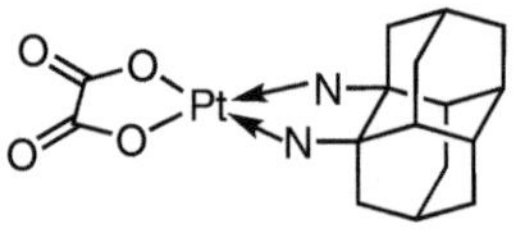

(R,R)-**DIAPTOX**

relatively strongly to cytochrome P450, where it is metabolized to mono and dialcohols as well as to ketones, with the major metabolite being 4-hydroxydiamantane (**4OHDIA**) [12]. What is more, the binding of the hydrocarbon to this enzyme is so efficient [13] that it is used as an enzyme inhibitor in order to quench the metabolism of other substrates in the cytochrome P450 active site [14]. Due to such strong and selective interaction of the diamantane cage with the enzyme active site [15], 3-azidodiamantane was developed as a photolabile, carbene-generating probe for cytochrome P450 [16]. Upon photolysis of the tritium-labeled probe (T on the apical C-9 position of the diamantane cage), the generated carbene immediately binds to the neighboring amino acid residues and makes a radiolabeled photoaffinity probe suitable for studying the cytochrome P450 binding site region [17]. Some other diamantane derivatives also show good binding capabilities [18] and therefore demonstrate the broad applicability of this cage moiety in this system [19]. As a side note related to the topic of metabolic regulations, the explored adamantane and diamantane aniline derivatives were reported to possess hypobetalipoproteinemic activity, meaning they can reduce lipoproteins and affect the lipoprotein-cholesterol ratio in rats [20].

The diamantane scaffold is also present in the structure of pharmacophore TD550, a retinoid agonist and antagonist (Figure 9.5) [21]. Retinoids are important for cell differentiation and proliferation but also modulate inflammatory and immune system cells. Thus, therapeutic agents that regulate retinoids are used for the treatment of inflammatory diseases like psoriasis and rheumatoid arthritis. TD550 was also explored for its antiproliferative potential but afforded only moderate efficacy [23]. Regulation of estrogen receptor binding maintains a delicate balance of the endocrine system, and one class of molecules that can bind

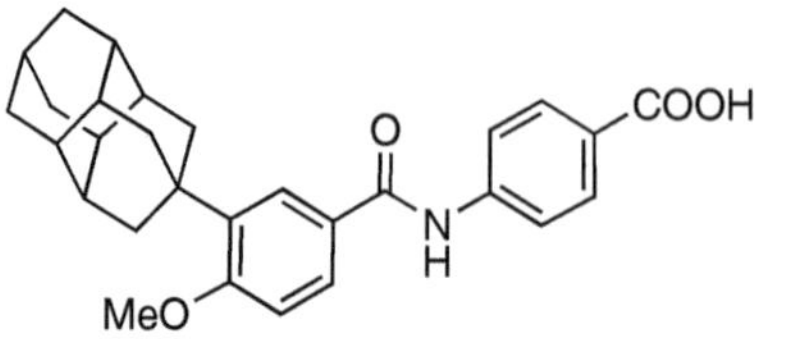

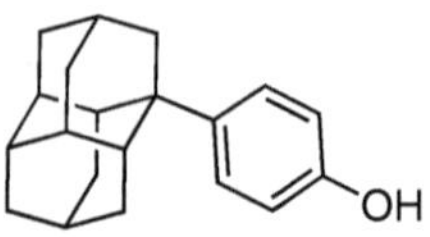

Figure 9.5 Structure of TD550, a retinoid agonist and antagonist (left) and p-(1-diamantyl) phenol capable of estrogenic-type activity (right). Source: From Refs. [21, 22].

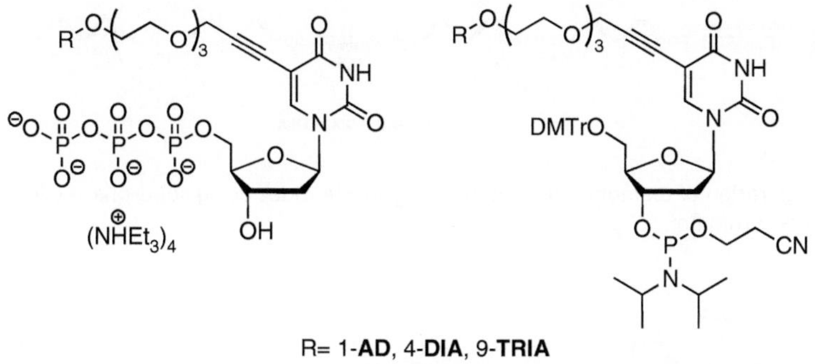

R= 1-**AD**, 4-**DIA**, 9-**TRIA**

Figure 9.6 DNA nucleobase analogues containing adamantane, diamantane, and triamantane subunits connected with a linker. DMTr = 4,4'- dimethoxytrityl protecting group. Source: From Ref. [24].

to these receptors with high affinity is composed of a phenolic ring connected to a hydrophobic group [22]. Among the explored bulky alkyl substituents was also the diamantane cage (Figure 9.5, right), though the corresponding adamantane derivative displayed slightly higher estrogenic activity. Indeed, the synthesized compounds behave as inhibitors of carcinoma cell growth (TD550) [23a] and display estrogenic activity.

Lastly, diamondoids have been used for the modification and functionalization of DNA helices by preparing nucleobase analogues decorated with diamondoid cages [24]. Note that a suitable site for functionalization is the part of the nucleobase that is not involved in hydrogen bonding of the Watson–Crick base pairs and thereby causes minimal disruptions in DNA duplex formation. Since diamondoids are sterically demanding substituents in this type of environment, the connection with the nucleobases was realized through the use of a linker moiety (Figure 9.6). Diamondoid subunits introduced at the C5 position of the pyrimidine backbone served as a bulky, lipophilic functional group, and the prepared hybrid thymidine-like structures were further incorporated into DNA chemically (automated solid-phase DNA synthesis of the phosphoramidites) and enzymatically (using DNA polymerase on the triphosphate series). The resulting oligonucleotides were characterized by thermal denaturation studies and circular dichroism spectroscopy, showing that there was little difference in duplex DNA conformation when comparing the modified oligonucleotides with the behavior of the unmodified ones. Moreover, diamondoid-modified thymidines were well accommodated in the DNA's major groove while at the same time being sufficiently flexible so as not to interfere with the DNA duplex formation, as evidenced by the characteristic profiles of their recorded CD spectra. Thermal denaturation measurements also highlighted comparable stabilities of all the studied duplexes, with only minor destabilization for the modified double strands.

Another way of incorporating diamondoids in DNA helices is by using "click" chemistry on solid support (Figure 9.7) [25]. Namely, diamondoid azido compounds were successfully coupled with an alkyne linker that was generated by H-phosphonate chemistry [26], and the subsequent cycloaddition click reaction

Figure 9.7 Preparation of diamondoid-decorated oligonucleotides using solid-phase click chemistry. Source: From Ref. [25].

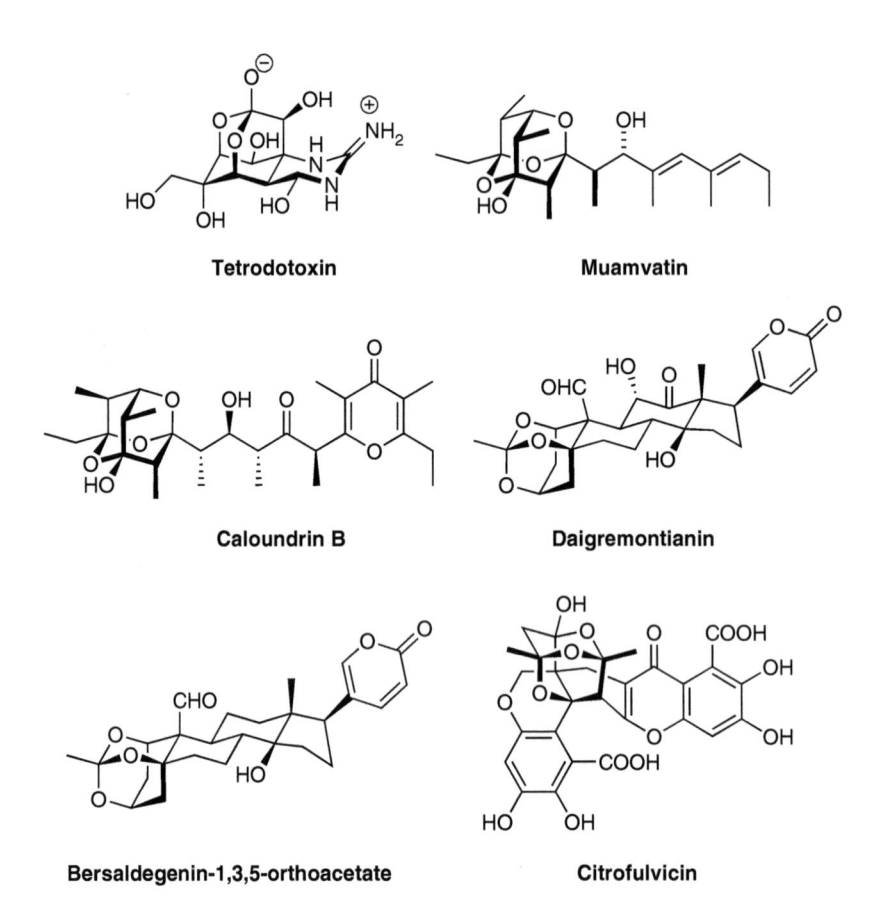

Figure 9.8 Examples of natural products that contain the trioxaadamantane core.

afforded the triazole moiety in the target oligonucleotides. The copper(I)-catalyzed [3 + 2] azide-alkyne cycloaddition [27] is an elegant and widely used method for DNA functionalization, but one drawback is the use of a copper catalyst that is not biocompatible. However, a variant of the reaction performed on solid support enables efficient removal of the copper catalyst by simple filtration. The recorded CD spectra of the prepared oligonucleotides revealed that the secondary DNA structure was not significantly perturbed, and thermal denaturation experiments confirmed that a larger size of the present diamondoid cage correlates with an increase in DNA duplex stability.

Terengganesine A   **Panacosmine**   **Nareline**

Scholarisine H   X = H, daphnezomine A   **Dapholdhamine B**
                       X = Me, daphnezomine B

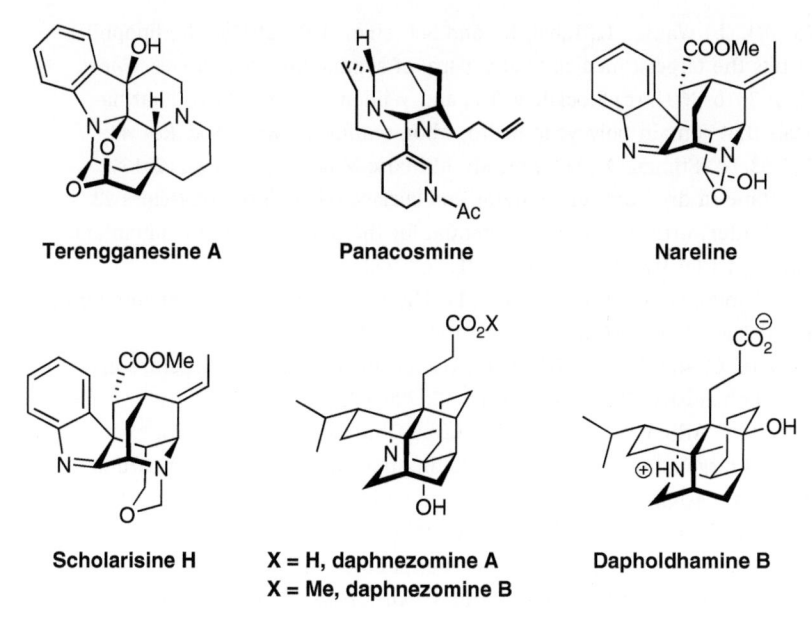

Figure 9.9 Examples of natural products that contain the azaadamantane core.

Based on the relatively scarce literature on diamondoid derivatives beyond adamantanes that show potential as bioactive scaffolds, it can be concluded that many opportunities for future discoveries exist. A possible outlook on the application of diamondoids as pharmacophores comes in the form of heterodiamondoids, cage derivatives consisting of one or more heteroatoms in the molecular scaffold. While such compounds isolated from marine organisms, plants, and fungi, e.g., tetrodotoxin [28], muamvatin [29], caloundrin B [30], daigremontianin [31], bersaldegenin-1,3,5-orthoacetate [32], and citrofulvicin [33] (Figure 9.8) display extraordinary bioactive properties (the extremely potent neurotoxin tetrodotoxin comes to mind first), derivatives resembling larger cages remain unexplored to date.

Among natural products, the azaadamantane core (Figure 9.9) is found in alkaloid terengganesine A [34a–d], isolated from the bark of *Kopsia terengganensis*, panacosmine [35a, b], isolated from the seeds of *Acosmium panamense*, nareline [36a, b] from the leaf extract of *Alstonia scholaris*, and scholarisine H [37] from long-term stored *A. scholaris* genus, as well as daphnezomine A and B [38a, b] and dapholdhamine B [39] isolated from genus *Daphniphyllum* trees.

Potentially useful compounds for future studies are also thiadiamondoids that occur naturally in mature petroleum and can therefore be exploited [40] and synthetically available oxa- [41] as well as azadiamondoids [42].

References

1 (a) Joubert, J., Geldenhuys, W.J., Van der Schyf, C.J. et al. (2012). Polycyclic cage structures as lipophilic scaffolds for neuroactive drugs. *ChemMedChem*

7: 375–384. (b) Wanka, L., Iqbal, K., and Schreiner, P.R. (2013). The lipophilic bullet hits the targets: medicinal chemistry of adamantane derivatives. *Chem. Rev.* 113: 3516–3604. (c) Stockdale, T.P. and Williams, C.M. (2015). Pharmaceuticals that contain polycyclic hydrocarbon scaffolds. *Chem. Soc. Rev.* 44: 7737–7763. (d) Štimac, A., Šekutor, M., Mlinarić-Majerski, K. et al. (2017). Adamantane in drug delivery systems and surface recognition. *Molecules* 22: 297. (e) Butterworth, R.F. (2021). Potential for the repurposing of adamantane antivirals for COVID-19. *Drugs R. D.* 21: 267–272.

2 Liu, J., Obando, D., Liao, V. et al. (2011). The many faces of the adamantyl group in drug design. *Eur. J. Med. Chem.* 46: 1949–1963.

3 Lamoureux, G. and Artavia, G. (2010). Use of the adamantane structure in medicinal chemistry. *Curr. Med. Chem.* 17: 2967–2978.

4 (a) Horvat, Š., Mlinarić-Majerski, K., Glavaš-Obrovac, L. et al. (2006). Tumor-cell-targeted methionine-enkephalin analogues containing unnatural amino acids: design, synthesis, and in vitro antitumor activity. *J. Med. Chem.* 49: 3136–3142. (b) Roščić, M., Sabljić, V., Mlinarić-Majerski, K., and Horvat, Š. (2008). In vitro enzymatic stabilities of methionine-enkephalin analogues containing an adamantane-type amino acid. *Croat. Chem. Acta* 81: 637–640. (c) Ivleva, E.A., Zaborskaya, M.S., Shiryaev, V.A., and Klimochkin, Y.N. (2023). One pot synthesis of bridgehead amino alcohols from diamandoid hydrocarbons. *Synth. Commun.* 476–491.

5 Schwertfeger, H., Würtele, C., Serafin, M. et al. (2008). Monoprotection of diols as a key step for the selective synthesis of unequally disubstituted diamondoids (nanodiamonds). *J. Org. Chem.* 73: 7789–7792.

6 Kapoerchan, V.V., Knijnenburg, A.D., Niamat, M. et al. (2010). An adamantyl amino acid containing gramicidin S analogue with broad spectrum antibacterial activity and reduced hemolytic activity. *Chem. Eur. J.* 16: 12174–12181.

7 Li, B., Wanka, L., Blanchard, J. et al. (2010). Neurotrophic peptides incorporating adamantane improve learning and memory, promote neurogenesis and synaptic plasticity in mice. *FEBS Lett.* 584: 3359–3365.

8 Müller, J., Kirschner, R.A., Berndt, J.P. et al. (2019). Diamondoid amino acid-based peptide kinase a inhibitor analogues. *ChemMedChem* 14: 663–672.

9 Wang, J.J., Wang, S.S., Lee, C.F. et al. (1997). In vitro antitumor and antimicrobial activities of N-substituents of maleimide by adamantane and diamantane. *Chemotherapy* 43: 182–189.

10 (a) Wang, J.J., Chang, Y.F., Chern, Y.T., and Chi, C.W. (2003). Study of in vitro and in vivo effects of 1,6-bis 4-(4-amino-3-hydroxyphenoxy)phenyl diamantane (DPD), a novel cytostatic and differentiation inducing agent, on human colon cancer cells. *Br. J. Cancer* 89: 1995–2003. (b) Wang, J.J., Huang, K.T., and Chern, Y.T. (2004). Induction of growth inhibition and G(1) arrest in human cancer cell lines by relatively low-toxic diamantane derivatives. *Anticancer Drugs* 15: 277–286. (c) Chang, Y.F., Chi, C.W., Chern, Y.T., and Wang, J.J. (2005). Effects of 1,6-Bis 4-(4-amino-3-hydroxyphenoxy)phenyl diamantane (DPD), a reactive oxygen species and apoptosis inducing agent, on human leukemia cells in vitro and in vivo. *Toxicol. Appl. Pharmacol.* 202: 1–12. (d) Ikediobi, O.N., Reimers, M.,

Durinck, S. et al. (2008). In vitro differential sensitivity of melanomas to phenothiazines is based on the presence of codon 600 BRAF mutation. *Mol. Cancer Ther.* 7: 1337–1346. (e) Wang, J.J., Hung, H.F., Huang, M.L. et al. (2012). Role of p21 as a determinant of 1,6-Bis 4-(4-amino-3-hydroxyphenoxy)phenyl diamantane response in human HCT-116 colon carcinoma cells. *Oncol. Rep.* 27: 529–534.

11 Bakhonsky, V.V., Pashenko, A.A., Becker, J. et al. (2020). Synthesis and antiproliferative activity of hindered, chiral 1,2-diaminodiamantane platinum(II) complexes. *Dalton Trans.* 49: 14009–14016.

12 Hodek, P., Janščak, P., Anzenbacher, P. et al. (1988). Metabolism of diamantane by rat liver microsomal cytochromes P-450. *Xenobiotica* 18: 1109–1118.

13 Hodek, P., Burkhard, J., and Janku, J. (1995). Probing the cytochrome P-450 2B1 active site with diamantoid compounds. *Gen. Physiol. Biophys.* 14: 225–239.

14 Stiborova, M., Hansikova, H., and Frei, E. (1996). Cytochromes P450 2B1 and P450 2B2 demethylate *N*-nitrosodimethylamine and *N*-nitrosomethylaniline *in vitro*. *Gen. Physiol. Biophys.* 15: 211–221.

15 Hodek, P., Sopko, B., Antonovič, L. et al. (2004). Evaluation of comparative cytochrome P450 2B4 model by photoaffinity labeling. *Gen. Physiol. Biophys.* 23: 467–488.

16 Hodek, P. and Smrček, S. (1999). Evaluation of 3-azidiamantane as photoaffinity probe of cytochrome P450. *Gen. Physiol. Biophys.* 18: 181–198.

17 Hodek, P., Karabec, M., Šulc, M. et al. (2007). Mapping of cytochrome P450 2B4 substrate binding sites by photolabile probe 3-azidiamantane: identification of putative substrate access regions. *Arch. Biochem. Biophys.* 468: 82–91.

18 Hodek, P., Borek-Dohalska, L., Sopko, B. et al. (2005). Structural requirements for inhibitors of cytochromes P450 2B: assessment of the enzyme interaction with diamondoids. *J. Enzyme Inhib. Med. Chem.* 20: 25–33.

19 (a) Borek-Dohalska, L., Hodek, P., and Stiborova, M. (2000). New selective inhibitors of cytochrome P450 2B4 and an activator of cytochrome P450 3A6 in rabbit liver microsomes. *Collect. Czech. Chem. Commun.* 65: 122–132. (b) Stiborova, M., Borek-Dohalska, L., Hodek, P. et al. (2002). New selective inhibitors of cytochromes P4502B and their application to antimutagenesis of tamoxifen. *Arch. Biochem. Biophys.* 403: 41–49. (c) Šulc, M., Hudeček, J., Stiborova, M., and Hodek, P. (2008). Structural analysis of binding of a diamantoid substrate to cytochrome P450 2B4: possible role of Arg 133 in modulation of function and activity of this enzyme. *Neuro Endocrinol. Lett.* 29: 722–727.

20 (a) Lednicer, D., Heyd, W.E., Emmert, D.E. et al. (1979). Hypobetalipoproteinemic agents. 2. Compounds related to 4-(1-adamantyloxy)aniline. *J. Med. Chem.* 22: 69–77. (b) Heyd, W.E., Bell, L.T., Heystek, J.R. et al. (1982). Hypobetalipoproteinemic agents. 3. Variation of the polycyclic portion of 4-(1-adamantyloxy)aniline. *J. Med. Chem.* 25: 1101–1103.

21 Kagechika, H., Kawachi, E., Fukasawa, H. et al. (1997). Inhibition of IL-1-induced IL-6 production by synthetic retinoids. *Biochem. Biophys. Res. Commun.* 231: 243–248.

22 Yamakoshi, Y., Otani, Y., Fujii, S., and Endo, Y. (2000). Dependence of estrogenic activity on the shape of the 4-alkyl substituent in simple phenols. *Biol. Pharm. Bull.* 23: 259–261.

23 (a) Sun, S.Y., Yue, P., Dawson, M.I. et al. (1997). Differential effects of synthetic nuclear retinoid receptor-selective retinoids on the growth of human non-small cell lung carcinoma cells. *Cancer Res.* 57: 4931–4939. (b) Sun, S.Y., Yue, P., Mao, L. et al. (2000). Identification of receptor-selective retinoids that are potent inhibitors of the growth of human head and neck squamous cell carcinoma cells. *Clin. Cancer Res.* 6: 1563–1573.

24 Wang, Y., Tkachenko, B.A., Schreiner, P.R., and Marx, A. (2011). Diamondoid-modified DNA. *Org. Biomol. Chem.* 9: 7482–7490.

25 Crumpton, J.B. and Santos, W.L. (2012). Site-specific incorporation of diamondoids on DNA using click chemistry. *Chem. Commun.* 48: 2018–2020.

26 Bouillon, C., Meyer, A., Vidal, S. et al. (2006). Microwave assisted "click" chemistry for the synthesis of multiple labeled-carbohydrate oligonucleotides on solid support. *J. Org. Chem.* 71: 4700–4702.

27 Gramlich, P.M.E., Wirges, C.T., Manetto, A., and Carell, T. (2008). Postsynthetic DNA modification through the copper-catalyzed azide-alkyne cycloaddition reaction. *Angew. Chem. Int. Ed.* 47: 8350–8358.

28 Bane, V., Lehane, M., Dikshit, M. et al. (2014). Tetrodotoxin: chemistry, toxicity, source, distribution and detection. *Toxins* 6: 693–755.

29 (a) Roll, D.M., Biskupiak, J.E., Mayne, C.L., and Ireland, C.M. (1986). Muamvatin, a novel tricyclic spiro ketal from the Fijian mollusk *Siphonaria normalis*. *J. Am. Chem. Soc.* 108: 6680–6682. (b) Paterson, I. and Perkins, M.V. (1993). Total synthesis of the marine polypropionate (+)-muamvatin. A configurational model for siphonariid metabolites. *J. Am. Chem. Soc.* 115: 1608–1610.

30 (a) Becerril-Jimenez, F. and Ward, D.E. (2012). On the origin of siphonariid polypropionates: Total synthesis of caloundrin B and its isomerization to siphonarin B. *Org. Lett.* 14: 1648–1651. (b) Blanchfield, J.T., Brecknell, D.J., Brereton, I.M. et al. (1994). Caloundrin B and Funiculatin A: new polypropionates from siphonariid limpets. *Aust. J. Chem.* 47: 2255–2269.

31 Wagner, H., Fischer, M., and Lotter, H. (1985). Isolation and structure determination of daigremontianin, a novel bufadienolide from *Kalanchoe daigremontiana*. *Planta Med.* 51: 169–170.

32 Moniuszko-Szajwaj, B., Pecio, L., Kowalczyk, M., and Stochmal, A. (2016). New bufadienolides isolated from the roots of *Kalanchoe daigremontiana* (Crassulaceae). *Molecules* 21: 243.

33 (a) Chen, Y., Jiang, N., Wei, Y.J. et al. (2018). Citrofulvicin, an antiosteoporotic polyketide from *Penicillium velutinum*. *Org. Lett.* 20: 3741–3744. (b) Doytchinova, K. and Winkler, J.D. (2021). Synthesis of the core ring system of the antiosteoporotic Citrofulvicin. *Org. Lett.* 23: 4575–4578.

34 (a) Uzir, S., Mustapha, A.M., Hadi, A.H.A. et al. (1997). Terengganensines A and B, dihydroeburnane alkaloids from *Kopsia terengganensis*. *Tetrahedron Lett.* 38: 1571–1574. (b) Li, G., Piemontesi, C., Wang, Q., and Zhu, J.P. (2019). Stereoselective total synthesis of eburnane-type alkaloids enabled by conformation-directed cyclization and rearrangement. *Angew. Chem. Int. Ed.* 58: 2870–2874. (c) Zhou, Q.L., Dai, X., Song, H. et al. (2018). Concise syntheses of eburnane indole alkaloids. *Chem. Commun.* 54: 9510–9512. (d) Piemontesi, C., Wang, Q., and Zhu, J.P. (2016). Enantioselective total synthesis of (−)-terengganensine A. *Angew. Chem. Int. Ed.* 55: 6556–6560.

35 (a) Trevisan, T.C., Silva, E.A., Dall'Oglio, E.L. et al. (2008). New quinolizidine and diaza-adamantane alkaloids from *Acosmium dasycarpum* (Vog.) Yakovlev-Fabaceae. *Tetrahedron Lett.* 49: 6289–6292. (b) Nuzillard, J.M., Connolly, J.D., Delaude, C. et al. (1999). Computer-assisted structural elucidation. Alkaloids with a novel diaza-adamantane skeleton from the seeds of *Acosmium panamense* (Fabaceae). *Tetrahedron* 55: 11511–11518.

36 (a) Kam, T.S., Nyeoh, K.T., Sim, K.M., and Yoganathan, K. (1997). Alkaloids from *Alstonia scholaris*. *Phytochemistry* 45: 1303–1305. (b) Morita, Y., Hesse, M., Schmid, H. et al. (1977). *Alstonia scholaris*: Struktur des Indolalkaloides Narelin. *Helv. Chim. Acta* 60: 1419–1434.

37 Yang, X.W., Luo, X.D., Lunga, P.K. et al. (2015). Scholarisines H-O, novel indole alkaloid derivatives from long-term stored *Alstonia scholaris*. *Tetrahedron* 71: 3694–3698.

38 (a) Xu, G.P., Wu, J.B., Li, L.Y. et al. (2020). Total synthesis of (−)-daphnezomines A and B. *J. Am. Chem. Soc.* 142: 15240–15245. (b) Hu, J.P., Chen, W.Q., Jiang, Y.Y., and Xu, J. (2023). Synthesis of tetracyclic core structure of daphnezomines A and B. *Chinese J. Org. Chem.* 43: 171–177.

39 Guo, L.D., Hou, J.P., Tu, W.T. et al. (2019). Total synthesis of dapholdhamine B and dapholdhamine B lactone. *J. Am. Chem. Soc.* 141: 11713–11720.

40 (a) Wei, Z., Moldowan, J.M., Fago, F. et al. (2007). Origins of thiadiamondoids and diamondoidthiols in petroleum. *Energy Fuel* 21: 3431–3436. (b) Wei, Z.B., Mankiewicz, P., Walters, C. et al. (2011). Natural occurrence of higher thiadiamondoids and diamondoidthiols in a deep petroleum reservoir in the Mobile Bay gas field. *Org. Geochem.* 42: 121–133. (c) Wei, Z.B., Walters, C.C., Moldowan, J.M. et al. (2012). Thiadiamondoids as proxies for the extent of thermochemical sulfate reduction. *Org. Geochem.* 44: 53–70. (d) Gvirtzman, Z., Said-Ahmad, W., Ellis, G.S. et al. (2015). Compound-specific sulfur isotope analysis of thiadiamondoids of oils from the Smackover formation, USA. *Geochim. Cosmochim. Acta* 167: 144–161. (e) Cai, C.F., Xiao, Q.L., Fang, C.C. et al. (2016). The effect of thermochemical sulfate reduction on formation and isomerization of thiadiamondoids and diamondoids in the lower Paleozoic petroleum pools of the Tarim Basin, NW China. *Org. Geochem.* 101: 49–62. (f) Zhu, G.Y., Zhang, Y., Zhang, Z.Y. et al. (2018). High abundance of alkylated diamondoids, thiadiamondoids and thioaromatics in recently discovered sulfur-rich LS2 condensate in the Tarim Basin. *Org. Geochem.* 123: 136–143.

41 (a) Fokin, A.A., Zhuk, T.S., Pashenko, A.E. et al. (2009). Oxygen-doped nanodiamonds: synthesis and functionalizations. *Org. Lett.* 11: 3068–3071. (b) Gunchenko, P.A., Li, J., Liu, B.F. et al. (2018). Aerobic oxidations with *N*-hydroxyphthalimide in trifluoroacetic acid. *Molec. Catal.* 447: 72–79.

42 (a) Krishnamurthy, V.V. and Fort, R.C. (1981). Heteroadamantanes. 2. Synthesis of 3-heterodiamantanes. *J. Org. Chem.* 46: 1388–1393. (b) Fokin, A.A., Reshetylova, O.K., Bakhonsky, V.V. et al. (2022). Synthetic doping of diamondoids through skeletal editing. *Org. Lett.* 24: 4845–4849. (c) Suslov, E.V., Ponomarev, K.Y., Volcho, K.P., and Salakhutdinov, N.F. (2021). Azaadamantanes, a new promising scaffold for medical chemistry. *Russ. J. Bioorg. Chem.* 47: 1133–1154.

10

Supramolecular Architectures

Diamondoids found use in supramolecular chemistry mostly as strongly binding guest molecules for spherical molecular containers, especially cyclodextrins (CDs, for recent reviews, see [1]) and cucurbiturils (CB[n]s), with many potential applications [2]. In general, strongly bound supramolecular complexes between guests (ligands) and hosts (receptors) are valuable systems for biotechnology and chemical sensing [3]. One naturally occurring example that often comes to mind is the extremely tightly bound avidin•biotin complex that found application in enzyme-linked immunosorbent assays, detection and purification of proteins and nucleic acids, as well as in immobilization chemistry on surfaces [4]. The quest for strong complexation is determined not only by practical issues but also adds to the understanding of host–guest interactions, and diamondoid derivatives are some of the strongest supramolecular guests known to date [5].

Arguably the most well-known diamondoid supramolecular systems comprise adamantane derivatives non-covalently bound to CD hosts that are also useful for diamondoid isolation from crude oil [6]. CDs are naturally occurring cyclic oligosaccharides (Figure 10.1) that are water-soluble and structurally resemble a truncated cone (torus) [1]. CDs typically contain glucose monomers ranging from six to eight units in a ring, constituting α-CD, β-CD, and γ-CD, respectively. The hydroxyl groups of the sugars are oriented toward the upper and lower rims of the cone, and the resulting central cavity containing the glycosidic oxygen is hydrophobic [8]. Since the spherical adamantane cage has a diameter of about 5 Å (Figure 10.1), it matches perfectly with the cavity of β-CD [9], forming 1 : 1 inclusion complexes with high association constants (typically between 10^4 and 10^5 M^{-1}).

Such a tight fit upon binding into the host cavity [10] is a demonstration of the "hydrophobic" effect [11], which, however, is difficult to separate from contributions from the van der Waals interactions, in particular, attractive London dispersion (LD) [12]. Among the plethora of explored derivatives featuring the adamantyl backbone, some representative examples include β-CD complexes with **1COOHAD** [9, 13], **1NH2AD** and **2NH2AD** [14], **1OHAD** [15], **1BRAD** [16], **1CH2OHAD** [17], CD-adamantane conjugates [18], supramolecular adamantane hydrogels and polymers [19], adamantane porphyrin derivatives [20], adamantane modified cytochrome [21], and many more. The application of adamantane-based supramolecular complexes with CDs goes beyond simple binary systems, and one

The Chemistry of Diamondoids: Building Blocks for Ligands, Catalysts, Materials, and Pharmaceuticals, First Edition. Andrey A. Fokin, Marina Šekutor, and Peter R. Schreiner.
© 2024 WILEY-VCH GmbH. Published 2024 by WILEY-VCH GmbH.

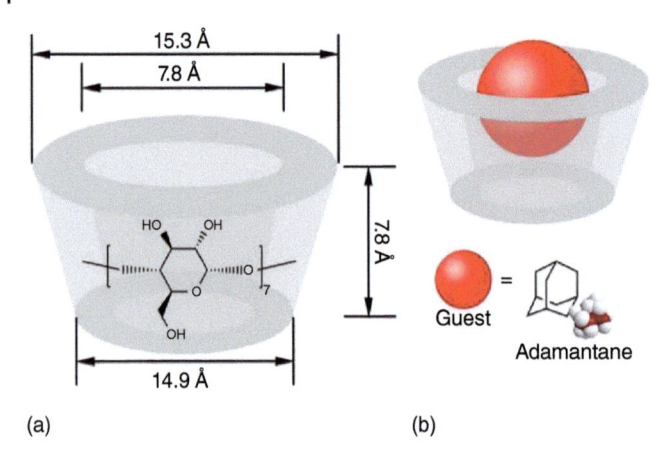

Figure 10.1 (a) Structure of β-cyclodextrin and (b) adamantane as guest molecule forming an inclusion complex with the β-CD host. Source: Reproduced from Ref. [7] with permission from the American Chemical Society, 2014.

such example is nanoparticles self-assembled from a ruthenium-β-CD complex and a peptide functionalized with adamantane [22]. These highly stable nanostructures containing a tumor-targeting peptide (cyclic Arg-Gly-Asp) can induce cell death in certain carcinoma cells while at the same time displaying quantitative drug loading.

A detailed description of every class of adamantane derivatives engaging in binding interactions with the CD hosts is beyond the scope of this work and instead, we will focus our attention on other diamondoid derivatives, mostly diamantanes. For example, saturated cage hydrocarbons can form nonbinary complexes that are useful as long-lived phosphorescence systems. Ternary complexes of naphthalene and phenanthrene with β-CD also incorporating **AD** [23] and **DIA** [24] are capable of phosphorescence at room temperature, whereby the arene molecules act as signaling units. The observed long-lasting arene phosphorescence is possible because of the shielding effect of the host and the second guest, thus facilitating protection from the deleterious influence of the dissolved oxygen that acts as a quencher. However, it appears that the efficiency of such triple complexes acting as phosphorescence centers is also correlated with the size of the saturated hydrocarbon. Since binary arene β-CD complexes do not exhibit long-lived phosphorescence even after the removal of oxygen from the sample and, moreover, since the addition of the cage molecule to the system results in phosphorescence, it follows that the saturated hydrocarbon upon binding to the remaining free portal of the host shields the arene from the environment. Also, poor solubility of the components and/or the resulting complexes and subsequent formation of molecular aggregates constitute another condition for long-lasting phosphorescence and further point to the importance of limited exposure of the arene to its surroundings. Lastly, as is often the case in supramolecular assemblies, a tight fit between the host and the guest is a decisive factor in governing the overall strength of the interaction and thereby the applicability of the system. Another study dealing with diamondoids in CDs explored inclusion complexes of adamantane, diamantane, and triamantane derivatives with β- and γ-CDs [25]. Since

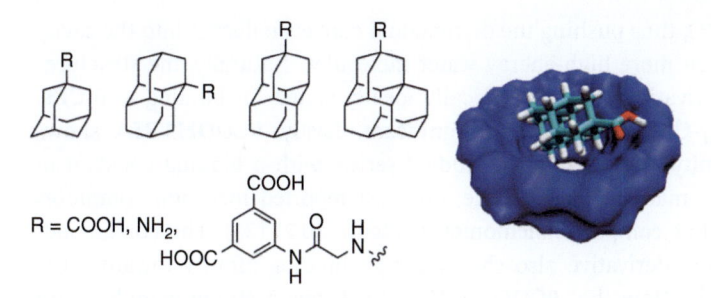

Figure 10.2 Diamondoid acid guests for β- and γ-CDs (left) and the computed γ-CD•**9COOHTRIA** complex (right).

γ-CD is more flexible than β-CD, the replacement of high-energy water molecules from the host cavity with a hydrophobic guest is more challenging and therefore a good test system for the binding of bulky diamondoid cages. Diamondoid carboxylic acids (Figure 10.2) were chosen as starting materials due to the combination of the hydrophobic cage backbone suitable in size and properties for the CD cavity with a substituent capability of forming hydrogen bonds with the CD rim, thus increasing the chances for tight binding.

Further derivatives under consideration include the even more hydrophilic diamondoid derivatives that incorporate 5-aminoisophthalic acid (Figure 10.2). As can be expected, all studied diamondoid derivatives interact with β-CD, but only the bulkiest triamantane compounds and medially substituted diamantane 5-isophthalic acid derivatives engage in binding with γ-CD, forming 1:1 stoichiometry complexes. When comparing the binding constants of complex formation between β-CD and guests **1COOHAD** and **1COOHDIA**, the diamantane derivative binds markedly stronger ($K_a = 2.0 \times 10^4$ M^{-1} vs. $K_a = 1.2 \times 10^5$ M^{-1}, respectively), owing to a better fit of the bigger cage to the CD cavity. B97-D3/def2-TZVPP//B97-D3/def-SV(P) computations verified this result by finding the β-CD•**1COOHDIA** complex to be more stable than the corresponding adamantane analogue [25]. Note, however, that the binding strength drops for the diamantane 5-aminoisophthalic acid derivative ($K_a = 4.1 \times 10^4$ M^{-1}), probably due to the larger acid moiety hindering the complete insertion of the guest into the cavity. Despite this, relatively strong binding still occurs and the isophthalic acid part engages in hydrogen bonding with the CD rim, as indicated by the pronounced tilt of the guest inside the CD that was observed in the computed structure of the complex. As for γ-CD complexes of the same diamantane derivative, there is a much more pronounced entropy contribution to the binding, meaning that changes in solvation and release of high-energy water molecules from the cavity play a significant role in complexation. On the other hand, apically substituted **4COOHDIA** showed a remarkably high binding constant with β-CD ($K_a = 2.1 \times 10^5$ M^{-1}), owing to the ability of the diamantane rod-like structure to immerse itself deeply into the hydrophobic cavity of the host. Similarly, the analogous isophthalic acid derivative demonstrates the same effect, and its binding constant is even slightly higher due to the beneficial hydrogen bond formation of the acid moiety with the rim

($K_a = 2.8 \times 10^5$ M^{-1}), thus pushing the diamondoid core even deeper into the cavity and displacing even more high-energy water molecules. Naturally, the structures of these apical derivatives are geometrically ideally suited for binding to β-CDs, and binding to γ-CDs is therefore insignificant. Lastly, **9COOHTRIA** shows the strongest affinity for γ-CD in the studied series, with a binding constant of $K_a = 5.0 \times 10^5$ M^{-1}, making it one of the strongest reported inclusion complexes with γ-CD for a 1:1 complex stoichiometry (Figure 10.2) [25]. The corresponding isophthalic acid derivative also shows a high binding affinity toward γ-CD ($K_a = 3.9 \times 10^5$ M^{-1}). Note that **9COOHTRIA** also forms a strong complex with β-CD ($K_a = 1.5 \times 10^5$ M^{-1}) by somewhat expanding the central cavity of the host, even though the penetration of the diamondoid core is not as efficient as for the γ-CD host.

Further exploration of the complex formation between CDs and diamondoid derivatives included adamantane, diamantane, and triamantane amines (Figure 10.2, R = NH$_2$) [26]. In general, binding constants for the adamantane and diamantane amines were somewhat smaller when compared to carboxylic acids described above due to stronger hydrogen bonding of the guest carboxylate group with the hydroxyl groups at the rim of the CDs. The binding to a smaller β-CD host can also lead to a 2:1 complex stoichiometry where two β-CDs can simultaneously accommodate a large triamantane cage as a guest [26a].

The ability of diamondoid cages to engage in strong interactions with CDs was also applied when designing supramolecular assemblies consisting of amphiphilic CD molecules that mimic the lipid bilayer and serve as a model for biological cell membranes (Figure 10.3) [26a]. The formed CD vesicles were decorated with inclusion complexes wherein the guests were functionalized with diamondoid scaffolds (adamantane, diamantane, and triamantane) for anchoring purposes

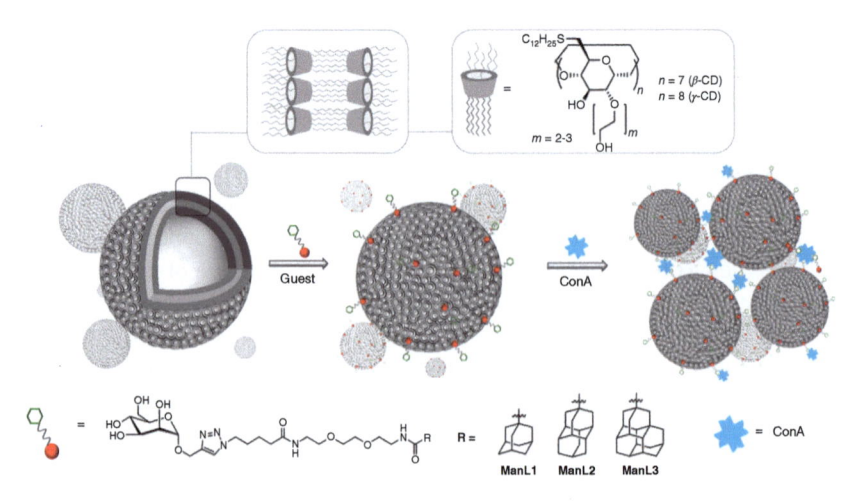

Figure 10.3 Structure of D-mannose-diamondoid conjugates and CD vesicles and a schematic representation of the agglutination of D-mannose-decorated CD vesicles induced by the lectin concanavalin A (ConA). Source: Reproduced from Ref. [26a], Wiley-VCH, 2017.

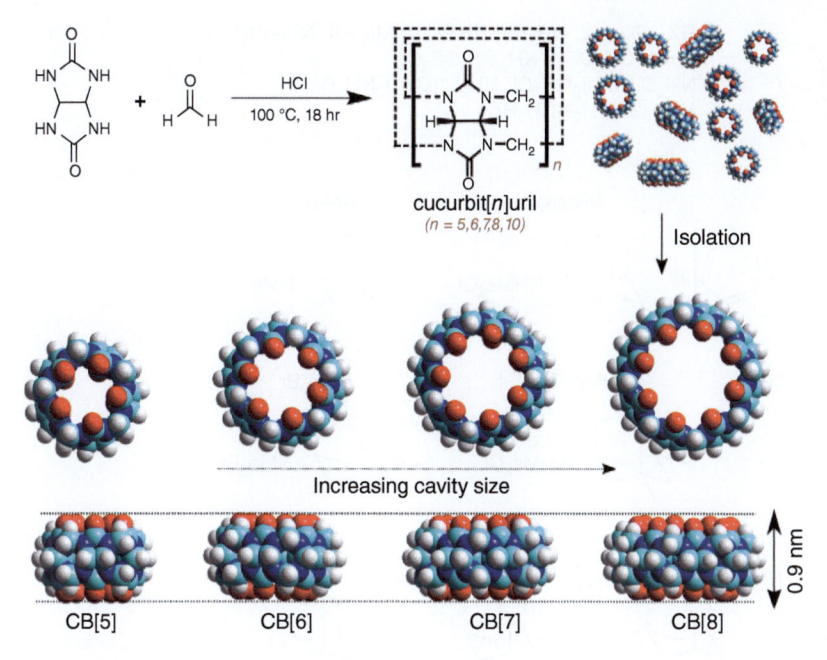

Figure 10.4 *Top*: The reaction between glycoluril and formaldehyde gives a mixture of cucurbiturils. *Bottom*: Space-filling models of CB[5]–[8] demonstrating the increasing width yet constant height of the macrocycles. Source: Reproduced from Ref. [29b] with permission from the American Chemical Society, 2015.

with carbohydrate moieties (D-mannose) that can bind to proteins where multiple bindings are necessary for cellular recognition and adhesion. Agglutination experiments performed on these self-assembled soft materials with lectin concanavalin A (ConA) revealed a correlation between the strength of complex association and the agglutination efficiency. Namely, the adamantane conjugate showed lower agglutination with β-CD vesicles than the diamantane and triamantane conjugates and almost no agglutination with γ-CD vesicles, whereas the latter two demonstrated a high agglutination capability with γ-CD vesicles. Overall, the ability of diamondoids to form strong complexes with CDs has many practical applications beyond simple host–guest chemistry and is a useful tool in, e.g., chromatographic separations [27], drug delivery [28], and many more.

Going to the next important class of supramolecular host molecules, we will now turn our attention to cucurbiturils, named so due to their resemblance to a pumpkin (*cucurbitaceae*) family [29]. The CB[n] class of molecular containers comprises a series of macrocyclic methylene-bridged glycoluril oligomers that are composed of two symmetry-equivalent ureidyl carbonyl portals connected with a hydrophobic cavity (Figure 10.4).

The carbonyl portals possess a highly negative electrostatic potential and are therefore suitable for cation binding. In 1986 it was first demonstrated that ligands with both a cationic and a hydrophobic unit bind to CB[6] with high affinity and selectivity [30]. While CB[n]s are able to host unsaturated hydrocarbons due to strong π-π

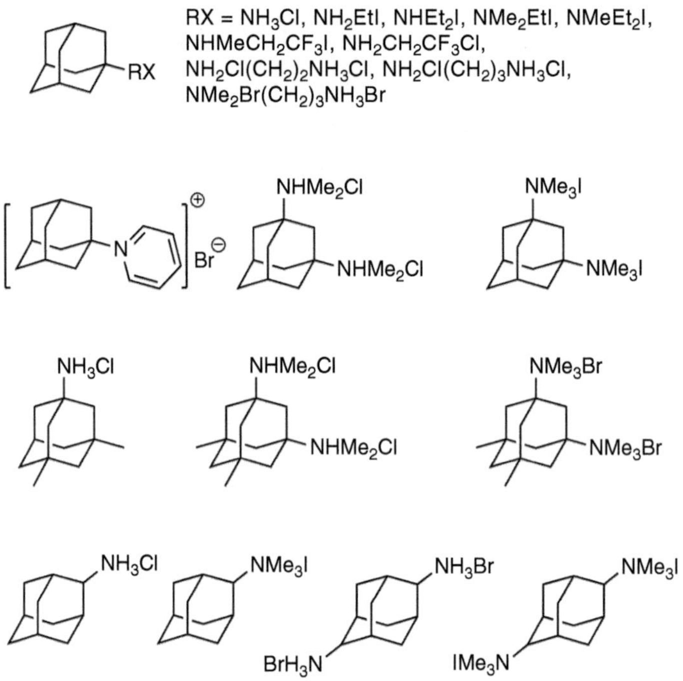

Figure 10.5 Examples of adamantane-containing amines that were studied as ligands with high affinity for cucurbituril hosts.

and σ-π attractions [31], their hosting of saturated hydrocarbons was also shown [32]. The ^{1}H NMR competition experiments demonstrated that CB[7] displays ultrahigh affinity for cationic adamantane derivatives (Figure 10.5), e.g., $K_a = 4.2 \times 10^{12}$ M^{-1} for **1NH₃AD+** in a water buffer was measured [33]. Following this observation, the next logical step was to test the affinity of diamantane derivatives for CB[n] hosts and that proved to be a truly fruitful endeavor.

At first, the binding results were not encouraging, as 4,9-diaminodiamantane (**49NH2DIA**) showed only a modest affinity toward CB[7] ($K_a = 1.3 \times 10^{11}$ M^{-1}) [34], and only its affinity toward *nor-seco*-cucurbit[10]uril was reported at the time [35].

Thus, dication **1NH2+CH2CH2NH3+AD** continued to hold the record for the tightest binding CB[7]•guest complex ever reported with $K_a = 5 \times 10^{15}$ M^{-1} in pure water and $K_a = 2.4 \times 10^{13}$ M^{-1} in a water buffer, even though only one carbonyl portal engaged in ion–dipole interactions [36]. However, everything changed with the groundbreaking report of the diamantane quaternary ammonium salt **49NME3IDIA** (Figure 10.6) [37] and its reported binding affinity [5]. The ultra-tight complex of **49NME3IDIA** with CB[7] amounted to $K_a = 7.2 \times 10^{17}$ M^{-1} in pure water and $K_a = 1.9 \times 10^{15}$ M^{-1} in a water buffer, making it 143-fold (water) and 79-fold (buffer) tighter than the analogous complex with **1NH2+CH2CH2NH3+AD** dication measured under identical conditions. This new record holder with an attomolar dissociation constant ($K_d = 1.4 \times 10^{-18}$ M) [5] therefore even exceeds the

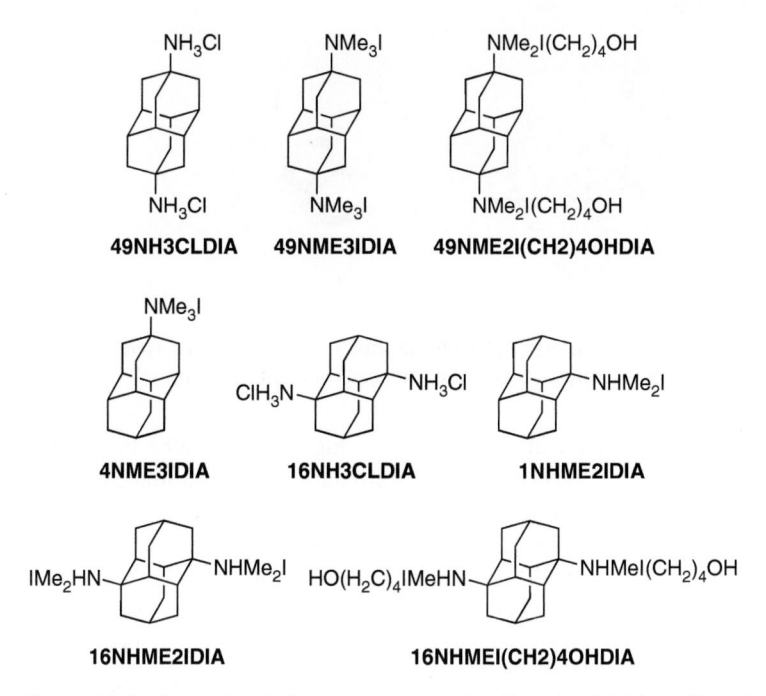

Figure 10.6 Examples of diamantane-containing ligands with high affinity for cucurbituril hosts.

strength of the avidin•biotin complex found in nature ($K_d \approx 10^{-15}$ M) [38], blurring the line between synthetic and natural receptors.

The X-ray crystal structure of the CB[7]•**49NME3DIA** dication complex (Figure 10.7) revealed that the observed binding affinity is a consequence of the hydrophobicity of the diamantane cage that interacts with the CB[7] core, maximizing the ion–dipole interactions between the methyl groups of the guest and the two carbonyl portals of the host (seven interactions per portal), co-linearity of the C_7-axis of CB[7] with the axial orientation of the diamondoid scaffold, and the expulsion of high-energy water molecules upon complex formation [5, 39].

Such a perfect fit of the designed **49NME3DIA** dication to CB[7] and the importance of the hydrophobic diamondoid residue for binding were subsequently underlined in a study using naphthalene ammonium salts [34]. The flatter 2,6-disubstituted naphthalene derivative analogous to **49NME3IDIA** formed a significantly weaker complex with CB[7] ($K_a = 1.7 \times 10^{11}$ M^{-1} in a water buffer), confirming that the bulky diamantane core indeed makes a significant contribution to the overall binding affinity. Reasons for this heightened interaction are numerous non-covalent close contacts between the ligand cage and the host hydrophobic core, as well as the ability of the bulkier diamantane backbone to extrude more of the high-energy water molecules from the host cavity upon binding [34]. This comes as no surprise since it is known that in an aqueous environment, a significant contribution to the inclusion of nonpolar ligands into cavities of spherical receptor molecules comes from the hydrophobic effect [11a]. Thus, the desolvation of the

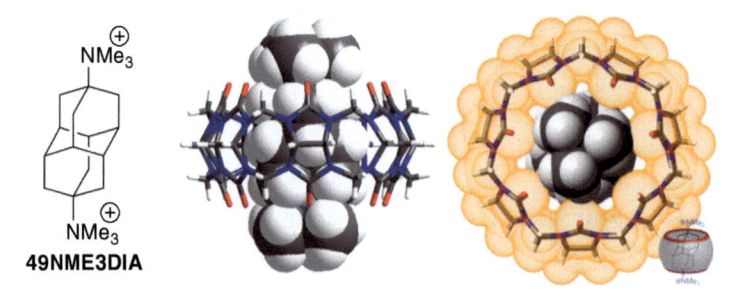

Figure 10.7 Side and top view of the CB[7]•**49NME3DIA** dication complex based on the X-ray single crystal structure analysis. Source: Reproduced from Ref. [5], Wiley-VCH, 2014.

hydrophobic guest combined with the release of high-energy water molecules from the host plays a part in the overall binding strength of the resulting complex. Note, however, that for bulky diamondoid ligands, LD interactions need to be taken into account as well, since they can also be an important modulator in the binding process, as was confirmed computationally [40]. The advantage of computational approaches lies in their predictive power for the binding strengths of compounds not yet measured as well as for gaining a deeper understanding of how small structural changes affect the shape and stability of the complexes under consideration. Since CB[n]•ligand structures are computationally demanding due to their size, it is especially useful when in silico studies can use available X-ray structures as a starting point, as was shown recently [40].

To further highlight the beneficial factors taken into account when designing **49NME3IDIA** with the goal to form an exceptionally stable complex with CB[7], it is illustrative to compare the structure of that complex with two other CB[7]•ligand pairs of similar structural features [39]. The first to consider is the CB[7]•**4NME3IDIA**, a system completely identical to the record-holding complex apart from one crucial difference: it has only one permethylated ammonium group to engage in interactions with the carbonyl rim (Figure 10.8(b)). The resulting complex is therefore markedly weaker ($K_a = 8.0 \times 10^{11}$ M^{-1} in a water buffer) due to the loss of one-half of ion–dipole interactions when compared to CB[7]•**49NME3IDIA** rim (Figure 10.8(a)). On the other hand, the adamantane subunit in CB[7]•**1NME3IAD** (Figure 10.8(c)) appears to sink more deeply into the host cavity in order to maximize the beneficial LD interactions acting on a smaller adamantane cage and thereby completely surrounding it with a nonpolar environment. This type of stabilization is directly reflected in the value of the binding constant for CB[7]•**1NME3IAD** ($K_a = 1.7 \times 10^{12}$ M^{-1} in a water buffer) that slightly exceeds the complex stability of the apical diamantane mono derivative **4NME3IDIA** under the same conditions.

In addition to apical diamantane ammonium salts, medially substituted diamantane derivatives (Figure 10.6) were also explored, and selectivity for the larger CB[8] host was found [39]. Moreover, **16NHME2IDIA** and **16NHMEI(CH2)4OHDIA**

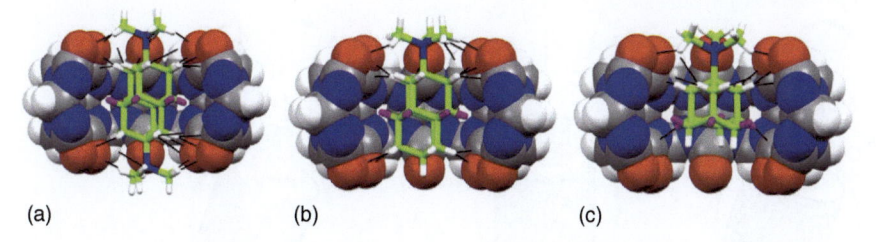

Figure 10.8 Cutaway representations rendered from the X-ray crystal structures of CB[7]•49NME3IDIA (left), CB[7]•4NME3IDIA (middle), and CB[7]•1NME3IAD (right). Source: Reproduced from Ref. [39] with permission from the American Chemical Society, 2017.

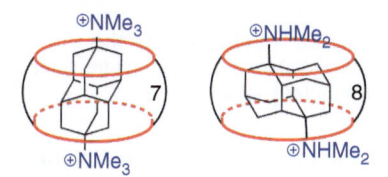

Figure 10.9 Schematic representations of the diamantane cucurbituril complexes that are characterized by record binding affinities. *Left*: The CB[7] complex with **49NME3DIA** dication ($K_a = 1.9 \times 10^{15}$ M^{-1}). *Right*: The CB[8] complex with the **16NHME2DIA** dication ($K_a = 5.7 \times 10^{14}$ M^{-1}). Source: Reproduced from Ref. [39] with permission from the American Chemical Society, 2017.

were identified as the strongest known tight binding guests for CB[8] with association constants in a water buffer amounting to 5.7×10^{14} M^{-1} and 9.2×10^{14} M^{-1}, respectively. The observed slight increase in affinity of **16NHMEI(CH2)4OHDIA** toward CB[8] could be due to loop formation via the long sidearm bearing the OH group that could engage in additional hydrogen bonding with the carbonyl portal, as supported by computations [39]. The orientation of the guest **16NHME2DIA** dication in CB[8] is clearly different from that of the **49NME3DIA** dication in CB[7] (Figure 10.9).

In general, the binding of diamondoid ammonium salts to CB[n] hosts can be subdivided into three distinct modes [40], depending on the nature of the salt (Figure 10.10). The first type, *the primary ammonium binding mode*, is characteristic of unsubstituted amines and can be easily recognized by a tilt of the ligand substituent toward the rim of the host's portal. Upon tilting, hydrogen bonds are generated between the amine moiety and the rim oxygens, while the central host cavity remains filled with the hydrophobic cage scaffold without deformation (Figure 10.10, left). Such binding mode cannot realize the full potential of ion–dipole interactions available from the host geometry, and consequently, the remaining oxygen atoms on the portal do not engage in binding interactions, resulting in lower association constants.

The next binding mode, *the quaternary ammonium binding mode*, occurs with permethylated amines and is characterized by substituents that are coaxial to the

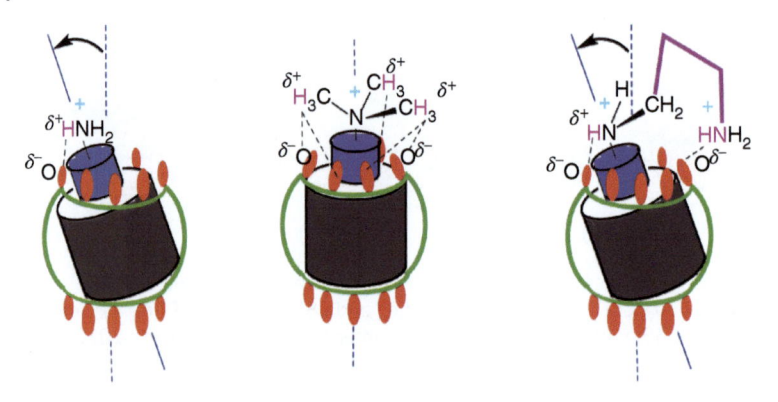

Figure 10.10 Binding modes of diamondoid ammonium salts with CB[n] hosts: the primary ammonium binding mode (left), the quaternary ammonium binding mode (middle), and the loop binding mode (right). Source: Reproduced from [40], Wiley-VCH, 2016.

host's C_n-axes, engaging in multiple electrostatic interactions between the methyl protons and the neighboring rim oxygens (Figure 10.10, middle). This binding mode consequently affords stronger complexes when compared to the first one. The third binding mode termed *the loop binding mode* (Figure 10.10, right) is found for amines with an alkyl side arm that terminates with another primary ammonium moiety. It is again characterized by a tilt of the ligand with respect to the receptor but also includes the terminal NH_3 group engaging in binding interactions during complex formation. The classification of CB[n]•cationic diamondoid complexes into these three binding modes based on specific host–guest interactions is consistently reflected in the values of the measured binding constants (Table 10.1) and lends credibility to the proposed generalization.

Since some selectivity of diamondoid derivatives toward CB[n] hosts of different sizes was observed, interest was sparked in the research community about whether cucurbiturils could be used for selective uptake of the parent diamondoid hydrocarbons as well [41]. The rationale was that cucurbiturils are good solubilizing agents for hydrocarbons in water due to their hydrophobic cavity, and since it was previously confirmed that CB[6] preferentially incorporates branched and cyclic hydrocarbons [32a], different diamondoid cage sizes could be a sufficient distinguishing factor for separation of single diamondoids from a hydrocarbon mixture. Indeed, the binding constant of CB[8] with **DIA** amounted to 1.5×10^7 M^{-1} while it was only 3.7×10^5 M^{-1} for **AD** under the same conditions [41], providing a proof-of-concept. Moreover, the authors also successfully extracted a mixture of large hydrocarbons from crude oil into an aqueous solution, which, in combination with computational modeling [32b], opens the way to future applicability of cucurbiturils in selective hydrocarbon separation under the conditions that lead to effective guest retrieval.

Table 10.1 Binding constants (K_a, M^{-1}) measured for the interaction between CB[7], CB[8], and various diamondoid cationic guests in 50 mM NaOAc buffer at pH 4.74.

Cationic guest	CB[7]	CB[8]	Cationic guest	CB[7]	CB[8]
1-ADNH$_3$	$(4.2 \pm 1.0) \times 10^{12}$	$(8.2 \pm 1.8) \times 10^8$	Me$_2$-1,3-AD(NHMe$_2$)$_2$	$(5.7 \pm 0.9) \times 10^5$	—
1-ADNH$_2$Et	$(8.7 \pm 2.0) \times 10^{11}$	$(1.4 \pm 0.4) \times 10^9$	Me$_2$-1,3-AD(NMe$_3$)$_2$	$(1.8 \pm 0.3) \times 10^6$	—
1-ADNHEt$_2$	$(1.2 \pm 0.3) \times 10^{11}$	$(3.1 \pm 0.8) \times 10^9$	2-ADNH$_3$	$(1.3 \pm 0.3) \times 10^{12}$	$(6.3 \pm 1.6) \times 10^9$
1-ADNMe$_2$Et	$(7.0 \pm 1.6) \times 10^{11}$	$(2.8 \pm 0.7) \times 10^{11}$	2-ADNMe$_3$	$(3.7 \pm 0.9) \times 10^{12}$	$(3.6 \pm 0.9) \times 10^{11}$
1-ADNMeEt$_2$	$(3.2 \pm 0.8) \times 10^{11}$	$(1.1 \pm 0.3) \times 10^{11}$	2,6-AD(NH$_3$)$_2$	$(1.9 \pm 0.4) \times 10^{12}$	$(4.7 \pm 1.2) \times 10^8$
1-ADNMe$_3$	$(1.7 \pm 0.4) \times 10^{12}$	$(9.7 \pm 2.5) \times 10^{10}$	2,6-AD(NMe$_3$)$_2$	$(3.3 \pm 0.9) \times 10^{13}$	$(5.2 \pm 1.4) \times 10^{10}$
1-ADNH$_2$(CH$_2$)$_2$NH$_3$	$(2.4 \pm 0.6) \times 10^{13}$	$(2.2 \pm 0.6) \times 10^{10}$	4,9-DIA(NH$_3$)$_2$	$(1.3 \pm 0.3) \times 10^{11}$	$(8.3 \pm 2.3) \times 10^{11}$
1-ADNH$_2$(CH$_2$)$_3$NH$_3$	$(1.5 \pm 0.4) \times 10^{13}$	—	4,9-DIA(NMe$_3$)$_2$	$(1.9 \pm 0.4) \times 10^{15}$	$(2.0 \pm 0.6) \times 10^{12}$
1-ADNMe$_2$(CH$_2$)$_3$NH$_3$	$(6.8 \pm 1.6) \times 10^{12}$	$(1.7 \pm 0.4) \times 10^{12}$	4,9-DIA(NMe$_2$(CH$_2$)$_4$OH)$_2$	$(1.9 \pm 0.4) \times 10^{15}$	$(1.3 \pm 0.3) \times 10^{13}$
1-ADNHMeCH$_2$CF$_3$	$(1.1 \pm 0.3) \times 10^{11}$	$(1.3 \pm 0.3) \times 10^9$	4-DIANMe$_3$	$(8.0 \pm 1.9) \times 10^{11}$	$(2.7 \pm 0.7) \times 10^{12}$
1-ADNH$_2$CH$_2$CF$_3$	$(5.9 \pm 1.4) \times 10^{11}$	$(1.0 \pm 0.3) \times 10^9$	1,6-DIA(NH$_3$)$_2$	2030	$(3.3 \pm 0.8) \times 10^{13}$
1-ADPy	$(2.0 \pm 0.4) \times 10^{12}$	$(2.0 \pm 0.5) \times 10^9$	1,6-DIA(NHMe$_2$)$_2$	686	$(5.7 \pm 1.5) \times 10^{14}$
Me$_2$-1-ADNH$_3$	$(2.5 \pm 0.4) \times 10^4$	$(4.3 \pm 1.1) \times 10^{11}$	1,6-DIA(NHMe(CH$_2$)$_4$OH)$_2$	194	$(9.2 \pm 2.4) \times 10^{14}$
1,3-AD(NHMe$_2$)$_2$	$(1.2 \pm 0.2) \times 10^6$	$(1.3 \pm 0.3) \times 10^{11}$	1-DIANHMe$_2$	643	$(7.8 \pm 0.8) \times 10^{13}$
1,3-AD(NMe$_3$)$_2$	$(6.4 \pm 1.0) \times 10^4$	$(1.1 \pm 0.3) \times 10^{11}$		—	—

Source: Adapted from Ref. [39].

262 | *10 Supramolecular Architectures*

(a) **FeCp₂OH**

(b) **1-AdOH**

(c) **4-DAOH**

(d) **4,9-DA(OH)₂**

(e) **3,9-TA(OH)₂**

cucurbit[7]uril — cucurbit[8]uril

No inclusion complex formed with CB7

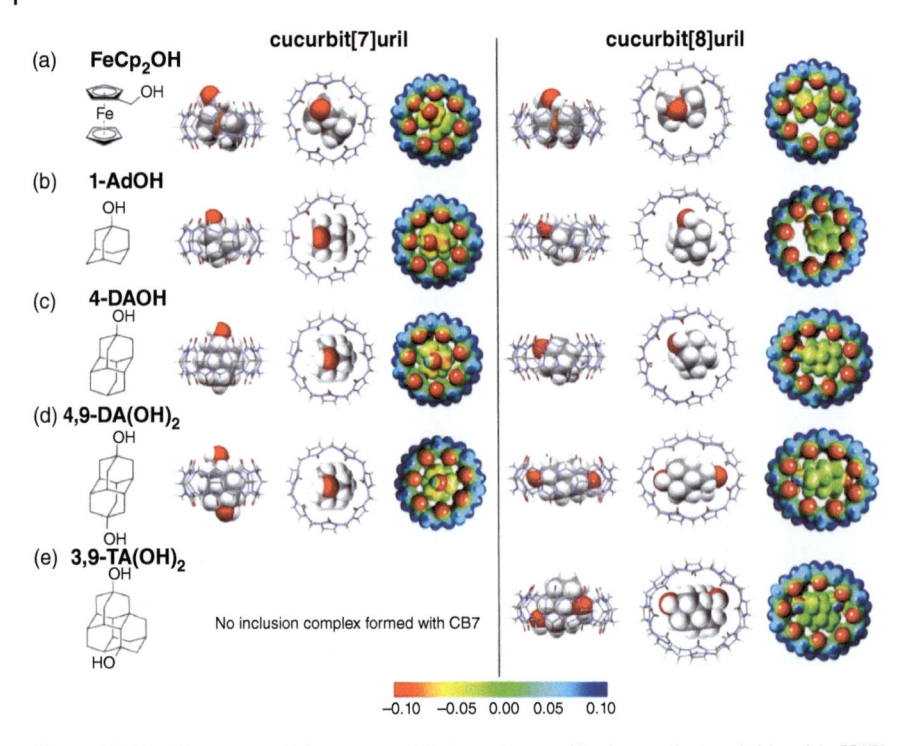

−0.10 −0.05 0.00 0.05 0.10

Figure 10.11 Structures of the most stable complexes of hydroxy-diamondoids with CB[7] (left) and CB[8] (right) computed at r²SCAN-3c/DCOSMO-RS and their corresponding electrostatic potential maps. Source: Reproduced from Ref. [42], Wiley-VCH, 2022.

Hydroxy-diamondoids (Figure 10.11) provide the best possible compromise between rigidity and symmetry, and by being reasonably soluble in water, are ideal model substrates for strong complexation with CB[n]. The only exception is **915OH2TRIA**, which was not soluble enough to allow calorimetric measurements. Good correlations between computed and experimental binding free energies were found with the r²SCAN-3c computational approach [43]. This model allowed to disclose driving forces for host–guest complexation where, however, the computed LD energies are rather constant for most of the complexes. Much larger guest/host-dependent variations were found for the electronic energy contributions. This is difficult to rationalize based on electrostatic potential plots because of different complexation topologies, especially for the complexation with CB[8] (Figure 10.11). Thus, neither LD nor electronic energies or entropies are among the selectivity-controlling factors for the complexation of hydroxy-diamondoids with CB[n]s. This only left peculiar host-related solvation effects, i.e., the solvation free energy differences between the host–guest complexes and between the unbound host and guest, responsible for complexation differences.

Another recent study on CB[n]•triamantane inclusion complexes focused on cationic triamantane derivatives (quaternary ammonium salts) and their binding affinities [44]. In principle, the triamantane skeleton is too voluminous to be comfortably encapsulated inside CB[7] and can be much more easily placed

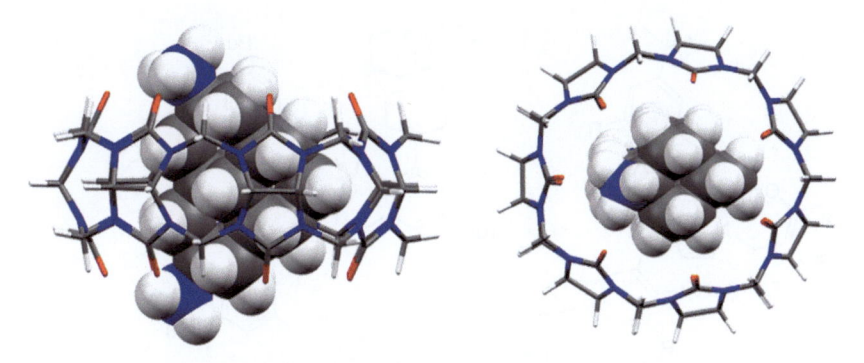

Figure 10.12 Side and top views of CB[7]• **915NH3CLTRIA**, an inclusion complex that to date incorporates the largest number of heavy atoms into the CB[7] cavity.

inside CB[8]. Thus, dihydroxy-triamantanes could not be successfully accommodated into the smaller CB[7] cavity [42]. However, the presence of cationic groups on **915NH3CLTRIA** provides sufficient ion–dipole interactions to make the inclusion of the triamantane framework inside CB[7] possible. The resulting CB[7]•**915NH3CLTRIA** complex, with its 18-carbon triamantane skeleton fully immersed inside the cavity, is the representative of the largest number of heavy (non-hydrogen) atoms incorporated into the CB[7] host cavity to date (Figure 10.12). Note that even larger **915NME3ITRIA** forms an exclusion complex with CB[7], showcasing the limit of how many heavy atoms can be incorporated. Despite this remarkable observation, cationic triamantanes still showed enhanced binding affinity toward CB[8] since the smaller CB[7] cavity could not easily accommodate such a large guest without incurring substantial energetic penalties due to over-packing [44]. As a result, the binding constants for **915NME3ITRIA** vary substantially (6.73×10^5 vs. 1.1×10^{14} M^{-1} with CB[7] and CB[8], respectively).

Up to now, studies of diamondoid derivatives as ligands for supramolecular hosts have been mostly limited to CD and CB[n] containers, and only a few literature examples exist for other types of host molecules binding diamondoids other than **AD**. For example, **DIA** was used as a guest for a few synthetic receptors [45], but mainly as a proof-of-concept molecule in order to determine the effect of a bulky ligand on the binding abilities of the receptor under study. The situation is similar for examples dealing with the incorporation of the diamondoid scaffold into the framework of the potential host molecule. As was the case for numerous instances of adamantane complexes with CDs, an exhaustive review of supramolecular receptors containing the adamantane moiety is out of the scope of this work. Nevertheless, just to briefly highlight a few recent examples, the adamantane scaffold can be incorporated into various cation and anion receptors and sensors, like adamantane derivatives of crown ethers and cryptands [46], thioethers [47], pyrromethanes [48], ureas [49], guanidines [50], and many more. One of the rare examples dealing with higher diamondoids is the diamantane-functionalized chelating triazacyclononane ligand capable of stabilizing uranium(III) ions [51]. Studied ligands possessing only

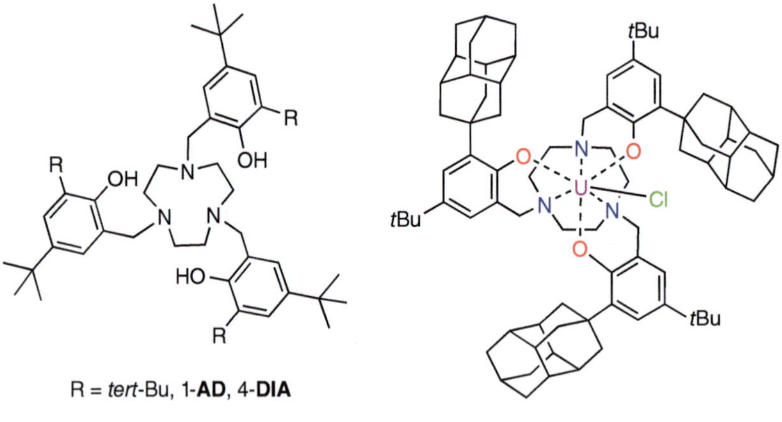

R = *tert*-Bu, 1-**AD**, 4-**DIA**

Figure 10.13 Chelating ligands for uranium coordination chemistry and structure of the chloro-complex with the **DIA** ligand after oxidation of the uranium center.

tert-butyl substituents (Figure 10.13) had a much shallower central cavity, and the introduction of diamantane substituents markedly improved the needed isolation of the reactive U(III) center from the environment [52]. The next step in the design was the introduction of a diamantane moiety into the triazacyclononane ligand [51] framework to increase the steric bulk even more, and the obtained ligand cavity was indeed even deeper. These second-generation complexes also had the characteristic deep hydrophobic pocket owing to the presence of diamantane cages. The broad potential for such structures lies in the field of complex-mediated alkane activations and functionalizations.

Peanut-shaped nanostructures comprising guest molecules surrounded by a polyaromatic shell were prepared and characterized in solution [53]. The double capsule **CAPS** (Figure 10.14) was constructed by tetramerization of *meta*-bis(10-bromo-9-anthryl)benzene (**ANTRBNZ**) in the presence of a palladium (II) salt. This host is able to encapsulate various guests, such as fullerene C_{60}, phenanthrene (**PHAN**), as well as **DIA**. Mixed (**DIA/2PHAN@CAPS**) and purely diamondoid complexes **2DIA@CAPS** form quantitatively upon slight heating of the guests with **CAPS** in H_2O/CH_3CN. The ability to host such diverge guests is determined by the high flexibility of **CAPS,** where cooperative changes in the volume of the linked cavities upon guest encapsulation are observed.

Diamondoid cages may serve as hosts when assembled in water [54]. For instance, a micellar adamantane-based capsule $(\mathbf{ADA})_n$ quantitatively formed in water from **ADA** (Figure 10.15(a)). Such capsules can take up a wide variety of spherical molecules in water (Figure 10.15(b)). While the capsules formed from **ADA** display superior binding ability if compared to aromatic (**AA**) or aliphatic (**CHA**) host-forming building blocks (Figure 10.15(c)), remarkable selectivity of $(\mathbf{ADA})_n$ toward binding of diamondoids was observed. Host–guest complexes $(\mathbf{ADA})_n \cdot m\,\mathbf{AD}$ and $(\mathbf{ADA})_n \cdot m\,\mathbf{DIA}$ were obtained as colorless aqueous solutions, and the concentrations of the incorporated **AD** and **DIA** were estimated by ^{1}H-NMR.

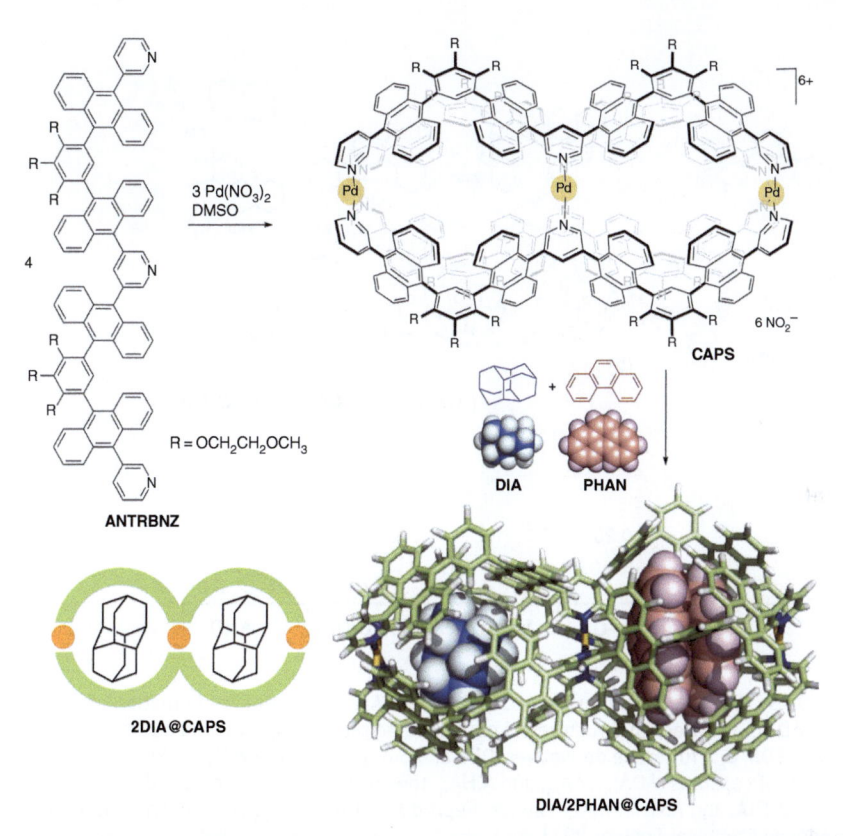

Figure 10.14 Schematic representation of the formation of double capsule **CAPS** and incorporation of one **DIA** and two phenanthrene (**PHAN**) molecules to form peanut-shaped nanostructure **DIA/2PHAN@CAPS** (bottom, right) and schematic representation of a complex with two **DIA** molecules **2DIA@CAPS** (bottom, left). Source: Reproduced from Ref. [53] with permission from Springer Nature, 2017.

Another example of diamondoid architectures serving as host molecules is diadamantyl ether **ADOAD** [55]. This molecule serves as a template model for docking and forms complexes with benzene and fluorinated benzenes, whose structures were studied by rotational spectroscopy. The experimentally observed complexes form largely due to LD interactions with the hydrogen-terminated surface of **ADOAD** (Figure 10.16). While the B3LYP-D3(BJ)/def2-TZVP computed interaction energies in the complex with benzene (**ADOADC6H6**) and perfluorobenzene (**ADOADC6F6**) are almost identical, the structure of the complexes varies due to different contributions from C – H•••π, C – H•••F, and C – H•••H – C interactions.

Supramolecular chemistry studies are not limited only to host–guest dynamics, as supramolecular architectures can also emerge via different modes and mechanisms of self-assembly. Very unusual supramolecular binding [56] was observed when 1-hydroxyadamantane (**1OHAD**) was reduced with alkali metals. With an excess of metal (M = Li, Na, or K) **1OHAD** forms stable clusters of formula **M(OAD)(1OHAD)2**, i.e., where two unreduced **1OHAD** clusters with

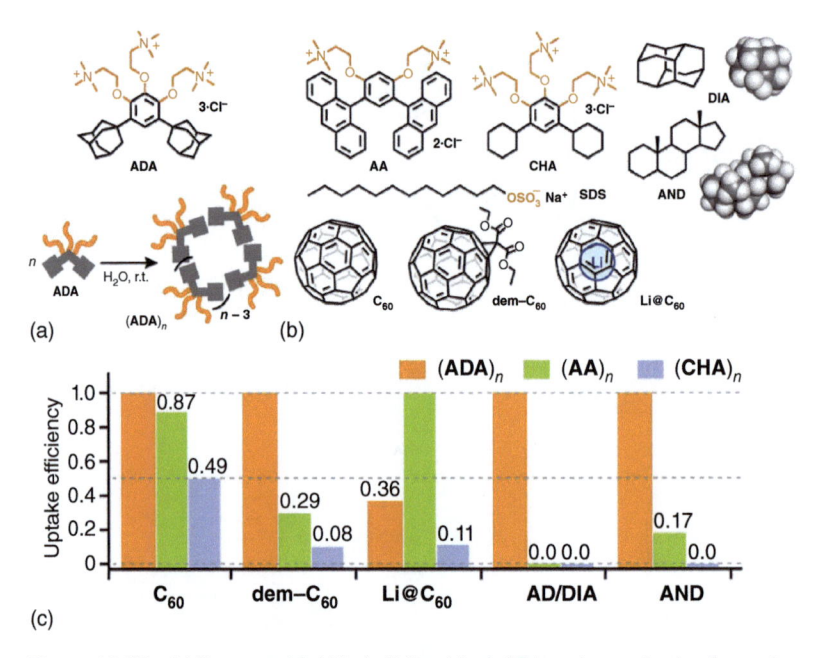

Figure 10.15 (a) Bent amphiphilic building block **ADA** and quantitative formation of capsule **(ADA)**$_n$ in water; (b) Structures of host building blocks **AA**, **CHA**, and guest molecules **SDS**, **C**$_{60}$, fullerene derivatives, **DIA**, and 5α-androstane (**AND**); (c) Relative uptake efficiencies of capsules **(ADA)**$_n$, **(AA)**$_n$, and **(CHA)**$_n$ toward guest molecules **C**$_{60}$, **dem−C**$_{60}$, **Li@C**$_{60}$, **AD**, **DIA**, and **AND** in water. Source: Reproduced from Ref. [54] with permission from the American Chemical Society, 2021.

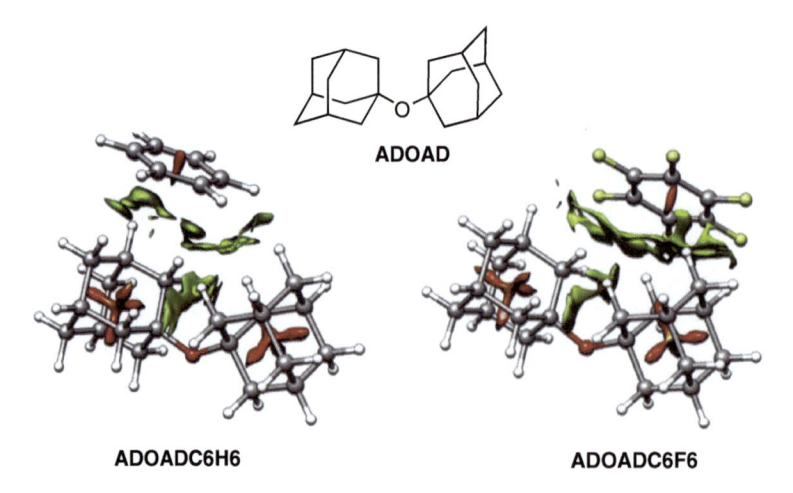

Figure 10.16 Structures of the complexes of diadamantyl ether (**ADOAD**) with benzene (**ADOADC6H6**) and perfluorobenzene (**ADOADC6F6**) from rotational spectroscopy data. Green surfaces represent attractive interactions and red indicate repulsive non-covalent interactions. Source: Reproduced from Ref. [55], Wiley-VCH, 2021.

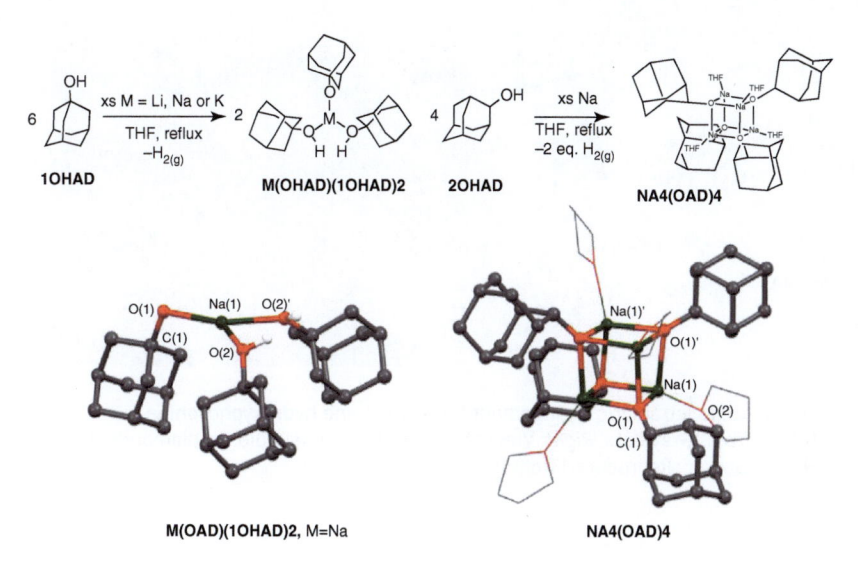

Figure 10.17 Isomeric hydroxy adamantanes **1OHAD** and **2OHAD** react with alkali metals differently. While **1OHAD** even with an excess of metal reacts incompletely and gives the LD-stabilized cluster **M(OAD)(1OHAD)2**, more sterically congested **2OHAD** forms the expected cubane-like alkoxide cluster **NA4(OAD)4** stabilized by complexation with THF (shown as a wireframe for clarity). Source: Reproduced from Ref. [56], Wiley-VCH, 2022.

an alkoxide AdOM fragment (Figure 10.17, left). This is the first example of an alkali metal reduction of an alcohol that is prevented from completion due to the complexation with the alcohol substrate. The X-ray crystal structure geometry displays that the extra stability of **M(OAD)(1OHAD)2** may arise from several attractive CH•••HC LD interactions between the cages. In contrast, the reduction of 2-hydroxyadamantane (**2OHAD**) gives the expected tetrameric cubane-type alkoxide **NA4(OAD)4** (Figure 10.17, right), where the cages are positioned distantly and are separated by the solvent (THF).

Another such example is the generation of linear nanopolymers consisting of diamondoid cages inside carbon nanotubes [57]. These one-dimensional diamondoid nanowires form upon annealing the suitable substrate in vacuo at high temperatures. Good precursor molecules for diamantane nanopolymer fusion in the confined environment of carbon nanotubes were found to be **49DICOOHDIA** [57b] and **49DIBRDIA** [57c]. The precursors were spontaneously pulled inside the nanotubes due to capillary forces and the starting molecules were simultaneously stabilized, compressed, and suitably oriented for 1D-polymerization (for more details, see Chapter 6). An example of nanowire design by supramolecular self-assembly demonstrates the importance of intermolecular LD interactions between the substrates that are sufficient to drive the assembly process [58]. LD thereby helps pre-organize diamondoid substrates for their subsequent reactivity [59].

The primary hydroxyphosphine **4OH9PH2DIA** (Figure 10.18) precipitated from the vapor phase with formation of a crystalline phase where cavities (channels) were found based on X-ray crystallography [60]. This template was used for low-temperature chemical vapor deposition of Pd to give nanocomposites with

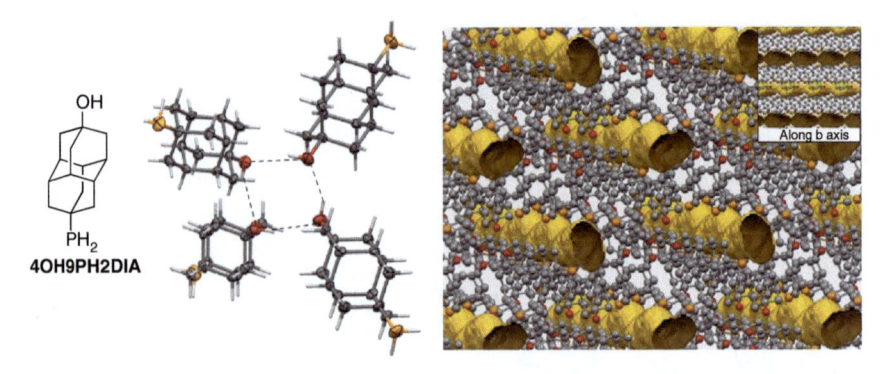

Figure 10.18 *Left*: Tetrameric arrangement for diamantane hydroxyphosphine **4OH9PH2DIA** from X-ray data. *Right*: View of accessible channel voids in diamondoid **4OH9PH2DIA**. Source: Reproduced from [60], Wiley-VCH, 2019.

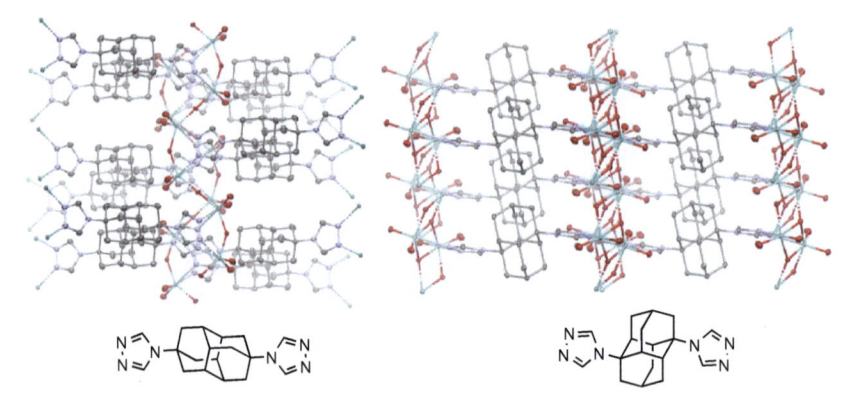

Figure 10.19 Bitopic 1,2,4-triazole diamondoid ligands used in the construction of 3D metal oxide–organic frameworks (MOOFs). Molybdenum(VI) oxide complex with 4,9-bis(1,2,4-triazol-4-yl)diamantane (left) and with 1,6-bis(1,2,4-triazol-4-yl)diamantane (right).

high structural uniformity and homogeneity. Due to the presence of channels, both **4OH9PH2DIA** and **Pd@4OH9PH2DIA** are useful for the construction of sp^3-carbon-based sensors for detection of NO_2 (down to 50 ppb) and NH_3 (25–100 ppm) [60].

Another example of supramolecular assemblies with diamondoid subunits are metal oxide–organic frameworks (MOOFs) constructed from molybdenum(VI) oxide and diamondoid 1,2,4-triazole ligands (Figure 10.19) [61]. Of special interest in the context of this chapter are bis-medial and bis-apical diamantane derivatives. The found 3D-structural motifs reflect the dual nature of the organic ligands as connectors for two adjacent coordination octahedra and as linkers between one-dimensional subtopologies.

Using a similar design rationale, metal–organic frameworks (MOFs) were recently constructed using metal salts and diamondoid thiols [62]. The resulting hybrid metal-diamondoid chalcogenide nanowires display solid inorganic cores

10 Supramolecular Architectures | **269**

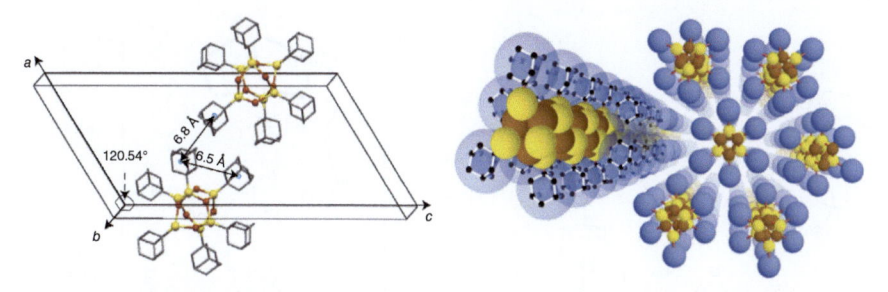

Figure 10.20 Unit cell of MOF crystals grown from the copper salt and **1SHAD** (left) and its structure viewed along the chain elongation direction (right). Source: Reproduced from Ref. [62] with permission from Springer Nature, 2017.

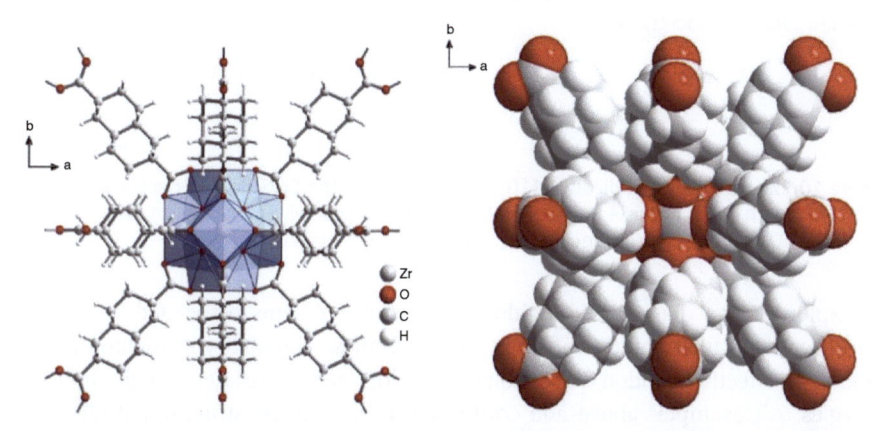

Figure 10.21 Different views of diamantane-4,9-dicarboxylate (L) zirconium Zr-MOF [$Zr_6O_4(OH)_4(L)_6$], which displays higher selectivity toward adsorption of CH_4 vs. CO_2 and H_2 than the respective terephthalic anolog UiO-66. Source: Reproduced from Ref. [63] with permission from the Royal Society of Chemistry, 2022.

with a three-atom cross-section and are at the same time uniformly insulated with the diamondoid substructures, enabling them to retain band-like electronic properties (Figure 10.20).

4,9-Diamantane dicarboxylate (**49COOH2DIA**) was tested [63] as a building block for the construction of zirconium MOFs with the general formula [$Zr_6O_4(OH)_4(L)_6$] (Figure 10.21). The diamantane-containing Zr-MOF displays preference toward the adsorption of CH_4 vs. CO_2 and H_2 compared to archetypal terephthalate-based UiO-66 [64].

Diamondoid-decorated Sn/S clusters were prepared [65] through condensation of diamondoid carboxylates with a keto-functionalized double-decker-like complex [$(RSn)_4S_6$] (**1**, Figure 10.22). This resulted in hybrid organic chalcogenido-metalates that are of great interest due to their (photo)catalytic, photoluminescent, and thermoelectric properties, as well as their outstanding capability of Li-ion storage.

The findings presented in this chapter demonstrate that the field of diamondoid supramolecular assemblies is on the rise, and there are still many aspects left to

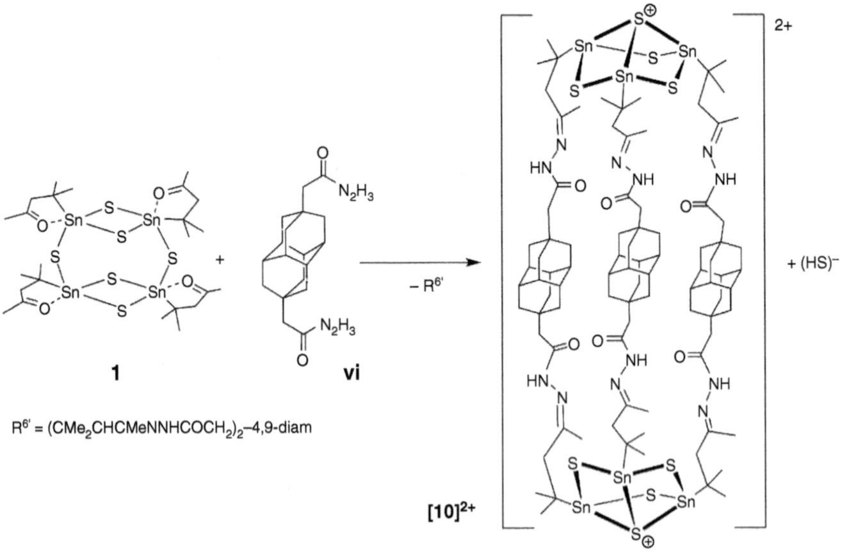

Figure 10.22 Formation of dication [10]$^{2+}$ by reaction of the Sn/S double-decker-like complex **1** with diamantane-4,9-dicarboxylic acid dihydrazide **vi**. Source: Reproduced from Ref. [65] with permission from the American Chemical Society, 2014.

be explored. The diamondoid rigidity eliminates problems associated with the conformational uncertainties of the guest molecules. The ability of diamondoids to serve as effective London dispersion energy donors often results in high binding energies. All examples above add confidence to host–guest structural studies of container molecules with diamondoids as guests. Although up to now the main focus of the community has been on exploring and applying host–guest interactions, recently the interest has also shifted to other types of supramolecular assemblies, e.g., flexible and rigid 3D frameworks, promising exciting discoveries yet to come in this ever-growing area of diamondoid research.

References

1 (a) Zhang, Y.M., Liu, Y.H., and Liu, Y. (2020). Cyclodextrin-based multistimuli-responsive supramolecular assemblies and their biological functions. *Adv. Mater.* 32: 1806158. (b) Liu, Z.J., Ye, L., Xi, J.N. et al. (2021). Cyclodextrin polymers: structure, synthesis, and use as drug carriers. *Prog. Polym. Sci.* 118: 101408. (c) Narayanan, G., Shen, J.L., Matai, I. et al. (2022). Cyclodextrin-based nanostructures. *Prog. Mater. Sci* 124: 100869.

2 (a) Chernikova, E.Y. and Berdnikova, D.V. (2020). Cucurbiturils in nucleic acids research. *Chem. Commun.* 56: 15360–15376. (b) Cicolani, R.S., Souza, L.R.R., Dias, G.B.D. et al. (2021). Cucurbiturils for environmental and analytical chemistry. *J. Incl. Phenom. Macrocycl.* 99: 1–12. (c) Das, D., Assaf, K.I., and Nau, W.M. (2019). Applications of cucurbiturils in medicinal chemistry and chemical

biology. *Front. Chem.* 7: 619. (d) Funk, S. and Schatz, J. (2020). Cucurbiturils in supramolecular catalysis. *J. Incl. Phenom. Macrocycl.* 96: 1–27. (e) Liu, Y.H., Zhang, Y.M., Yu, H.J., and Liu, Y. (2021). Cucurbituril-based biomacromolecular assemblies. *Angew. Chem. Int. Ed.* 60: 3870–3880. (f) Tominaga, M., Hyodo, T., Hikami, Y., and Yamaguchi, K. (2021). Solvent-dependent alignments and halogen-related interactions in inclusion crystals of adamantane-based macrocycle with pyridazine moieties. *CrstEngComm* 23: 436–442.

3 (a) Atwood, J.L., Gokel, G.W., and Barbour, L. (2017). *Comprehensive Supramolecular Chemistry II*, 2nd ed. Elsevier Ltd p. 4568. (b) Beer, P., Borendt, T.R., and Lim, J.Y.C. (2022). *Supramolecular Chemistry: Fundamentals and Applications*. Oxford University Press p. 192.

4 (a) Wilchek, M. and Bayer, E.A. (1990). *Avidin-biotin technology*, vol. 184. Academic Press p. 746. (b) Laitinen, O.H., Nordlund, H.R., Hytonen, V.P., and Kulomaa, M.S. (2007). Brave new (strept)avidins in biotechnology. *Trends Biotechnol.* 25: 269–277. (c) Ostojic, G.N. and Hersam, M.C. (2012). Biomolecule-directed assembly of self-supported, nanoporous, conductive, and luminescent single-walled carbon nanotube scaffolds. *Small* 8: 1840–1845.

5 Cao, L.P., Šekutor, M., Zavalij, P.Y. et al. (2014). Cucurbit[7]uril·guest pair with an attomolar dissociation constant. *Angew. Chem. Int. Ed.* 53: 988–993.

6 Huang, L., Zhang, S.C., Wang, H.T. et al. (2011). A novel method for isolation of diamondoids from crude oils for compound-specific isotope analysis. *Org. Geochem.* 42: 566–571.

7 Hu, Q.D., Tang, G.P., and Chu, P.K. (2014). Cyclodextrin-based host-guest supramolecular nanoparticles for delivery: from design to applications. *Acc. Chem. Res.* 47: 2017–2025.

8 Saenger, W.R., Jacob, J., Gessler, K. et al. (1998). Structures of the common cyclodextrins and their larger analogues – beyond the doughnut. *Chem. Rev.* 98: 1787–1802.

9 Eftink, M.R., Andy, M.L., Bystrom, K. et al. (1989). Cyclodextrin inclusion complexes: studies of the variation in the size of alicyclic guests. *J. Am. Chem. Soc.* 111: 6765–6772.

10 Harries, D., Rau, D.C., and Parsegian, V.A. (2005). Solutes probe hydration in specific association of cyclodextrin and adamantane. *J. Am. Chem. Soc.* 127: 2184–2190.

11 (a) Biedermann, F., Nau, W.M., and Schneider, H.J. (2014). The hydrophobic effect revisited-studies with supramolecular complexes imply high-energy water as a noncovalent driving force. *Angew. Chem. Int. Ed.* 53: 11158–11171. (b) Biela, A., Nasief, N.N., Betz, M. et al. (2013). Dissecting the hydrophobic effect on the molecular level: the role of water, enthalpy, and entropy in ligand binding to thermolysin. *Angew. Chem. Int. Ed.* 52: 1822–1828.

12 (a) Haberhauer, G., Woitschetzki, S., and Bandmann, H. (2014). Strongly underestimated dispersion energy in cryptophanes and their complexes. *Nat. Commun.* 5: 3542. (b) Sure, R. and Grimme, S. (2015). Comprehensive benchmark of association (free) energies of realistic host–guest complexes. *J. Chem. Theory Comput.* 11: 3785–3801. (c) Yang, L.X., Adam, C., Nichol, G.S., and Cockroft, S.L.

(2013). How much do van der Waals dispersion forces contribute to molecular recognition in solution? *Nat. Chem.* 5: 1006–1010.

13 (a) Cromwell, W.C., Bystrom, K., and Eftink, M.R. (1985). Cyclodextrin-adamantanecarboxylate inclusion complexes: studies of the variation in cavity size. *J. Phys. Chem.* 89: 326–332. (b) Perry, C.S., Charman, S.A., Prankerd, R.J. et al. (2006). The binding interaction of synthetic ozonide antimalarials with natural and modified beta-cyclodextrins. *J. Pharm. Sci.* 95: 146–158. (c) Harrison, J.C. and Eftink, M.R. (1982). Cyclodextrin–adamantanecarboxylate inclusion complexes: a model system for the hydrophobic effect. *Biopolymers* 21: 1153–1166.

14 (a) Gelb, R.I. and Schwartz, L.M. (1989). Complexation of adamantane-ammonium substrates by beta-cyclodextrin and its *O*-methylated derivatives. *J. Incl. Phenom. Macrocycl. Chem.* 7: 537–543. (b) Palepu, R. and Reinsborough, V.C. (1990). β-Cyclodextrin inclusion of adamantane derivatives in solution. *Aust. J. Chem.* 43: 2119–2123.

15 Czugler, M., Eckle, E., and Stezowski, J.J. (1981). Crystal and molecular structure of a 2,6-tetradeca-*O*-methyl-β-cyclodextrin–adamantanol 1:1 inclusion complex. *J. Chem. Soc. Chem. Commun.* 1291–1292.

16 Jaime, C., Redondo, J., Sanchezferrando, F., and Virgili, A. (1990). Solution geometry of β-cyclodextrin-1-bromoadamantane host-guest complex as determined by ^{1}H[^{1}H] intermolecular NOE and MM2 calculations. *J. Org. Chem.* 55: 4772–4776.

17 Hamilton, J.A. (1985). β-Cyclodextrin 1-hydroxymethyl-adamantane undecahydrate. *Carbohydr. Res.* 142: 21–37.

18 Tran, D.N., Colesnic, D., de Beaumais, S.A. et al. (2014). Cyclodextrin-adamantane conjugates, self-inclusion and aggregation versus supramolecular polymer formation. *Org. Chem. Front.* 1: 703–706.

19 (a) Miyamae, K., Nakahata, M., Takashima, Y., and Harada, A. (2015). Self-healing, expansion-contraction, and shape-memory properties of a preorganized supramolecular hydrogel through host-guest interactions. *Angew. Chem. Int. Ed.* 54: 8984–8987. (b) Nakahata, M., Takashima, Y., and Harada, A. (2016). Highly flexible, tough, and self-healing supramolecular polymeric materials using host-guest interaction. *Macromol. Rapid Commun.* 37: 86–92. (c) Ohga, K., Takashima, Y., Takahashi, H. et al. (2005). Preparation of supramolecular polymers from a cyclodextrin dimer and ditopic guest molecules: control of structure by linker flexibility. *Macromolecules* 38: 5897–5904. (d) Davis, M.E., Zuckerman, J.E., Choi, C.H.J. et al. (2010). Evidence of RNAi in humans from systemically administered siRNA via targeted nanoparticles. *Nature* 464: 1067–1071. (e) Sandier, A., Brown, W., Mays, H., and Amiel, C. (2000). Interaction between an adamantane end-capped poly(ethylene oxide) and a beta-cyclodextrin polymer. *Langmuir* 16: 1634–1642. (f) Evenou, P., Rossignol, J., Pembouong, G. et al. (2018). Bridging beta-cyclodextrin prevents self-inclusion, promotes supra-molecular polymerization, and promotes cooperative interaction with nucleic acids. *Angew. Chem. Int. Ed.* 57: 7753–7758. (g) Kakuta, T., Takashima, Y., Nakahata, M. et al. (2013). Preorganized hydrogel: self-healing properties of

supramolecular hydrogels formed by polymerization of host-guest-monomers that contain cyclodextrins and hydrophobic guest groups. *Adv. Mater.* 25: 2849–2853.

20 Fathalla, M., Neuberger, A., Li, S.C. et al. (2010). Straightforward self-assembly of porphyrin nanowires in water: harnessing adamantane/β-cyclodextrin interactions. *J. Am. Chem. Soc.* 132: 9966–9967.

21 Fragoso, A., Caballero, J., Almirall, E. et al. (2002). Immobilization of adamantane-modified cytochrome c at electrode surfaces through supramolecular interactions. *Langmuir* 18: 5051–5054.

22 Xue, S.S., Tan, C.P., Chen, M.H. et al. (2017). Tumor-targeted supramolecular nanoparticles self-assembled from a ruthenium-beta-cyclodextrin complex and an adamantane-functionalized peptide. *Chem. Commun.* 53: 842–845.

23 Nazarov, V.B., Avakyan, V.G., Alfimov, M.V., and Vershinnikova, T.G. (2003). Long-lived room temperature phosphorescence of a naphthalene-β-cyclodextrin-adamantane complex in the presence of oxygen. *Russ. Chem. Bull.* 52: 916–922.

24 Nazarov, V.B., Avakyan, V.G., Bagrii, E.I. et al. (2005). Long-lived phosphorescence of arenes in complexes with cyclodextrins – 2. Room-temperature phosphorescence of ternary complexes of naphthalene and phenanthrene with beta-cyclodextrin and adamantane derivatives in the presence of oxygen. *Russ. Chem. Bull.* 54: 2752–2756.

25 Voskuhl, J., Waller, M., Bandaru, S. et al. (2012). Nanodiamonds in sugar rings: an experimental and theoretical investigation of cyclodextrin-nanodiamond inclusion complexes. *Org. Biomol. Chem.* 10: 4524–4530.

26 (a) Schibilla, F., Voskuhl, J., Fokina, N.A. et al. (2017). Host-guest complexes of cyclodextrins and nanodiamonds as a strong non-covalent binding motif for self-assembled nanomaterials. *Chem. Eur. J.* 23: 16059–16065. (b) Alešković, M., Roca, S., Jozepović, R. et al. (2022). Unravelling binding effects in cyclodextrin inclusion complexes with diamondoid ammonium salt guests. *New J. Chem.* 46: 13406–13414.

27 (a) Yashkin, S.N. and Ageeva, Y.A. (2013). Gas-chromatographic studies of the sorption thermodynamics of adamantanes on a carbon adsorbent modified with polyethylene glycol with beta-cyclodextrin additives. *Russ. J. Phys. Chem. A* 87: 1921–1928. (b) Bazilin, A.V., Yashkina, E.A., and Yashkin, S.N. (2016). Chromatographic study of complex formation of adamantane derivatives with beta-cyclodextrin. *Russ. Chem. Bull.* 65: 103–109.

28 Štimac, A., Šekutor, M., Mlinarić-Majerski, K. et al. (2017). Adamantane in drug delivery systems and surface recognition. *Molecules* 22: 297.

29 (a) Freeman, W.A., Mock, W.L., and Shih, N.Y. (1981). Cucurbituril. *J. Am. Chem. Soc.* 103: 7367–7368. (b) Barrow, S.J., Kasera, S., Rowland, M.J. et al. (2015). Cucurbituril-based molecular recognition. *Chem. Rev.* 115: 12320–12406.

30 Mock, W.L. and Shih, N.Y. (1986). Structure and selectivity in host-guest complexes of cucurbituril. *J. Org. Chem.* 51: 4440–4446.

31 Barrow, S.J., Assaf, K.I., Palma, A. et al. (2019). Preferential binding of unsaturated hydrocarbons in aryl-bisimidazolium cucurbit[8]uril complexes furbishes evidence for small-molecule π–π interactions. *Chem. Sci.* 10: 10240–10246.

32 (a) Florea, M. and Nau, W.M. (2011). Strong binding of hydrocarbons to cucurbituril probed by fluorescent dye displacement: a supramolecular gas-sensing ensemble. *Angew. Chem. Int. Ed.* 50: 9338–9342. (b) Assaf, K.I., Florea, M., Antony, J. et al. (2017). HYDROPHOBE challenge: a joint experimental and computational. Study on the host guest binding of hydrocarbons to cucurbiturils, allowing explicit evaluation of guest hydration free-energy contributions. *J. Phys. Chem. B* 121: 11144–11162.

33 Liu, S.M., Ruspic, C., Mukhopadhyay, P. et al. (2005). The cucurbit[n]uril family: prime components for self-sorting systems. *J. Am. Chem. Soc.* 127: 15959–15967.

34 Cao, L.P., Škalamera, Đ., Zavalij, P.Y. et al. (2015). Influence of hydrophobic residues on the binding of CB[7] toward diammonium ions of common ammonium⋯ammonium distance. *Org. Biomol. Chem.* 13: 6249–6254.

35 Huang, W.H., Liu, S.M., Zavalij, P.Y., and Isaacs, L. (2006). Nor-seco-cucurbit[10]uril exhibits homotropic allosterism. *J. Am. Chem. Soc.* 128: 14744–14745.

36 Moghaddam, S., Yang, C., Rekharsky, M. et al. (2011). New ultrahigh affinity host–guest complexes of cucurbit[7]uril with bicyclo[2.2.2]octane and adamantane guests: thermodynamic analysis and evaluation of M2 affinity calculations. *J. Am. Chem. Soc.* 133: 3570–3581.

37 Šekutor, M., Molčanov, K., Cao, L.P. et al. (2014). Design, synthesis, and X-ray structural analyses of diamantane diammonium salts: guests for cucurbit[n]uril (CB[n]) hosts. *Eur. J. Org. Chem.* 2014: 2533–2542.

38 Green, N.M. and Avidin. (1963). 3. The nature of the biotin-binding site. *Biochem. J.* 89: 599–609.

39 Sigwalt, D., Šekutor, M., Cao, L. et al. (2017). Unraveling the structure-affinity relationship between cucurbit[n]urils (n = 7, 8) and cationic diamondoids. *J. Am. Chem. Soc.* 139: 3249–3258.

40 Hostas, J., Sigwalt, D., Šekutor, M. et al. (2016). A nexus between theory and experiment: non-empirical quantum mechanical computational methodology applied to cucurbit[n]uril·guest binding interactions. *Chem. Eur. J.* 22: 17226–17238.

41 Lu, X.Y. and Isaacs, L. (2016). Uptake of hydrocarbons in aqueous solution by encapsulation in acyclic cucurbit[n]uril-type molecular containers. *Angew. Chem. Int. Ed.* 55: 8076–8080.

42 Grimm, L.M., Spicher, S., Tkachenko, B. et al. (2022). The role of packing, dispersion, electrostatics, and solvation in high-affinity complexes of cucurbit[n]urils with uncharged polar guests. *Chem. Eur. J.* 28: e202200529.

43 (a) Grimme, S., Hansen, A., Ehlert, S., and Mewes, J.M. (2021). r^{2}SCAN-3c: a "Swiss army knife" composite electronic-structure method. *J. Chem. Phys.* 154: 064103. (b) Gasevic, T., Stuckrath, J.B., Grimme, S., and Bursch, M. (2022). Optimization of the r^{2}SCAN-3c composite electronic-structure method for use with slater-type orbital basis sets. *J. Phys. Chem. A* 126: 3826–3838.

44 King, D., Šumanovac, T., Murkli, S. et al. (2023). Cucurbit[8]uril forms tight inclusion complexes with cationic triamantanes. *New J. Chem.* 47: 5338–5346.

45 (a) Ronson, T.K., League, A.B., Gagliardi, L. et al. (2014). Pyrene-edged $Fe^{II}_4L_6$ cages adaptively reconfigure during guest binding. *J. Am. Chem. Soc.* 136: 15615–15624. (b) Hof, F., Nuckolls, C., Craig, S.L. et al. (2000). Emergent conformational preferences of a self-assembling small molecule: structure and dynamics in a tetrameric capsule. *J. Am. Chem. Soc.* 122: 10991–10996. (c) Aoki, S., Shiro, M., and Kimura, E. (2002). A cuboctahedral supramolecular capsule by 4:4 self-assembly of tris(Zn^{II}-cyclen) and trianionic trithiocyanurate in aqueous solution at neutral pH (cyclen=1,4,7,10-tetraazacyclododecane). *Chem. Eur. J.* 8: 929–939. (d) Lledo, A. and Rebek, J. (2010). Self-folding cavitands: structural characterization of the induced-fit model. *Chem. Commun.* 46: 1637–1639.

46 (a) Marchand, A.P., Kumar, K.A., McKim, A.S. et al. (1997). Synthesis and alkali metal picrate extraction capabilities of novel cage-functionalized 17-crown-5 and 17-crown-6 ethers. *Tetrahedron* 53: 3467–3474. (b) Mlinarić-Majerski, K. and Kragol, G. (2001). Design, synthesis and cation-binding properties of novel adamantane- and 2-oxaadamantane-containing crown ethers. *Tetrahedron* 57: 449–457. (c) Mlinarić-Majerski, K. and Šumanovac Ramljak, T. (2002). Synthesis and alkali metal binding properties of novel *N*-adamantylaza-crown ethers. *Tetrahedron* 58: 4893–4898. (d) Supek, F., Šumanovac Ramljak, T., Marjanović, M. et al. (2011). Could LogP be a principal determinant of biological activity in 18-crown-6 ethers? Synthesis of biologically active adamantane-substituted diaza-crowns. *Eur. J. Med. Chem.* 46: 3444–3454. (e) Šumanovac Ramljak, T., Mlinarić-Majerski, K., and Bertoša, B. (2012). Alkali metal ion complexation of adamantane functionalized diaza-bibracchial lariat ethers. *Croat. Chem. Acta* 85: 559–568. (f) Šumanovac Ramljak, T., Despotović, I., Bertoša, B., and Mlinarić-Majerski, K. (2013). Synthesis and alkali metal complexation studies of novel cage-functionalized cryptands. *Tetrahedron* 69: 10610–10620.

47 (a) Vujasinović, I., Veljković, J., Molčanov, K. et al. (2008). Thiamacrocyclic lactones: new Ag(I)-ionophores. *J. Org. Chem.* 73: 9221–9227. (b) Vujasinović, I., Mlinarić-Majerski, K., Bertoša, B., and Tomić, S. (2009). Influence of the rigid spacer to macrocyclization of poly(thialactones): synthesis and computational analysis. *J. Phys. Org. Chem.* 22: 431–437.

48 (a) Renić, M., Basarić, N., and Mlinarić-Majerski, K. (2007). Adamantane-dipyrromethanes: novel anion receptors. *Tetrahedron Lett.* 48: 7873–7877. (b) Alešković, M., Halasz, I., Basarić, N., and Mlinarić-Majerski, K. (2009). Synthesis, structural characterization, and anion binding ability of sterically congested adamantane-calix[4]pyrroles and adamantane-calixphyrins. *Tetrahedron* 65: 2051–2058. (c) Alešković, M., Basarić, N., Mlinarić-Majerski, K. et al. (2010). Anion recognition through hydrogen bonding by adamantane-dipyrromethane receptors. *Tetrahedron* 66: 1689–1698. (d) Alešković, M., Basarić, N., Halasz, I. et al. (2013). Aryl substituted adamantane-dipyrromethanes: chromogenic and fluorescent anion sensors. *Tetrahedron* 69: 1725–1734. (e) Alešković, M., Basarić, N., Došlić, N. et al. (2014). HSO_4^- sensing based on proton transfer in H-bonding complexes. *Supramol. Chem.* 26: 850–855.

49 (a) Blažek, V., Bregović, N., Mlinarić-Majerski, K., and Basarić, N. (2011). Phosphate selective alkylenebisurea receptors: structure-binding relationship. *Tetrahedron* 67: 3846–3857. (b) Blažek, V., Mlinarić-Majerski, K., Qin, W.W., and Basarić, N. (2012). Photophysical study of the aggregation of naphthyl-, anthryl-and pyrenyl-adamantanebisurea derivatives. *J. Photochem. Photobiol. A* 229: 1–10. (c) Blažek, V., Molčanov, K., Mlinarić-Majerski, K. et al. (2013). Adamantane bisurea derivatives: anion binding in the solution and in the solid state. *Tetrahedron* 69: 517–526. (d) Blažek Bregović, V., Halasz, I., Basarić, N., and Mlinarić-Majerski, K. (2015). Anthracene adamantylbisurea receptors: switching of anion binding by photocyclization. *Tetrahedron* 71: 9321–9327.

50 Šekutor, M. and Mlinarić-Majerski, K. (2014). Adamantyl aminoguanidines as receptors for oxo-anions. *Tetrahedron Lett.* 55: 6665–6670.

51 Lam, O.P., Heinemann, F.W., and Meyer, K. (2010). A new diamantane functionalized tris(aryloxide) ligand system for small molecule activation chemistry at reactive uranium complexes. *C. R. Chim.* 13: 803–811.

52 (a) Castro-Rodriguez, I., Nakai, H., Gantzel, P. et al. (2003). Evidence for alkane coordination to an electron-rich uranium center. *J. Am. Chem. Soc.* 125: 15734–15735. (b) Castro-Rodriguez, I. and Meyer, K. (2006). Small molecule activation at uranium coordination complexes: control of reactivity via molecular architecture. *Chem. Commun.* 1353–1368.

53 Yazaki, K., Akita, M., Prusty, S. et al. (2017). Polyaromatic molecular peanuts. *Nat. Commun.* 8: 15914.

54 Katagiri, Y., Tsuchida, Y., Matsuo, Y., and Yoshizawa, M. (2021). An adamantane capsule and its efficient uptake of spherical guests up to 3 nm in water. *J. Am. Chem. Soc.* 143: 21492–21496.

55 Quesada-Moreno, M.M., Pinacho, P., Perez, C. et al. (2021). Do docking sites persist upon fluorination? The diadamantyl ether-aromatics challenge for rotational spectroscopy and theory. *Chem. Eur. J.* 27: 6198–6203.

56 Mears, K.L., Stennett, C.R., Fettinger, J.C. et al. (2022). Inhibition of alkali metal reduction of 1-adamantanol by London dispersion effects. *Angew. Chem. Int. Ed.* 61: e202201318.

57 (a) Zhang, J.Y., Feng, Y.Q., Ishiwata, H. et al. (2012). Synthesis and transformation of linear adamantane assemblies inside carbon nanotubes. *ACS Nano* 6: 8674–8683. (b) Zhang, J., Zhu, Z., Feng, Y. et al. (2013). Evidence of diamond nanowires formed inside carbon nanotubes from diamantane dicarboxylic acid. *Angew. Chem. Int. Ed.* 52: 3717–3721. (c) Nakanishi, Y., Omachi, H., Fokina, N.A. et al. (2015). Template synthesis of linear-chain nanodiamonds inside carbon nanotubes from bridgehead-halogenated diamantane precursors. *Angew. Chem. Int. Ed.* 54: 10802–10806.

58 (a) Ebeling, D., Šekutor, M., Stiefermann, M. et al. (2017). London dispersion directs on-surface self-assembly of [121]tetramantane molecules. *ACS Nano* 11: 9459–9466. (b) Ebeling, D., Šekutor, M., Stiefermann, M. et al. (2018). Assigning the absolute configuration of single aliphatic molecules by visual inspection. *Nat. Commun.* 9: 2420.

59 Gao, H.Y., Šekutor, M., Liu, L.C. et al. (2019). Diamantane suspended single copper atoms. *J. Am. Chem. Soc.* 141: 315–322.
60 Moncea, O., Casanova-Chafer, J., Poinsot, D. et al. (2019). Diamondoid nanostructures as sp^3-carbon-based gas sensors. *Angew. Chem. Int. Ed.* 58: 9933–9938.
61 Lysenko, A.B., Senchyk, G.A., Lincke, J. et al. (2010). Metal oxide-organic frameworks (MOOFs), a new series of coordination hybrids constructed from molybdenum(VI) oxide and bitopic 1,2,4-triazole linkers. *J. Chem. Soc. Dalton Trans.* 39: 4223–4231.
62 Yan, H., Hohman, J.N., Li, F.H. et al. (2017). Hybrid metal-organic chalcogenide nanowires with electrically conductive inorganic core through diamondoid-directed assembly. *Nat. Mater.* 16: 349–357.
63 Gvilava, V., Vieten, M., Oestreich, R. et al. (2022). A diamantane-4,9-dicarboxylate based UiO-66 analogue: challenging larger hydrocarbon cage platforms. *CrstEngComm* 24.
64 Kandiah, M., Nilsen, M.H., Usseglio, S. et al. (2010). Synthesis and stability of tagged UiO-66 Zr-MOFs. *Chem. Mater.* 22: 6632–6640.
65 Barth, B.E.K., Tkachenko, B.A., Eussner, J.P. et al. (2014). Diamondoid hydrazones and hydrazides: sterically demanding ligands for Sn/S cluster design. *Organometallics* 33: 1678–1688.

11

Diamondoid Oligomers

Although nature offers a large variety of diamondoids (at least up to undecamantane, see Chapter 2) [1], the isolation of large diamondoids is cumbersome and expensive. Cyclohexamantane is the largest diamondoid that has been isolated and fully characterized [2], but higher diamondoids were only identified or isolated in trace amounts insufficient for full characterization. Stitching diamondoids together is currently the only realistic tactic to access large diamondoid molecules. This can be achieved through sp^3–sp^3 or sp^2–sp^2 C–C-couplings of readily available smaller diamondoids. The sp^3–sp^3 coupled structures perfectly reproduce segments of diamond lattice, as shown in Figure 11.1 for the simplest sp^3–sp^3 coupled diamondoid 1-adamantyl-1-adamantane (**1ADAD**), and sp^2–sp^2 coupling leads to 2-(2-adamantylidene)adamantane (**AD=AD**) that models nanodiamond particles with "defects" relating back to unsaturation (Figure 11.1) [4].

Such defects are usually present not only in partially graphitized nanodiamond particles but also on the surface of diamond [5]. The electronic properties of such materials are determined by low levels of sp^2 non-diamond-like "impurities" [6]. Due to the presence of a very sterically encumbered C—C bond, **AD=AD** displays a number of very unusual properties that will be discussed in detail below. Note that the dimension (ca. 0.8 nm) of dimers **1ADAD** and **AD=AD** is already larger (Figure 11.1) than that of [121]tetramantane (**121TET**). The growing interest in **1ADAD** derivatives is based on their potential as insulating materials [7], organic capacitors [8], high-resistance polymers [9], and dehydrogenase inhibitors [10]. The coupling of higher diamondoids [11] provides access to much larger particles (currently up to 2 nm for the triamantane dimer, see below).

11.1 Saturated Diamondoid Oligomers

The simplest representative, **1ADAD**, was first prepared in 1957 by Wurtz coupling 1-bromo adamantane (**ADHAL**, Hal = Br, Scheme 11.1) with sodium in ether [12]. The formation of **1ADAD** was documented in many organometallic transformations

---

The Chemistry of Diamondoids: Building Blocks for Ligands, Catalysts, Materials, and Pharmaceuticals,
First Edition. Andrey A. Fokin, Marina Šekutor, and Peter R. Schreiner.
© 2024 WILEY-VCH GmbH. Published 2024 by WILEY-VCH GmbH.

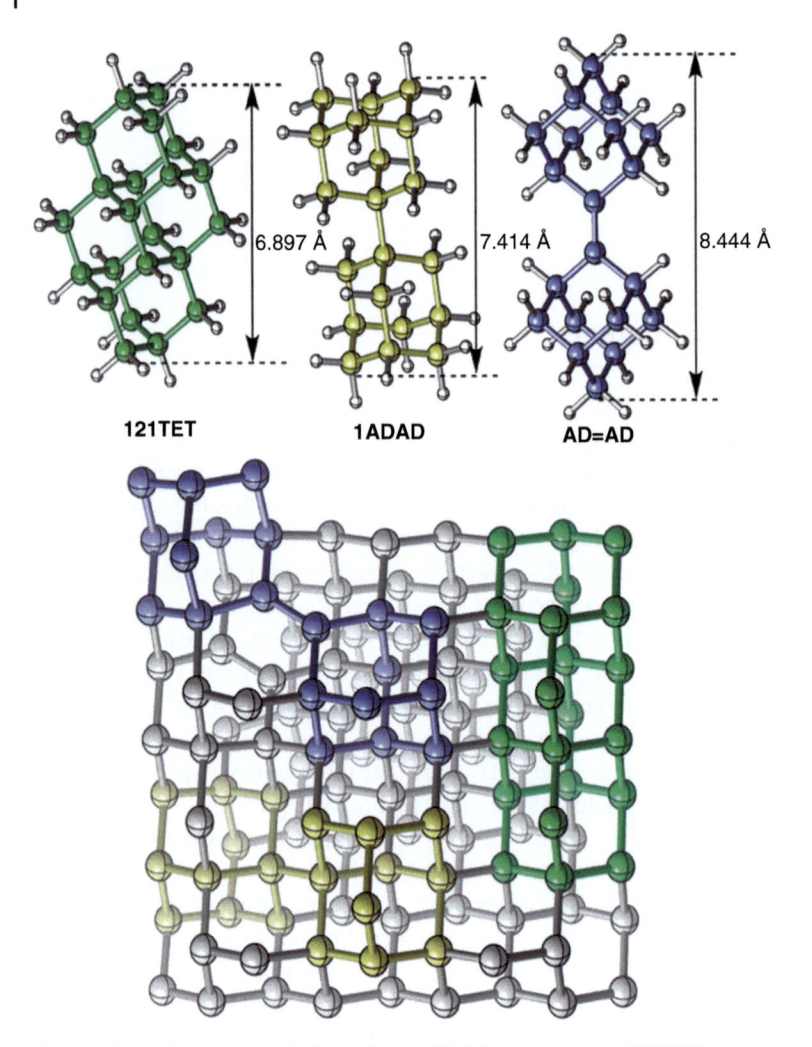

Figure 11.1 Structures and dimensions of [121]tetramantane (**121TET**), 1-adamantyl-1-adamantane (**1ADAD**), and 2-(2-adamantylidene) adamantane (**AD=AD**), and their representation as part of the diamond lattice. Source: Reproduced from Ref. [3] with permission from the American Chemical Society, 2015.

of halogenated adamantanes [13] in the presence of metallic magnesium [14], lithium [15], lanthanum [16], or with *tert*-butyl lithium (Scheme 11.1) [17].

Alternatively, **1ADAD** was obtained through recombination [18] of adamantyl radicals formed from the single-electron reduction of 1-iodo adamantane (**ADHAL**, Hal = I) [19], thermolysis [20], photolysis [21], or oxidation [22] of azo-1-adamantane (**ADN=NAD**). Minor amounts of **1ADAD** were detected in the reaction mixture of phenyl-1-adamantyl ketone (**PHCOAD**) photolysis (Scheme 11.1) [23].

Due to its high symmetry **1ADAD** packs well in the solid and has a very high melting point (296 °C) as well as extraordinary thermal stability [24]. Such stability

11.1 Saturated Diamondoid Oligomers

Scheme 11.1 The preparations of 1-adamantyl-1-adamantane (**1ADAD**) by various coupling methods.

is not surprising as the central C—C bond in **1ADAD** is only slightly elongated (1.578 Å) [25] as compared to the typical sp^3–sp^3 C—C bond length in alkanes (1.54 Å); the adamantyl moieties are nearly undistorted (other C—C-bonding distances in **1ADAD** are around 1.53 ± 0.01 Å). The low strain energy of **1ADAD** [26] is also in accord with the fact that the experimental activation enthalpy for the dissociation of **1ADAD** to form two adamantyl radicals ($\Delta H^{\ddagger}$ = 69 kcal mol^{-1}) is close to that for the dissociation of hexamethyl ethane (2,2,3,3-tetramethylbutane) into two *tert*-butyl radicals ($\Delta H^{\ddagger}$ = 70 kcal mol^{-1}) [27]. The situation changes dramatically for dimers constructed from larger diamondoids. For example, the Wurtz coupling of 1-bromodiamantane (**1BRDIA**) at the sterically hindered medial position gives the adduct **1DIADIA** that displays a long central C—C bond of 1.647 Å (Scheme 11.2) [28].

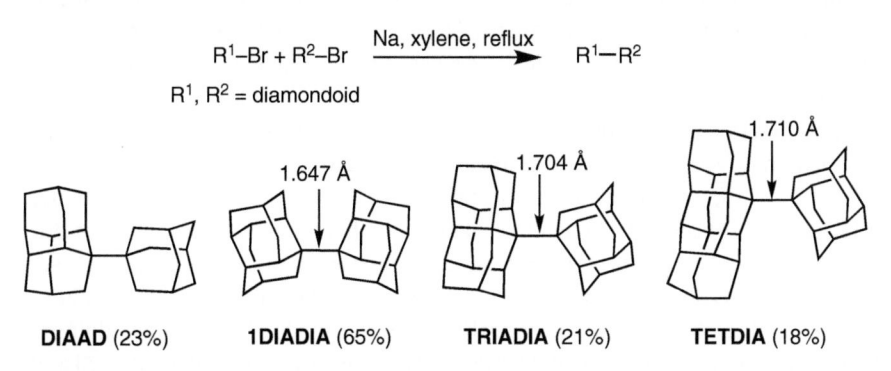

Scheme 11.2 Wurtz couplings of higher diamondoid bromo-derivatives give dimers with extremely long central C—C bonds. Source: From Refs. [11, 28].

Various lower and higher diamondoids were coupled, and their dimers were characterized [11]. They are more sterically congested, and, consequently, even longer C—C bonds are present in their structures (1.704 and 1.710 Å in **TRIADIA** and **TETDIA**, respectively). The latter currently has the longest value ever observed in a fully saturated hydrocarbon [11, 29]. Surprisingly, these dimers, despite having

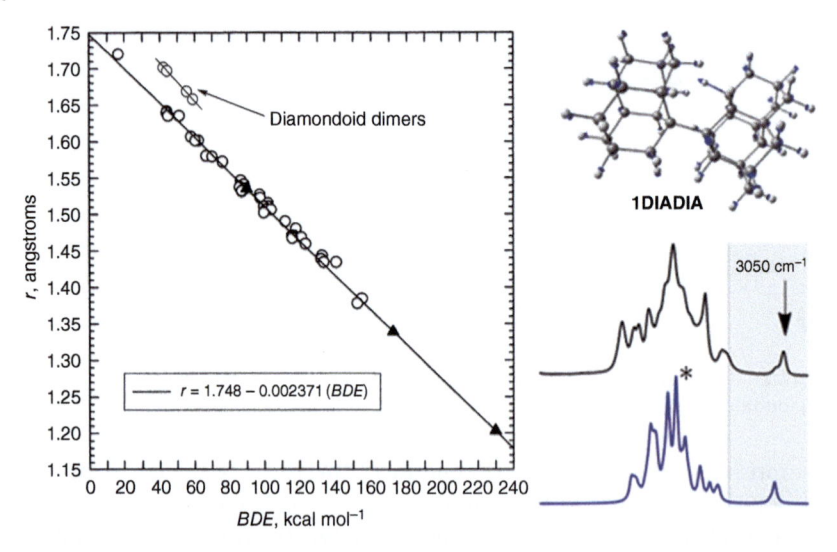

Figure 11.2 *Left*: Positions of diamondoid dimers in a bond lengths/bond energy linear correlation plot (source: Adopted from Ref. [33] with permission from the American Chemical Society, 2003); *Right*: CH-stretching region in the experimental (black) and computed (blue) vibrational spectra of **1DIADIA** with a characteristic band at 3050 cm^{-1} assigned to motions of the inward hydrogens (source: Adopted from Ref. [34] with permission from Elsevier, 2020).

very long bonds, are stable upon heating and melt without decomposition at temperatures above 200 °C. Experiments and computations [11, 28] demonstrate that such elongated bonds in diamondoid dimers are strengthened due to numerous stabilizing LD interactions between the diamondoid moieties [30]. Another source of stabilization that reduces the bond dissociation energies (BDEs) are the small relaxation energies of the ensuing radicals compared to less rigid acyclic hydrocarbons [31]. Additionally, electrostatic interactions between diamondoid fragments stabilize the dimers [32]. Because of such extra-stabilization, these large diamondoid dimers do not fit into conventional linear bond lengths/bond energy relationships [33] (Figure 11.2, left). For instance, the dissociation energy of the central C—C bond of **1DIADIA** (65 kcal mol^{-1}) is ca. 15 kcal mol^{-1} higher than correlation predicts for a bond length of 1.65 Å. The LD and electrostatic contributions increase with molecular size and this opens possibilities to prepare much larger diamondoid assemblies that are highly stable and suitable for robust material applications [11]. In addition, diamondoid dimers serve as models regarding the nature of covalent bonds [35] and as benchmark molecules in testing the performance of newly developed or improved computational methods [36].

Strong through-space interactions between diamondoid moieties are reflected in the vibrational spectra of dimers **1DIADIA**, **TRIADIA**, and **TETDIA**, in which specific C—H stretching bands of the inward hydrogens shift to 3050 cm^{-1} (Figure 11.2, right) [34]. Such a "confined" vibration is absent in essentially unstrained **1ADAD**, for which the highest experimental C—H stretching mode is observed at "normal" positions for saturated hydrocarbons (around 2950 cm^{-1}). Thus, the high-frequency

Figure 11.3 HOMO and HOMO−1 of diamantane (**DIA**) and the sp³−sp³ diamantane dimer (**1DIADIA**). Source: Reproduced from Ref. [37] with permission from the American Institute of Physics, 2013.

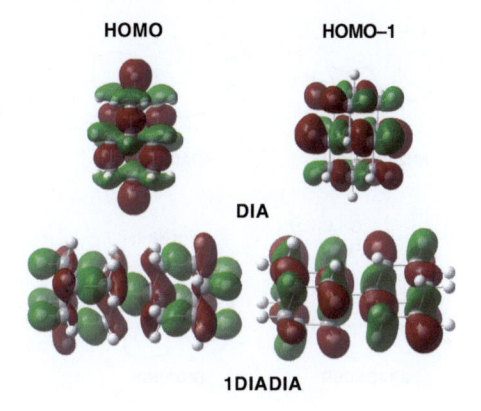

C—H stretching modes may be used as a direct marker of strong intramolecular H/H interactions.

Valence photoelectron spectroscopy was used to compare diamondoid monomers and their sp³–sp³ dimers [37]. The overall electronic structures of the dimers were determined by the superposition of the bonding partner orbitals (Figure 11.3). For instance, both cages of **1DIADIA** contribute equally to the HOMO and HOMO−1, and their shapes are similar to orbital pictures of monomeric diamondoid particles. The electronic properties of diamondoid dimers depend strongly on their sizes, i.e., they display pronounced quantum confinement effects. For example, while the experimental adiabatic IP for **DIA** is 8.79 eV, this value for **1DIADIA** is only 8.19 eV [37]. For nonsymmetric dimers, the HOMO is unequally distributed, and the quantum confinement is less pronounced as compared to the homo-dimers.

Although the diamondoid dimer HOMOs are highly delocalized and the nature of the central bond has only little impact on the electronic structure itself, relaxation after ionization to the radical cations leads to dissociation of the central bond. For example, the electrooxidation of the *all-tert*-hexamethyl derivative of **1ADAD** in acetonitrile leads to 3,5,7-trimethyladamantyl 1-acetamide almost exclusively [38].

Possible applications of the diamondoid assemblies for surface modifications in the construction of diamond-like materials require their functionalizations with various groups (OH, SH, COOH, etc.) that serve as surface attachment points. In contrast to **AD**, where incorporation of the first functional group substantially alters its overall reactivity, the cage moieties in **1ADAD** react independently, and disubstitution such as bromination [39] and carboxylation [40] can readily be achieved (Scheme 11.3) [39, 40]. Even the hexabromination of **1ADAD** is possible [41] in the presence of Lewis acids to give **BR3ADADBR3** in 70% yield [39a]. The latter was used for the preparation of hexaaryl derivatives **BRARADADARBR,** which are useful as microporous polymer building blocks [42] as well as for the construction of superhydrophobic materials [43]. The exhaustive methylation [44] of the *tert*-CH positions of **ADAD** using AlCl₃ and tetramethylsilane as the methylation agent gives the hexamethyl derivative **ME3ADADME3** in 68% yield (Scheme 11.3).

Mono-substitution of **1ADAD** could be achieved only at low conversion. The bromination with bromine diluted in CCl₄ gave monobromide **1ADADBR** in

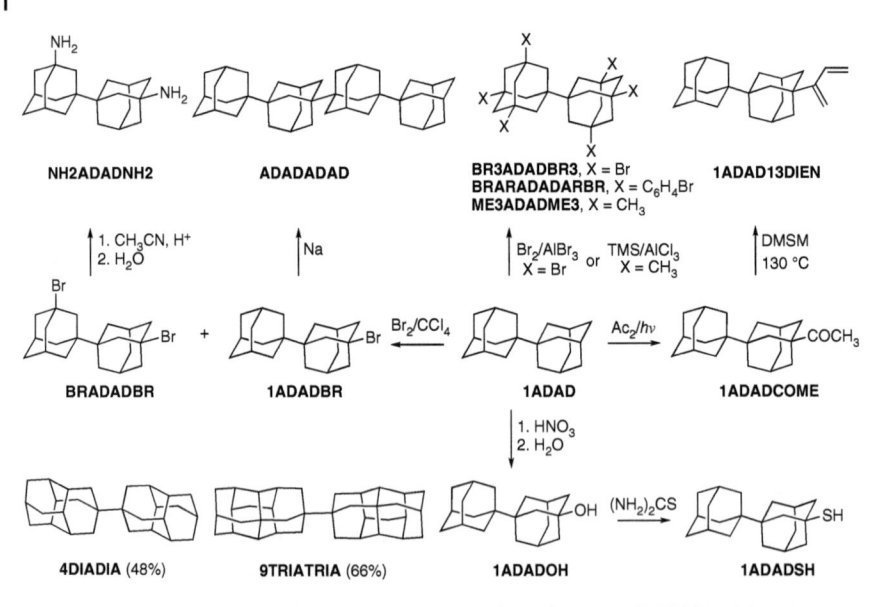

Scheme 11.3 Functionalizations of 1-adamantyl-1-adamantane (**1ADAD**) with electrophilic and radical reagents. The preparation of poorly soluble tetramer **ADADADAD** and representatives of higher apically coupled diamondoid dimers **4DIADIA** and **9TRIATRIA** that display satisfactory solubility in organic solvents but whose chemistry remains unexplored (yields are shown for their preparation through Wurtz couplings of respective apical bromides).

58% yield, but is still accompanied by the formation of substantial amounts of 3,3′-dibromo derivative **BRADADBR** [45]. Only moderate yields of **1ADADCOME** and **1ADADOH** were reported in, respectively, monoacetylation of **1ADAD** with diacetyl and nitroxylation with 100%-nitric acid, followed by hydrolysis [46]. Functional group exchanges of **1ADAD** derivatives under mild reaction conditions retain the cage structure and were subjected to amination [46, 47], thiolation (**1ADADSH**) [48], and diolefination (**1ADAD13DIEN**) [49]. In contrast, in the presence of strong Lewis acids, **1ADAD** rearranges to the thermodynamically more stable 2-adamantyl-1-adamantane and 2-adamantyl-2-adamantane (**2ADAD**, see below) isomers [50]. Monobromide **1ADADBR** was tested as a precursor for assembling larger adamantane oligomers that mimic one-dimensional diamond nanotube topology. In particular, tetramer **ADADADAD**, whose dimension already reaches 2 nm, was obtained by Wurtz coupling of **1ADADBR** with sodium in octane (Scheme 11.3). However, the obtained white powder was not fully characterized [45] due to its insolubility in all common organic solvents, which is the primary drawback of oligoadamantane chemistry. For instance, while diamine **NH2ADADNH2** could be used for controlled interparticle spacing and self-assembly of Ru nanoparticles [51], the corresponding diacid is not useful due to its generally low solubility. Luckily, the linear dimers of higher diamondoids of similar dimensions **4DIADIA** and **9TRIATRIA** (Scheme 11.3) [11] are quite soluble in organic solvents. The chemistry of large diamondoid dimers still remains largely unexplored, and only recently [52] the first functionalization of diamantane

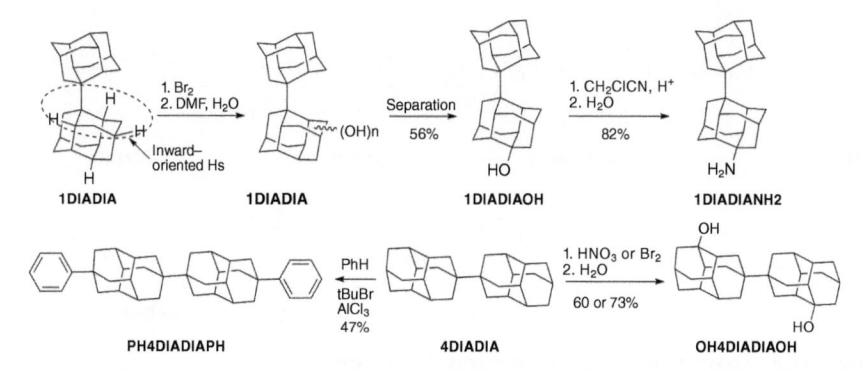

Scheme 11.4 Mono-functionalization of sterically-congested medial diamantane dimer **1DIADIA** and bis-functionalizations of apical **4DIADIA**. Source: From Refs. [52, 53].

dimers was reported (Scheme 11.4). Bromination of **1DIADIA** occurs at the medial position, which is the most distant from the central C—C-bond [54]. This gives a mixture of bromo derivatives, whose hydrolysis followed by chromatographic separations gave the hydroxy derivative **1DIADIAOH** as the main product. The reactivity of **1DIADIA** is remarkable as it contains seven inequivalent *tert*-CH positions, but only one is predominantly attacked owing to steric unavailability of the inward-oriented C—H bonds (Scheme 11.4). Alcohol **1DIADIAOH** was then transferred to its amino-derivative **1DIADIANH2** through an effective two-step procedure. Bromination and nitroxylation of **4DIADIA** gave a mixture of C—H substitution products where the medial positions of both cages are predominantly attacked, and, after hydrolysis, the most symmetric C_{2h}-diol **HO4DIADIAOH** was isolated. We conclude that the differences in reactivity of **1DIADIA** and **4DIADIA** are because many of the *tert*-C—H positions in very crowded **1DIADIA** are blocked against the attack of an electrophile. This provides viable opportunities for selective functionalizations of other sterically congested diamondoid dimers.

In contrast to the kinetically controlled reactions with electrophiles, the thermodynamically controlled di-substitutions occur mostly at the apical positions of **4DIADIA** due to the generally higher stability of the apical diamondoid derivatives [55]. The arylation of **4DIADIA** in the presence of Lewis acids gives 9,9-diphenyl-4,4-bis-diamantyl (**PH4DIADIAPH**) in good yield (Scheme 11.4). The latter molecule can be used for the construction of extended diamondoid-based gyroscopes in analogy to those previously prepared based on 4,9-diphenyldiamantane.

As will be discussed in Chapter 12, doping dramatically changes the electronic properties of diamondoids. The preparation of diamondoid assemblies doped with heteroatoms is currently limited to adamantane and diamantane derivatives only. To include heteroatoms, two different approaches were developed. The first is based on the subsequent fragmentation and reconstruction of the cages and was used for the preparation of the oxaadamantane dimer **OADADO** (Scheme 11.5) [56] The transformation involves the tetrabromination of **1ADAD** to **BR2ADADBR2** followed by double fragmentation to the keto-olefin, which was ozonized to

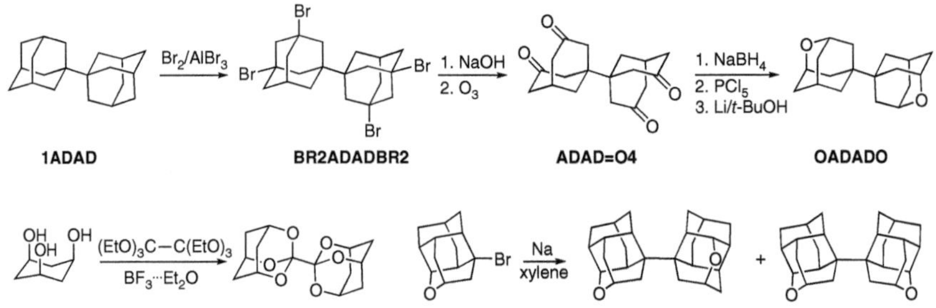

Scheme 11.5 Oxygen-doped adamantane and diamantane dimers.

tetraketone **ADAD=O4**. Transannular cyclization and reduction gave **OADADO** in ca. 8% overall yield.

A very simple and elegant one-step approach to **O3ADADO3** (Scheme 11.5) was achieved based on the reaction of *all-cis* 1,3,5-trihydroxy cyclohexane with hexaethoxyethane [57] in 58% yield. Such a cage construction approach, however, is difficult to achieve for higher diamondoid assemblies, for which Wurtz coupling is still the most useful, for instance, in the preparation of 3-oxadiamantane dimer **ODIADIAO** [58] through the coupling of readily available [59] oxabromide **ODIABR**. This resulted in the formation of a diastereomeric mixture of C_1- (**ODIADIAOC1**) and C_2- (**ODIADIAOC2**) isomers (Scheme 11.5). As **ODIADIAOC2** has a much larger permanent dipole moment than **1DIADIA**, its gas phase structure was successfully elucidated by microwave spectroscopy. In combination with gas electron diffraction (GED), this allowed comparisons of the gas phase and condensed state geometries of large molecules serving as benchmarks for the development of some theoretical methods [58].

2-Adamantyl-2-adamantane (**2ADAD**, Scheme 11.6) may also be seen as a structural motif of a hydrogen-terminated diamond lattice and, remarkably, is even less strained than **1ADAD** [50, 60]. It contains two almost undistorted adamantane units connected with the "normal" C—C bond of length 1.542 Å without substantial intramolecular repulsions between the inward hydrogens [34]. First prepared by the Kolbe reaction of 2-adamantane carboxylic acid [61], **2ADAD** is now easily accessible through the hydrogenation of **AD=AD** (Scheme 11.6) [62].

While direct functionalization of **2ADAD** may be problematic because of the presence of similarly reactive inequivalent C—H bonds, its derivatives are easily accessible from ketone **2ADAD=O**. The latter was obtained from **AD=O** [63], followed by acid-catalyzed rearrangement [64]. Chiral alcohol (**2ADADOH**) and amine (**2ADADNH2**) are interesting sterically encumbered pre-ligands and recently became available in optically active forms through subsequent chlorination of **AD=AD**, followed by hydrolysis, esterification, and cholesterol esterase hydrolysis to give optically active (−)-**AD=ADOH** (Scheme 11.6). The latter rearranges to (+)-**2ADAD=O** with conservation of optical purity through an acid-catalyzed [1,4]-hydride shift [63b].

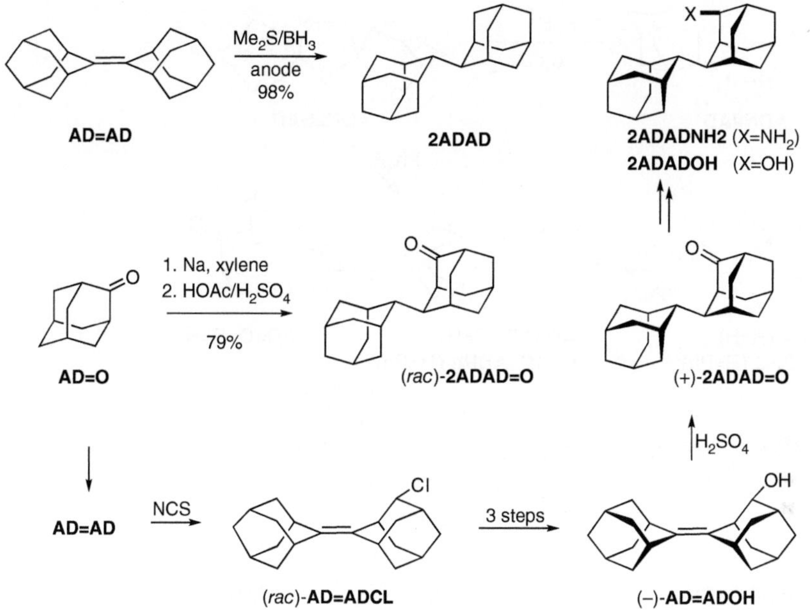

Scheme 11.6 The preparation of 2-adamantyl-2-adamantane (**2ADAD**) and functional derivatives **2ADADX** useful as optically active pre-ligands.

11.2 Unsaturated Oligomers

The incorporation of defect states is a powerful way to change the electronic properties of diamond by introducing additional electronic transitions into the band structure [65]. While the sp^3/sp^2-ratios are difficult to control in nanodiamond materials [66], diamondoid oligomers with unsaturated "defects" represent nanodiamond particles with well-controlled shapes and sp^2-defects that are of interest for nanoelectronics. The simplest representative of this class of molecules is **AD=AD**, which was first prepared by the low-yielding photodecarbonylation of the 2-adamantylketene dimer [67]. Alternatively, coupling of *gem*-dibromoadamantane (**ADBR2**, R = H) in the presence of a Zn/Cu couple gave **AD=AD** in 75% yield (Scheme 11.7) [68].

As the latter reaction appeared to be very sensitive to the reaction conditions [69], many other approaches to **AD=AD** were established. The cationic rearrangement of spiro[adamantane-2,4′-homoadamantan-5′-ol] led to a mixture of hydrocarbons with very low **AD=AD** content [70]. The first high-yielding synthesis of **AD=AD** was developed in 1973 [69] and involves the sulfurization of hydrazone **ADN2AD** to thiadiazine **ADN2SAD**; melting with triphenylphosphine gave **AD=AD** in 65% yield (Scheme 11.7). However, after McMurry's [71] ketone coupling discovery in the presence of Ti(III)-reagents [72], this new reaction became the most powerful tool for the preparation of **AD=AD** and its sterically even more hindered derivatives [73]. McMurry coupling also allows the preparation of larger unsaturated

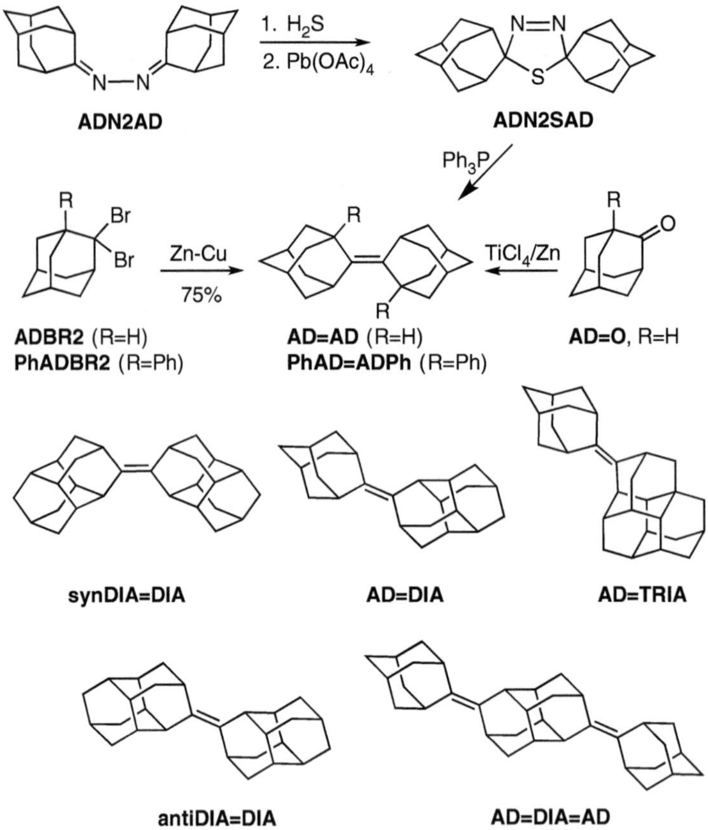

Scheme 11.7 Preparations of 2-(2-adamantylidene)adamantane (**AD=AD**) and higher unsaturated diamondoid dimers synthesized through McMurry coupling of the corresponding diamondoid ketones.

diamondoid assemblies [3] such as **AD=DIA**, *syn*- and *anti*-diastereomers of **DIA=DIA**, as well as **AD=DIA=AD** that represent large nanodiamond particles with sp²-defects (Scheme 11.7). However, this approach is difficult to use for the preparation of highly strained and twisted **AD=AD** derivatives, for which the condensation of the corresponding *gem*-dibromides is still useful (as, for instance, in the preparation of **PhAD=ADPh**, Scheme 11.7) [74].

Despite significant crowding, the C—C bond in unsubstituted **AD=AD** displays a planar arrangement as derived from X-ray analysis [75] and DFT computations [76]. The AIM analysis indicates only two close H•••H-contacts (between β-hydrogens), in accord with their short distance of ca. 1.85 Å [77]. In contrast, steric hindrance in **PhAD=ADPh** results in 23° twisting of the C—C bond [74]. The ionization of **AD=AD** leads to the unusually stable and long-lived radical cation **AD=ADRC** with a high degree of spin/charge delocalization over the cage moieties [78]. This radical cation was generated through protic acid-promoted oxidations of **AD=AD** [79] as well as in the reactions of **AD=AD** with nitrogen oxides [80] or through its electrochemical oxidation [81]. In contrast to the adamantane radical cation **ADRC** (see Chapters 3 and 4 for more details), which undergoes fast proton/hydrogen losses, **AD=ADRC** is persistent both in the gas phase and in condensed states and only very slowly eliminates H_2 [82]. The persistence of **AD=ADRC** is controlled by steric crowding caused by the cage moieties that are additionally able to participate in charge/spin delocalization [76]. This enabled measurements of the optical [83] as well as ESR and ENDOR spectra of **AD=ADRC** under ambient conditions [84].

Olefins such as **AD=AD** and larger analogues like **DIA=DIA** form highly delocalized ionized states not only at the single-molecule level but also within their van der Waals clusters [3]. It was experimentally demonstrated that in the gas phase, the self-exchange electron transfer between the ionized and neutral states of **AD=AD** ($k = [7.85 \pm 0.94] \times 10^{-9}$ molec^{-1} cm^3 s^{-1}) is comparable to the rate of electron exchange in the **AD/ADRC** pair ($k = [1.50 \pm 0.15] \times 10^{-10}$ molec^{-1} cm^3 s^{-1}) and is at the same order of magnitude as the collision rate [82]. From this point, the unsaturated diamondoid dimers are indeed suitable materials for mimicking the electronic properties of H-terminated diamond surfaces, as electron/hole transfer processes in **AD=AD** clusters are expected to be effective [3]. As both the LUMOs and HOMOs of unsaturated diamondoid dimers [85] are localized on the double bond and excitation causes only little changes of the cage moieties [3], their surface self-assembled monolayers, as well as van der Waals crystals [85b], are expected to be resistant against irradiation. The absorption and emission spectra of diamondoid dimers are red-shifted with respect to pristine diamondoids [86].

The peculiar structure and reactivity of **AD=AD** allowed the isolation of the first cyclic bromonium salt, **AD=ADBR4**, (Scheme 11.8) [87], which was postulated by Roberts and Kimball in 1937 [88] as the key intermediate in olefin bromination reactions. Relatively stable **AD=ADBR4** forms in the reaction of **AD=AD** with bromine as a yellow crystalline material, whose X-ray structure [89] was reproduced well by computations [90]. In solution, the formation of the bromonium ion is fast and reversible [91] and occurs through a charge-transfer π-complex that is stabilized due to substantial LD interactions [92]. The bromonium and iodonium salts derived from **AD=AD** with triflate as the counterion are even more stable [93] and were also characterized through single-crystal X-ray diffraction [94]. The chloronium salt with $SbCl_6^-$ was also isolated [95] and structurally characterized [96]. Bromination under phase-transfer conditions in the presence of NH_4VO_3 or Na_2MoO_4 led to **AD=ADBR$^+$** with the corresponding metal-containing counteranions [91b]. The extraordinary persistence of **AD=ADBR$^+$** is due to significant structural crowding that prevents nucleophilic attack [89], and it can therefore readily be used as a brominating reagent. The kinetics of the bromine transfer from **AD=ADBR$^+$** to olefins were studied [97], and **AD=ADBR$^+$** together with the weakly coordinating counterion tetrakis[3,5-bis(trifluoromethyl)phenyl] borate, was recently used as a halogenating agent for olefins and heteroatomic organic compounds [98]. Recently, stable seleniranium and telluriranium salts were also isolated and characterized, utilizing **AD=AD** as a sterically encumbered olefin [99].

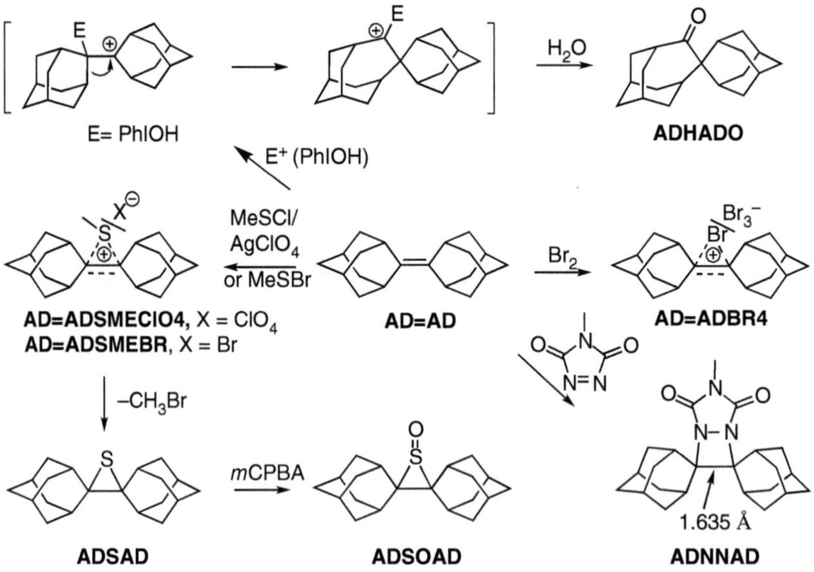

Scheme 11.8 Isolation of stable intermediates resulting from the attack of electrophiles onto the double bond of 2-(2-adamantylidene)adamantane **AD=AD**.

The reaction of **AD=AD** with Koser's PhIOH reagent is accompanied by skeletal rearrangement to spiro-derivative **ADHADO** (Scheme 11.8); a stable thiiranium salt **AD=ADSMEClO4** forms in MeSCl/AgClO$_4$ [100]. With more nucleophilic counterions (Br$^-$) **AD=ADSMEBR** may undergo demethylation, affording very stable thiirane **ADSAD**. This is a very unusual reaction in organosulfur chemistry, as thiiranum salts generally decompose to ring-opened or desulfurization products. The oxidation of **ADSAD** gave **ADSOAD** [101], which may be used as a clean source of sulfur monoxide [101, 102].

An effective nonionic addition to the **AD=AD** double bond was observed only for very reactive species like triazolinedione, which at room temperature gives the formal [2 + 2]cycloaddition [103] product **ADNNAD** through an aziridinium imide intermediate (Scheme 11.8) [104]. The structure of **ADNNAD** displays an unusually long C—C bond connecting two adamantyl units (1.635 Å) [105]. A shorter C—C bond was detected in the dioxazolidine adduct formed in the photochemical reaction of **AD=AD** with 1,3,5-trinitrobenzene [106]. Only a few more examples of additions to the otherwise very unreactive **AD=AD** double bond are known and include cyclopropanation with Me$_3$Al–CH$_2$I$_2$ [107], SO$_3$ addition [108], and hydrogenation with the borane-methyl sulfide (BMS) complex mentioned above (Scheme 11.6) [62].

While the reaction of **AD=AD** with Br$_2$ results in **AD=ADBR$^+$**, the reaction with *N*-bromosuccinimide (NBS) gives C—H-substituted 4-bromo derivative **AD=AD4BR** in 97% yield [109]. The ionic nature of this reaction is supported by the fact that the model radical reaction with the Br$_3$C$^\bullet$ generated under phase transfer catalysis (PTC) conditions [110] occurs exclusively at the tertiary position to give **AD=AD1BR** in 52% yield (Scheme 11.9).

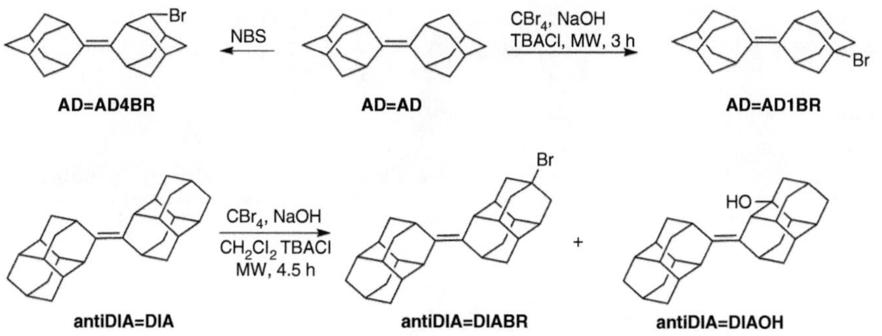

Scheme 11.9 Functionalizations of unsaturated diamondoid dimers with radical reagents.

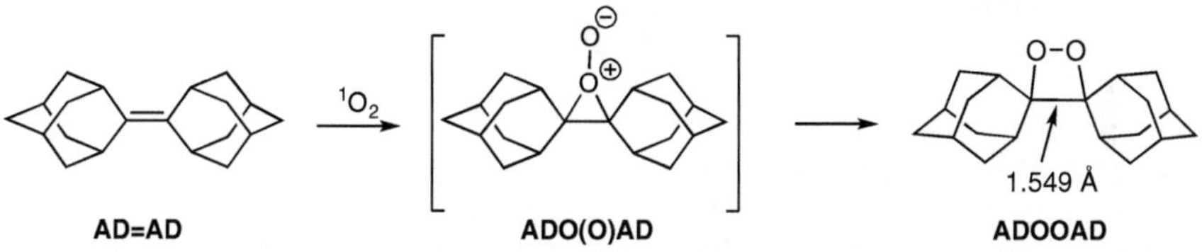

Scheme 11.10 Formation of unusually stable cyclic peroxide **ADOOAD**.

As several inequivalent C—H positions are present in larger diamondoid dimers such as **antiDIA=DIA**, a mixture of tertiary derivatives (**antiDIA=DIABR** and **antiDIA=DIAOH**) formed in 23 and 47% yields, respectively, upon PTC bromination with CBr_4 (Scheme 11.9). The availability of both apically (**antiDIA=DIABR**) and medially (**antiDIA=DIAOH**) substituted **DIA=DIA** allows for the construction of surface SAMs of different topologies.

Under oxidative conditions in the presence of oxygen, **AD=AD** forms unusually stable dioxetane **ADOOAD** (Scheme 11.10). Surprisingly, the single crystal X-ray diffraction structure displays a C—C bond of normal length (1.549 Å) [111] that is shorter than in the nitrogen analogue **ADNNAD**. Oxetane **ADOOAD** was prepared through the reaction with singlet oxygen generated upon sensitization with dyes [112] or metalloporphyrines [113], as well as via catalytic decomposition of hydrogen peroxide [114]. The oxidation proceeds through a perepoxide intermediate [115], **ADO(O)AD**, which behaves as a mild oxidant and is able to transfer oxygen to sulfoxides [116], phosphites [117], and fullerenes [118].

The reaction of **AD=AD** with oxygen is faster in the presence of tetracyanoethylene [119] and in protic solvents [120]; however, the yields of **ADOOAD** are only moderate. The formation of **ADOOAD** in the reaction of **AD=AD** with ground-state triplet oxygen requires single-electron transfer (SET) initiation. The radical cation **AD=ADRC**, generated electrochemically [81, 121], or through SET to nitronium, nitrosonium [122], aminium [123], or pyrilium [124] salts readily reacts with oxygen to give **ADOOAD**, usually in high yields.

Recent interest in **ADOOAD** arises from its high antiplasmodium activity [125], which is at least two orders of magnitude higher than that of the antimalarial drug artemisinin [126]. Owing the combination of high stability with the ability to

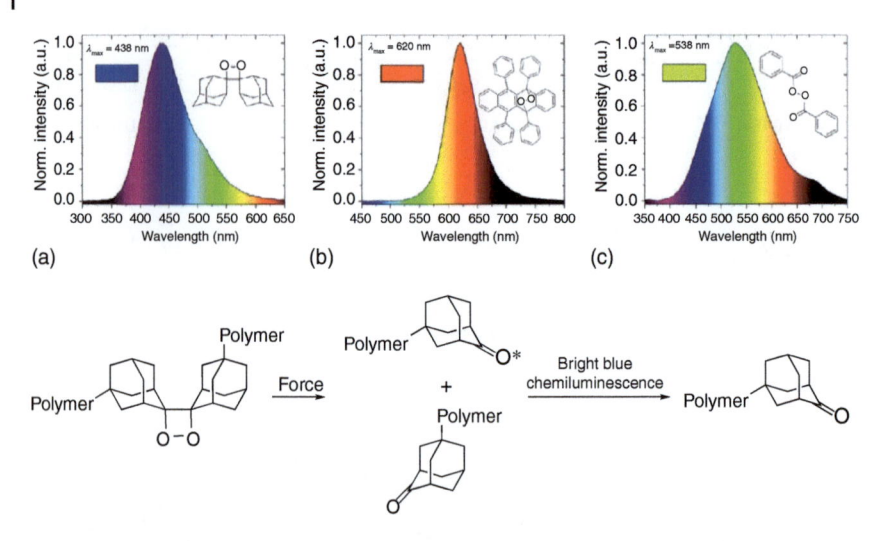

Figure 11.4 *Top*: Solid-state thermochemiluminescence spectra of some organic peroxides: bis(adamantyl)-1,2-dioxetane (a), rubrene endoperoxide (b), and benzoyl (c) (Source: Reproduced from Ref. [129] with permission from Springer Nature, 2019); *Bottom*: Mechanically induced decomposition of a polymeric dioxetane that results in chemiluminescence where the thus formed ketone relaxes from its excited state to the ground state producing blue light (Source: From Ref. [130a]).

produce under thermolysis two molecules of **AD=O**, one of which forms in the excited or triplet electronic state [127], **ADOOAD** derivatives are useful for the design of new luminescent materials [128]. In the solid state, **ADOOAD** displays unique thermochemiluminescence [129] that is blue-shifted relative to other classes of peroxides (Figure 11.4). The decomposition of **ADOOAD** produces an excited ketone that, in combination with an electron acceptor [131], efficiently produces blue light. Polymers based on **ADOOAD** exhibit mechanically induced chemiluminescence [130], i.e., an optical response to mechanical stress. Consequently, **ADOOAD** was used as a cross-linker in the photopolymerization of methyl acrylate to give mechanochemiluminescent materials, which allows real-time monitoring of chain-scission events with high temporal resolution. This is useful for the study of stress distribution in polymers and sensing and mapping of deformations and damages. The ability of **ADOOAD**-based polymers to emit in focal spots of ultrasonic irradiation is potentially useful for noninvasively triggering cellular responses in optogenetics [131]. **ADOOAD** substituted with a phthalimide function in the *tert*-C—H position of adamantane moiety exists in diastereomeric *syn*- and *anti*-forms. The latter displays contrastive chemiluminescence in the crystalline state, with packing determining the emission pattern [132].

Sesquihomoadamantene (**SHAD**, Figure 11.5) was first prepared together with its isomer **AD=AD** through the acid-catalyzed rearrangement of pinacol **ADOH2AD**, which exists exclusively in a *trans*-conformation [63a, 134]. The structure of **SHAD** [135] displays an almost undistorted planar olefin moiety and serves, together with isomer **AD=AD**, as a model for olefin oxidations. In contrast to radical

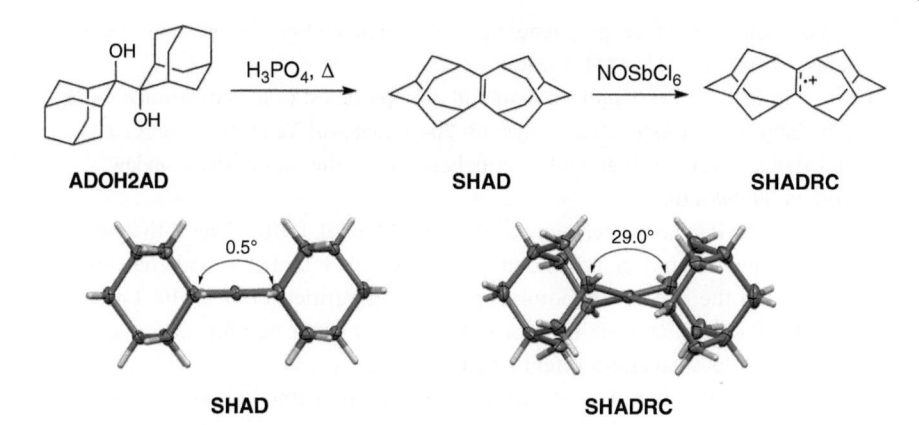

Figure 11.5 Preparation of sesquihomoadamantene (**SHAD**), its radical cation (**SHADRC**) and their X-ray structures (from Ref. [133], Wiley, 2000).

cations generated from aliphatic olefins that are unstable due to H-loss from the allylic positions, such a reaction is prevented for the rigid **AD=AD** and **SHAD** cages. Both radical cations **AD=ADRC** and **SHADRC** are almost strain-free and display similar distortions, bond lengths, and angles. However, **AD=ADRC** is too short-lived (see above) to be isolated, despite being ca. 9.6 kcal mol^{-1} more stable than isolable **SHADRC** [76]. The outstanding kinetic stability of **SHADRC** is due to the C—H-bond protective shielding of the double bond (Figure 11.5) that sterically prevents nucleophilic attack. The X-ray crystal structure [133] of radical cation **SHADRC** derived from **SHAD** and nitrosonium salt NO$^+$SbCl$_6^-$ displays ca. 29° twisting of the double bond. This provided the first unequivocal confirmation of Mulliken's suggestion that olefinic radical cations should be twisted and must show "shallow minima at perhaps about φ=30°" [136].

References

1 Dahl, J.E., Liu, S.G., and Carlson, R.M.K. (2003). Isolation and structure of higher diamondoids, nanometer-sized diamond molecules. *Science* 299: 96–99.
2 (a) Dahl, J.E.P., Moldowan, J.M., Peakman, T.M. et al. (2003). Isolation and structural proof of the large diamond molecule, cyclohexamantane (C$_{26}$H$_{30}$). *Angew. Chem. Int. Ed.* 42: 2040–2044. (b) Fokin, A.A., Tkachenko, B.A., Fokina, N.A. et al. (2009). Reactivities of the prism-shaped diamondoids [1(2)3]tetramantane and [12312]hexamantane (Cyclohexamantane). *Chem. Eur. J.* 15: 3851–3862.
3 Zhuk, T.S., Koso, T., Pashenko, A.E. et al. (2015). Toward an understanding of diamond sp^2-defects with unsaturated diamondoid oligomer models. *J. Am. Chem. Soc.* 137: 6577–6586.
4 Georgakilas, V., Perman, J.A., Tucek, J., and Zboril, R. (2015). Broad family of carbon nanoallotropes: classification, chemistry, and applications of fullerenes,

carbon dots, nanotubes, graphene, nanodiamonds, and combined superstructures. *Chem. Rev.* 115: 4744–4822.

5 (a) Holt, K.B. (2010). Undoped diamond nanoparticles: origins of surface redox chemistry. *Phys. Chem. Chem. Phys.* 12: 2048–2058. (b) Yang, N.J., Foord, J.S., and Jiang, X. (2016). Diamond electrochemistry at the nanoscale: a review. *Carbon* 99: 90–110.

6 Cinkova, K., Batchelor-McAuley, C., Marton, M. et al. (2016). The activity of non-metallic boron-doped diamond electrodes with sub-micron scale heterogeneity and the role of the morphology of sp^2 impurities. *Carbon* 110: 148–154.

7 Harada, T. (2009). Organic resin materials for insulating film for semiconductor devices. In: 2009 Japanese Patent JP 2009295972.

8 Cao, Z., Cao, Y., Chen, J. et al. (2014). Manufacture method of organic capacitive touch screen having copper-doped adamantane-contg. prepolymer as electrode material. In: 2014 Chinese Patent CN 103744570 A.

9 (a) Kubo, Y., Iwato, K., and Inabe, H. Film forming compositions based on compounds and(or) polymers of compounds having adamantane groups for use in electronic devices. In: 2008 US Patent US 2009004842. (b) Kato, T., Oki, H., and Ishisone, T. (2005). Heat-resistant styrene polymers and their efficient manufacture. In: Japanese Patent JP 2006096988. (c) Kobayashi, S., Matsuzawa, T., Matsuoka, S.I. et al. (2006). Living anionic polymerizations of 4-(1-adamantyl) styrene and 3-(4-vinylphenyl)-1,1′-biadamantane. *Macromolecules* 39: 5979–5986.

10 Eckhardt, M., Hamilton, B.S., and Himmelsbach, F. (2010). 1,1′-Diadamantyl carboxylic acids, medicaments containing such compounds and their use. In: WO Patent WO 2010010174.

11 Fokin, A.A., Chernish, L.V., Gunchenko, P.A. et al. (2012). Stable alkanes containing very long carbon-carbon bonds. *J. Am. Chem. Soc.* 134: 13641–13650.

12 Landa, S. and Hala, S. (1957). Über einige neue Derivate des Adamantans und ein neues Verfahren zur Isolierung des Adamantans aus Erdöl. *Angew. Chem. Int. Ed.* 69: 684.

13 Stepanov, F.N. and Baklan, V.F. (1964). On the reactivity of adamantane bromides with metals. *Zh. Obsh. Khim.* 34: 579–584.

14 (a) Dubois, J.E., Molle, G., Tourillon, G., and Bauer, P. (1979). XPS study of surface changes in magnesium subjected to attack by an alkyl halide. *Tetrahedron Lett.* 5069–5072. (b) Molle, G., Bauer, P., and Dubois, J.E. (1982). Formation of cage-structure organomagnesium compounds. Influence of the degree of adsorption of the transient species at the metal surface. *J. Org. Chem.* 47: 4120–4128.

15 Molle, G., Dubois, J.E., and Bauer, P. (1978). Cage structure organometallic compounds: 1-Diamantyl, 1-twistyl, 1-triptycyl and 2-adamantyl lithium compounds. Synthesis and reactivity. *Tetrahedron Lett.* 3177–3180.

16 Nishino, T., Watanabe, T., Okada, M. et al. (2002). Reduction of organic halides with lanthanum metal: a novel generation method of alkyl radicals. *J. Org. Chem.* 67: 966–969.

17 Wieringa, J.H., Strating, J., and Wynberg, H. (1972). 1-Lithioadamantane. *Synth. Commun.* 2: 191–195.

18 DeCosta, D.P., Bennett, A., Pincock, A.L. et al. (2000). Photochemistry of aryl *tert*-butyl ethers in methanol: the effect of substituents on an excited state cleavage reaction. *J. Org. Chem.* 65: 4162–4168.

19 (a) Rondinini, S., Mussini, P.R., Muttini, P., and Sello, G. (2001). Silver as a powerful electrocatalyst for organic halide reduction: the critical role of molecular structure. *Electrochim. Acta* 46: 3245–3258. (b) Paddon, C.A., Bhatti, F.L., Donohoe, T.J., and Compton, R.G. (2007). Electrocatalytic reduction of alkyl iodides in tetrahydrofuran at silver electrodes. *J. Phys. Org. Chem.* 20: 115–121. (c) Paddon, C.A., Bhatti, F.L., Donohoe, T.J., and Compton, R.G. (2007). Electrosynthetic reduction of 1-iodoadamantane forming 1,1′-biadamantane and adamantane in aprotic solvents: Insonation switches the mechanism from dimerisation to exclusive monomer formation. *Ultrason. Sonochem.* 14: 502–508.

20 Engel, P.S., Chae, W.K., Baughman, S.A. et al. (1983). A product study of 1-adamantyl and 1-bicyclo[2.2.2]octyl radicals in hydrocarbon solvents. An unusually large hydrogen isotope effect. *J. Am. Chem. Soc.* 105: 5030–5034.

21 Engel, P.S., Lee, W.K., Marschke, G.E., and Shine, H.J. (1987). The reactions of 1-adamantyl radicals with acetonitrile and their bearing on the oxidative decomposition of 1,1′-azoadamantane. *J. Org. Chem.* 52: 2813–2817.

22 Bae, D.H., Engel, P.S., Hoque, A. et al. (1985). Oxidative decomposition of 1,1′-azoadamantane by thianthrene cation radical. Carbocationic chemistry from a free radical source. *J. Am. Chem. Soc.* 107: 2561–2562.

23 Turro, N.J. and Tung, C.H. (1980). Photochemistry of phenyladamantyl ketone in homogeneous organic and in micellar solution. *Tetrahedron Lett.* 21: 4321–4322.

24 Karpushenkava, L.S., Kabo, G.J., Bazyleva, A.B. et al. (2007). Thermodynamic properties of 1,1′-biadamantane. *Thermochim. Acta* 459: 104–110.

25 (a) Alden, R.A., Kraut, J., and Traylor, T.G. (1968). Concerning the mechanism of single-bond shortening. Evidence from the crystal structures of 1-biapocamphane, 1-binorbornane, and 1-biadamantane. *J. Am. Chem. Soc.* 90: 74–82. (b) Alden, R., Kraut, J., and Traylor, T.G. (1967). Hybridization, conjugation, and bond lengths. An experimental test. *J. Phys. Chem.* 71: 2379–2380.

26 Engler, E.M., Andose, J.D., and Schleyer, P.v.R. (1973). Critical evaluation of molecular mechanics. *J. Am. Chem. Soc.* 95: 8005–8025.

27 Beckhaus, H.D., Flamm, M.A., and Rüchardt, C. (1982). Thermolabile hydrocarbons. 15. The thermolysis of 1,1′-biadamantane. *Tetrahedron Lett.* 23: 1805–1808.

28 Schreiner, P.R., Chernish, L.V., Gunchenko, P.A. et al. (2011). Overcoming lability of extremely long alkane carbon-carbon bonds through dispersion forces. *Nature* 477: 308–311.

29 (a) Mandal, N. and Datta, A. (2020). Molecular designs for expanding the limits of ultralong C–C bonds and ultrashort H•••H non-bonded contacts. *Chem. Commun.* 56: 15377–15386. (b) Shimajiri, T., Kawakami, Y., Kawaguchi, S. et al. (2023). Ultralong C(sp^3)-C(sp^3) single bonds shortened and stabilized by London dispersion. *Synlett* 34: 1147–1152.

30 Wagner, J.P. and Schreiner, P.R. (2015). London dispersion in molecular chemistry-reconsidering steric effects. *Angew. Chem. Int. Ed.* 54: 12274–12296.

31 Cho, D., Ikabata, Y., Yoshikawa, T. et al. (2015). Theoretical study of extremely long yet stable carbon–carbon bonds: effect of attractive C•••H interactions and small radical stabilization of diamondoids. *Bull. Chem. Soc. Jpn.* 88: 1636–1641.

32 Wan, X.J., He, X., Li, M. et al. (2022). The synergetic and multifaceted nature of carbon-carbon rotation reveals the origin of conformational barrier heights with bulky alkane groups. *J. Phys. Org. Chem.* 36: e4352.

33 Zavitsas, A.A. (2003). The relation between bond lengths and dissociation energies of carbon-carbon bonds. *J. Phys. Chem. A* 107: 897–898.

34 Tyborski, C., Huckstaedt, T., Gillen, R. et al. (2020). Vibrational signatures of diamondoid dimers with large intramolecular London dispersion interactions. *Carbon* 157: 201–207.

35 (a) Lobato, A., Salvado, M.A., Recio, J.M. et al. (2021). Highs and lows of bond lengths: is there any limit? *Angew. Chem. Int. Ed.* 60: 17028–17036. (b) Delgado, A.A.A., Humason, A., Kalescky, R. et al. (2021). Exceptionally long covalent CC bonds-a local vibrational mode study. *Molecules* 26: 950.

36 (a) Allinger, N.L., Lii, J.H., and Schaefer, H.F. III, (2016). Molecular mechanics (MM4) studies on unusually long carbon-carbon bond distances in hydrocarbons. *J. Chem. Theory Comput.* 12: 2774–2778. (b) Morgante, P. and Peverati, R. (2021). CLB18: a new structural database with unusual carbon-carbon long bonds. *Chem. Phys. Lett.* 765. (c) Yost, S.R. and Head-Gordon, M. (2018). Efficient implementation of NOCI-MP2 using the resolution of the identity approximation with application to charged dimers and long C–C bonds in ethane derivatives. *J. Chem. Theory Comput.* 14: 4791–4805. (d) Fokin, A.A. (2023). Long but strong C—C single bonds: challenges for theory. *Chem. Rec.* 24: e202300170.

37 Zimmermann, T., Richter, R., Knecht, A. et al. (2013). Exploring covalently bonded diamondoid particles with valence photoelectron spectroscopy. *J. Chem. Phys.* 139: 084310.

38 Edwards, G.J., Jones, S.R., and Mellor, J.M. (1977). Anodic oxidation of substituted adamantanes. *J. Chem. Soc. Perkin Trans.* 2: 505–510.

39 (a) Lai, X., Guo, J., Fu, S., and Zhu, D. (2016). Synthesis of rigid cores based on 1,1′-biadamantane. *RSC Adv.* 6: 8677–8680. (b) Reinhardt, H.F. (1962). Biadamantane and some of its derivatives. *J. Org. Chem.* 27: 3258–3261.

40 Butenko, L.N., Protopopov, P.A., Derbisher, V.E., and Khardin, A.P. (1984). Synthesis of functional derivatives of tricyclic hydrocarbons. *Synth. Commun.* 14: 113–119.

41 Rezaei-Seresht, E., Balali, H., and Maleki, B. (2022). Synthesis of novel rigid structures derived from 1,1′-biadamantane. *Fuller. Nanotub. Carbon Nanostructures* 1–4.

42 Guo, J.W., Lai, X.F., Fu, S.Q. et al. (2017). Microporous organic polymers based on hexaphenylbiadamantane: synthesis, ultra-high stability and gas capture. *Mater. Lett.* 187: 76–79.

43 (a) Li, X., Guo, J.W., Tong, R. et al. (2018). Microporous frameworks based on adamantane building blocks: synthesis, porosity, selective adsorption and functional application. *React. Funct. Polym.* 130: 126–132. (b) Li, L.J., Deng, J.H., Guo, J.W., and Yue, H.B. (2020). Synthesis and properties of microporous organic polymers based on adamantane. *Prog. Chem.* 32: 190–203.

44 Bonsir, M., Davila, C., Kennedy, A.R., and Geerts, Y. (2021). Exhaustive one-step bridgehead methylation of adamantane derivatives with tetramethylsilane. *Eur. J. Org. Chem.* 2021: 5227–5237.

45 Ishizone, T., Tajima, H., Matsuoka, S., and Nakahama, S. (2001). Synthesis of tetramers of 1,3-adamantane derivatives. *Tetrahedron Lett.* 42: 8645–8647.

46 Chernish, L.V., Gunchenko, P.A., Barabash, A.V. et al. (2008). Selective synthesis of mono-derivatives of 1,1-diadamantane. *J. Org. Pharm. Chem.* 6: 48–51.

47 Senchyk, G.A., Lysenko, A.B., Krautscheid, H., and Domasevitch, K.V. (2011). "Fluoride molecular scissors": a rational construction of new Mo(VI) oxofluorido/1,2,4-triazole MOFs. *Inorg. Chem. Commun.* 14: 1365–1368.

48 Tkachenko, B.A., Fokina, N.A., Chernish, L.V. et al. (2006). Functionalized nanodiamonds part 3: Thiolation of tertiary bridgehead alcohols. *Org. Lett.* 8: 1767–1770.

49 Fokin, A.A., Butova, E.D., Chernish, L.V. et al. (2007). Simple preparation of diamondoid 1,3-dienes via oxetane ring opening. *Org. Lett.* 9: 2541–2544.

50 Slutsky, J., Engler, E.M., and Schleyer, P.v.R. (1973). Equilibration of biadamantane isomers. *J. Chem. Soc. Chem. Commun.* 685–686.

51 Min, Y.Y., Nasrallah, H., Poinsot, D. et al. (2020). 3D Ruthenium nanoparticle covalent assemblies from polymantane ligands for confined catalysis. *Chem. Mater.* 32: 2365–2378.

52 Gunchenko, P.A., Chernish, L.V., Tikhonchuk, E.Y. et al. (2020). Functionalization of diamantane dimers. *J. Org. Pharm. Chem.* 18: 16–22.

53 Karlen, S.D., Ortiz, R., Chapman, O.L., and Garcia-Garibay, M.A. (2005). Effects of rotational symmetry order on the solid state dynamics of phenylene and diamantane rotators. *J. Am. Chem. Soc.* 127: 6554–6555.

54 Fokin, A.A., Tkachenko, B.A., Gunchenko, P.A. et al. (2005). Functionalized nanodiamonds Part I. An experimental assessment of diamantane and computational predictions for higher diamondoids. *Chem. Eur. J.* 11: 7091–7101.

55 Johnston, D.E., Rooney, J.J., and McKervey, M.A. (1972). Equilibration of diamantan-1-ol and diamantan-4-ol. Conformational enthalpy of the hydroxy-group, and an unusual example of how entropy and symmetry factors can influence relative thermodynamic stabilities. *J. Chem. Soc. Chem. Commun.* 29–30.

56 Teager, D.S. and Murray, R.K. (1993). Synthesis of 5,5′-bi-2-oxaadamantane. *J. Org. Chem.* 58: 5560–5561.

57 Stetter, H. and Hunds, A. (1984). Über Verbindungen mit Urotropin-Struktur, LX. Orthooxalsäure- und Orthokohlensäurederivate rnit Adamantanstruktur. *Liebigs Ann. Chem.* 1577–1590.

58 Fokin, A.A., Zhuk, T.S., Blomeyer, S. et al. (2017). Intramolecular London dispersion interaction effects on gas-phase and solid-state structures of diamondoid dimers. *J. Am. Chem. Soc.* 139: 16696–16707.

59 Fokin, A.A., Zhuk, T.S., Pashenko, A.E. et al. (2009). Oxygen-doped nanodiamonds: synthesis and functionalizations. *Org. Lett.* 11: 3068–3071.

60 Mielke, K., Fry, J.L., Finnen, D.C., and Pinkerton, A.A. (1994). 2,2′-Biadamantane. *Acta Crystallogr. C* 50: 267–269.

61 Vanzorge, J.A., Strating, J., and Wynberg, H. (1970). A comparison of behaviour of (1-adamantyl)- and (2-adamantyl)-carboxylic acids during Kolbe electrolysis. *Recl. Trav. Chim. Pays-Bas* 89: 781–789.

62 Rathore, R., Weigand, U., and Kochi, J.K. (1996). Efficient hydrogenation of sterically hindered olefins with borane-methyl sulfide complex. *J. Org. Chem.* 61: 5246–5256.

63 (a) Gill, G.B. and Hands, D. (1971). Reactions of highly hindered spiroadamantanes. *Tetrahedron Lett.* 181–184. (b) Ganga-Sah, Y., Leznoff, D.B., and Bennet, A.J. (2019). Synthesis of sterically congested 2,2′-bi(adamantyl)-based alcohol and amines. *J. Org. Chem.* 84: 15276–15282.

64 Chou, D.T.H., Huang, X., Batchelor, R.J. et al. (1998). Rearrangement of a homoallylic alcohol via an acid-catalyzed 1,4-hydride shift yields a saturated ketone. *J. Org. Chem.* 63: 575–581.

65 Williams, O.A., Nesladek, M., Daenen, M. et al. (2008). Growth, electronic properties and applications of nanodiamond. *Diamond Relat. Mater.* 17: 1080–1088.

66 Osswald, S., Yushin, G., Mochalin, V. et al. (2006). Control of sp^2/sp^3 carbon ratio and surface chemistry of nanodiamond powders by selective oxidation in air. *J. Am. Chem. Soc.* 128: 11635–11642.

67 Strating, J., Scharp, J., and Wynberg, H. (1970). Ketenes with the adamantane skeleton. *Rec. Trav. Chim.* 89: 23–31.

68 Geluk, H.W. (1970). Synthesis of adamantylideneadamantane. *Synthesis* 652–653.

69 Schaap, A.P. and Faler, G.R. (1973). Convenient synthesis of adamantylideneadamantane. *J. Org. Chem.* 38: 3061–3062.

70 Boelema, E., Wynberg, H., and Strating, J. (1971). Cationic rearrangements of spiro[adamantane-2,4′-homoadamantan-5′-ol]. *Tetrahedron Lett.* 4029–4032.

71 McMurry, J.E. (1983). Titanium-induced dicarbonyl-coupling reactions. *Acc. Chem. Res.* 16: 405–411.

72 (a) McMurry, J.E. and Fleming, M.P. (1976). Improved procedures for the reductive coupling of carbonyls to olefins and for the reduction of diols to olefins. *J. Org. Chem.* 41: 896–897. (b) Fleming, M.P. and McMurry, J.E. (1981). Reductive coupling of carbonyls to alkenes: Adamantylideneadamantane. Tricyclo[3.3.1.1^{3,7}]decane, tricyclo[3.3.1.1^{3,7}]decylidene. *Org. Synth.* 60: 113–117. (c) Tolstikov, G.A., Lerman, B.M., and Belogaeva, T.A. (1991). A convenient synthesis of adamantylideneadamantane. *Synth. Commun.* 21: 877–879.

73 Lenoir, D., Frank, R.M., Cordt, F. et al. (1980). Sterisch gehinderte Olefine, V. Synthese und Röntgenstrukturanalyse von *trans*-1-Ethyl-2-(1-ethyl-2-adamantyliden)adamantan, einem hochgespannten Ethylen, Vergleich mit Kraftfeld-Rechnungen. *Chem. Ber.* 113: 739–749.

74 Okazaki, T., Ogawa, K., Kitagawa, T., and Takeuchi, K. (2002). Trans-2,2′-bi(1-phenyladamantylidene): the most twisted biadamantylidene. *J. Org. Chem.* 67: 5981–5986.

75 Swen-Walstra, S.C. and Visser, G.J. (1971). X-ray crystal and molecular structure of adamantylideneadamantane. *J. Chem. Soc. Chem. Commun.* 82–83.

76 Rathore, R., Lindeman, S.V., Zhu, C.J. et al. (2002). Steric hindrance as a mechanistic probe for olefin reactivity: variability of the hydrogenic canopy over the isomeric adamantylideneadamantane/sesquihomoadamantene pair (a combined experimental and theoretical study). *J. Org. Chem.* 67: 5106–5116.

77 Abboud, J.-L.M., Alkorta, I., Davalos, J.Z. et al. (2016). The thermodynamic stability of adamantylideneadamantane and its proton- and electron-exchanges. Comparison with simple alkenes. *Bull. Chem. Soc. Jpn.* 89: 762–769.

78 Nelsen, S.F. and Kessel, C.R. (1979). Adamantylideneadamantane radical cation, a long-lived one-electron, two-center, π-bonded species. *J. Am. Chem. Soc.* 101: 2503–2504.

79 Rathore, R., Zhu, C.J., Lindeman, S.V., and Kochi, J.K. (2000). Spontaneous oxidation of organic donors to their cation radicals using Bronsted acids. Identification of the elusive oxidant. *J. Chem. Soc. Perkin Trans.* 2: 1837–1840.

80 Bosch, E. and Kochi, J.K. (1996). Catalytic epoxidation of hindered olefins with dioxygen. Fast oxygen atom transfer to olefin cation radicals from nitrogen oxides. *J. Am. Chem. Soc.* 118: 1319–1329.

81 Nelsen, S.F., Kapp, D.L., Akaba, R., and Evans, D.H. (1986). Cyclic voltammetry study of the cation radical catalyzed oxygenation of tetraalkyl olefins to dioxetanes. *J. Am. Chem. Soc.* 108: 6863–6871.

82 Guerrero, A., Herrero, R., Quintanilla, E. et al. (2010). Single-electron self-exchange between cage hydrocarbons and their radical cations in the gas phase. *ChemPhysChem* 11: 713–721.

83 Clark, T., Teasley, M.F., Nelsen, S.F., and Wynberg, H. (1987). Optical absorption spectra of tetraalkyl olefin cation radicals. *J. Am. Chem. Soc.* 109: 5719–5724.

84 (a) Nelsen, S.F., Kapp, D.L., Gerson, F., and Lopez, J. (1986). Dioxetane radical cations in solution. An ESR and cyclic voltammetry study. *J. Am. Chem. Soc.* 108: 1027–1032. (b) Gerson, F., Lopez, J., Akaba, R., and Nelsen, S.F. (1981). Alkyl group stabilization of monoolefin radical cations. An ESR, ENDOR, and cyclic voltammetry study. *J. Am. Chem. Soc.* 103: 6716–6722.

85 (a) Meinke, R., Richter, R., Merli, A. et al. (2014). UV resonance Raman analysis of trishomocubane and diamondoid dimers. *J. Chem. Phys.* 140: 034309. (b) Tyborski, C., Meinke, R., Gillen, R. et al. (2017). From isolated diamondoids to a van-der-Waals crystal: a theoretical and experimental analysis of a trishomocubane and a diamantane dimer in the gas and solid phase. *J. Chem. Phys.* 147: 044303.

86 Banerjee, S., Stueker, T., and Saalfrank, P. (2015). Vibrationally resolved optical spectra of modified diamondoids obtained from time-dependent correlation function methods. *Phys. Chem. Chem. Phys.* 17: 19656–19669.

87 Strating, J., Wieringa, J.H., and Wynberg, H. (1969). The isolation of a stabilized bromonium ion. *J. Chem. Soc. Chem. Commun.* 907-908.

88 Roberts, I. and Kimball, G.E. (1937). The halogenation of ethylenes. *J. Am. Chem. Soc.* 59: 947-948.

89 Slebockatilk, H., Ball, R.G., and Brown, R.S. (1985). The question of reversible formation of bromonium ions during the course of electrophilic bromination of olefins. 2. The crystal and molecular structure of the bromonium ion of adamantylideneadamantane. *J. Am. Chem. Soc.* 107: 4504–4508.

90 Islam, S.M. and Poirier, R.A. (2008). Addition reaction of adamantylideneadamantane with Br_2 and $2Br_2$: A computational study. *J. Phys. Chem. A* 112: 152–159.

91 (a) Bellucci, G., Bianchini, R., Chiappe, C. et al. (1993). A dynamic NMR investigation of the adamantylideneadamantane/Br_2 system. Kinetic and thermodynamic evidence for reversible formation of the bromonium ion/Br_n^- pairs. *J. Org. Chem.* 58: 3401–3406. (b) Bortolini, O., Chiappe, C., Conte, V., and Carraro, M. (1999). Direct synthesis of stable adamantylideneadamantane bromonium salts. *Eur. J. Org. Chem.* 3237–3239.

92 Chiappe, C., Detert, H., Lenoir, D. et al. (2003). Polarizability effects and dispersion interactions in alkene-Br_2 π-complexes. *J. Am. Chem. Soc.* 125: 2864–2865.

93 Bennet, A.J., Brown, R.S., McClung, R.E.D. et al. (1991). An unprecedented rapid and direct Br^+ ion transfer from the bromonium ion of adamantylideneadamantane to acceptor olefins. *J. Am. Chem. Soc.* 113: 8532–8534.

94 Brown, R.S., Nagorski, R.W., Bennet, A.J. et al. (1994). Stable bromonium and iodonium ions of the hindered olefins adamantylideneadamantane and bicyclo[3.3.1]nonylidenebicyclo[3.3.1]nonane. X-ray structure, transfer of positive halogens to acceptor olefins, and *ab initio* studies. *J. Am. Chem. Soc.* 116: 2448–2456.

95 Nugent, W.A. (1980). Unusual reactions of adamantylideneadamantane with metal oxidants. Isolation of stable chloronium salts. *J. Org. Chem.* 45: 4533–4534.

96 Mori, T., Rathore, R., Lindeman, S.V., and Kochi, J.K. (1998). X-ray structure of bridged 2,2′-bi(adamant-2-ylidene) chloronium cation and comparison of its reactivity with a singly-bonded chloroarenium cation. *Chem. Commun.* 927–928.

97 Neverov, A.A. and Brown, R.S. (1996). Br^+ and I^+ transfer from the halonium ions of adamantylideneadamantane to acceptor olefins. Halocyclization of 1,ω-alkenols and alkenoic acids proceeds via reversibly formed intermediates. *J. Org. Chem.* 61: 962–968.

98 (a) Ascheberg, C., Bock, J., Buss, F. et al. (2017). Stable bromiranium ions with weakly-coordinating counterions as efficient electrophilic brominating agents. *Chem. Eur. J.* 23: 11578–11586. (b) Bock, J., Daniliuc, C.G., and Hennecke, U. (2019). Stable bromiranium ion salts as reagents for biomimetic indole terpenoid cyclizations. *Org. Lett.* 21: 1704–1707.

99 (a) Poleschner, H. and Seppelt, K. (2018). Seleniranium and telluriranium salts. *Chem. Eur. J.* 24: 17155–17161. (b) Poleschner, H. and Seppelt, K. (2021).

Attempts to synthesize a thiirane, selenirane, and thiirene by dealkylation of chalcogeniranium and thiirenium salts. *Chem. Eur. J.* 27: 649–659.
100 Bolster, J. and Kellogg, R.M. (1978). Preparation of some *S*-methyl thiiranium salts of adamantylideneadamantane. Demethylation and desulphurization with nucleophiles. *J. Chem. Soc. Chem. Commun.* 630–631.
101 AbuYousef, I.A. and Harpp, D.N. (1997). Effective precursors for sulfur monoxide formation. *J. Org. Chem.* 62: 8366–8371.
102 (a) Abuyousef, I.A. and Harpp, D.N. (1995). A useful precursor for sulfur monoxide transfer. *Tetrahedron Lett.* 36: 201–204. (b) Nakayama, J., Tajima, Y., Xue-Hua, P., and Sugihara, Y. (2007). [1+2] Cycloadditions of sulfur monoxide (SO) to alkenes and alkynes and [1+4] cycloadditions to dienes (polyenes). Generation and reactions of singlet SO? *J. Am. Chem. Soc.* 129: 7250–7251.
103 Seymour, C.A. and Greene, F.D. (1980). Mechanism of triazolinedione-olefin reactions. Ene and cycloaddition. *J. Am. Chem. Soc.* 102: 6384–6385.
104 (a) Nelsen, S.F. and Klein, S.J. (1997). Addition of *N*-methyltriazolinedione to biadamantylidene. *J. Phys. Org. Chem.* 10: 456–460. (b) Nelsen, S.F. and Kapp, D.L. (1985). Direct observation of the aziridinium imide intermediates in the reaction of biadamantylidene with triazolinediones. *J. Am. Chem. Soc.* 107: 5548–5549.
105 Cheng, C.C., Seymour, C.A., Petti, M.A. et al. (1984). Reaction of electrophiles with unsaturated systems: Triazolinedione-olefin reactions. *J. Org. Chem.* 49: 2910–2916.
106 Okada, K., Saito, Y., and Oda, M. (1992). Photochemical reaction of polynitrobenzenes with adamantylideneadamantane: the X-ray structure analysis and chemical properties of the dispiro *N*-(2,4,6-trinitrophenyl)-1,3,2-dioxazolidine product. *J. Chem. Soc. Chem. Commun.* 1731–1732.
107 Ramazanov, I.R., Kadikova, R.N., Zosim, T.P. et al. (2016). Cyclopropanation of [2,2′]biadamantylidene with $Me_3Al\text{-}CH_2I_2$ reagent. *Mendeleev Commun.* 26: 434–436.
108 Bakker, B.H., Cerfontain, H., and Tomassen, H.P.M. (1989). Aliphatic sulfonation. 3. Reactions of adamantylidenealkanes and cyclopropylidenealkanes with sulfur trioxide. *J. Org. Chem.* 54: 1680–1684.
109 Meijer, E.W., Kellogg, R.M., and Wynberg, H. (1982). Rearrangements in the halogenation of tetraalkylethylenes with *N*-halosuccinimides and *tert*-butyl hypochlorite. *J. Org. Chem.* 47: 2005–2009.
110 Schreiner, P.R., Lauenstein, O., Butova, E.D. et al. (2001). Selective radical reactions in multiphase systems: phase-transfer halogenations of alkanes. *Chem. Eur. J.* 7: 4996–5003.
111 Hess, J. and Vos, A. (1977). Adamantylideneadamantane peroxide, a stable 1,2-dioxetane. *Acta Crystallogr. B* 33: 3527–3530.
112 (a) Wieringa, J.H., Wynberg, H., Strating, J., and Adam, W. (1972). Adamantylideneadamantane peroxide, a stable 1,2-dioxetane. *Tetrahedron Lett.* 169–172. (b) Jefford, C.W. and Boschung, A.F. (1976). The dye-sensitized photo-oxygenation of biadmantylidene. *Tetrahedron Lett.* 4771–4774.

(c) Meijer, E.W. and Wynberg, H. (1981). 1,2-Dioxetanes as chemiluminescent intermediates in the triplet oxygen oxygenation of olefins. *Tetrahedron Lett.* 22: 785–788. (d) Jefford, C.W., Estrada, M.J., and Barchietto, G. (1987). The reaction of adamantylideneadamantane with singlet oxygen mediated by rose bengal and charge transfer complexes. *Tetrahedron* 43: 1737–1745. (e) Jefford, C.W., Estrada, M.J., Barchietto, G. et al. (1990). The reaction of singlet oxygen with adamantylideneadamantane mediated by rose bengal. *Helv. Chim. Acta* 73: 1653–1658. (f) Jefford, C.W. and Boschung, A.F. (1977). The reaction of biadamantylidene with singlet oxygen in the presence of dyes. *Helv. Chim. Acta* 60: 2673–2685.

113 Akasaka, T., Haranaka, M., and Ando, W. (1993). The novel formation of metalloporphyrin-oxo species in singlet oxygen oxidation of adamantylideneadamantane. *J. Am. Chem. Soc.* 115: 7005–7006.

114 Aubry, J.M. and Bouttemy, S. (1997). Preparative oxidation of organic compounds in microemulsions with singlet oxygen generated chemically by the sodium molybdate hydrogen peroxide system. *J. Am. Chem. Soc.* 119: 5286–5294.

115 (a) Clennan, E.L., Chen, M.F., and Xu, G. (1996). New potent trapping agents for the peroxidic intermediates formed in the reactions of singlet oxygen. *Tetrahedron Lett.* 37: 2911–2914. (b) Clennan, E.L., Stensaas, K.L., and Rupert, S.D. (1998). Trapping of peroxidic intermediates with sulfur and phosphorus centered electrophiles. *Heteroat. Chem.* 9: 51–56.

116 Schaap, A.P., Recher, S.G., Faler, G.R., and Villasenor, S.R. (1983). Evidence for a perepoxide intermediate in the 1,2-cycloaddition of singlet oxygen to adamantylideneadamantane: nucleophilic oxygen atom transfer to sulfoxides. *J. Am. Chem. Soc.* 105: 1691–1693.

117 Akaba, R., Aihara, S., Sakuragi, H., and Tokumaru, K. (1987). Triphenylpyrylium salt as sensitizer for electron transfer oxygenation not involving superoxide anion. *J. Chem. Soc. Perkin Trans. 2* 1262–1263.

118 Maeda, Y., Niino, Y., Kondo, T. et al. (2011). Oxygen atom transfer from peroxide intermediates to fullerenes. *Chem. Lett.* 40: 1431–1433.

119 Akaba, R., Sakuragi, H., and Tokumaru, K. (1984). Photooxidation of adamantylideneadamantane in the presence of tetracyanoethylene. A new photoepoxidation process. *Chem. Lett.* 1677–1680.

120 Akaba, R., Sakuragi, H., and Tokumaru, K. (1984). Autoxidation of adamantylideneadamantane in a protic solvent. Facile proton-induced electron transfer from the olefin to oxygen. *Tetrahedron Lett.* 25: 665–668.

121 Clennan, E.L., Simmons, W., and Almgren, C.W. (1981). Electroorganic synthesis. 1. Electrode-catalyzed synthesis of a dioxetane. *J. Am. Chem. Soc.* 103: 2098–2099.

122 Nelsen, S.F. and Akaba, R. (1981). Oxygenation reactions of the adamantylideneadamantane cation radical. *J. Am. Chem. Soc.* 103: 2096–2097.

123 Lopez, L., Farinola, G.M., Nacci, A., and Sportelli, S. (1998). Monodeoxygenation of spiro adamantane-1,2-dioxetanes induced by aminium salt. *Tetrahedron* 54: 6939–6946.

124 (a) Akaba, R., Sakuragi, H., and Tokumaru, K. (1991). Triphenylpyrylium-salt-sensitized electron transfer oxygenation of adamantylideneadamantane. Product, fluorescence quenching, and laser flash photolysis studies. *J. Chem. Soc. Perkin Trans. 2* 291–297. (b) Clennan, E.L. and Liao, C. (2014). Synthesis, characterization, photophysics and photochemistry of pyrylogen electron transfer sensitizers. *Photochem. Photobiol.* 90: 344–357.

125 Silva, A.F., Oliveira, V.X. Jr., Silva, L.S. et al. (2016). Antiplasmodial activity of alkyl-substituted 1,2-dioxetanes against plasmodium falciparum. *Bioorg. Med. Chem. Lett.* 26: 5007–5008.

126 Klayman, D.L. (1985). *Qinghaosu* (Artemisinin): an antimalarial drug from China. *Science* 228: 1049–1055.

127 (a) Schuster, G.B., Turro, N.J., Steinmetzer, H.C. et al. (1975). Adamantylideneadamantane-1,2-dioxetane. An investigation of the chemiluminescence and decomposition kinetics of an unusually stable 1,2-dioxetane. *J. Am. Chem. Soc.* 97: 7110–7118. (b) Hohne, G., Schmidt, A.H., and Lechtken, P. (1976). Thermoanalyse von kristallinem Adamantylideneadamantan-1,2-dioxetan. *Tetrahedron Lett.* 3587–3590.

128 Sagara, Y., Yamane, S., Mitani, M. et al. (2016). Mechanoresponsive luminescent molecular assemblies: an emerging class of materials. *Adv. Mater.* 28: 1073–1095.

129 Schramm, S., Karothu, D.P., Lui, N.M. et al. (2019). Thermochemiluminescent peroxide crystals. *Nat. Commun.* 10: 997.

130 (a) Chen, Y., Spiering, A.J.H., Karthikeyan, S. et al. (2012). Mechanically induced chemiluminescence from polymers incorporating a 1,2-dioxetane unit in the main chain. *Nat. Chem.* 4: 559–562. (b) Clough, J.M. and Sijbesma, R.P. (2014). Dioxetane scission products unchanged by mechanical force. *ChemPhysChem* 15: 3565–3571.

131 Kim, G., Lau, V.M., Halmes, A.J. et al. (2019). High-intensity focused ultrasound-induced mechanochemical transduction in synthetic elastomers. *Proc. Natl. Acad. Sci. U. S. A.* 116: 10214–10222.

132 Matsuhashi, C., Ueno, T., Uekusa, H. et al. (2020). Isomeric difference in the crystalline-state chemiluminescence property of an adamantylideneadamantane 1,2-dioxetane with a phthalimide chromophore. *Chem. Commun.* 56: 3369–3372.

133 Kochi, J.K., Rathore, R., Zhu, C.J., and Lindeman, S.V. (2000). Structural characterization of novel olefinic cation radicals: X-ray crystallographic evidence of s-p hyperconjugation. *Angew. Chem. Int. Ed.* 39: 3671–3674.

134 Wynberg, H., Boelema, E., Wieringa, J.H., and Strating, J. (1970). Adamantylidene adamantane glycol. *Tetrahedron Lett.* 3613–3614.

135 Watson, W.H. and Nagl, A. (1987). Structure of bis(homoadamantane). *Acta Crystallogr. C* 43: 2465–2466.

136 Mulliken, R.S. (1959). Conjugation and hyperconjugation: a survey with emphasis on isovalent hyperconjugation. *Tetrahedron* 5: 253–274.

12

Doped Diamondoids

Pristine diamondoids display poor conductivity and are viewed as insulators [1] with dielectric constants of 2.5–2.7 fm^{-1}, which is even lower than that of pure type-IIa diamond ($\kappa = 5.66$ fm^{-1}). It is well established that doping changes the electron transport properties of diamond materials. Typical p-dopants include boron, and such type-IIb-doped diamond is also present in nature, though in very small quantities [2]. n-Doping requires the presence of heavier elements like phosphorus or nitrogen, and nitrogen-doped colorless type-Ia and yellow type-Ib diamonds are very abundant in nature (>98% of all diamond material) [2, 3]. Doping affects the electron emission and conduction properties of diamond drastically. While boron-doped diamond displays good conductivity but poor emission, the opposite is true for nitrogen-doped type-Ia diamond [4]. The strong emitting properties of artificially phosphorus-doped chemical vapor deposition (CVD) diamond were confirmed recently [5]; however, the emission is difficult to control as it is very sensitive to the surface topology [6].

Even though microelectronic applications of naturally occurring and artificially doped diamond continue to attract great attention [7], their utilization for the construction of nanoelectronic devices is still problematic because the available nanodiamond materials suffer from inhomogeneities, wide particle size distributions, polycrystallinity, and the presence of various impurities. These factors significantly affect their electronic properties [8], and the statement that "…the more defective nanodiamond particles are, the more efficiently they emit" [9] is only valid at the macroscopic level. Since the encouraging proposal of a blend of diamond and diamondoids in the construction of electron emitters predicted theoretically [10] was later confirmed experimentally [11], many attempts have been made to understand the influence of doping on the properties of diamondoids with the hope to create new materials with well-controlled and tailored properties. All three ways, namely the placing of the dopant inside the cage (@-doping), replacement of carbons by heteroatoms (internal doping), and C—H-substitution with functional groups (external doping), were suggested for fine-tuning the electronic properties of diamondoids (Fig. 12.1). Very recently, the studies of the electronic properties of diamondoid clusters [12] that may be termed as "dispersion doping" (d) emerged.

The Chemistry of Diamondoids: Building Blocks for Ligands, Catalysts, Materials, and Pharmaceuticals,
First Edition. Andrey A. Fokin, Marina Šekutor, and Peter R. Schreiner.
© 2024 WILEY-VCH GmbH. Published 2024 by WILEY-VCH GmbH.

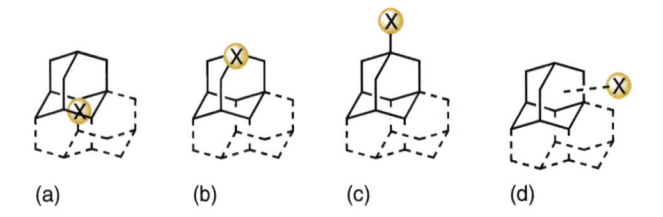

Figure 12.1 Examples of @- (a), internal (b), external (c), and "dispersion" (d) doping of diamondoids, where "X" symbolizes an atom or a functional group.

12.1 @-Doping

The incorporation of host atoms and molecules within carbon cavities attracted attention after the seminal detection of the fullerene-lanthanum inclusion complex La@C$_{60}$, formed in the thermolysis of lanthanum-impregnated graphite [13]. In general, placing the host atoms inside fullerenes is a reasonably simple task due to the large cavity sizes. Many such X@C60(C70) endohedral complexes were obtained through "physical synthesis" as above, where carbon materials form fullerenes in the presence of host atom sources under harsh reaction conditions [14]. Alternatively, incorporation of host atoms was achieved through cage opening followed by fullerene core reconstruction, termed "molecular surgery" [15]. However, the latter approach has been seen as too elaborate for practical large-scale synthesis. Although placing host atoms inside cage molecules smaller than fullerenes is energetically unfavorable, the topic attracted much attention after the successful preparation of the endohedral dodecahedrane complex He@C$_{20}$H$_{20}$ [16]. This complex is stable for weeks even though the incorporation of He inside the cage is ca. 34 kcal mol^{-1} endothermic [17]. The formation of endohedral complexes of much smaller **AD** is even less favorable [18] due to high Pauli repulsion. Placing a proton inside the cage causes the migration of the extra proton to the surface [19]. The He + **AD** → He@**AD** reaction is 160 kcal mol^{-1} endothermic at B3LYP/6-31G(d) [20] and 158 kcal mol^{-1} at MP2(full)/6-311++G(2d,2p) levels of theory [21]. However, **AD** is the smallest cage that is structurally able to incorporate various guest atoms like He, Li$^+$, Be^{2+}, Mg^{2+}, and Ne, although the associated encapsulation processes are all highly endothermic. Among metal cations, only Li$^+$ displays a true transition structure for the insertion/extrusion of the metal ion with the conservation of the cage structure, while all other cations lead to cage destruction. The incorporation of charged species is somewhat less favorable, especially in the presence of counterions [22]; expectedly, the (in)stability of the Li$^+$@**AD** complex depends strongly on the nature of the counter anion (Fig. 12.2). While in the presence of fluoride, the transition structure **TS**$_F$ for Li$^+$ extrusion from Li$^+$@**AD** is located 174.2 kcal mol^{-1} above the starting materials, the corresponding barrier for the reaction with Br$^-$ as the counterion through **TS**$_{Br}$ amounts to 198.0 kcal mol^{-1}. That is, none of these reactions are experimentally feasible.

The endohedral complexes of **AD** with larger 3d-transition metals (M@**AD**, M = Cr, Mn, Fe, Co, and Ni) display strong C•••C interactions, although the

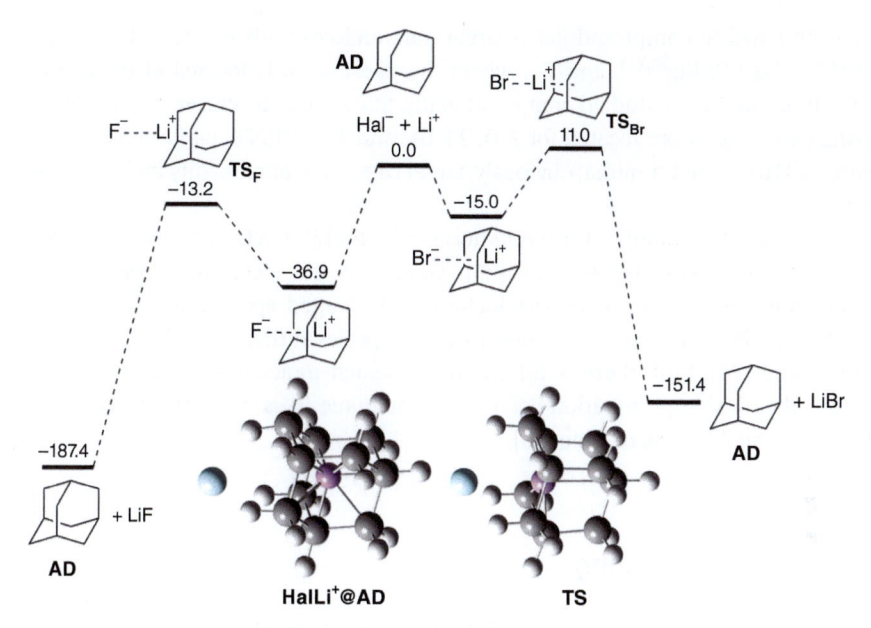

Figure 12.2 Relative MP2/6-311++G(3df,2p) energies for a lithium cation crossing the adamantane (**AD**) cavity in the presence of halide anions and the transition structures (**TS**) for LiHal extrusion from **HalLi$^+$@AD**. Source: Reproduced in part from Ref. [22] with permission from the American Chemical Society, 2011.

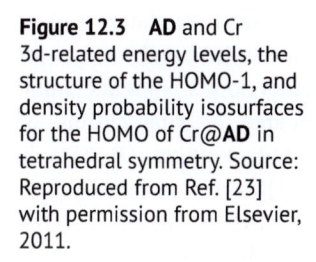

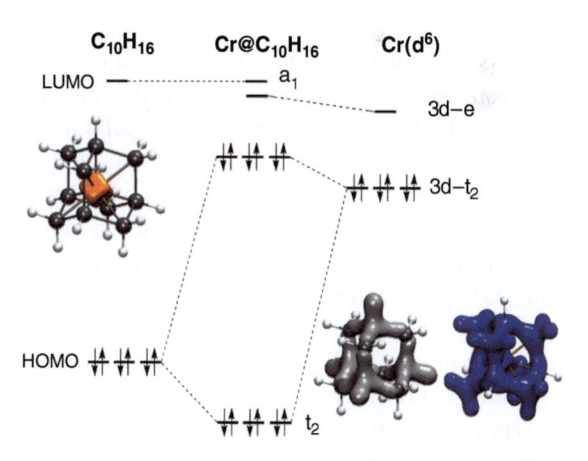

Figure 12.3 AD and Cr 3d-related energy levels, the structure of the HOMO-1, and density probability isosurfaces for the HOMO of Cr@AD in tetrahedral symmetry. Source: Reproduced from Ref. [23] with permission from Elsevier, 2011.

formation energies are even more endothermic (8–10 eV), but still lower than the inclusion energies of transition metals into diamond (12–13 eV) [23]. Doping with transition metals changes the electronic structures and substantially decreases the HOMO–LUMO gaps due to contributions from the 3d orbitals, as shown in Fig. 12.3 for Cr@**AD**. Doping with transition metals leaves the lowest unoccupied molecular orbital (LUMO) of **AD** almost unaffected, in good agreement with the fact that this only increases the occupied 3d-related states in transition-metal-doped artificial diamond materials [24].

The endohedral complexations of larger diamondoids with H$^+$, He, Ne, Li, Li$^+$, Be$^{(0,+,2+)}$, Na$^{(0,+)}$, Mg$^{(0,2+)}$, and F$^-$, where the guest atom is located at the center of the molecule, were studied computationally. Stationary structures of such endohedral complexes were located for **AD**, **TRIA**, and **1212PENT**, while those in the center of **DIA** are not minima; obviously, larger cages can more readily include guest atoms.

For instance, the complexation endothermicities for Li$^+$ + **AD** and Li$^+$ + **123PENT** are 2.64 and 2.08 eV, respectively [25]. Generally, the electronic properties of X@diamondoid are altered by the inclusion of charged species and depend on the charge, size, and type of the encapsulated species rather than the size of the diamondoid [25]. Still, there is little hope that such molecules can be prepared, as even the "inside protonation" of 1-azaadamantane is ca. 80–90 kcal mol^{-1} less favorable than from the outside [26].

12.2 Internal Doping

In contrast to @-doped diamondoids, internal doping is much more realistic from a practical viewpoint, as some X-hetero diamondoids (X = O, N, S, B, Si) are available synthetically as well as naturally. Most of the theoretical findings noted below are still far from experimental verifications due to the limited accessibility of doped diamondoids. The experimentally available structures are restricted mostly to adamantane (Scheme 12.1) and diamantane derivatives. Oxaadamantane (**O_AD**) was first obtained (Scheme 12.1) as the dimethyl derivative [27] and then as an unsubstituted cage from readily available 7-methylenebicyclo[3.3.1]nonane-3-one (**KETMET**) [28]. Thiaadamantane (**S_AD**) was prepared by ring closure of 2,6-bicyclo[3.3.1]nonane derivatives [29], and 1-boraadamantane (**B_AD**) is available from 3-propyl-7-methylene-3-borabicyclo[3.3.1]nonane obtained from the condensation of allylborane with allene followed by hydrogenation [30, 31]. 1-Azaadamantane (**N_AD**) was obtained either from 1,3,5-derivatives of cyclohexane [32] or through ring closure of aza-bicyclononane derivatives [33]. Multistep exchange of boron in 1-boraadamantane for nitrogen is also possible [34]. Phosphaadamantane (**P_AD**) was prepared [35] through multistep synthesis involving the closure of the bicyclo[3.3.1]nonane system.

Very recently, the preparation of 2,4,10-trioxaadamantane derivatives (**3O_AD**) [36] from *cis*-phloroglucin was adapted to the preparation [37] of hexaoxadiamantane derivatives (**6O_DIA**) from naturally occurring *myo*-inositol (Scheme 12.2). Alternatively, the oxygen-doped structures were obtained through the ring closure of bicyclo[3.3.1]nona-2,6-diene [38] and various bicyclo[3.3.1]nonane derivatives [39], cage opening and re-closure [40], as well as oxygen insertions into the diamondoid ketones through a "double Criegee rearrangement" in the presence of trifluoroperacetic acid [41]. Such "skeletal editing" [42], i.e., the replacement of the cage group CH$_2$ by a heteroatom, provides access to a large variety of O-doped diamondoids, including higher diamondoids (Scheme 12.3) [43].

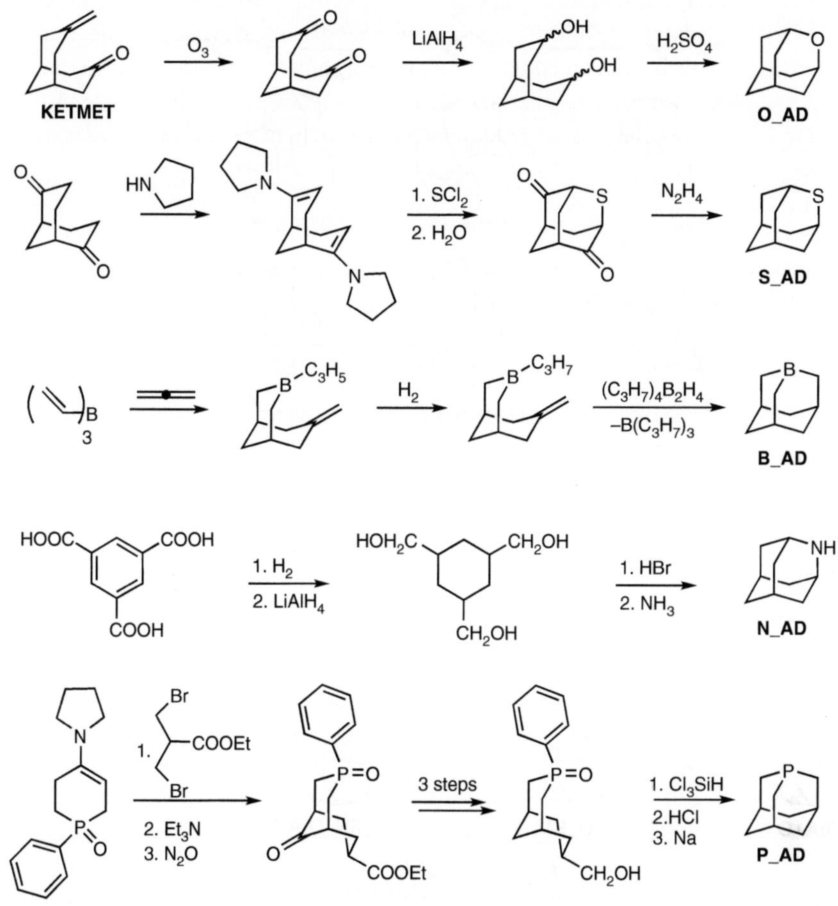

12.2 Internal Doping | 309

Scheme 12.1 Synthetic routes to internally doped adamantanes (heteroadamantanes).

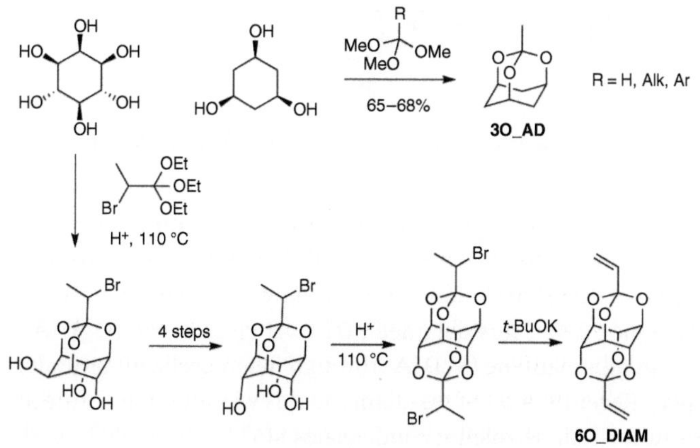

Scheme 12.2 Construction of trioxaadamantane and hexaoxadiamantane cages.

Scheme 12.3 "Skeletal editing" for the replacement of the cage CH_2-fragment by a heteroatom.

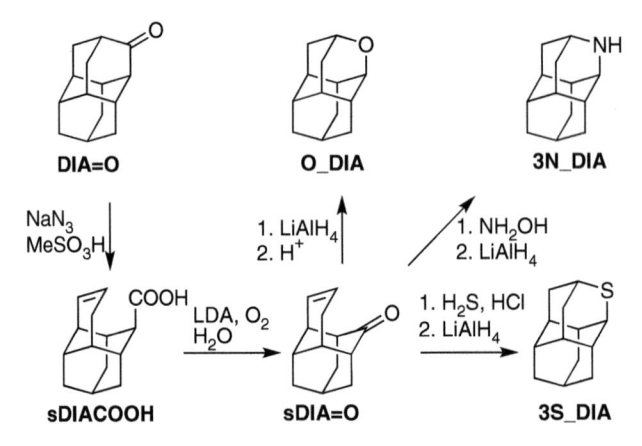

Scheme 12.4 Preparation of heterodiamantanes through unsaturated acid **sDIACOOH**. Source: From Ref. [44].

The general way to O-, N-, and S-doped diamantanes is based on seco-ketone **sDIA=O** prepared from diamantanone **DIA=O** through fragmentation to the acid **sDIACOOH** and subsequent decarboxylation under strongly basic conditions in the presence of oxygen (Scheme 12.4). Thus obtained **sDIA=O** gives 3-oxa- (**O_DIA**), 3-aza- (**3N_DIA**), and 3-thiadiamantane (**S_DIA**) through heterocyclizations under reductive conditions [44]. Recently, a set of thiadiamondoids was found in nature as a minor component of sulfur-rich petroleum condensates [45].

Alternatively, effective skeletal editing was achieved through two retro-Barbier fragmentations followed by cage reconstruction in the presence of a dopant [42b].

Phenylhydroxy diamantane (**PHOHDIA**) upon treatment with Br$_2$ undergoes cage opening to ester **nBROCOPHDIA** as the key intermediate (Scheme 12.5). The latter was transferred to **S_DIA** through subsequent iodination to **nBRIDIA** and sulfur insertion with Na$_2$S. The reduction of **nBROCOPHDIA** to **nBROHDIA** followed by oxidation to the corresponding ketone and reaction with hydroxylamine gave oxime **nNOHBRDIA** and 3-azadiamantane (**3N_DIA**) after reduction.

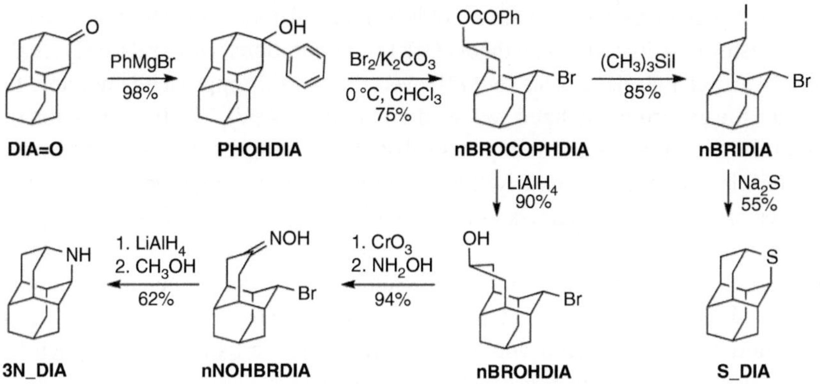

Scheme 12.5 Synthesis of 3-aza- (**3N_DIA**) and thia- (**S_DIA**) diamantanes utilizing retro-Barbier fragmentation of phenylhydroxy diamantane (**PHOHDIA**) to intermediate **sBROCOPHDIA** followed by cage reconstruction in the presence of respective nucleophiles. Source: From Ref. [42b].

High yields and preparative ease make the above scheme viable for large-scale synthesis of heterodiamondoids that was extended for the first preparation of thiatriamantane utilizing a slightly modified procedure (Scheme 12.6). The first retro-Barbier fragmentation of **PHOHTRIA** in the I$_2$/Pb(OAc)$_4$ system gives

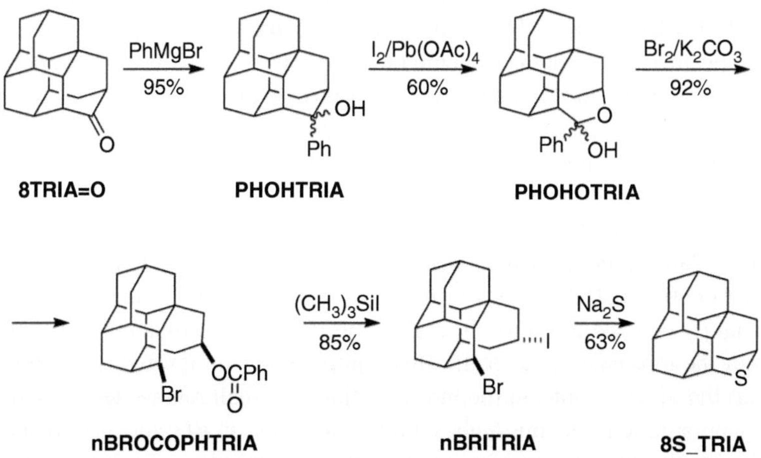

Scheme 12.6 First preparation of thiatriamantane (**8S_TRIA**) through double retro-Barbier fragmentation followed by cage reconstruction with a sulfur source (Na$_2$S).

intermediate hemiketal **PHOHOTRIA**, which undergoes a second retro-Barbier type fragmentation with Br_2/K_2CO_3 to give bromoester **sBROCOPHTRIA**. The latter, after iodination, gives target 8-thiatriamantane (**8S_TRIA**) in the presence of Na_2S.

The current interest in internally doped diamondoids is determined by their use as seeds for CVD incorporation of color centers into microscopic diamond [46] as well as by their potential as nanoelectronic materials. The general trends in changing the electronic properties of diamondoids by internal doping may be qualitatively analyzed using density functional theory (DFT) computations with a relatively small double-ζ valence polarized basis set [47], even though this approach is not able to reproduce the experimental band gap of adamantane (9.4 eV). The incorporation of boron and nitrogen as n- and p-dopants substantially reduces the band gap by ca. 2 eV. This is in agreement with early diffusion Monte–Carlo simulations of electronic properties of silicon nanoclusters [48] and with the general trends where doping with boron most significantly alters the electronic properties of adamantane [47]. Consequently, most of the work on the electronic properties of doped diamondoids is associated with these two elements. Despite pronounced changes in the shapes of the frontier orbitals, silicon doping [49] influences the band gap only slightly, i.e., from 7.49 eV for **AD** to 6.34 eV (B3LYP/6-31+G(d,p)) [50] for its silicon analog $Si_{10}H_{16}$, which is the smallest repeating fragment of a solid silicon lattice and is experimentally known only as the fully TMS-protected form $Si_{10}(SiMe_3)_{16}$ [51].

The most well-known representative of a massively n-doped diamondoid is 1,3,5,7-tetraazaadamantane (urotropine), whose electronic properties are available [52]; its 2,4,6,10-C_{3v}-isomer was synthesized recently and also studied experimentally [53]. The availability of high-resolution spectral data for adamantane and urotropine [52] makes these two molecules excellent benchmarks for computational modeling of the electronic properties of doped diamondoids [53a, 54]. While the general agreement with the experiment is good, the main sources of computational errors arise from temperature effects that lead to level broadening (ionization spectra) and also from the highly diffuse nature of unoccupied Rydberg states (absorption spectra) [54a]. Doping of adamantane with nitrogen reduces the optical gap by ca. 1 eV relative to **AD** due to the high energy of the valence electrons on nitrogen (Fig. 12.4). In contrast to doped diamond, where nitrogen-vacancy centers typically cause visible luminescence, the incorporation of nitrogen into **AD** completely quenches the luminescence (Fig. 12.4).

Nitrogen doping reduces the symmetry of the adamantane HOMO, which represents an active Jahn–Teller (JT) system that strongly distorts upon ionization [55]. The fully symmetric a_1-HOMO of urotropine is localized on the nitrogen atoms and only slightly coupled to the triply degenerate t_2 HOMO-1, thus reducing the JT activity of the molecule. Quasi-static DFT computations on ionized urotropine confirm that the JT distortions in the radical cation are small relative to adamantane [56]. Computations on the highest-occupied states of **AD** and urotropine agree well with the low-energy parts of the experimental photoelectron spectra of these molecules. Large discrepancies in optical spectra simulations between the many-body perturbation theory methods and time-dependent DFT in the

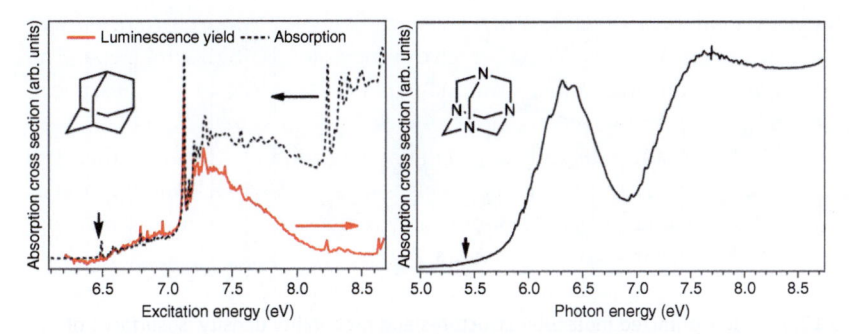

Figure 12.4 *Left*: Luminescence yield of **AD** compared to its optical absorption. The optical gap at 6.5 eV is marked by an arrow. *Right*: Optical absorption of urotropine. The optical gap at 5.42 eV is marked by an arrow. Source: Reproduced from Ref. [53a] with permission from the American Physical Society, 2009.

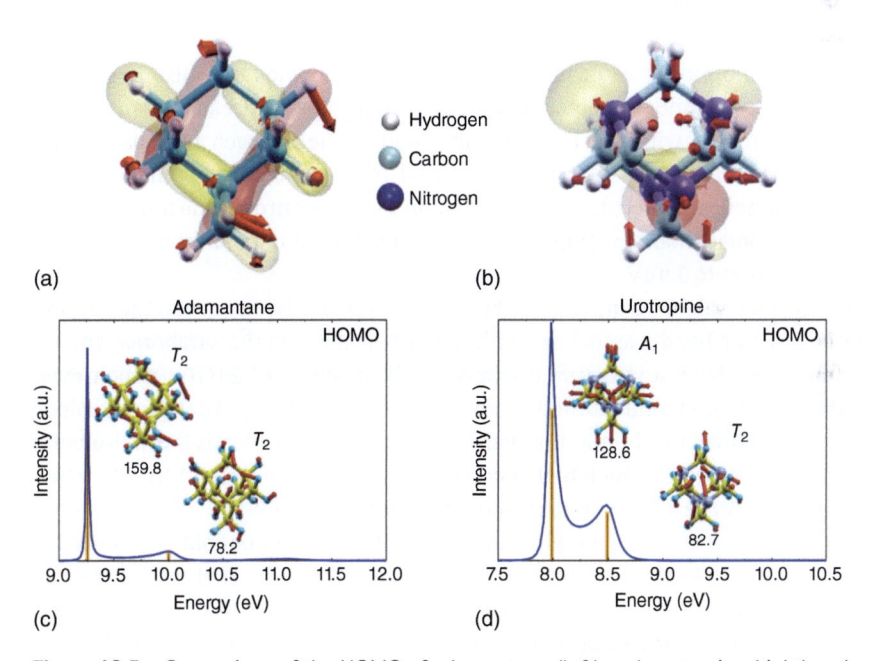

Figure 12.5 Comparison of the HOMO of adamantane (left) and urotropine (right) and calculated vibronic spectral functions for the labeled occupied states for which positive/negative isosurfaces are depicted by red/yellow lobes on the ball-and-stick model. Source: Reproduced from Ref. [56] with permission from Springer Nature, 2016.

adiabatic local density approximation were found but favor the former. Within the Green's function and the screened Coulomb interaction (G_0W_0) and G_0W + BSE approximations, the quasiparticle gap corrections can exceed thrice the DFT gap. The computed G_0W_0 quasiparticle levels agree well with the corresponding experimental vertical ionization energies (Fig. 12.5), where nuclear dynamics contribute significantly to the ionization process [54a].

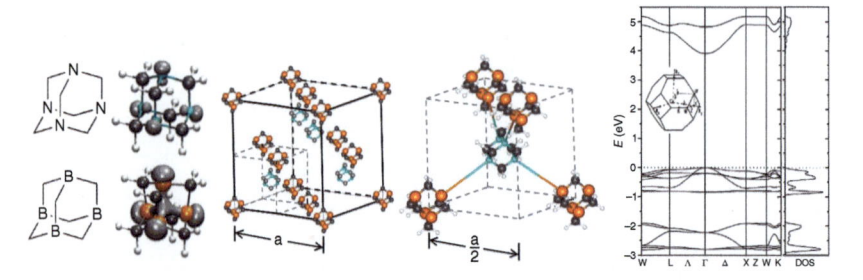

Figure 12.6 *Left*: Optimized molecular structures and probability density isosurfaces of the HOMO and LUMO of urotropine and 1,3,5,7-tetraboraadamantane. *Middle*: Schematic representation of the molecular crystal, composed of 1,3,5,7-tetraboraadamantane and urotropine (N: blue, B: orange). *Right*: Energy band structure along several high-symmetry directions and density of states of the molecular crystal. The inset shows the first Brillouin zone and the respective high-symmetry points. Source: Reproduced from Ref. [57c] with permission from the American Physical Society, 2009.

The electronic properties of internally doped diamondoids may also be altered by intermolecular donor–acceptor combinations. The most noticeable arrangement involves urotropine with the HOMO, comprised of the nitrogen lone pairs, and its boron analog with the LUMO, constructed from the empty boron p-orbitals [57]. A hypothetical molecular crystal based on the donor urotropine and acceptor 1,3,5,7-tetraboraadamantane (Fig. 12.6) was modeled computationally; its electronic band gap shrinks to 3.9 eV.

The top of the valence band is described as B ← N dative bonding, and the bottom of the conduction band is associated with the carbon atoms in the tetraboraadamantane fragments. Such a structure displays a bulk modulus of 20 GPa, which may provide high stability and stiffness at room temperature [57c]. Analogous molecular crystals formed by di-bora-di-azaadamantane are even stiffer as they have bulk moduli of 43 GPa with a much higher cohesive energy of 3.69 eV (cf. 1.81 eV for the urotropine/1,3,5,7-tetraboraadamantane molecular crystal) [58].

Internal doping changes the electronic properties of diamondoids not only by introducing extra electronic states into the band gap but also by slightly shifting the electronic levels (Fig. 12.7) [59]. While p-doping with boron (**B_DIA**) has essentially no influence on the valence band of **DIA**, the low-energy levels appear in the conduction band, thereby lowering the band gap to ca. 4 eV. Doping with nitrogen (**4N_DIA**) reduces the band gap due to lowering the energy of the conductance level.

Because of many similarities between hydrogen-terminated cubic boron nitride [60] and bulk diamond, studies were extended to boron nitrides up to the size of heptamantane [57a, 61]. DFT computations show that, in contrast to their all-carbon counterparts, the band gap almost vanishes in going from BN-adamantane to BN-heptamantane, thereby revealing strong quantum confinement effects. The HOMO–LUMO gaps of BN-adamantane, BN-triamantane, and BN-[121]tetramantane are smaller than the band gap of cubic boron nitride [61], although direct comparison of diamondoids of different sizes is not strictly correct because of different stoichiometries. For instance, BN-diamantane and

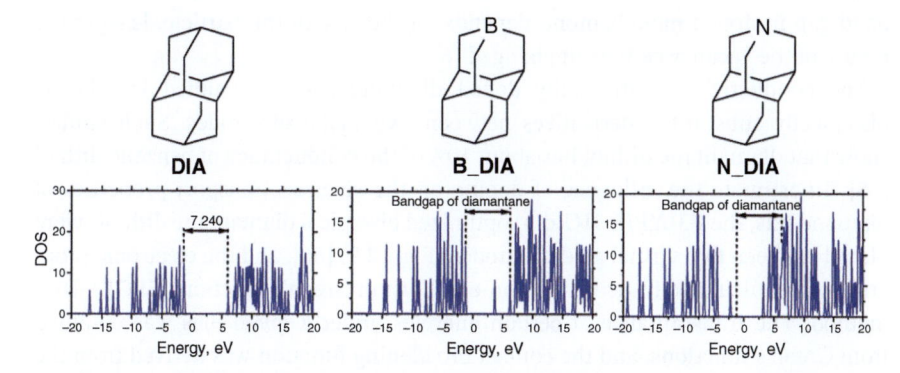

Figure 12.7 Density of states of diamantane (**DIA**, left), 4-boradiamantane (**B_DIA**, middle), and 4-azadiamantane (**4N_DIA**, right). Source: Reproduced from Ref. [59] with permission from the American Physical Society, 2004.

BN-tetramantane are isoelectronic with their corresponding C-diamondoids, whereas BN-triamantane and BN-adamantane are not. Due to such an imbalance, the ionization energies and electron affinities of BN-diamondoids do not follow the same trends found for diamondoids [61]. Additional stabilization arises from the 3D-aromaticity of molecules with $[4n+2]\pi$-electrons [62]. In general, the HOMO–LUMO gaps of BN-diamondoids are smaller than the band gap of bulk cubic boron nitride (ca. 6.1 eV) and may be varied to reach the semiconductor level [57b].

While doping with oxygen only slightly affects the electronic properties of adamantane [54a], with powerful n-dopants such as sulfur, the valence band level rises substantially relative to the undoped molecule due to contributions to the HOMO from the sulfur lone pairs [63]. The size of the molecule has almost no influence on the conduction band structure of thiadiamondoids. The HOMO level is largely size-independent in thiadiamondoids; however, the valence band density increases due to the energy increase of the lower occupied HOMO-n states. For doped triamantanes (Fig. 12.8), the position of the dopant has a negligible influence on the B3PW91/6-31G(d,p) computed HOMO–LUMO gaps. Due to the dominant contributions of the atomic orbitals of the dopant into the HOMOs and the LUMOs, the narrowing of the gap is determined only by the nature of the dopant, rather than by the size of the cage. Such trends agree well with the observations that the

Figure 12.8 For doped triamantanes, the nature rather than the position of the dopant has an influence on computed HOMO–LUMO gaps. Source: From Ref. [63].

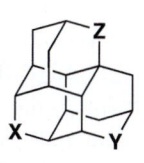

X	Y	Z	HOMO–LUMO gap, eV
CH_2	CH_2	CH_2	8.7
O	CH_2	CH_2	8.1
S	CH_2	CH_2	6.8
CH_2	O	CH_2	8.1
CH_2	S	CH_2	6.8
CH_2	CH_2	O	8.2
CH_2	CH_2	S	6.7

band gap in doped nanodiamond depends on the size of the particle, leaving the nature of the vacancy centers unchanged [64].

The conductivity of internally doped diamondoids was studied by placing bis-apically substituted derivatives between two metal electrodes. Such studies show that diamantane dithiol has about 10% of the conductance of benzene dithiol [65]. To estimate the influence of doping on the electron transport properties of diamondoids, the B3LYP/6-31G(d,p) optimized bis-apical diamantane dithiols were placed between two virtual gold electrodes (Fig. 12.9) [66], and the electronic properties were calculated by using the non-equilibrium Green's function (NEGF)–DFT method. The transmittance at certain energies under a given bias was obtained from Green's functions, and the contact broadening function was derived from the electrode's self-energy. The transmittance spectra reflect the electron tunneling probability at a given energy level and the transmittance at the Fermi level provides information about conduction. The HOMO of **DIA** confined between two electrodes is far away from the Fermi level, which is shifted to zero, resulting in poor conductivity (Fig. 12.9a). Although the HOMO of 1-phosphadiamantane is mostly comprised of the diffuse phosphorus 3p-orbitals, the transmittance is still too low to increase conductance. In contrast, the high-lying HOMO of 1-azadiamantane is located much closer to the Fermi level, causing relatively high conductivity, while 1-boradiamantane has a low-lying LUMO located too far from the Fermi level, and this molecule again behaves as a poor conductor.

Theoretical studies of electric junctions constructed from doped diamondoids were extended to larger systems, namely to **121TET** bis-apical di-thiols [67]. As an alternative to thiol functions, an N-heterocyclic carbene was considered as the connecting point for the junction. The di-thiol linker is more efficient as it introduces additional electron transport paths. N-doping leads to a considerable increase in conductance because of the asymmetric nature of the I–V curves within a very small range of bias voltages [67]. These results are in agreement with the influence of doping on the tunnel magnetoresistance (TMR) of dithiolated diamondoids placed between two ferromagnetic nickel electrodes. DFT computations predict that boron doping considerably decreases while nitrogen doping increases the TMR ratio [65]. Again, the enhanced conducting properties of nitrogen-doped diamantanes are attributed to the shift of the HOMO closer to the Fermi level.

Computationally, N-doped lower and higher diamondoids exergonically form very tight sandwich-like complexes with alkali metal ions Li^+, Na^+, K^+, and benzene (Fig. 12.10) [68]. Generally, complexations involving lithium ions are the strongest. The diamondoid size affects the complex stability only within a few kcal mol^{-1} and better correlates with the diamondoid HOMO level. The stability of the complexes toward dissociation results from the combination of covalent, electrostatic, and dispersion interactions and provides up to 80 kcal mol^{-1} energy gain for the formation of nitrogen-doped pentamantane **PENT_N_Li_PhH** sandwich complex. Surprisingly, contributions from London dispersion to the stability of such complexes are relatively small, as B3LYP, B97X-D, and M06-2X gave similar complexation energies (74.2, 79.5, and 79.2 kcal mol^{-1}, respectively), where the inclusion of dispersion corrections gives only ca. 5 kcal mol^{-1} energy gain.

Figure 12.9 *Top*: Computational modeling of diamantane bis-apical dithiol placed between two gold electrodes as a model of a doped conductive junction. *Middle*: The transmittance spectra of diamantane (a) and 1-phosphadiamantane (b); *Bottom*: The transmittance spectra of 1-azadiamantane (c) and 1-boradiamantane (d). Relatively high conductivity is predicted only for 1-azadiamantane due to HOMO located much closer to the Fermi level. The blue lines indicate the energy level of the projected self-consistent Hamiltonian states of the HOMO and LUMO. Source: Reproduced from Ref. [66] with permission from Elsevier, 2015.

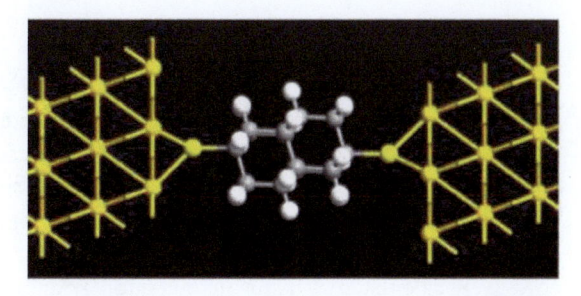

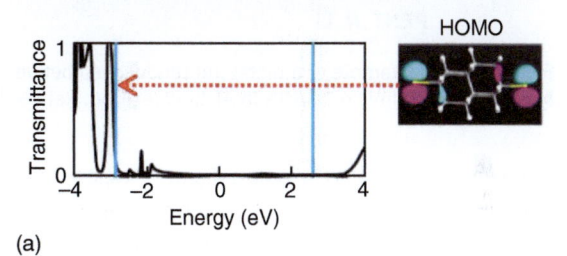

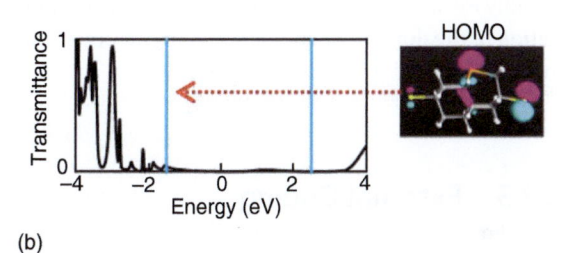

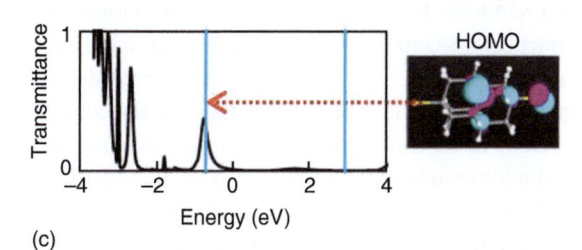

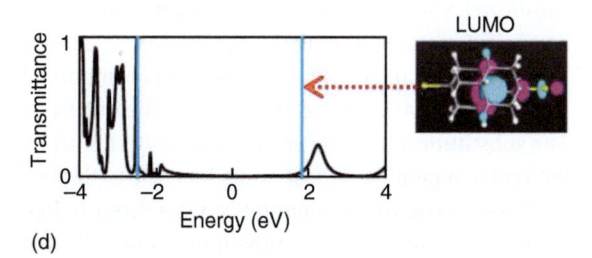

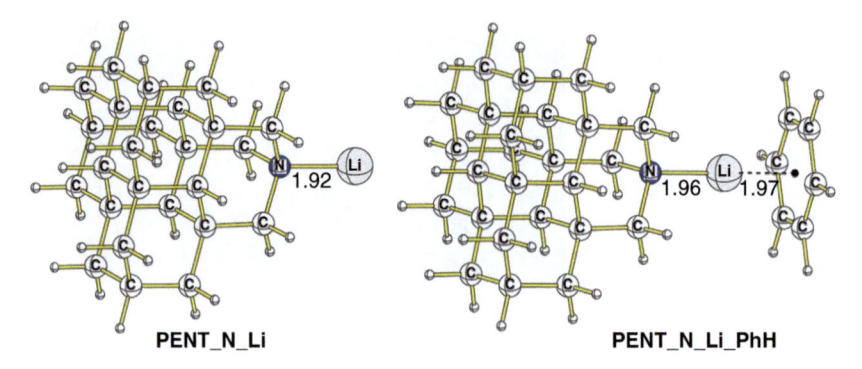

Figure 12.10 Example of a metal ion sandwiched between N-doped **1(2,3)4PENT** and benzene (interatomic distances in Å). Source: From Ref. [68].

Recently [69], the reorientations of nitrogen, boron, and oxygen-doped diamondoids up to C_{198} in size were modeled computationally with the possibility of a heavy atom quantum tunneling mechanism. The relative tunneling contribution decreases rapidly for smaller cages, but the addition of an external electric field causes reorientation of the dopant. Such phenomena can potentially be useful for the construction of atomic memory devices.

12.3 External Doping

C—H bond functionalizations allow effective tuning of the electronic properties of diamondoids [63, 70], as such "external doping" is the most common way for the practical utilization of diamondoids, and many diamondoid derivatives of this type are available [71]. As in the case of internal doping, the frontier orbitals of externally doped diamondoids display substantial contributions from the cage atoms and thereby affect the HOMO–LUMO gaps substantially. The most prominent external p-doping comes with highly electropositive alkali metals (Li [72], Na [72b, 73], and K [74]) that lead to structures with high-lying (carbanion-type) HOMOs and substantially reduced band gaps. Adamantanes $C_{10}H_{16-n}Na_n$ where tertiary hydrogens are substituted with sodium atoms computationally reduce the band gap to a semimetallic or even metallic level in the gas phase and in the bulk (Fig. 12.11) [73]. A significant reduction in HOMO–LUMO gaps is computed for alkali metal substitution in **DIA**, where **C14H19K** has the lowest value of 1.64 eV (cf. 8.88 eV for pristine **DIA**). The substitution with alkali metals leads to a shift in absorption from ultraviolet to the visible region, and these shifts increase with the size of the metal [75].

Although sodium adamantyls are far from being realistic and practical due to their extreme lability, some other organometallic diamondoids are quite stable. For instance, 1-adamantylzinc bromide is even available commercially.

The opposite extreme case of external doping is fluoroalkyl and fluorine substitutions that increase the HOMO–LUMO gaps, thereby also increasing the insulating

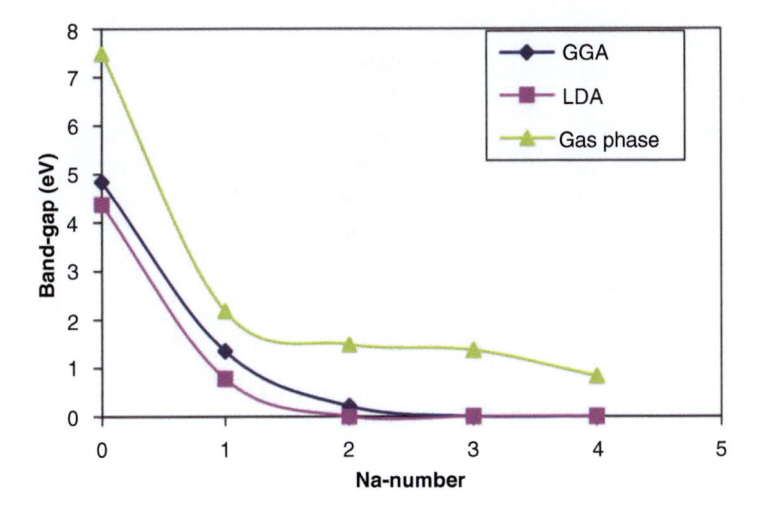

Figure 12.11 Computed band gap of adamantane depending on the number of sodium atoms replacing hydrogens at the tertiary positions. Source: Reproduced from Ref. [73] with permission from Elsevier, 2010.

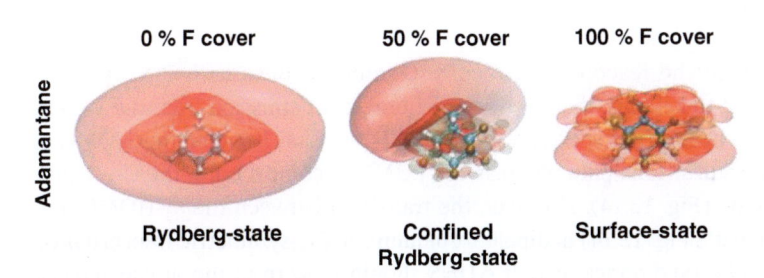

Figure 12.12 LUMO states of fluorinated adamantane at 0%, 50%, and 100% fluorine coverage. Source: Reproduced from Ref. [78] with permission from the American Chemical Society, 2008.

properties of diamondoids [63, 76]. Upon electron-acceptor substitution with the highly electronegative fluorine atoms, the HOMO energy is reduced [77] while the LUMO level remains nearly constant; this increases the HOMO–LUMO gap. When approximately 50% of all C—H bonds of adamantane are replaced by fluorine [78], the LUMO adopts Rydberg-type surface states (Fig. 12.12).

Hopes are associated with the combination of fluorine and alkali metal substitutions of the diamondoid cages. A systematic study on alkali metal-substituted fluoroadamantanes was conducted up to $LiC_{10}F_{15}$ [79]. Such a combination shifts the absorption wavelengths within UV to IR regions, where the increase in the number of fluorine substitutions causes the bathochromic shift that is desirable for photovoltaics (Fig. 12.13).

Remarkably, external C—S doping leads to absorptions in the visible region and makes such diamondoids relevant for optoelectronic applications [80]. Red–orange crystalline adamantanethione (**AD=S**) and orange 2,6-adamantanedithione were

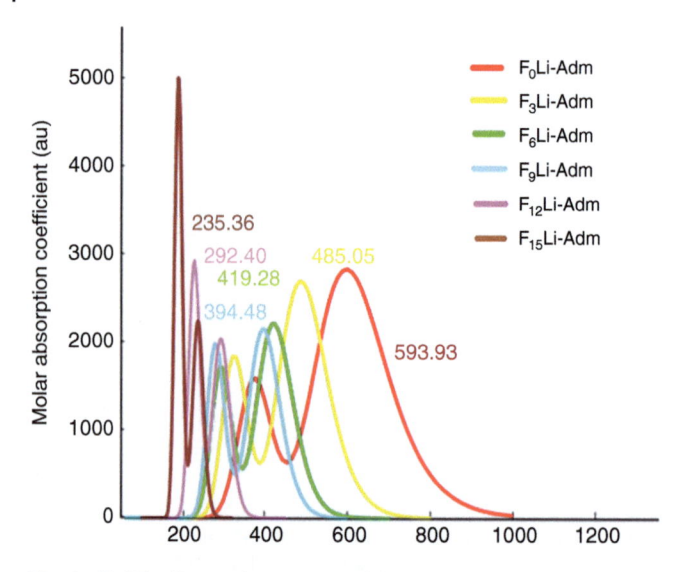

Figure 12.13 Absorption spectra of fluorinated Li-adamantanes obtained with CAM-B3LYP/aug-cc-pVDZ level of DFT. Source: Reproduced from Ref. [79] with permission from Elsevier, 2019.

prepared through the reaction of the corresponding ketones with P_4S_{10} [81]. In marked contrast to ketones, there is a strong fluorescent emission from the $S_2 \rightarrow S_0$ transition. The adsorption spectrum of **AD=S** is unusual because the HOMO is located on the sulfur lone pair, whereas the LUMO is comprised of the C—S antibonding π^* state (Fig. 12.14). However, the transition between the b_2-HOMO and b_1-LUMO (n → π^*, Fig. 12.14) is dipole forbidden for C_{2v}-symmetric thiocarbonyls [83], and the observed optical gap of **AD=S** should arise from the dipole-allowed transition from the HOMO to the sulfur 4s-Rydberg LUMO+1 state (the TD-DFT computed value is 5.03 eV [82]). Together with other symmetry-allowed transitions from HOMO-1 and HOMO-2 to π^*, these absorptions are observed in the UV spectrum (223 and 240 nm) of **AD=S** both in solution and in the gas phase [81a, 84]. However, these transitions are too high in energy to be responsible for the experimentally visible light absorption of **AD=S** [81a, 84, 85]. Such a low-energy symmetry forbidden n → π^* transition in C_{2v}-symmetric **AD=S** may be rationalized based on the vibronic interaction mechanism where orbitally forbidden transitions may exchange energy with the nearest dipole-allowed states [83a]. This explains the observable but very low-intense absorption of **AD=S** at 488 nm (ε = 11.3) and 554 nm (ε = 2.7), while the intensity of symmetry-allowed absorption in the UV region is much higher (ε = 12.700) [81a].

The addition of more than one thione group to adamantane leads to a further decrease in the optical gap. For adamantane-2,4-dithione and 2,6-dithione, the weakly allowed HOMO → LUMO (2.39 eV) and HOMO − 1(HOMO) → LUMO (LUMO + 1) (3.34 eV) transitions lead to strong absorptions in the visible region (Fig. 12.14d) [54b], with the extreme case being tetrathione, which displays an optical absorption near the RGB primary colors [80].

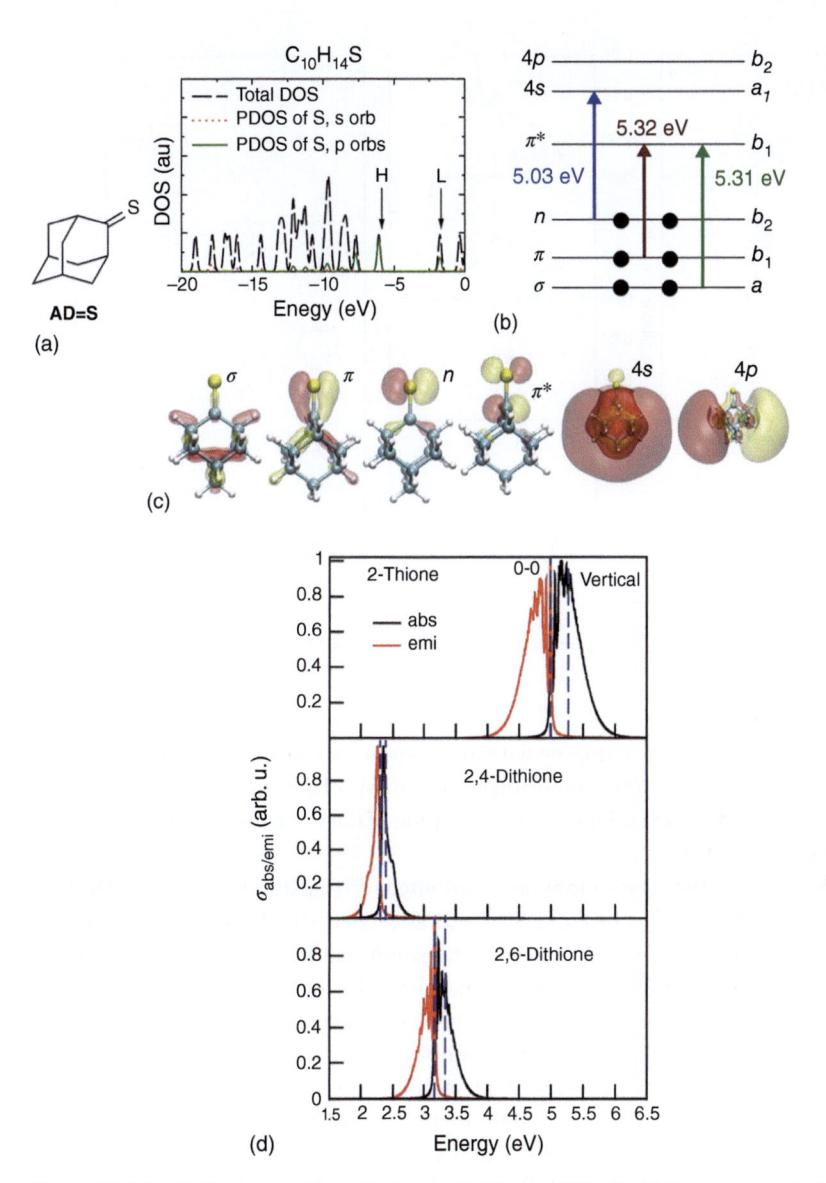

Figure 12.14 Optical properties of adamantanethione (**AD=S**). (a) Total and projected density of states (PDOS) of **AD=S**, where H and L denote the HOMO and LUMO energies. (b) Allowed transitions in **AD=S** with corresponding excitation energies. (c) Selected molecular orbitals of **AD=S**. Source: Reproduced from Ref. [82] with permission from the American Physical Society, 2005. (d) Vibronic absorption and emission spectra of the diamondoid thiones computed at the B3LYP/TZVP level of theory. The dashed blue lines represent the 0–0 and vertical transition energies for all three plots. Source: Reproduced from Ref. [54] with permission from the Royal Society of Chemistry, 2015.

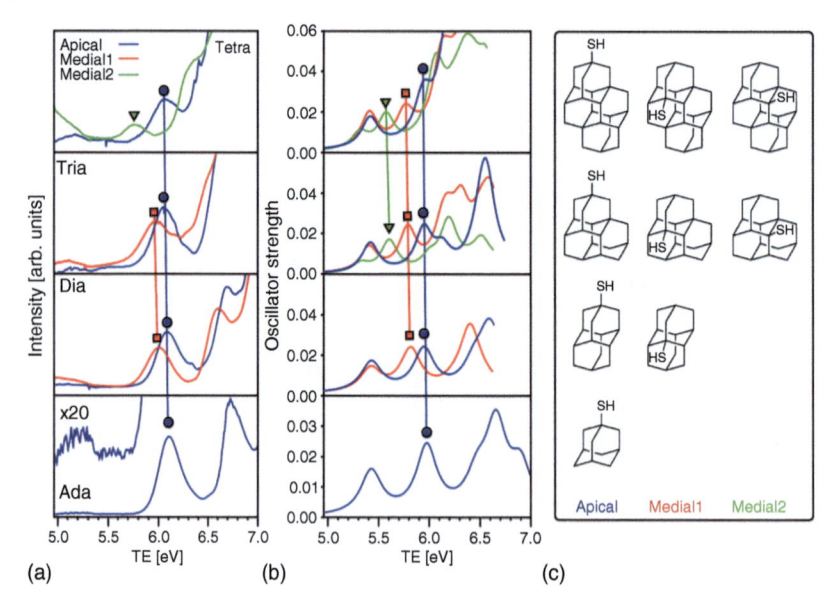

Figure 12.15 Comparison of experimental (a) and computed (b) absorption spectra of diamondoid thiols (c). Source: Reproduced from Ref. [86] with permission from AIP Publishing, 2010.

The influence of external doping with an SH-group on the electronic properties of diamondoids was also studied experimentally [86]. As observed before, their absorptions in the deep UV depend more on the position of the thiol group than on diamondoid size (Fig. 12.15).

In most cases, the absorptions are attributed to transitions from the HOMO to the SH (first excited state) and SC (second excited state) antibonding orbitals (Fig. 12.16). The lowest energy n → σ_{SH}^* transition that leads to dissociation of the SH bond is almost independent of diamondoid size [87], whereas the n → σ_{SC}^* transition depends not only on the size of the molecule but slightly also on the position of the SH group in the diamondoid (Fig. 12.16). The σ → S transition belongs to an excited Rydberg state of adamantane thiol; however, upon increasing the size of the diamondoid, this cage-originating transition shifts to lower energies with an energetic crossover of σ → S and n → σ_{SH}^* transitions for thiols around pentamantane (Fig. 12.16, right). Beyond this point, diamondoid-like transitions will be the energetically lowest transitions in diamondoid thiols, and the excitations of the diamondoid cage will dominate. This not only allows band gap tuning in the higher diamondoid thiols but also predicts that higher diamondoid thiols should display UV photoluminescence.

The incorporation of a phosphine group (PH_2) is another example of external n-doping of diamondoids. The stability of the ensuing primary phosphines toward air is usually quite low and is determined by the SOMO level of the respective radical cations presumably formed en route in the reaction with oxygen [88]. The higher stability of diamondoid phosphines is determined additionally by the steric bulkiness of the cage and the nature of the HOMO, which is highly

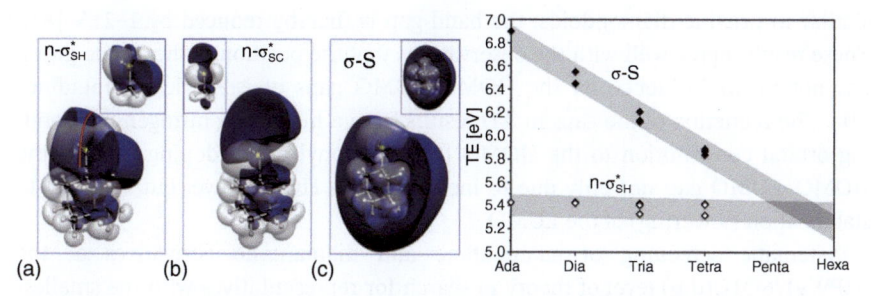

Figure 12.16 *Left*: Electron density difference between excited states and the ground states of adamantane thiol, corresponding to n → σ_{SH}^*, n → σ_{SC}^* and σ → S transitions. *Right*: Size dependence of n → σ_{SH}^* and σ → S transition energies for various diamondoid thiols. While the n → σ_{SH}^* is size-independent, the energy of the σ → S transition decreases with the cage size reaching the crossover point around pentamantane. Source: Reproduced from Ref. [86] with permission from AIP Publishing, 2010.

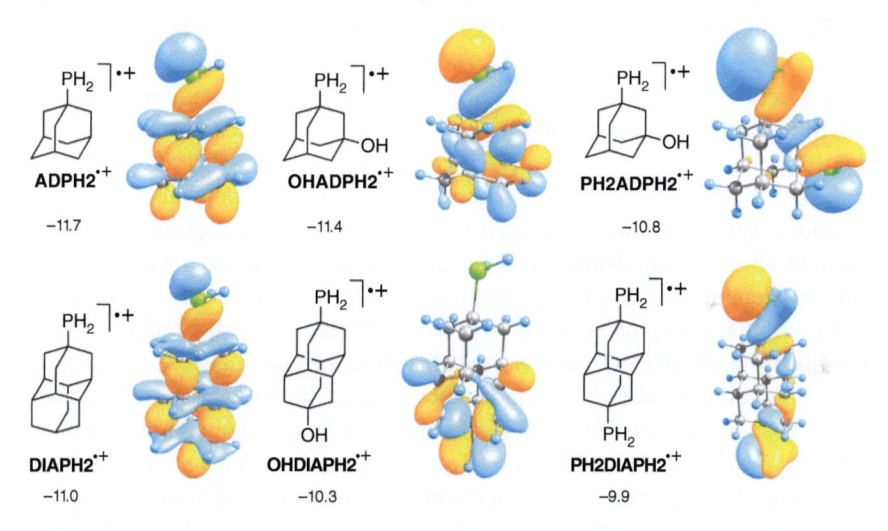

Figure 12.17 Shapes and energies (in eV) of the SOMOs for the B3LYP/6-311+G(d,p) optimized structures of the radical cations derived from selected primary diamondoid phosphines. Source: Reproduced from Ref. [89] with permission from the American Chemical Society, 2016.

delocalized over the entire molecule [89]. The experimentally observed stabilities [89] of primary diamantyl phosphines toward oxygen compared to their adamantyl [90] analogs agree well with the SOMO energy levels of the corresponding phosphine radical cations obtained by DFT computations. The unusual stability of 9-hydroxydiamantyl-4-phosphine (**OHDIAPH2**) is due to the location of the SOMO predominantly on the oxygen atom rather than on the phosphorus atom (Fig. 12.17).

The HOMO of externally NH_2-doped diamondoids is more localized on the nitrogen lone pair, and the LUMO is completely represented by the NH_2-group. In contrast to thiols, where excitation quenches the photoluminescence due to SH bond breaking, external doping with an NH_2 group retains the UV emission similar to pristine diamondoids; the band gap is thereby reduced by 1–2 eV [87]. These results agree well with the observations that the position of the amino group has not much influence on the HOMO–LUMO gaps of isomeric rimantadines [91]. The reduction of the gaps in **AD** results from a high-lying nitrogen nonbonding orbital contribution to the HOMO [91]. Carbonyl group doping reduces the HOMO–LUMO gap not only due to increasing the HOMO level but also due to stabilization (lowering) of the LUMO.

Systematic screening of adamantane and diamantane derivatives at the B3PW91/6-31G(d,p) level of theory in search for representatives with the smallest and highest HOMO–LUMO energy gaps considered 10 functionalization sites [76] and resulted in HOMO–LUMO gaps ranging from 2.42 to 10.63 eV. The maximal values were obtained by doping with F- and CF_3-groups (Fig. 12.18).

The lowest gaps are associated with push-pull substitution, with the push effect being due to sulfur or nitrogen dopants and the pull effect due to the nitro-group (Fig. 12.18, top). The latter is responsible for lowering the LUMO level, while the internal doping with nitrogen atoms increases the HOMO level. The HOMO–LUMO gaps varying in a very wide range (2.94–10.08 eV) were predicted for externally doped structures (Fig. 12.18, bottom). Earlier, very low band gaps were only achieved in diamondoid betaine structures (1.8 eV for 4-amino-diamantane-9-carboxylic acid) [63]. This effect was recently confirmed through TD-DFT computations at PBE1PBE/6-31+G(d,p) level of theory, where it was found that the vertical excitations were substantially reduced by the electron-acceptor groups [92]. The push-pull effect is also distinctive for functionalized adamantanes with aromatic electron accepting and donating groups. The combination of benzonitrile (acceptor) and carbazole (donor) moieties reduces the experimental optical band gap [93] of such 4-benzonitrile (**Cz_AD_PhCN**) to 3.52 eV (Fig. 12.19a) [93].

DFT computations show that the frontier orbitals of **Cz_AD_PhCN** are completely separated, as the HOMO is located on the electron-rich carbazole moiety while the LUMO is located exclusively on the electron-deficient benzonitrile group (Fig. 12.19a). Such a combination effectively hampers intramolecular charge transfer and is useful for the construction of electrophosphorescent light-emitting devices. The bis[(4,6-difluorophenyl)pyridinato-N,C2'](picolinato)iridium (III) (FIrpic)-based OLED device hosted by **Cz_AD_PhCN** exhibits the highest current efficiency among blue phosphorescent OLEDs due to the high energy of the triplet state, which confines excitons to the emitting layer [93]. Another combination of adamantane-containing rigid donor 10H-spiro[acridine-9,2'-adamantane] with 2,4,6-triphenyl-1,3,5-triazine as an acceptor in **TPT_AD_AC** leads to an unusual blue OLED device (Fig. 12.19b) [94]. The rigid adamantane core not only stabilizes the excited state but also induces the formation of two conformeric forms of **TPT_AD_AC**. This causes a deep-blue conventional fluorescence in combination with sky-blue thermally activated delayed fluorescence with a quantum efficiency that is the highest reported for OLEDs based on dual-conformation emission. The adamantane-bridged phenanthroimidazole **AD_BPI** with two twisted chromophores (Fig. 12.19c) and strong HOMO and LUMO overlap (Fig. 12.19d) displays effective hole and electron transport capabilities and gives a material that

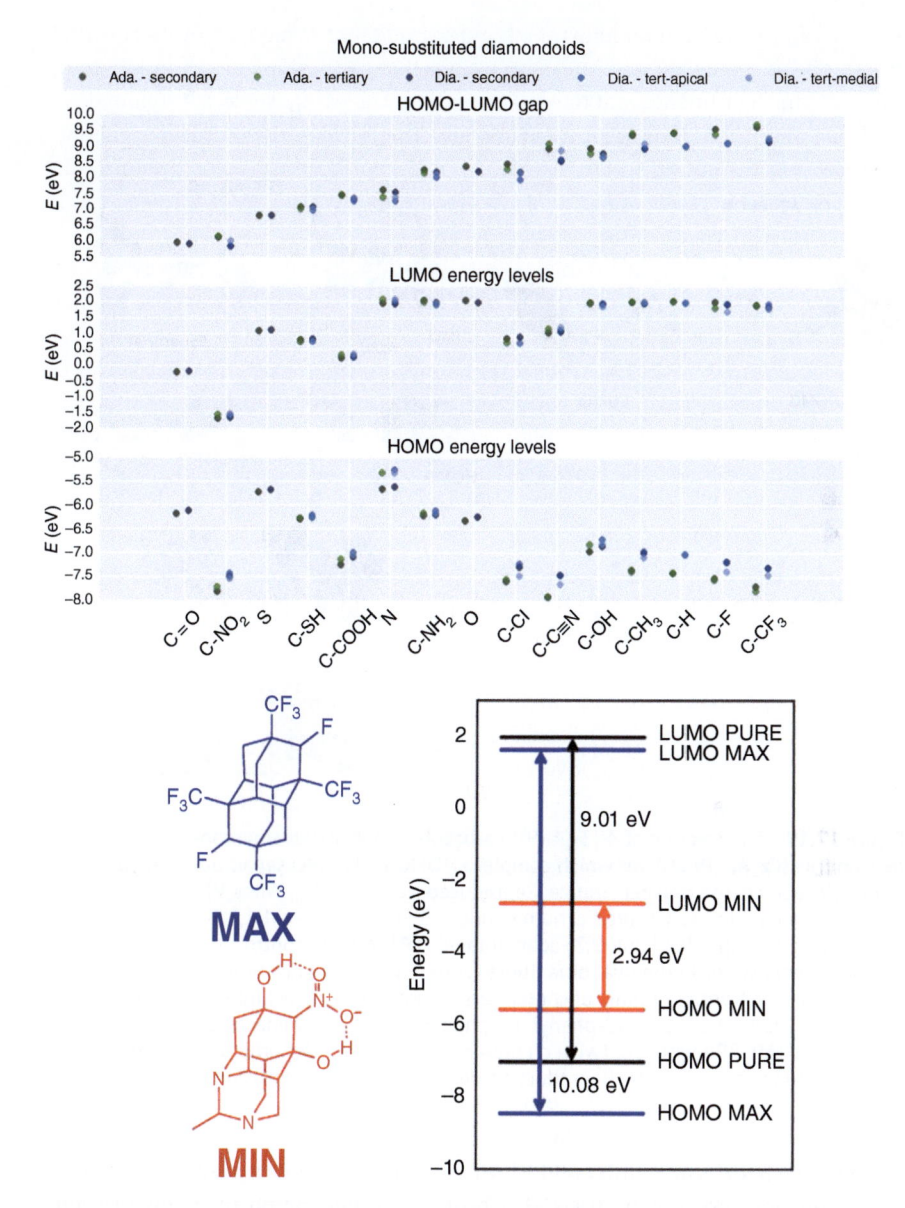

Figure 12.18 *Top*: HOMO–LUMO gaps and frontier molecular orbital levels (eV) for monosubstituted diamondoids. *Bottom*: Frontier molecular orbital levels of pristine diamantane (black) and minimal (red) and maximal (blue) HOMO–LUMO energy gaps of selected externally doped structures (bottom left). Source: Reproduced from Ref. [76] with permission from the American Chemical Society, 2017.

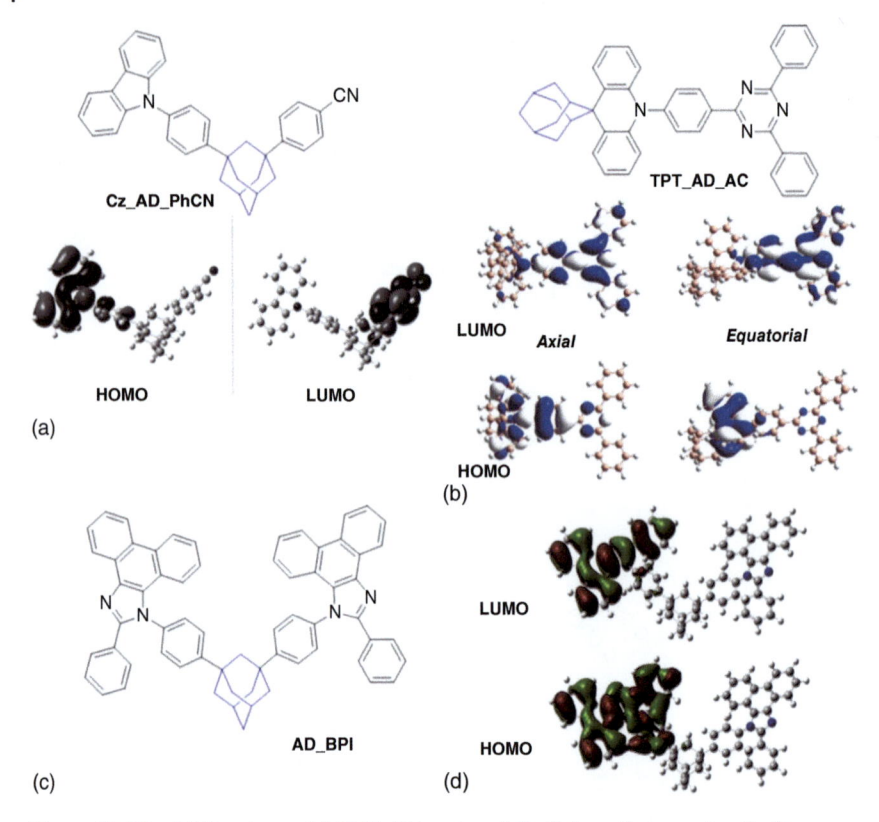

Figure 12.19 (a) Structure of 4-[3-[4-(9H-carbazol-9-yl)phenyl]adamantan-1-yl] benzonitrile (**Cz_AD_PhCN**), for which complete HOMO and LUMO separation hampers the intramolecular charge transfer. Source: Reproduced from Ref. [93], Wiley-VCH, 2015. (b) Rigid adamantane-substituted acridine donor 10-(4-(4,6-diphenyl-1,3,5-triazin-2-yl)phenyl)-10H-spiro[acridine-9,2′-adamantane] (**TPT_AD_AC**) controls the conformational states in the dual blue fluorescence emission OLED device and the structures of the frontier orbitals of conformers. Source: Reproduced from Ref. [94], Wiley-VCH, 2019. (c) 1,3-bis(4-(2-phenyl-1H-phenanthro[9,10-d]imidazol-1-yl)phenyl) adamantane (**AD_BPI**) with two twisted chromophores; (d) frontier orbitals of **AD_BPI** display strong overlap. Source: Reproduced from Ref. [95] with permission from Elsevier, 2020.

may be used both as an emitter and a host. It not only exhibits a deep blue emission at 422 nm but may also be used as a host in the fabrication of highly efficient red/yellow/green phosphorescent OLEDs.

The perylene-3,4,9,10-tetracarboxylic bis-imide moiety (Fig. 12.20, left) exhibits chemical and photochemical stability as well as high fluorescence and, when connected to bulky groups, can be influenced by noncovalent interactions that cause hypsochromic shifts in the UV/vis absorption and fluorescence spectra [96]. As the quantum yields remain unaffected such dyes are useful for fine-tuning of UV/vis spectra [96]. The energy transfer in donor/acceptor perylene dyads constructed around the rigid diamantane core in **PERYLEN_DIA** (Fig. 12.20, right) displays highly efficient fluorescence with quantum yields >90%. Notably, the

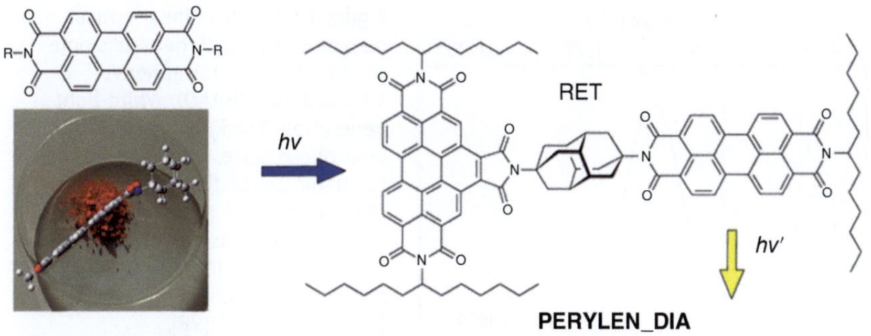

Figure 12.20 *Left*: Adamantane-containing perylene dye, whose spectral hypsochromic shifts are caused by noncovalent interactions. Source: From Ref. [96]. *Right*: Resonance energy transfer (RET) across the orthogonal perylene moieties connected by diamantane fragments in **PERYLEN_DIA** that resemble a mirror placed at 45° toward the direction of the incoming light. Source: From Ref. [97].

energy transfer occurs in a system with two orthogonal chromophores, contradicting fluorescent (Förster) resonance energy transfer (RET) theory, which requires effective dipole–dipole interactions. The fact that two chromophores are linked by a rigid, saturated cage indicates that energy transfer does not require p-orbital overlap between donor and acceptor. It was suggested [97] that vibronic coupling is responsible for the efficient transition between the orthogonal fragments of **PERYLEN_DIA**. As a result of the orthogonal position of the chromophores, the incoming light undergoes a 90° flip, i.e., the molecule resembles a mirror placed at 45° toward the direction of the incoming light. Such unusual energy transfer may find applications in light collection and handling optoelectronic devices.

Recently, it was found that single crystals of 1,3,5,7-tetraphenyladamantane (**PH4AD**) exhibit narrow-band second-harmonic light generation upon irradiation with IR laser light (Fig. 12.21) [98]. The emission covers nearly the entire visible spectrum similar to the one observed for inorganic Sn/S-adamantane analog **PH4SNSAD** [98]. The calculations of the nonlinear optical response reveal an overall enhancement of the optical susceptibilities for strongly dispersion-bonded dimeric **PH4AD** structures [99]. Modifications of the **PH4AD** structure by introducing substituents on the aromatic moieties cause a blue shift in the emission maximum [100]. A study of the influence of cage bromine substitution in phenyl adamantanes (AdBr$_x$Ph$_{4-x}$) reveals the key mechanistic features of this type of optical response [101]. While irradiation has only little influence on the crystal of **PH4AD**, for the brominated derivatives the Coulomb forces result in an electronic excitation that causes debromination.

As diamondoids are effective dispersion energy donors [102], they may form strongly bound clusters with other donor molecules, a finding that may be considered as "dispersion" doping. The latter remains largely undisclosed in diamondoid chemistry. Recently, very fast charge transfer in perylene tetracarboxylic acid diimide (PDI) vdW complexes with higher diamondoids (up to C$_{190}$H$_{110}$) was predicted (Fig. 12.22) [12]. Due to the size and rigidity of both entities, the electron

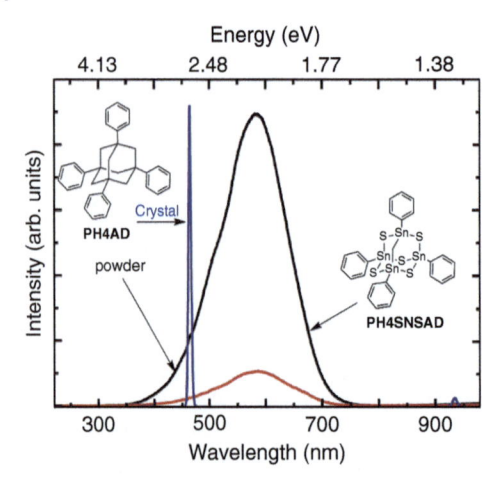

Figure 12.21 Nonlinear optical response of the powder and single crystal of 1,3,5,7-tetraphenyl-adamantane (**PH4AD**). White-light generation (black) is observed in the amorphous state, while second-harmonic generation is characteristic of the crystalline state (blue). The nonlinear response of respective Sn/S adamantane analog (**PH4SNSAD**) powder (red) is given as a reference. Source: From Ref. [98].

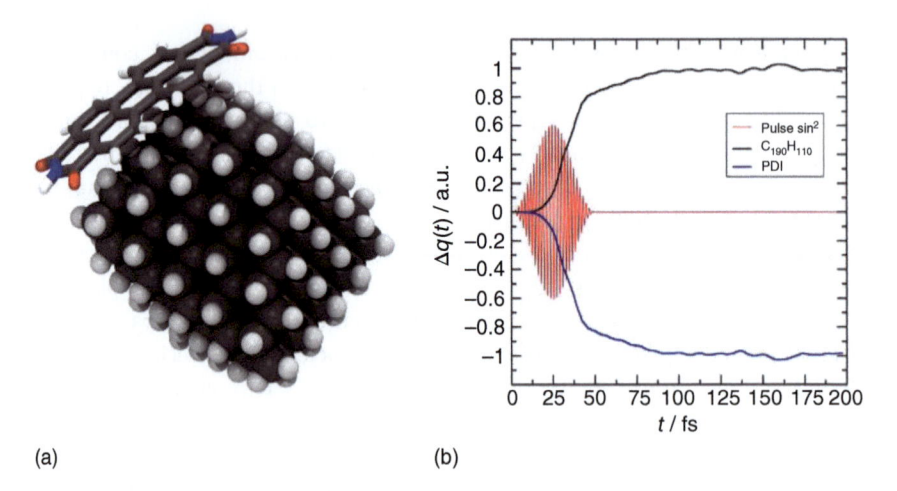

Figure 12.22 (a) Example of "dispersion" doping of diamondoid represented by the complex of the diamondoid $C_{190}H_{110}$ with diimide of perylene tetracarboxylic acid; (b) charge as a function of time for donor (black) and acceptor (blue). The squared-sinusoidal pulse perturbation in tune with the HOMO–LUMO transition of the acceptor is represented in red. Source: Reproduced from Ref. [12] with permission from the American Chemical Society, 2018.

transfer within this dispersion-dimer system occurs without polaron formation or nuclear relaxation, reaching a steady state originating only from electronic density redistributions. After charge transfer, the system demonstrates high stability and avoids charge recombination as the diamondoid plays the role of an effective hole acceptor. This is in full agreement with the ability of diamondoids to form positive ionized states with a high degree of spin/charge delocalization [103].

Very recently [104], the geometric, electronic, and nonlinear optical responses of adamantane complexes with alkali metal oxides M_3O (M = Li, Na, and K) were computed at DFT. Such externally doped $M_3O\bullet\bullet\bullet AD$ vdW-clusters substantially reduce the energy gaps (down to 2.19–3.06 eV) and shift the absorption maxima to the near-infrared region.

References

1 Clay, W.A., Sasagawa, T., Kelly, M. et al. (2008). Diamondoids as low-κ dielectric materials. *Appl. Phys. Lett.* 93: 172901.
2 Gaillou, E. and Rossman, G.R. (2013). Color in natural diamonds: the beauty of defects. *Rocks Miner.* 89: 66–75.
3 Liu, F.K., Guo, Y., Zhao, B., and Li, X. (2022). The color origin and evaluation of natural colored diamonds. *Sci. Technol. Adv. Mater.* 14: 243–256.
4 (a) Geis, M.W., Twichell, J.C., Efremow, N.N. et al. (1996). Comparison of electric field emission from nitrogen-doped, type Ib diamond, and boron-doped diamond. *Appl. Phys. Lett.* 68: 2294–2296. (b) Okano, K., Koizumi, S., Silva, S.R.P., and Amaratunga, G.A.J. (1996). Low-threshold cold cathodes made of nitrogen-doped chemical-vapour-deposited diamond. *Nature* 381: 140–141.
5 Yamada, T., Masuzawa, T., Mimura, H., and Okano, K. (2016). Electron emission from conduction band of heavily phosphorus doped diamond negative electron affinity surface. *J. Phys. D* 49: 045102.
6 (a) Diederich, L., Kuttel, O., Aebi, P., and Schlapbach, L. (1998). Electron affinity and work function of differently oriented and doped diamond surfaces determined by photoelectron spectroscopy. *Surf. Sci.* 418: 219–239. (b) Diederich, L., Kuttel, O.M., Ruffieux, P. et al. (1998). Photoelectron emission from nitrogen- and boron-doped diamond (100) surfaces. *Surf. Sci.* 417: 41–52.
7 (a) Kalish, R. (1998). Doping diamond for electronic applications. *Isr. J. Chem.* 38: 41–50. (b) Kalish, R. (1999). Doping of diamond. *Carbon* 37: 781–785. (c) Barnard, A.S., Russo, S.P., and Snook, I.K. (2005). Simulation and bonding of dopants in nanocrystalline diamond. *J. Nanosci. Nanotechnol.* 5: 1395–1407. (d) Ristein, J. (2006). Surface transfer doping of diamond. *J. Phys. D* 39: R71–R81. (e) Koizumi, S. and Suzuki, M. (2006). n-Type doping of diamond. *Phys. Status Solidi A* 203: 3358–3366. (f) Nesladek, M. (2005). Conventional n-type doping in diamond: state of the art and recent progress. *Semicond. Sci. Technol.* 20: R19–R27. (g) Kraft, A. (2007). Doped diamond: a compact review on a new, versatile electrode material. *Int. J. Electrochem. Sci.* 2: 355–385. (h) Williams, O.A. (2006). Ultrananocrystalline diamond for electronic applications. *Semicond. Sci. Technol.* 21: R49–R56. (i) Yang, N.J., Foord, J.S., and Jiang, X. (2016). Diamond electrochemistry at the nanoscale: a review. *Carbon* 99: 90–110. (j) Evtukh, A., Hartnagel, H.L., Yilmazoglu, O. et al. (2015). Carbon-based quantum cathodes. In: *Vacuum Nanoelectronic Devices: Novel Electron Sources and Applications*, p314–p374. Chennai: Wiley-VCH. (k) Giubileo, F., Di Bartolomeo, A., Iemmo, L. et al. (2018). Field emission from carbon nanostructures. *Appl. Sci.* 8: 526. (l) Rodgers, L.V.H., Hughes, L.B., Xie, M.Z. et al. (2021). Materials challenges for quantum technologies based on color centers in diamond. *MRS Bull.* 46: 623–633.

8 Kalish, R. (2007). Diamond as a unique high-tech electronic material: difficulties and prospects. *J. Phys. D* 40: 6467–6478.
9 Velardi, L., Valentini, A., and Cicala, G. (2017). UV photocathodes based on nanodiamond particles: effect of carbon hybridization on the efficiency. *Diam. Relat. Mater.* 76: 1–8.
10 Drummond, N.D., Williamson, A.J., Needs, R.J., and Galli, G. (2005). Electron emission from diamondoids: a diffusion quantum Monte Carlo study. *Phys. Rev. Lett.* 95: 096801.
11 Yang, W.L., Fabbri, J.D., Willey, T.M. et al. (2007). Monochromatic electron photoemission from diamondoid monolayers. *Science* 316: 1460–1462.
12 Medrano, C.R. and Sanchez, C.G. (2018). Trap-door-like irreversible photoinduced charge transfer in a donor-acceptor complex. *J. Phys. Chem. Lett.* 9: 3517–3524.
13 Heath, J.R., Obrien, S.C., Zhang, Q. et al. (1985). Lanthanum complexes of spheroidal carbon shells. *J. Am. Chem. Soc.* 107: 7779–7780.
14 Bethune, D.S., Johnson, R.D., Salem, J.R. et al. (1993). Atoms in carbon cages: the structure and properties of endohedral fullerenes. *Nature* 366: 123–128.
15 Vougioukalakis, G.C., Roubelakis, M.M., and Orfanopoulos, M. (2010). Open-cage fullerenes: towards the construction of nanosized molecular containers. *Chem. Soc. Rev.* 39: 817–844.
16 Cross, R.J., Saunders, M., and Prinzbach, H. (1999). Putting helium inside dodecahedrane. *Org. Lett.* 1: 1479–1481.
17 Jimenez-Vazquez, H.A., Tamariz, J., and Cross, R.J. (2001). Binding energy in and equilibrium constant of formation for the dodecahedrane compounds He@$C_{20}H_{20}$ and Ne@$C_{20}H_{20}$. *J. Phys. Chem. A* 105: 1315–1319.
18 Wang, S.-G., Qiu, Y.-X., and Schwarz, W.H.E. (2009). Bonding or nonbonding? Description or explanation? "Confinement bonding" of He@adamantane. *Chem. Eur. J.* 15: 6032–6040.
19 Camacho-Mojica, D.C., Ha, J.K., Min, S.K. et al. (2022). Proton affinity and gas phase basicity of diamandoid molecules: diamantane to $C_{131}H_{116}$. *Phys. Chem. Chem. Phys.* 24: 3470–3477.
20 Moran, D., Woodcock, H.L., Chen, Z.F. et al. (2003). The viability of small endohedral hydrocarbon cage complexes: X@C_4H_4, X@C_8H_8, X@C_8H_{14}, X@$C_{10}H_{16}$, X@$C_{12}H_{12}$, and X@$C_{16}H_{16}$. *J. Am. Chem. Soc.* 125: 11442–11451.
21 Bader, R.F.W. and Fang, D.C. (2005). Properties of atoms in molecules: caged atoms and the Ehrenfest force. *J. Chem. Theory Comput.* 1: 403–414.
22 Trujillo, C., Sanchez-Sanz, G., Alkorta, I., and Elguero, J. (2011). Simultaneous interactions of anions and cations with cyclohexane and adamantane: aliphatic cyclic hydrocarbons as charge insulators. *J. Phys. Chem. A* 115: 13124–13132.
23 Garcia, J.C., Machado, W.V.M., Assali, L.V.C., and Justo, J.F. (2011). Transition metal atoms encapsulated in adamantane molecules. *Diam. Relat. Mater.* 20: 1222–1224.
24 (a) Chanier, T., Pryor, C., and Flatte, M.E. (2012). Chemical trends of substitutional transition-metal dopants in diamond: an *ab initio* study. *Phys. Rev. B* 86: 085203. (b) Assali, L.V.C., Machado, W.V.M., and Justo, J.F. (2011). 3d transition

25 Marsusi, F. and Mirabbaszadeh, K. (2009). Altering the electronic properties of diamondoids through encapsulating small particles. *J. Phys. Condens. Matter* 21: 215303.

26 Beshara, C. and Shustov, G. (2019). Inside-protonated 1-azaadamantane: computational studies on the structure, stability, and generation. *Can. J. Chem.* 97: 169–177.

27 (a) Stetter, H. and Mayer, J. (1959). Über Verbindungen mit Urotropin-Struktur, XV. Synthese des 2-Oxa-adamantan-Ringsystems. *Chem. Ber.* 92: 2664–2666. (b) Stetter, H. and Mayer, J. (1959). Neue Synthese des Adamantan-und 2-Oxa-adamantan-Ringsystems. *Angew. Chem. Int. Ed.* 71: 430–430.

28 (a) Stetter, H. and Tacke, P. (1962). Eine Fragmentierung in der Adamantan-Reihe. *Angew. Chem. Int. Ed.* 74: 354–355. (b) Schaefer, J.P. and Honig, L.M. (1968). Bicyclo[3.3.1] nonanes. IV. Dehydration of the bicyclo[3.3.1]nonane-2,6-diols. *J. Org. Chem.* 33: 2655–2659.

29 Stetter, H., Held, H., and Schulte-Oestrich, A. (1962). Über Verbindungen mit Urotropin-Struktur, XXIII. Synthese des 2-Thia-adamantans. *Chem. Ber.* 95: 1687–1681.

30 Mikhailov, B.M. and Smirnov, V.N. (1973). 1-Boraadamantane and its transformation into 1-adamantanol. *Izv. Akad. Nauk, Ser. Khim.* 2165–2166.

31 Mikhailov, B.M. (1980). The chemistry of boron-cage compounds. *Pure Appl. Chem.* 52: 691–704.

32 (a) Fusco, R. and Bianchetti, G. (1953). Synthesis of azaadamantane. *Atti Accad. Naz. Lincei, Rend. Cl. Sci. Fis. Mat.* 15: 420–421. (b) Lukes, R. and Galik, V. (1954). Synthesis of 1-azaadamantane. *Collect. Czechoslov. Chem. Commun.* 19: 712.

33 Speckamp, W.N., Dijkink, J., and Huisman, H.O. (1970). 1-Azaadamantanes. *J. Chem. Soc. D* 197-198.

34 (a) Mikhailov, B.M. and Shagova, E.A. (1983). Organoboron compounds. 406. Conversion of 1-boraadamantane to 1-azaadamantane. *J. Organomet. Chem.* 258: 131–136. (b) Bubnov, Y.N., Gursky, M.E., and Pershin, D.G. (1991). A novel method of synthesis of 1-azaadamantane from 1-boraadamantane. *J. Organomet. Chem.* 412: 1–8. (c) Bubnov, Y.N., Gurskii, M.E., and Pershin, D.G. (1990). A new synthesis for 1-azaadamantane from 1-boraadamantane. *Bull. Acad. Sci. USSR, Div. Chem. Sci.* 39: 857–857.

35 Meeuwissen, H.J., Vanderknaap, T.A., and Bickelhaupt, F. (1983). Synthesis of phosphaadamantane. *Phosphorus Sulfur Silicon Relat. Elem.* 18: 109–112.

36 Stetter, H. and Steinacker, K.H. (1953). Über Verbindungen mit Urotropin-Struktur, V: Über die Maskierung von Carboxygruppen durch Orthoesterbildung mit *cis*-Phloroglucit. *Chem. Ber.* 86: 790–793.

37 Ikeya, K., Okamoto, S., and Sudo, A. (2021). Synthesis of a divinyl-functionalized diamantane-analogue from naturally occurring *myo*-inositol and

its application to polymer synthesis via the thiol-ene reaction. *Results Chem.* 3: 100176.
38 Stetter, H. and Schwartz, E.F. (1968). Compounds with urotropine structure. 40. Ring closure reactions of bicyclo[3.3.1]nonadi-2,6-ene. *Chem. Ber.* 101: 2464–2467.
39 (a) Averina, N.V., Zefirov, N.S., Kadzyauskas, P.P., and Sadovaya, N.K. (1975). Skeleton and polycyclic compounds. 8. Synthesis of isomeric oxatricyclodecanes – 2-oxaadamantane and 2-oxatwistane. *Zh. Org. Khim.* 11: 77–85. (b) Fisch, M., Smallcombe, S., Gramain, J.C. et al. (1970). On the conformation of *endo*-bicyclo[3.3.1]nonan-3-ol. A new synthesis of oxaadamantane. *J. Org. Chem.* 35: 1886–1890.
40 Suginome, H. and Yamada, S. (1986). The replacement of the carbonyl group of adamantanone by an oxygen or sulfur atom and the one-step transformation of 2-methyladamantan-2-ol into 2-oxa-adamantane; an efficient new synthesis of 2-oxa- and 2-thiaadamantane. *Synthesis* 741–743.
41 (a) Krasutsky, P.A., Kolomitsyn, I.V., Kiprof, P. et al. (2000). Observation of a stable carbocation in a consecutive Criegee rearrangement with trifluoroperacetic acid. *J. Org. Chem.* 65: 3926–3933. (b) Marchand, A.P., Kumar, V.S., and Hariprakasha, H.K. (2001). Synthesis of novel cage oxaheterocycles. *J. Org. Chem.* 66: 2072–2077.
42 (a) Kennedy, S.H., Dherange, B.D., Berger, K.J., and Levin, M.D. (2021). Skeletal editing through direct nitrogen deletion of secondary amines. *Nature* 593: 223–227. (b) Fokin, A.A., Reshetylova, O.K., Bakhonsky, V.V. et al. (2022). Synthetic doping of diamondoids through skeletal editing. *Org. Lett.* 24: 4845–4849. (c) Li, Y., Cheng, S.H., Tian, Y. et al. (2022). Recent ring distortion reactions for diversifying complex natural products. *Nat. Prod. Rep.* 39: 1970–1992. (d) Zippel, C., Seibert, J., and Brase, S. (2021). Skeletal editing-nitrogen deletion of secondary amines by anomeric amide reagents. *Angew. Chem. Int. Ed.* 60: 19522–19524.
43 Fokin, A.A., Zhuk, T.S., Pashenko, A.E. et al. (2009). Oxygen-doped nanodiamonds: synthesis and functionalizations. *Org. Lett.* 11: 3068–3071.
44 Krishnamurthy, V.V. and Fort, R.C. (1981). Heteroadamantanes. 2. Synthesis of 3-heterodiamantanes. *J. Org. Chem.* 46: 1388–1393.
45 Zhu, G.Y., Zhang, Y., Zhang, Z.Y. et al. (2018). High abundance of alkylated diamondoids, thiadiamondoids and thioaromatics in recently discovered sulfur-rich LS2 condensate in the Tarim Basin. *Org. Geochem.* 123: 136–143.
46 (a) Nizovtsev, A.P., Kilin, S.Y., Pushkarchuk, A.L. et al. (2020). Hyperfine characteristics of quantum registers NV-^{13}C in diamond nanocrystals formed by seeding approach from isotopic aza-adamantane and methyl-aza-adamantane. *Semiconductors* 54: 1689–1691. (b) Nizovtsev, A.P., Pushkarchuk, A.L., Kilin, S.Y. et al. (2021). Hyperfine interactions in the NV-^{13}C quantum registers in diamond grown from the azaadamantane seed. *Nanomaterials* 11: 1303.
47 Adhikari, B. and Fyta, M. (2015). Towards double-functionalized small diamondoids: selective electronic band-gap tuning. *Nanotechnology* 26: 035701.

48 Williamson, A.J., Grossman, J.C., Hood, R.Q. et al. (2002). Quantum Monte Carlo calculations of nanostructure optical gaps: application to silicon quantum dots. *Phys. Rev. Lett.* 89: 196803.

49 Fotooh, F.K. and Atashparvar, M. (2019). Theoretical study of the effect of simultaneous doping with silicon, on structure and electronic properties of adamantane. *Russ. J. Phys. Chem. B* 13: 1–8.

50 Marsusi, F., Mirabbaszadeh, K., and Mansoori, G.A. (2009). Opto-electronic properties of adamantane and hydrogen-terminated sila- and germa-adamantane: a comparative study. *Phys. E Low Dimens. Syst. Nanostruct.* 41: 1151–1156.

51 Fischer, J., Baumgartner, J., and Marschner, C. (2005). Synthesis and structure of sila-adamantane. *Science* 310: 825–825.

52 Schmidt, W. (1973). Photoelectron spectra of σ-bonded molecules, part 4. Photoelectron spectra of diamondoid molecules: adamantane, silamantane and urotropine. *Tetrahedron* 29: 2129–2134.

53 (a) Landt, L., Kielich, W., Wolter, D. et al. (2009). Intrinsic photoluminescence of adamantane in the ultraviolet spectral region. *Phys. Rev. B* 80: 205323. (b) Semakin, A.N., Sukhorukov, A.Y., Nelyubina, Y.V. et al. (2014). Urotropine isomer (1,4,6,10-tetraazaadamantane): synthesis, structure, and chemistry. *J. Org. Chem.* 79: 6079–6086.

54 (a) Demjan, T., Voros, M., Palummo, M., and Gali, A. (2014). Electronic and optical properties of pure and modified diamondoids studied by many-body perturbation theory and time-dependent density functional theory. *J. Chem. Phys.* 141: 064308. (b) Banerjee, S., Stüker, T., and Saalfrank, P. (2015). Vibrationally resolved optical spectra of modified diamondoids obtained from time-dependent correlation function methods. *Phys. Chem. Chem. Phys.* 17: 19656–19669.

55 (a) Fokin, A.A., Schreiner, P.R., Gunchenko, P.A. et al. (2000). Oxidative single-electron transfer activation of σ-bonds in aliphatic halogenation reactions. *J. Am. Chem. Soc.* 122: 7317–7326. (b) Patzer, A., Schütz, M., Möller, T., and Dopfer, O. (2012). Infrared spectrum and structure of the adamantane cation: direct evidence for Jahn-Teller distortion. *Angew. Chem. Int. Ed.* 51: 4925–4929.

56 Gali, A., Demjan, T., Voros, M. et al. (2016). Electron-vibration coupling induced renormalization in the photoemission spectrum of diamondoids. *Nat. Commun.* 7: 11327.

57 (a) Fyta, M. (2014). Stable boron nitride diamondoids as nanoscale materials. *Nanotechnology* 25: 365601. (b) Miranda, W., Moreira, E., Tavares, M.S. et al. (2021). BN adamantane isomers: an optical absorption spectrum study. *Appl. Phys. A Mater. Sci. Process.* 127: 32. (c) Garcia, J.C., Justo, J.F., Machado, W.V.M., and Assali, L.V.C. (2009). Functionalized adamantane: building blocks for nanostructure self-assembly. *Phys. Rev. B* 80: 125421.

58 Garcia, J.C., Justo, J.F., Machado, W.V.M., and Assali, L.V.C. (2010). Boron and nitrogen functionalized diamondoids: a first principles investigation. *Diam. Relat. Mater.* 19: 837–840.

59 McIntosh, G.C., Yoon, M., Berber, S., and Tomanek, D. (2004). Diamond fragments as building blocks of functional nanostructures. *Phys. Rev. B* 70: 045401.

60 Karlsson, J. and Larsson, K. (2010). Hydrogen-induced de/reconstruction of the c-BN(100) surface. *J. Phys. Chem. C* 114: 3516–3521.

61 Gao, W.W., Hung, L.D., Ogut, S., and Chelikowsky, J.R. (2018). The stability, electronic structure, and optical absorption of boron-nitride diamondoids predicted with first-principles calculations. *Phys. Chem. Chem. Phys.* 20: 19188–19194.

62 Fokin, A.A., Kiran, B., Bremer, M. et al. (2000). Which electron count rules are needed for four-center three-dimensional aromaticity? *Chem. Eur. J.* 6: 1615–1628.

63 Fokin, A.A. and Schreiner, P.R. (2009). Band gap tuning in nanodiamonds: first principle computational studies. *Mol. Phys.* 107: 823–830.

64 Hu, W., Li, Z.Y., and Yang, J.L. (2013). Surface and size effects on the charge state of NV center in nanodiamonds. *Comput. Theor. Chem.* 1021: 49–53.

65 Matsuura, Y. (2016). Tunnel magnetoresistance of diamondoids. *Chem. Phys. Lett.* 663: 21–26.

66 Matsuura, Y. (2015). Electronic transport properties of diamondoids. *Comput. Theor. Chem.* 1074: 131–135.

67 Adhikari, B., Sivaraman, G., and Fyta, M. (2016). Diamondoid-based molecular junctions: a computational study. *Nanotechnology* 27: 485207.

68 Sharma, H., Saha, B., and Bhattacharyya, P.K. (2017). Sandwiches of N-doped diamondoids and benzene via lone pair-cation and cation-pi interaction: a DFT study. *New J. Chem.* 41: 14420–14430.

69 Sedgi, I. and Kozuch, S. (2023). A playground for heavy atom tunnelling: neutral substitutional defect rearrangement from diamondoids to diamonds. *Chem. Eur. J.* 29: e202300673.

70 Maier, F.C., Sarap, C.S., Dou, M.F. et al. (2019). Diamondoid-functionalized nanogaps: from small molecules to electronic biosensing. *Eur. Phys. J.-Spec. Top.* 227: 1681–1692.

71 Zhou, Y.J., Brittain, A.D., Kong, D.Y. et al. (2015). Derivatization of diamondoids for functional applications. *J. Mater. Chem. C* 3: 6947–6961.

72 (a) Ranjbar, A., Khazaei, M., Venkataramanan, N.S. et al. (2011). Chemical engineering of adamantane by lithium functionalization: a first-principles density functional theory study. *Phys. Rev. B* 83: 115401. (b) Song, Y.D., Wang, L., and Wu, L.M. (2016). The structures and nonlinear optical responses of Li/Na doped adamantane: a density functional study. *Optik* 127: 10825–10837.

73 Hamadanian, M., Khoshnevisan, B., and Fotooh, F.K. (2010). Structure and electronic properties of Na-doped adamantane crystals. *J. Mol. Struct.* 961: 48–54.

74 Krongsuk, S., Shinsuphan, N., and Amornkitbumrung, V. (2019). Effect of the alkali metal (Li, Na, K) substitution on the geometric, electronic and optical properties of the smallest diamondoid: first principles calculations. *Chin. J. Chem. Eng.* 27: 476–482.

75 Khan, P., Mahmood, T., Ayub, K. et al. (2021). Turning diamondoids into nonlinear optical materials by alkali metal substitution: a DFT investigation. *Opt. Laser Technol.* 142.

76 Teunissen, J.L., De Proft, F., and De Vleeschouwer, F. (2017). Tuning the HOMO-LUMO energy gap of small diamondoids using inverse molecular design. *J. Chem. Theory Comput.* 13: 1351–1365.

77 Marsusi, F. (2018). Nuclear dynamic effects on electronic properties of functionalized diamondoids. *Phys. E Low Dimens. Syst. Nanostruct.* 103: 435–443.

78 Szilvasi, T. and Gali, A. (2014). Fluorine modification of the surface of diamondoids: a time-dependent density functional study. *J. Phys. Chem. C* 118: 4410–4415.

79 Elavarasi, S.B., Mariam, D., Momeen, M.U. et al. (2019). Effect of fluorination on bandgap, first and second order hyperpolarizabilities in lithium substituted adamantane: a time dependent density functional theory. *Chem. Phys. Lett.* 715: 310–316.

80 Santos, A.M.S., Moreira, E., Meiyazhagan, A., and Azevedo, D.L. (2021). Bond order effects on the optoelectronic properties of oxygen/sulfur functionalized adamantanes. *J. Mol. Graph.* 105: 107869.

81 (a) Greidanus, J.W. and Schwalm, W.J. (1969). Adamantanethione and its reduction to 2-adamantanethiol. *Can. J. Chem.* 47: 3715–3716. (b) Bernhard, S. and Belser, P. (1996). Synthesis of new rigid, bridging ligands for the study of energy and electron-transfer reactions. *Synthesis* 192–194.

82 Voros, M., Demjen, T., Szilvasi, T., and Gali, A. (2012). Tuning the optical gap of nanometer-size diamond cages by sulfurization: a time-dependent density functional study. *Phys. Rev. Lett.* 108: 267401.

83 (a) Steer, R.P. (1981). Structure and decay dynamics of electronic excited states of thiocarbonyl compounds. *Rev. Chem. Intermed.* 4: 1–41. (b) Clouthier, D.J., Hackett, P.A., Knight, A.R., and Steer, R.P. (1981). Structures and decay dynamics of thiocarbonyl excited states. *J. Photochem. Photobiol.* 17: 319–326.

84 Falk, K.J. and Steer, R.P. (1988). Rydberg states in adamantanethione, thiofenchone, and thiocamphor. *Can. J. Chem.* 66: 575–577.

85 (a) Falk, K.J., Knight, A.R., Maciejewski, A., and Steer, R.P. (1984). Concerning the lifetime of the second excited singlet state of adamantanethione. *J. Am. Chem. Soc.* 106: 8292–8293. (b) Falk, K.J. and Steer, R.P. (1989). Photophysics and intramolecular photochemistry of adamantanethione, thiocamphor, and thiofenchone excited to their second excited singlet states: evidence for subpicosecond photoprocess. *J. Am. Chem. Soc.* 111: 6518–6524.

86 Landt, L., Bostedt, C., Wolter, D. et al. (2010). Experimental and theoretical study of the absorption properties of thiolated diamondoids. *J. Chem. Phys.* 132: 144305.

87 Sarap, C.S., Adhikari, B., Meng, S. et al. (2018). Optical properties of single- and double-functionalized small diamondoids. *J. Phys. Chem. A* 122: 3583–3593.

88 Stewart, B., Harriman, A., and Higham, L.J. (2011). Predicting the air stability of phosphines. *Organometallics* 30: 5338–5343.

89 Moncea, O., Gunawan, M.A., Poinsot, D. et al. (2016). Defying stereotypes with nanodiamonds: stable primary diamondoid phosphines. *J. Org. Chem.* 81: 8759–8769.

90 (a) Goerlich, J.R. and Schmutzler, R. (1995). Organophosphorus compounds with tertiary alkyl substituents. 6. A convenient method for the preparation of di-1-adamantylphosphine and di-1-adamantylchlorophosphine. *Phosphorus Sulfur Silicon Relat. Elem.* 102: 211–215. (b) Stetter, H. and Last, W.D. (1969). Über Verbindungen mit Urotropin-Struktur, XLIV. Über Adamantan-phosphonsäure-1-dichlorid. *Chem. Ber.* 102: 3364–3366.

91 Garcia, J.C., Justo, J.F., Machado, W.V.M., and Assali, L.V.C. (2010). Structural, electronic, and vibrational properties of amino-adamantane and rimantadine isomers. *J. Phys. Chem. A* 114: 11977–11983.

92 Jeneesh, K.K.M. and Padmanaban, R. (2018). Effects of functionalization on the electronic and absorption properties of the smaller diamondoids: a computational study. *J. Chem. Sci.* 130: 113.

93 Gu, Y., Zhu, L.P., Li, Y.F. et al. (2015). Adamantane-based wide-bandgap host material: blue electrophosphorescence with high efficiency and very high brightness. *Chem. Eur. J.* 21: 8250–8256.

94 Li, W., Cai, X.Y., Li, B.B. et al. (2019). Adamantane-substituted acridine donor for blue dual fluorescence and efficient organic light-emitting diodes. *Angew. Chem. Int. Ed.* 58: 582–586.

95 Guan, H.M., Hu, Y.X., Xiao, G.Y. et al. (2020). Novel adamantane-bridged phenanthroimidazole molecule for highly efficient full-color organic light-emitting diodes. *Dyes Pigments* 177: 108273.

96 Langhals, H., Dietl, C., Zimpel, A., and Mayer, P. (2012). Noncovalent control of absorption and fluorescence spectra. *J. Org. Chem.* 77: 5965–5970.

97 Langhals, H., Dietl, C., Dahl, J. et al. (2020). OrthoFRET in diamantane FRET in orthogonal stiff dyads; diamond restriction for frozen vibrations. *J. Org. Chem.* 85: 11154–11169.

98 Rosemann, N.W., Locke, H., Schreiner, P.R., and Chatterjee, S. (2018). White-light generation through nonlinear optical response of 1,3,5,7-tetraphenyladamantane: amorphous versus crystalline states. *Adv. Opt. Mater.* 6: 1701162.

99 Schwan, S., Achazi, A.J., Ziese, F. et al. (2023). Insights into molecular cluster materials with adamantane-like core structures by considering dimer interactions. *J. Comput. Chem.* 44: 843–856.

100 (a) Gowrisankar, S., Bernhardt, B., Becker, J., and Schreiner, P.R. (2021). Regioselective synthesis of *meta*-tetraaryl-substituted adamantane derivatives and evaluation of their white light emission. *Eur. J. Org. Chem.* 2021: 6806–6810. (b) Gowrisankar, S., Hosier, C.A., Schreiner, P.R., and Dehnen, S. (2022). Manipulating white-light generation in adamantane-like molecules via functional group substitution. *ChemPhotoChem* 6: e202200128.

101 Belz, J., Haust, J., Muller, M.J. et al. (2022). Adamantanes as white-light emitters: controlling the arrangement and functionality by external Coulomb forces. *J. Phys. Chem. C* 126: 9843–9854.

102 (a) Grimme, S., Huenerbein, R., and Ehrlich, S. (2011). On the importance of the dispersion energy for the thermodynamic stability of molecules. *ChemPhysChem* 12: 1258–1261. (b) Schreiner, P.R., Chernish, L.V., Gunchenko, P.A. et al. (2011). Overcoming lability of extremely long alkane carbon-carbon bonds through dispersion forces. *Nature* 477: 308–311. (c) Wagner, J.P. and Schreiner, P.R. (2015). London dispersion in molecular chemistry-reconsidering steric effects. *Angew. Chem. Int. Ed.* 54: 12274–12296. (d) Rummel, L. and Schreiner, P.R. (2024). Advances and prospects in understanding London dispersion interactions in molecular chemistry. *Angew. Chem. Int. Ed.* 63: e202316364.

103 Shubina, T.E. and Fokin, A.A. (2011). Hydrocarbon sigma-radical cations. *WIREs Comput. Mol. Sci.* 1: 661–679.

104 Bano, R., Ayub, K., Mahmood, T. et al. (2023). Diamondoid as potential nonlinear optical material by superalkali doping: a first principles study. *Diam. Relat. Mater.* 135: 109826.

13

Perspective

As we have hopefully amply outlined in this book, diamondoids are an incredibly fascinating class of molecules. They bear simplicity, beauty (through symmetry), exceptionally high thermodynamic stability plus a long list of unique properties such as negative electron affinity (and structural integrity upon one-electron oxidation) [1], high band gaps [2], and, depending on shape, nearly isotropic polarizability, only to name but a few. Most importantly: they are available in large quantities, at least as far as the lower diamondoids adamantane, diamantane, and triamantane are concerned [3]. They can be isolated from petroleum feedstocks or synthesized in the laboratory through thermodynamic rearrangement reactions from a variety of hydrocarbon precursors [4]. As such, they are sustainable feedstocks for future applications that can also be recovered or readily be degraded after use through microbial biotransformations [5].

The applications side is to a large degree uncharted territory. Even though adamantane is a moiety commonly found, for instance, in quite a few pharmaceutically active ingredients, the next higher diamondoids have virtually not been utilized in medicinal chemistry [6]. Moreover, the currently issued patents often do not even claim the higher diamondoids as potential building blocks, not to speak of their use in active ingredients. As diamondoids are non-toxic in pharmaceutically active concentrations but convey high lipophilicity [7], chemical stability, and structural rigidity, there is clearly a very large untapped potential.

Similar arguments apply to catalysis for which applications have been found for adamantane and diamantane [8]; to the best of our knowledge, there is only one example for the use of triamantane in catalysis [9], even though it is abundantly available. With the advent of recognizing that London dispersion [10], which is proportional to the polarizability of a structural entity [11], is an important design element for catalysis, this is a clear omission. Additionally, pendant groups attached to diamondoids can be placed in a defined geometric relationship, so that predictable architectures can be constructed [12].

A myriad of applications for diamondoids can be envisaged in the realm of material chemistry. This hypothesis is born out of the realization that nanoscale diamond has found many applications [13] and diamondoids bear many of the essential properties of diamond, with the added huge advantage that they are "knowable," i.e., the structures can be very well and atomistically resolved, characterized, and the

The Chemistry of Diamondoids: Building Blocks for Ligands, Catalysts, Materials, and Pharmaceuticals,
First Edition. Andrey A. Fokin, Marina Šekutor, and Peter R. Schreiner.
© 2024 WILEY-VCH GmbH. Published 2024 by WILEY-VCH GmbH.

compounds purified. In contrast to diamond, which is always a bulk material with very varied surface characteristics, diamondoids can be modified highly selectively so that well-designed and highly pure functionalized diamondoids are available [14]. One can immediately think of doped diamondoids [15] as the basis for doped diamonds, the use of diamondoids for surface coatings and modifications [16], as sound and heat propagators in electronic applications, and many more. We have just begun exploring these possibilities, and many tailor-made applications can be envisaged utilizing functionalized diamondoids as the key building blocks.

We are very much looking forward to many more groups joining the team of diamondoid explorers and hope this book has aroused your interest!

References

1 Yang, W.L., Fabbri, J.D., Willey, T.M. et al. (2007). Monochromatic electron photoemission from diamondoid monolayers. *Science* 316: 1460–1462.

2 Fokin, A.A. and Schreiner, P.R. (2009). Band gap tuning in nanodiamonds: First principle computational studies. *Mol. Phys.* 107: 823–830.

3 Dahl, J.E., Liu, S.G., and Carlson, R.M.K. (2003). Isolation and structure of higher diamondoids, nanometer-sized diamond molecules. *Science* 299: 96–99.

4 (a) Carrow, B.P. and Chen, L. (2017). Tri(1-adamantyl) phosphine: Exceptional catalytic effects enabled by the synergy of chemical stability, donicity, and polarizability. *Synlett* 28: 280–288. (b) Fort, R.C. and Schleyer, P.v.R. (1964). Adamantane: Consequences of the diamondoid structure. *Chem. Rev.* 64: 277–300.

5 Folwell, B.D., McGenity, T.J., and Whitby, C. (2020). Diamondoids are not forever: Microbial biotransformation of diamondoid carboxylic acids. *J. Microbial. Biotechnol.* 13: 495–508.

6 Hodek, P., Borek-Dohalska, L., Sopko, B. et al. (2005). Structural requirements for inhibitors of cytochromes P4502B: Assessment of the enzyme interaction with diamondoids. *J. Enzyme Inhib. Med. Chem.* 20: 25–33.

7 Wanka, L., Iqbal, K., and Schreiner, P.R. (2013). The lipophilic bullet hits the targets: Medicinal chemistry of adamantane derivatives. *Chem. Rev.* 113: 3516–3604.

8 Agnew-Francis, K.A. and Williams, C.M. (2016). Catalysts containing the adamantane scaffold. *Adv. Synth. Catal.* 358: 675–700.

9 Richter, H., Schwertfeger, H., Schreiner, P.R. et al. (2009). Thieme Chemistry Journal Awardees – where are they now? Synthesis of diamantane-derived N-heterocyclic carbenes and applications in catalysis. *Synlett* 193–197.

10 (a) Fokin, A.A., Chernish, L.V., Gunchenko, P.A. et al. (2012). Stable alkanes containing very long carbon-carbon bonds. *J. Am. Chem. Soc.* 134: 13641–13650. (b) Schreiner, P.R., Chernish, L.V., Gunchenko, P.A. et al. (2011). Overcoming lability of extremely long alkane carbon-carbon bonds through dispersion forces. *Nature* 477: 308–311. (c) Wagner, J.P. and Schreiner, P.R. (2015). London dispersion in molecular chemistry-reconsidering steric effects. *Angew. Chem. Int. Ed.* 54: 12274–12296.

11 Israelachvilli, J.N. (2011). *Intermolecular and Surface Forces*, 3rd ed. USA: Academic Press Waltham.

12 (a) Müller, C.E., Wanka, L., Jewell, K., and Schreiner, P.R. (2008). Enantioselective kinetic resolution of *trans*-cycloalkane-1,2-diols. *Angew. Chem. Int. Ed.* 47: 6180–6183. (b) Prochazkova, E., Kolmer, A., Ilgen, J. et al. (2016). Uncovering key structural features of an enantioselective peptide-catalyzed acylation utilizing advanced NMR techniques. *Angew. Chem. Int. Ed.* 55: 15754–15759.

13 Krüger, A. (2008). The structure and reactivity of nanoscale diamond. *J. Mater. Chem.* 18: 1485–1492.

14 Gunawan, M.A., Poinsot, D., Domenichini, B. et al. (2015). Nanodiamonds: emergence of functionalized diamondoids and their unique applications. In: *Chemistry of Organo-Hybrids: Synthesis and Characterization of Functional Nano-Objects* (ed. B. Charleux, C. Copéret, and E. Lacôte), 69–113. Hoboken, New Jersey: John Wiley & Sons, Inc.

15 (a) Fokin, A.A., Zhuk, T.S., Pashenko, A.E. et al. (2009). Oxygen-doped nanodiamonds: Synthesis and functionalizations. *Org. Lett.* 11: 3068–3071. (b) Fokin, A.A., Reshetylova, O.K., Bakhonsky, V.V. et al. (2022). Synthetic doping of diamondoids through skeletal editing. *Org. Lett.* 24: 4845–4849.

16 (a) Narasimha, K.T., Ge, C., Fabbri, J.D. et al. (2016). Ultralow effective work function surfaces using diamondoid monolayers. *Nat. Nanotechnol.* 11: 267–273. (b) Feng, K., Solel, E., Schreiner, P.R. et al. (2021). Diamantanethiols on metal surfaces: Spatial configurations, bond dissociations, and polymerization. *J. Phys. Chem. Lett.* 12: 3468–3475. (c) Jantayod, A., Doonyapisut, D., Eknapakul, T. et al. (2020). Resistive switching in diamondoid thin films. *Sci. Rep.* 10: 19009. (d) Lopatina, Y.Y., Vorobyova, V.I., Fokin, A.A. et al. (2019). Structures and dynamics in thiolated diamantane derivative monolayers. *J. Phys. Chem. C* 123: 27477–27482.

Index

a

α-(acylamino)acrylic derivatives 222
adamantane and diamantane amino acids 102
adamantane-containing perylene dye 327
adamantane-containing polymers 193
adamantane-containing SAMs on surfaces 137–146
adamantane modified cytochrome 251
adamantane porphyrin derivatives 251
adamantane radical cation formation 65, 66, 113, 115, 289
adamantane-stabilized metal nanoparticles 160–162
1-adamantanethiolate SAMs 139
adamantanethione 319, 321
1-adamantyl acetic acid 99
1-adamantyl-1-adamantane 193–197, 279–281, 284
2-adamantyl-1-adamantane 284
  functionalization 284
2-adamantyl-2-adamantane 284, 286, 287
1-adamantyl and 4-diamantyl complexes 219
2-(2-adamantylidene)adamantane 279, 280, 288, 290
AlBr$_3$/*tert*-BuBr sludge catalyst 10
alkyladamantanes 39, 118
all-cis 1,3,5-trihydroxycyclohexane 286
1-aminoadamantane 103, 118, 226
γ-aminoadamantane carboxylic acids 101

aminoadamantane dirhodium tetracarboxylate complex 226–227
amino alcohol derivative 105
3-aminodiamantane 103
4-aminodiamantane-9-carboxylic acid 324
4-aminodiamantane derivative 258
5-aminoisophthalic acid 253
anti-tetramantane 4
apical *bis*-dialcohols 88
apical 4-bromodiamantane 77
apical diamondoid thiols 150, 155
artificially phosphorus-doped CVD diamond 305
aryl ethers 218–220
1-azadiamantane 307, 308, 316–317
4-azadiamantane 315
3-azidodiamantane 242
azobis(isobutyronitrile) (AIBN) 100

b

Balaban–Schleyer nomenclature 4
Balaban–Schleyer system 4
bastardane 10
benzo-fused cyclobutanes 224
4-benzonitrile 324
bicyclo[3.3.1]nona-2,6-diene 308
2,6-bicyclo[3.3.1]nonane derivatives 308
bicyclo[3.3.1]nonane system 308
B3LYP/6-31G(d,p) optimized bis-apical diamantane dithiols 316
binding energies 36, 150, 270
bioactive diamondoid compounds 239

biphenyl diamondoid phosphine ligand 219
bis(adamantyl)-1,2-dioxetane 292
1,2-bis(alkylmethylphosphino)ethane 222
1,6-bis[4-(4-amino-3-hydroxyphenoxy) phenyl]diamantane 241
4,9-bis-4-(4-aminophenoxy)phenyl diamantane 205
bis-apical diamantane amino acid 239
bis-apical diamantane derivatives 204, 268
bis-apical 4,9-diaminodiamantane 205
bis-apical phosphonic acid dichloride 109
bis-diamondoidyl chlorophosphonates 109
4,9-bis[4(3,4-dicarboxyphenoxy) phenyl] diamantane dianhydride 205
bis[(4,6-difluorophenyl)pyridinato-N,C2′](picolinato)iridium(III) (FIrpic)-based OLED device 324
1,3-bis-(1,2,4-triazol-4-yl)adamantane 196
1,6-bis(1,2,4-triazol-4-yl)diamantane 268
4,9-bis(1,2,4-triazol-4-yl)diamantane 268
3,5-bis(trifluoromethyl)phenyl moiety 226
bitopic 1,2,4-triazole diamondoid ligands 268
black oil reservoirs 41
body-centered cubic structured gold nanocluster $Au_{38}S_2(AdS)_{20}$ 160
bond dissociation energies (BDEs) 282
1-boraadamantane 308
1-boradiamantane 316–317
4-boradiamantane 315
borane-methyl sulfide (BMS) complex 290
"bromine-free" diamondoid derivatives 88
"bromine-free" methods 83
1-bromoadamantane 63, 65, 181, 279
2-bromo-derivative 65
1-bromo-2-hydroxydiamantane 94
4-bromotoluene 217
1-bromo-3,5,7-tris(mercaptomethyl) adamantane 143
Buchwald–Hartwig amination reactions 215
bulky trialkylphosphines 215

C

camphene 172
carboxybenzyl-(Cbz) protected 1-aminoadamantane-3-carboxylic acid 226
carboxylic acids and their derivatives 98–103
cataCXium® 215–218
catalysis, diamondoids in 213
catalytically active gold(I) compounds 218
cationic triamantane derivatives 262
C–C bond forming methods 118
C60(C70) endohedral complexes 306
CD-adamantane conjugates 251
C–H-bond functionalization 109, 110, 318
chiral ligand 1,2-diaminodiamantane 241
chloroacetamide diamantane derivative 104
C–H-substituted 4-bromo derivative 290
cis-phloroglucin 308
citraconimide derivatives 241
citrofulvicin 245
cluster halogenation 66
combined distillation/crystallization technique 48
congressane 5, 10
copper (I)-catalyzed [3+2] azide-alkyne cycloaddition 244
Corey–Chaykovsky methylenation reaction 92
C_{3v}-tetramantane 2, 4, 8, 43, 67–68, 77, 86, 87
$[Cu_2(bipy)_2(DIENDIA)]^{2+}$ complex 120
cucurbiturils 251, 255, 260, 262
cyclic oligosaccharides 251
β-cyclodextrin 252
cyclodextrins 160, 251
cyclohexamantane 2, 25, 46, 48, 279

d

dangling bonds 35, 177
daphnezomine A and B 245
dapholdhamine B 245
([1231241(2)3]decamantane ($C_{35}H_{36}$) 5, 186
1,3-dehydroadamantane 90
dehydrogenated octahedral and cuboctahedral nanodiamond 20
3,6-dehydrohomoadamantane 90
deuterium-labeled diamondoids 45
1,7-diadamantane thioperylene-3,4,9,10-tetracarboxylic diimide 141
1,3-diadamantyl disubstituted imidazolium salt 224
diadamantyl ether 265, 266
dialkyl or alkylaryl phosphinic acid chlorides 110
diamantane 1
  cucurbituril complexes 259
  dichlorodicarboxyates 200
  functionalized chelating triazacyclononane ligand 264
  polyimides 204, 205
  quaternary ammonium salt 256
  radical cation 113, 116, 117
3-diamantane acrylates 198
diamantane 5-aminoisophthalic acid derivative 253
diamantane-based polymeric maleimides and acrylates 197
4,9-diamantane bis-carbaldehyde 107
diamantane-containing polyamides 202
diamantane 4,9-derivatives 182
4,9-diamantane dicarboxylate 269
diamantine-1,6-dicarboxylic acid 11, 201
diamantine-4,9-dicarboxylic acid 156, 182–184, 270
diamantane-4,9-dicarboxylic acid dihydrazide 270
1,6-diamantanedichlorodicarboxyate 200
1-diamantane thiol 146
4,9-diamantyldithiol 147
4-diamantyl acrylate 198
4-diamantyl methacrylate 198
1-diamantylmethyl ketone 92, 93
1-diamantyl triflate 112

1,2-diaminodiamantane 241
  based platinum complexes 241, 242
1,6-diaminodiamantane 205
4,9-diaminodiamantane 205, 256
diamond-3C-SiC microdomes 177
diamond growth inside nanotubes 180–186
diamond-like 2D-electronics 146
diamondoid acetic and propionic acids 100
diamondoid amine synthesis 103
diamondoid amino acids 239, 240
diamondoid carboxylates 225, 228, 269
diamondoid carboxylic acids 98–100, 253
diamondoid-containing p-σ-n type molecular rectifier 160
diamondoid-decorated oligonucleotides 244
diamondoid-decorated Sn/S clusters 269
diamondoid dirhodium tetracarboxylate complexes 228
diamondoid functionalizations
  alcohols and ketones 83–98
  as alkane CH activation models 63–69
  carboxylic acids and their derivatives 98–103
  halogenations 75–83
  nitrogen-containing compounds 103–108
  other diamondoid derivatives 118–121
  phosphorous-and sulfur-containing compounds 108–113
  single electron oxidations 113–118
diamondoid indices 40, 41
diamondoid-modified oligopeptide catalysts 228
diamondoid NHC ligands 224
diamondoid nitrates 83
diamondoid phosphine ligands 213, 218, 219
diamondoid phosphonate 7-dichlorophosphoryl-[1(2,3)4]pentamantane 177
diamondoid phosphonates 154, 177, 179, 180
diamondoid phosphonic acid dichlorides 109, 110

diamondoid polymers
  based on 1-adamantyl-1-adamantane 194–197
  based on difunctionalized diamantanes 198–207
  based on monofunctionalized diamantanes 197–198
diamondoid-promoted growth of diamond under HT-HP or CVD conditions 171–177
diamondoid self-assembly
  adamantane-containing SAMs on surfaces 137–146
  adamantane-stabilized metal nanoparticles 160–162
  applications of diamondoid SAM materials 160
  other applications of diamondoid SAM materials 156–160
  pristine diamondoids on surfaces 156
  for SAM formation 146–156
diamondoid thiols 50, 51, 52, 112, 146, 147, 150, 154–156, 158, 268, 322, 323
diamondoid-to-diamond conversion mechanism 173
diamondoid/Ru nanoparticles 229
diamondoids
  defined 1
  electron affinities 24–25
  ionization potentials 23–24
  molecular symmetry of 1
  nomenclature 2–6
  physical properties of 14–18
  preparation of 8–14
  spectroscopy of 18–23
  strain 5–9
  structures and symmetry of 2
  synthesis from adamantane 182–186
  vibrational spectroscopy 25–28
diamond planets 36
di-bora-di-azaadamantane 314
3,3′-dibromo derivative 284
4,9-dibromodiamantane 181–182
4,9-dichlorocarboxylates 201
7-dichlorophosphoryl-[1(2,3)4]pentamantane 44, 177, 178
diethylaminosulfur trifluoride (DAST) 82

1,3-diethynyladamantane 194, 200
3,3′-diethynyl-1,1-biadamantane 194
4,9-diethynyldiamantane 198, 200
4,9-dihalodiamantanes 182
4,9-dihydroxydiamantane 182
4-dimethylaminodiamantane 258
dioxiranes 90, 91
9,9-diphenyl-4,4-bis-diamantyl 285
4,9-diphenyldiamantane 285
dirhodium adamantyl glycine catalysts 228
dirhodium tetracarboxylate adamantane complex 225
dirhodium tetracarboxylate adamantyl complex 226
dispersion doping 305, 327, 328
1,3-disubstituted adamantane 1,3-phenoxycarboxylic acid 204
1,1-disubstituted alkenes 225
2,6-disubstituted naphthalene derivative 257
D-mannose-diamondoid conjugates 254
1-dodecanethiol 139, 140
doped diamondoids
  @-doping 306–307
  external doping 318–329
  internal doping 308–318
double Criegee rearrangement 308
dualist graph convention, for diamondoids nomenclature 5

e

electron affinities 24–25, 315
electronic photodissociation 114, 116
electrophilic nitronium species 83
energy diagram 152
ethanodiamantanes 40, 41
ethanotriamantanes 40, 41
external doping 305, 318–329

f

farnesol carbamate 226, 227
ferrocene-terminated alkane thiols 144
"flexible steric bulk" concept 232
fluorene 172
fullerene C_{60} 264

g

GC-GC-TOFMS method 39
gem-dibromoadamantane 287

Gramicidin S 240
graphite-cluster diamond (GCD) 175
guanosine monophosphate (GMP) 241

h

H-abstraction/substitution reactions 64
Hal$_3$C· steric factors 68
halogenations 65, 68, 69, 75–83
4-{3-[4-(9H-carbazol-9-yl)phenyl]
  adamantan-1-yl}benzonitrile 326
H-coupled electron transfer (HCET) 65, 66
Heck reaction 215, 216, 218
[121321]heptamantane 5
heteroadamantanes 309
heterodiamondoids 245, 311
[12312]hexamantane 5
hexaoxadiamantane cages 308–309
higher diamondoids for diamond
  nucleation 177–180
homodiamantane 91
10H-spiro[acridine-9,2′-adamantane]
  with 2,4,6-triphenyl-1,3,5-triazine
  324
hybrid diamantane/aryl polyamides
  202–203
hybrid diamond-SiC microdome
  structures 177
hydride transfer 64
hydrogen-coupled electron transfer
  (HCET) transition structures 67, 68
hydrogen-terminated diamonds 36
hydrophobic effect 251
β-hydroxy carboxylic acid derivatives
  106, 107
1-hydroxydiamantane 93, 94
4-hydroxydiamantane 242
2-hydroxy-2-(1-diamantyl)acetic acid 93
9-hydroxydiamantyl-4-phosphine 323
hydroxy-diamondoids 83, 85, 88, 262
9-hydroxyhomodiamantane 91
hydroxyphosphine 267, 268

i

ICl$_6^+$ heptahalogen (Hal$_7$)$^+$ electrophiles
  66
internal doping 18, 96, 305, 308–318, 324
internally-doped adamantanes 308–309

1-iodoadamantane 82, 280
ionization potentials 23–24, 27, 113, 114, 155
isomeric alkyl diamondoid derivatives
  40
isomeric γ-cations oxadiamant-6-yl 97
isomeric diamondoids 5, 41, 43
isomeric hydroxy adamantanes 267
isopropylidene diamantane 92
iso-tetramantane 86

k

kinetic isotope effect (KIE) 66
Koch–Haaf reaction 98, 99, 101
Koser's PhIOH reagent 290

l

least sterically crowded pentamantanes
  43
lectin concanavalin A (ConA) 254, 255
London dispersion (LD) interactions
  137, 215, 224, 251, 258, 262, 265, 267, 282, 289

m

maleimide derivatives 241
m-chloroperbenzoic acid (MCPBA) 105
Meerwein's ester 8, 9
meso-alkane-1,2-diols 230, 231
meta-bis(10-bromo-9-anthryl)benzene
  264
methyladamantane 40, 91
ω-(4′-methylbiphenyl-4-yl)ethane thiol
  141
ω-(4′-methylbiphenyl-4-yl)propane thiol
  141
methyldiamantane 40, 49
7-methylenebicyclo[3.3.1]nonane-3-one
  308
1-methylimidazole 88
micellar adamantane-based capsule 264
microwave-assisted nonionic surfactant
  extraction (MANSE) 50
microwave plasma (MP) CVD techniques
  173
molecular surgery 306
molecular symmetry of diamondoids 1
multicatalyst oligopeptide 230
myo-inositol 308

n

nanodiamonds 1
  dehydrogenated octahedral and cuboctahedral nanodiamond 20
  particles 2, 20, 25, 35, 36, 151, 171, 172, 175, 279, 287, 288, 305
nareline 245
naturally occurring diamondoids
  alternative natural sources of 48–50
  in the Earth's crust 38–40
  formation in the Earth's crust 42–44
  in geochemical studies 40–42
  large-scale isolation of 44–48
  other derivatives in nature 50–52
N-(1)-diamantylmaleimide 197
N-heterocyclic carbenes (NHC) 154, 223–224, 231
N-heterocyclic diamondoid carbenes 232
nitrogen-containing compounds 103–108
nitronium tetrafluoroborate 82
non-equilibrium Green's function (NEGF)–DFT method 316
non-proteinogenic adamantane amino acids 240
norbornene dimer 10, 42

o

octahedral crystals 44
oligomeric bis(2-ethylhexyl)-p-phenylenevinylene monomers 206
oligopeptide-based multicatalyst system 231
1D-diamond nanoscale materials 180
orange 2,6-adamantanedithione 319
organometallic diamondoid derivatives 121
oxaadamantane 96, 285, 308
3-oxadiamantane dimer 286
oxadiamondoids 96, 97
5-oxatriamant-2-yl 97
oxidative esterification 231
oxindoles 215, 217
oxygen-doped adamantane and diamantane dimers 286
ortho-methylated carbamoyl chlorides 215

p

palladium triadamantyl phosphine complex 221
panacosmine 245
paraffin wax 172
Pd-catalyzed Buchwald–Hartwig amination 217
Pd-catalyzed Sonogashira–Hagihara coupling 218
Pd-catalyzed Suzuki coupling 216
Pd-catalyzed synthesis of aryl ethers 219
peanut-shaped nanostructures 264
pentamantane 1, 5, 8, 21, 39, 43, 44, 46, 47, 51, 92, 146, 151, 177–180, 186, 316, 322, 323
[12(1)3]pentamantane 179
[1212]pentamantane 5, 92
[1(2,3)4]pentamantane chlorophosphonate 179
pentamantane thiols 146
peptide-based bioactive compounds 239
perylene-3,4,9,10-tetracarboxylic bis-imide moiety 326
perylene-3,4,9,10-tetracarboxylic diimide 141, 142
pharmacophore TD550 242
phenyl-1-adamantyl ketone 280
phenylboronic acid 217
phenylhydroxy diamantane 311
phosphaadamantane 308
1-phosphadiamantane 317
phosphorous-and sulfur-containing compounds 108–113
π-methyl histidine (Pmh) 229
poly(amide-imide) diamantane co-polymers 206
poly(amide-imide)s 206
polyaryene ether ketone triphenylphosphine oxide polymers 201
polyhalogen cations 66, 67
polyhalogen electrophiles 66
polymantanes 2, 3, 10, 15
polymeric maleimides 197
polymers based on 1-adamantyl-1-adamantane 194–197
polymers based on difunctionalized diamantanes 198–207

Index | 349

polymers based on monofunctionalized diamantanes 197–198
polyvinyl adamantane 193
pristine diamondoids 50, 289, 305, 324
on surfaces 156
3-propyl-7-methylene-3-borabicyclo [3.3.1]-nonane 308
protobranching 6, 9
pseudotetrahedral tetrahalogenated adamantane 81
pure type-IIa diamond 305

q
quantitative diamondoid analysis 45
quinuclidine radical cation 68

r
radical C–H-halogenations 78
red-orange crystalline adamantanethione 319
relatively stable primary diamondoid phosphines 223
rubrene endoperoxide 292
ruthenium-substituted polyoxometalates 91

s
saturated diamondoid oligomers 279–287
scholarisine H 245
Schotten–Baumann reactions 107
secondary 1-adamantyl and 4-diamantyl phosphine oxides 219–220
secondary phosphine oxides 219
self-assembled monolayers (SAMs) 25, 137–162, 289, 291
self-healing polymers 197
sesquihomoadamantene 292, 293
shock-wave diamond formation 36
single electron oxidations of diamondoids 113–118
singly occupied molecular orbital (SOMO) 114, 116, 153, 322, 323
skew-tetramantane 4
Sonogashira coupling 217, 224
spiro[adamantane-2,4′-homoadamantan-5′-ol] 287
spirooxindoles 215
stabilomeric synthesis 9, 11, 42

sterically-congested medial diamantane dimer 285
sterically crowded adamantane and diamantane amides 107
sterically demanding triadamantyl phosphine 221
substituted diamondoid phosphine derivatives 224
sulfamate ester 105
supramolecular adamantane hydrogels and polymers 251
supramolecular architectures
adamantane modified cytochrome 251
adamantane porphyrin derivatives 251
4-aminodiamantane mono derivative 258
5-aminoisophthalic acid 253
binding constants 260–261
biological cell membranes 254
1,6-bis(1,2,4-triazol-4-yl)diamantane 268
4,9-bis(1,2,4-triazol-4-yl)diamantane 268
bitopic 1,2,4-triazole diamondoid ligands 268
cationic triamantane derivatives 262
CD-adamantane conjugates 251
cucurbiturils 251
cyclic oligosaccharides 251
cyclodextrins 251
β-cyclodextrin 251–252
diadamantyl ether 265–266
diamantane 5-aminoisophthalic acid derivative 253
diamantane cucurbituril complexes 259
4,9-diamantane dicarboxylate 269
diamantane functionalized chelating triazacyclononane ligand 264
diamantane-4,9-dicarboxylic acid dihydrazide 270
4,9-diaminodiamantane 256
diamondoid carboxylic acids 253
diamondoid-decorated Sn/S clusters 269
4-dimethylaminodiamantane 258

supramolecular architectures (*contd.*)
   D-mannose-diamondoid conjugates 254
   fullerene C_{60} 264
   "hydrophobic" effect 251
   hydroxy-diamondoids 262
   hydroxyphosphine 267
   isomeric hydroxy adamantanes 267
   lectin concanavalin A (ConA) 255
   loop binding mode 260
   meta-bis(10-bromo-9-anthryl)benzene 264
   micellar adamantane-based capsule 264
   primary ammonium binding mode 260
   quaternary ammoniumbinding mode 260
   supramolecular adamantane hydrogels and polymers 251
   tetrameric cubane-type alkoxide 267
Suzuki–Miyaura couplings 217, 218
   of chloroarenes 221

t

Taft polarizability parameter 221
1,3,5,7-tetraboraadamantane 314
terengganesine A 245
terminal diamondoid 1,3-dienes 119
tert-butyldimethylsilyl (TBDMS) 88
tert-disubstituted diamantanes 76
tert-disubstituted triamantanes 75
tertiary and secondary diamantane amines 104
tertiary diamondoid phosphines 111
1,3,5,7-tetraazaadamantane 312
1,3,5,7-tetrabromoadamantane 79
1,2,4,5-tetracyanobenzene (TCB) 115–117
tetraester 8
tetrahydrodicyclopentadiene (THCPD) 9, 42
tetrakis[3,5-bis(trifluoromethyl)phenyl] borate 289
[1(2)3]tetramantane 4, 43
tetramantanes 21, 39, 40, 46, 47, 51, 86, 113, 156

[121]tetramantane 4, 5, 279, 280
[123]tetramantane 1–2, 4, 9, 25, 27, 75, 77, 78, 113, 156, 158
tetramantane thiols 146
1,3,5,7-tetramethyladamantane 7, 67
tetrapeptide (Boc-L-Pmh-^{A}Gly-L-Cha-L-Phe-OMe) 229
1,3,5,7-tetraphenyladamantane 327, 328
thiatriamantane 51, 311, 312
Tolman cone angle 213, 214
T_d-pentamantane 8
trans-1,2-alkanediols 230
trans-cycloalkane-1,2-diols 230
triadamantyl-1-phosphine 112
triamantane 1, 45, 100, 146, 155, 198, 243, 253–255, 262, 263, 279, 314, 339
triamantane thiols 146
1,3,5-triazine-2,4,6-triamine 141, 142
1,2,4-triazole linkers 196
1,3,5-tribromoadamantane 79
2,2,2-trifluoroethanol 88
trioxaadamantane 244, 308, 309
2,4,10-trioxaadamantane derivatives 308
2,4,6-trimethyl adamantane 44
3,5,7-trimethyladamantyl-1-acetamide 283

u

unnatural amino acids 103, 239
unsaturated oligomers 287–293
unusually stable cyclic peroxide 291
urea-functionalized adamantyl dirhodium tetracarboxylate complex 226–227
urotropine/1,3,5,7-tetraboraadamantane molecular crystal 314

v

valence photoelectron spectroscopy 27, 283
vibrational spectroscopy 15, 25–28
vinyl diamondoids 120

w

Wurtz coupling of 1-bromodiamantane 281